INSECT NEUROENDOCRINES

V.B. Awasthi
M.Sc., Ph.D.
Department of Zoology
JNV University, Jodhpur

SCIENTIFIC PUBLISHERS (INDIA)
P.O. BOX 91 JODHPUR

Published by :
PAWAN KUMAR
Scientific Publishers (India)
5-A, New Pali Road, P.O. Box 91,
Jodhpur - 342 001 India
Tel.: +91-291-2433323
Fax.: +91-291-2512580
E-mail: info@scientificpub.com
http://www.scientificpub.com

ISBN : 81-7233-390-0

Laser typeset : Rajesh Ojha
Printed in India

PREFACE

The neuroendocrine system of insects is of special importance, because it controls directly or indirectly, the activities of almost all the systems and a variety of biological phenomena cannot be explained without knowing the role played by the different endocrine glands. Every aspect of insect life is influenced and controlled by the secretions of the neuroendocrines.

On this super important subject, now a large number of authoritative books are available, which provide a comprehensive account on almost all aspects of the neuroendocrine system of insects but there is no book which deals with the ultrastructural illustrations, which are indeed very much needed for the proper understanding of the subject. Moreover, except Nayar's elementary book published three decade ago no one has tried to publish a book on this vital subject in India. As we know, most of the biological phenomena especially pertaining to the structure and functional correlations are based on the interpretation of the figures. Many structural details that are difficult to explain theoretically can be easily understood by having a glance of figures. For example, the "C" bodies can be understood only by going through the figures, otherwise it is difficult to grasp, likewise modifications of the endocrine glands and malformations of various cell organelles under different conditions. Hence, it was decided to select and compile the figures of the insect neuroendocrine system scattered in the research papers, representing some of the insect orders that have most thoroughly been studied together with the explanation of figures given by the authors. In the text, no specific reference of figures has therefore been made. The readers can findout themselves without difficulty the relevant information, going through them orderwise.

It is hoped that the theoretical account and a large number of illustrations of prominent pioneer authors included in the book will provide ample information to the graduate students and research scholars of insect neuroendocrinology as well as to those engaged in carrying out researches on comparative endocrinology, insect ethology, cell biology and developmental biology, and acquaint them well with almost all the components of the neuroendocrine system in the active, inactive and various developmental stages, alongwith ultrastructural changes occurring in the gland cells and their organelles, that are associated with specific functions.

The aim of the author has been to present an account of the neuroendocrine system of insects and also to provide information directly through the ultrastructural illustrations of the neurosecretory cells of the brain and other ganglia, the neurohaemal organs (corpora cardiaca, cephalic aorta, perisympathetic and other neurohaemal organs), the corpora allata, the prothoracic or ecdysial glands and Weismann's ring. The sites and manners of release of secretory products have also been well documented. The chemical nature of insect hormones has been discussed, and account of a large number of hormonally controlled phenomena is given.

I do hope that the readers will find the book informative, thought provoking and interesting, and enjoy the illustrations of prominent authors, which would be a rare occasion for the workers, to see them.

The illustrations of the neuroendocrine organs of insect orders that have been selected from a very large number of papers included here are : (1) Collembola, (2) Dictyoptera (Blattaria), (3) Phasmida, (4) Homoptera, (5) Heteroptera, (6) Orthoptera, (7) Dermaptera, (8) Diptera, (9) Coleoptera and (10) Lepidoptera. This arrangement has been made for the sake of convenience and not from taxonomic point of view. The appendix will facilitate the students, researchers, teachers to understand the morphological organisation of the neuroendocrine system of some common insects.

I most humbly extend my sincere thanks to a large number of authors whose illustrations have been included and to the editors and publishers of various journals. Their number is so vast that it is difficult to acknowledge them individually but it was due to them that this compilation work was possible, for which I am extremely grateful.

I take this opportunity to remember with profound regards and wish to place on record the valuable suggestions that I had received from time to time in my research career in the past more than three decades from Prof. V.B. Wigglesworth, Prof. B. Scharrer, Prof. G. Fraenkel, Prof. H.E. Hinton, Prof. M. Gersch, Prof. H. Mislin, Prof. W. Weber, Prof. M. Raabe, Prof. B. Johnson, Prof. L. I. Gilbert, Prof. Y. Kono, Prof. T.C. Normann and Dr. L. Strong.

I am grateful to the authorities of Scientific Publishers (India), Jodhpur, Shri Prahlad Rai Sharma and Shri Pawan Kumar Sharma for their keen interest in my work and with whole hearted support in every act of the publication of the book.

I am thankful to my friend Shri A.P. Jain, Botanical Survey of India, Jodhpur, for reading the proof and offering valuable suggestions. Thanks are also due to my family members viz. wife Sneh Lata, son Santosh Kumar, daughter Shubha, daughter-in-law Hem Lata and grand-daughter Richa for their whole hearted co-operation without which this work was not possible.

JODHPUR **V.B. Awasthi**

September, 2004

ABBREVIATIONS

AO - Aorta
AH - Activation hormone
AER - Agranular endoplasmic reticulum
AX - Axon
ARF - Anterior retraction factor
BH - Brain hormone
BM - Basement membrane
BR - Brain
*CA - Corpus allatum
CAH - Corpus allatum hormone
*CC - Corpus cardiacum
CH - Chromatin
CNS - Central nervous system
DH - Diapausing hormone
EG - Ecdysial gland
EH - Eclosion hormone
ER - Endoplasmic reticulum
ERC - Circular endoplasmic reticulum
ERN - Net like endoplasmic reticulum
F - Filamentous structure
FG - Frontal ganglion
FGC - Frontal ganglion connective
G - Golgi apparatus/complex
GER - Granular endoplasmic reticulum
H - Haemocoel
HG - Hypocerebral ganglion
JH - Juvenile hormone
JHA - Juvenile hormone analogue
M - Mitochondrion
MH - Moulting hormone
MHA - Moulting hormone analogue
MVB - Multi vesicular body
N - Nucleus
NC - Nucleolus
*NCA - Nervus corporis allati
NCA I - Nervus corporis allati I
NCA II - Nervus corporis alllati II
NCC I - Nervus corporis cardiaci I
NCC II - Nervus corporis cardiaci II
NCCIII - Nervus corporis cardiaci III
*NCC - Nervus corporis cardiaci (common nerve)
NES - Neuroendocrine system
NHO - Neurohaemal organ
NSC - Neurosecretory cell
NSG - Neurosecretory granule
NSM - Neurosecretory material
NSP - Neurosecretory pathway
PI - Pars intercerebralis
PM - Plasma membrane
PO - Perisympathetic organ
PG - Prothoracic gland
PIF - Pupariation immobalizing factor
PSF - Puparium stimulating factor
PTF - Purparium tanning factor
PNO - Perisympathetic neurohaemal organ
PTTH - Prothoracicotropic hormone
R - Ribosome
RER - Rough surfaced or rough endoplasmic reticulum
SOG - Suboesophageal ganglion
V - Vesicle/Vacuole

*Pl. : Corpora allata, Corpora cardiaca,
Nervi corporis allati, Nervi corporis cardiaci

CONTENTS

1

THE NEUROENDOCRINE SYSTEM OF INSECTS

The NEUROENDOCRINE system of insects consists of (1) the neurosecretory cells of the nervous system *viz.* brain, suboesophageal, thoracic, abdominal and other ganglia, (2) The neurohaemal organs which serve as the site at which the neurosecretory products escape into the blood stream, (3) the corpora allata of multiple functions and (4) the prothoracic or ecdysial glands which secrete moulting hormone or ecdysone.

Kopec (1912, 1917, 1922) suggested that metamorphosis in *Lymantria dispar* was controlled by the brain and observed that the larvae failed to pupate if their heads were tied off with a ligature or if brain removed. According to him brain acted as an endocrine organ. Reimplantation of brain caused moulting and metamorphosis in the brainless larvae and until this time, the concept of neurosecretion was not known which was proposed later by E. Scharrer (1928).

Scharrer, E. (1928) for the first time observed the secretory activity in the cells of the brain (preoptic nucleus) of a teleostean fish, *Phoxinus laevis*. He called these glandular nerve cells, showing cytological evidence of secretory activity as neurosecretory cells (NSCs) and their product as neurosecretion or neurosecretory material (NSM).

The concept of neurosecretion was more clearly understood in 1937, when E. and B. Scharrer, published a comprehensive review.

In insects the presence of NSCs in the protocerebrum, the source of brain hormone was first described by Weyer (1935) in the honeybee, *Apis mellifica*. Later the presence of the NSCs in the brain and the ganglia of the ventral nerve cord and their functional significance along with other endocrine glands were demonstrated by numerous workers which have been extensively reviewed by Vander Kloot (1960). Wigglesworth (1964, 1970), De Wilde (1964), De Wilde and De Loof (1973); Gilbert (1964), Gilbert and King (1973); Gabe (1966), Novak (1966, 1975); Doane (1973); Slama *et al.* (1974); Scharrer B. (1965), Bargmann and Scharrer (1970); Kerkut and Gilbert (1985); Raabe (1982, 1989), Gersch (1975); Rowell (1976); Highnam and Hill (1979); Prabhu (1981); Awasthi (1985b), and many others.

A considerable difference in histological, histochemical, ultrastructural and staining properties, and the variation in size, location of cells as well as specific functional differences clearly suggest that the neurosecretory cells of the brain may produce not only one but several compounds differing in physiological functions and chemical structure, which has been also clearly indicated by the presence of different and distinct types of cells and electron density of the granules. For the regulation of insect development and almost all vital processes the most important product or products, of the neurosecretory cells of the protocerebrum, appears to be the hormone which stimulates the secretory activity of other endocrine glands. As will be seen (*vide infra*) the neurosecretory hormone of the brain is released into the blood stream *via* corpora cardiaca, the principal neurohaemal organ of insects in general (see details elsewhere), either as such or by the modification of the secretions of the chromophil or intrinsic cells of the corpora cardiaca. The hormone secreted by the neurosecretory cells of the brain is termed as brain hormone (BH), prothoracicotropic hormone (PTTH), activation hormone (AH), etc. by different workers. But most frequently, the secretion of the neurosecretory cells of the brain, specially that of the pars intercerebralis of the protocerebrum is commonly known as brain hormone (BH), which in fact is a mixture a several hormones, this is because of the presence of distinct and different types of cells which is further substantiated by electron density of the granules. Most probably, each type of cell secretes a distinct type of hormone and this possibility can not be ruled out at the moment. It may be briefly stated that the developmental processes connected with physiological changes in insects are regulated by three hormones collectively referred to as developmental or metamorphosis hormones. One of them is a neurohormone released by the neurosecretory system (cells) while the second is released by the corpora allata and the third by the prothoracic or ecdysial glands (Slama *et al.*, 1974).

In the following pages only a brief account on the neuroendocrine system ie. neurosecretory cells of the various ganglia, neurohaemal organs i.e. the sites of storage and release of neurohormones, (corpora cardiaca and aorta); the corpora allata and ecdysial or prothoracic glands have been given to acquaint the readers with the subject. A comprehensive account has already been given in several authoritative books, as mentioned in the beginning.

2

THE NEUROSECRETORY CELLS

As already remarked, the neurosecretory cells are glandular nerve cells; secrete various types of neurohormones and control almost all life processes in insects either directly or through the secretions produced by other endocrine glands, which are non-nerous in nature.

STAINING METHODS AND CLASSIFICATION OF THE NEUROSECRETORY CELLS

The neurosecretory cells (NSCs) are stained with "specific" neurosecretory stains like - chrome alum haematoxylin phloxine (CHP), aldehyde fuchsin (AF), paraldehyde fuchsin (PF), performic acid resorcin fuchsin (PARF), performic acid victoria blue (PAVB), alcian blue phloxine (ABP) etc. The CHP, PF, AF and PARF staining techniques differentiate maximum types of cells and are used for the classification of the neurosecretory cells, others stain only one or two types of cells. On the basis of staining properties two to four types of cells have been classified viz A, B, C and D. (see: Nayar, 1955; Johansson, 1957; Highnam, 1961 etc.). Although in most of the cases the cells can be compared with similar types of cells of other insects but they may differ also due to their chemical nature. Hence, it is not always possible to compare the work done on other insects. For example Mitsuhashi (1963) distinguished A, B, C and D-types of cells in several lepidopteran insects but in *sensu stricto*, on the basis of staining properties all of them belong to one type i.e A type except the type C because A, B, and D-cells contain a small or a large amount of PF positive material. With CHP the A- cells stain blue black or blue gray and the B- cells pink or red (by Phloxine) whereas with PF, AF or PARF, the A-cells stain purple, purple red, violet, violet red and the B - cells green (by the counter stain - Halmi's mixture). Likewise, C- and D-cells are also distinguished. The two types of cells viz A-and B-types are found in majority of the insects whereas the C- and D- types are found only in a few insect orders. In certain cases, only A- cells are present in the brain. Some workers like Fletcher (1969) has classified NSCs from I to XIII types but all of them fall under 3 or 4 classical types with several subtypes on staining patterns but his system is not better than the alphabetical system of classification, hence not followed by most of the workers. It is noticeable that the visible secretion is only a carrier protein "neurophysin" to which the hormone molecule is attached. Thus, the carrier substance which is the stainable part, acts as a marker for the hormone. When the hormone is released, it becomes separated from the carrier and is then free to enter the blood stream (i.e. haemolymph).

HISTOCHEMISTRY OF THE NEUROSECRETORY CELLS

A number of histochemical tests have been performed on species belonging to main insect orders. The medial and lateral A-cells have shown the presence of carbohydrates, lipids and a large amount of proteins. The A-cells often contain sulphydril and disulphur groups. The SS and SH groups are present in large amounts in the A-cells of hemimetabolous insects of the order - Dictyoptera, Orthoptera, Heteroptera, etc. but are either absent or present in small amounts in holometabolous insect orders like - Diptera, Lepidoptera etc. The phenol groups also appear to be present in fairly large amounts in the A-cells of the medial groups (see : Raabe, 1982, 1989, Awasthi, 1985b; Awasthi and Singh, 1986b).

ULTRASTRUCTURE OF THE NEUROSECRETORY CELLS

Under the electron microscope also, several types of neurosecretory cells have been distinguished on the basis of the size of granules and electron density which has been clearly demonstrated in the figures. The A-cells always contain membrane bound granules 1000-3000Å or

1000-4000Å in diameter (Wigglesworth, 1970, 1974) and the matrix is electron dense. The granules of other types of cells are much smaller and less than 1000Å in diameter and the matrix may be electron dense, less electron dense, electron transparent, electron lucent etc. Evidence(s) gathered from both light and electron microscopic investigation leads to the conclusion that the neurosecretory substances are produced in the perikaryon. The ultrastructural events involved in the elaboration are virtually the same as those in other well documented cases of synthesis of proteinaceous glandular products. The neurosecretory material appears to be manufactured in the ergastoplasm ie. granular - endoplasmic reticulum of the perikaryon and transported to the Golgi bodies. In the Golgi bodies they are packaged and budded off in the form of membrane bound neurosecretory granules. Under certain circumstances the granules may accumulate in the perikaryon in the axons and axon-terminals (Slama *et al.,* 1974). This has been clearly shown in text figures. For details of cell-organelles see illustrations.

A mention may be made here as stated above, that the raw material is synthesized in the granular endoplasmic reticulum and from here it is transported to the Golgi cisternae. In the Golgi bodies elaboration of NSM takes place through the mediation of alkaline phosphatase (Tewari and Awasthi, 1968) and finally the membrane bound vesicles are budded off in the form of neurosecretory granules. These granules have been shown in several instances which are stored in the perikaryon and transported to the neurohaemal organs *via* axons or axonal pathways.

The neurosecretory granules are stored in the cytoplasm, apparently for varying periods of time until their content is released. According to Highnam (1967, 1971) the amount of stainable material present in the NSCs is a result of a balance between the synthesis and release of material. This may take place, at least in principle, at various levels of the neuron i.e. perikaryon. However, in most of the known neurosecretory centres, discharge occurs at the axon terminals (see : NHOs and Slama *et al.*, 1974). Neurosecretory granules reach this site by proximo-distal - axonal transport. Evidence for the occurrence of this "migration" is conclusively provided by the accumulation of neurosecretory material at the proximal stump of the surgically interrupted pathway and by the time sequence in the changing distribution of incorporated labelled precursors as determined by the autoradiographic techniques (Flament-Durand, 1966). Using radio-active tracer methods by measuring the rate of incorporation of labelled cystine, Highnam (1968) found that the neurosecretory activity is indeed much higher in females with developing ovaries than in those without. The velocity of axonal transport of neurosecretory material has been estimated to be approximately 0.8 to 1.6 mm per hour. Schooneveld (1970) suggested that the axonal transport may be regulated by nervous activity but the release of the material from axon terminals may be regulated by other factors. In addition to developmental changes there exist many factors which have more immediate influence on the amount of neurosecretion, especially in the neurohaemal organs. Among these factors are electrical irritations, intensive nervous stress as during mating or oviposition, ingestion of food or water after starvation or thirsting, etc. (Highnam and Hill, 1979).

A prominent characteristic of neuroendocrine system among insects as well as in other invertebrates is the aggregation of bulbous secretion filled terminals, around the vascular system, they form separate anatomical entities called "Neurohaemal organs".

THE DISTRIBUTION OF THE NEUROSECRETORY CELLS

The neurosecretory cells have been observed in the brain, suboesophageal, thoracic, abdominal as well as in the frontal and hypocerebral ganglia in both apterygote and pterygote insects.

The neurosecretory cells in the brain

Two to several paired groups of NSCs are found in the brain. In the majority of the insects, two medial and two lateral groups have been most frequently observed. In addition to these four groups, a pair of optic, a pair of posterior and a pair of ventral groups of NSCs have also been observed.

In the pars intercerebralis medialis of the protocerebrum on each side of the medial line (furrow) are found a cluster of NSCs in most of the insects. These constitute the medial groups of neurosecretory cells (MNCs). Each medial group of NSCs is composed of either. A or A- and B- or more types of cells as classified earlier. In some insects only A -

cells are present while in others A and B both eg. in the bug *Nezara viridula* the medial groups have only the A- cells, in the *Gryllodes sigillatus* (Orthoptera) *Utetheisa pulchella* (a moth) A, and B- cells, in *Philosamia ricini, Achoea janata, Amsacta collaris* (moths) A,B,C- cells are clearly seen. Four types of cells viz. A,B,C and D are present in the brain of *Schistocerca gregaria*. These cells on the basis of their size have been further classified into sub types: A1, A2, B1, B2, C1, C2, and so on.

The second paired group of neurosecretory cells is present in the lateral part of the protocerebrum and constitute the lateral group of neurosecretory cells (LNCs). The lateral groups are also composed of either only A or A- and B- both the types of cells. In some insects the lateral group is totally absent. As remarked above, in certain insects in addition to medial and lateral groups, optic, posterior and ventral groups of cells have also been observed. In Lepidoptera there are always present two posterior groups of NSCs, each of which is composed of two A- cells which has been considered as the characteristic feature of this order (Herman and Gilbert, 1965).

In apterygote, *Petrobius maritimus* (Leach), it is possible to distinguish neurosecretory cells belonging to the two principal cell types, A and B, previously described for the Insecta. The histological appearance of the secretion in type A cells differs according to whether the cells are in the brain or in the ventral nerve cord. There is, on the other hand, a uniformity of appearance of elements in the brain. In view of the similarity of the pathways of the nerves arising from these cells they can be homologized with similar structures seen in the Pterygota. The Gomori-positive material in these cells, whether in the brain or in the ventral nerve cord, undergoes quantitative, cyclical variations which seem to be related to the moulting cycle. During moulting the secretion of the cerebral neurons appears to be discharged, whereas that in the cells of the ventral nerve cord reaches its maximum concentration. Experimental ablation of the frontal complex and section of the internal cardiac nerves ie. NCC-I has been performed, but has failed to indicate the role of those secretions whose source has been thus removed from circulation (Bart, 1963).

THE NUMBER OF THE NEUROSECRETORY CELLS

The number of medial neurosecretory cells (MNCs) of the brain vary considerably from insect to insect from a few to several hundred (see table 1).

Table 1.

Order/Insect	MNCs	Technique	Reference
Ephemeroptera *Ecdyonurus* sp.	25+25	PF, CHP	Arvy & Gabe (1953)
Odonata *Aeschna* sp.	40 + 40	PSI*	Sterba & Hoheisel (1964)
Isoptera *Anacanthotermes*	a few A cells	PAVB	Awasthi, (1973a)
Odontotermes	10-13 + 10-13	PAVB	Bassaglingappa et al., (1984)
Orthoptera *Gryllotalpa*	300 + 300	PAVB	Dogra (1967)
Gryllodes	150+150	PF, PAVB	Awasthi (1969a)
Schistocerca	1000 + 1000	PF, CHP	Highnam (1961)
	604 + 604	EM	Rowell &Mason (1976)
Melanoplus	400+400	PAVB, CHP	Dogra & Ewen (1970)
Poecilocera	80+80	PF	Saini (1971)
Dictyoptera *Periplaneta*	45 + 45 or 50 + 50	CHP, PF	Fuller (1960)
Periptaneta	50 + 50	CHP, PF	Pipa (1962)
Epilampra	100+100 ca.	PAVB	Awasthi (1978)
Phasmida *Clitumnus*	40 + 40	—	Cazal (1948)
Dermaptera *Anisolabis*	10-20 + 10 – 20	AF, CHP	Ozeki (1958)
Labidura	8-10 + 8-10	PF, PARF, PAVB	Awasthi (1975c)
Euborellia	10-14 + 10-14	PF, PARF, PAVB	Awasthi (1976d)
Heteroptera *Rhodnius*	20 + 20	—	Wigglesworth (1940)
Rhodinus	65+65	CH, P, AZ	Baehr (1968)
Adelphocoris	12 + 12	PF, CHP, PAVB	Ewen (1962)
Oncopeltus	7+7	PF, CHP	Johansson (1957, 1958)
Iphita	16 + 16	PF, CHP	Nayar (1955)
Dysdercus	9 - 10 + 9 - 10	PF, PAVB	Dogra (1967a)
Dysdercus	9+ 9	PF, PAVB	Awasthi (1972b)
Nezara	5 + 5	PF, PAVB	Awasthi (1969b, 1972a)

* PSI- Pseudocyanin (a technique of fluorescence-microscopy)

Metochus	9+9	PF, PAVB	Awasthi (1973c)
Lygaeus	5 + 5	PARF, PAVB	Awasthi (1980a)
Leptocorisa	5 + 5	AF, CHP	Tiwari *et al.* (1975)
Belostoma	8 – 10 + 8 – 10	PF, PAVB	Dogra 1969
Sphaerodema	11 + 11	PAVB	Awasthi (1973b)
Ranatra	9-10 + 9-10	PAVB	Dogra (1967b)
Homoptera *Aphis*	4-6 + 4-6	PF, AZ, CHP	Johnson (1963)
Mecoptera *Panorpe*	14 + 14	—	Cazal (1948)
Lepidoptera *Achoea*	8 + 8	PF, PARF	Awasthi & Singh (1982b)
Amsacta	8 + 8	PF, PARF	Awasthi & Singh (1981a)
Utetheisa	8 + 8	PF, PARF	Awasthi & Singh (1985)
Philosamia	8 + 8	PF, PARF	Awasthi & Singh (1981b)
Philosamia	8 + 8	PF, PTH, CHP	Panov & Kind (1963)
Ephestia	6-8+6-8	AZ, IH*	Rhem (1951)
Galleria	38-47+38-47	PF, AZ, CHP	Delepine (1965)
Sphinx	8+8	PF, PTH, CHP	Panov & Kind (1963)
Bombyx	8+8	PF, PTH, CHP	Panov & Kind (1963)
Bombyx	10 + 10	PF, CHP	Kobayashi (1957)
Antheraea	8 + 8	PF, PTH, CHP	Panov & Kind (1963)
Hyalophora	13 + 13	PF, CHP	Herman & Gilbert (1965)
Vanessa	8 + 8	PF, PTH, CHP	Panov & Kind (1963)
Pieris	8 + 8	PF, PTH, CHP	Panov & Kind (1963)
Diptera *Aedes*	12 + 12	PF, EM	Meola & Lea (1972)
Calliphora	12 - 13 + 12 - 13	PF, CHP	Thomsen (1965)
Glossina	8-10 + 8-10	CHP	Longley (1965)
Sarcophaga	13 + 13	PF, PAVB	Dogra & Tandon (1966)
Sarcophaga	13 + 13	PF, PAVB	Awasthi (1972c)
Hymenoptera *Apis*	60 + 60	—	Weyer (1935)
Apis	65 + 65	CHP	Formigoni (1956)

** IH- Iron haematoxyline.

Coleoptera *Hydrous*	20 + 20	CHP, Giemsa, Feulgan	DeLerma (1954)
Leptinotarsa	80 - 80	PF, PTH, CHP	Schooneveld (1970)
Aulacophora	8-10 + 8-10	AF, CHP	Saini (1966)

The size of the neurosecretory cells also varies considerably from order to order. A large number of cells are present in Orthoptera and Dictyoptera but their size is small, while in Lepidoptera and Heteroptera a small number of cells are present but their size is many time larger than those of Orthoptera and dictyopteran insects. In Dermaptera and Diptera the size is nearly in between the above mentioned orders. There appears a close correlation between the NSC - size, number and production of the total amount of hormone, which may be equal because when the size is small their number is large and when the size is large their number is small. However their may be certain exception in the vast group of insects.

The Neurosecretory cells in the other ganglia

In the following pages only a brief account of the presence of the NSCs in the ganglia of the ventral nerve cord and frontal ganglion is given.

The suboesophageal ganglion consists of different types of neurosecretory cells. In Heteroptera, there are two prominent neurosecretory cells of A-type, which are located in the posterior ventral part of the ganglion. In addition to two conspicuous A - cells, there are numerous B- type of cells present in the suboesophageal ganglion. The A- and B- types of cells in staining properties and ultrastructural details are similar to the A and B cells of the brain as shown in figures. Similar types of cells have also been reported in other insects. All other ganglia of the ventral nerve cord viz - thoracic, and abdominal, consist of clusters of A - and B- types of cells. These secrete specific hormones which are stored in the perisympathetic organs (see : under NHOs). The frontal ganglion of Lepidoptera, Heteroptera and Dictyoptera (Blattaria) are also composed of A- and B- types of cells. According to Bell *et al.*, (1974) the frontal ganglion (FG) of the tobacco hornworm,

Manduca sexta, was found to contain two neurosecretory cells (NSCs) (max. dia. = 40-45 μm). The cytoplasmic inclusions of the NSCs were stained purple with paraldehyde fuchsin, and marked fluctuations in amounts of NSM in the perikarya were observed, depending upon the developmental status of the insect. The perikarya of NSCs in the frontal ganglia of starved larvae and diapause pupae contained large accumulations of NSM, whereas feeding larvae and developing pharate adults showed relatively low amounts of neurosecretion. Electron microscopy revealed large accumulations of NS granules (dia = 80-240 nm) in the frontal ganglia of diapause pupae, but only slight accumulations of granules were seen in the NSCs of developing larvae and pharate adults.

It was concluded that axonal transport and release but not synthesis is shut down during starvation and diapause, leading to accumulation of NSM, in the perikarya. It is also suggested that the failure of many investigators to differentiate NSCs in the frontal ganglion of various insects may have been due to the selection of very active stages when the amount of available NSM was too low to be visualized by conventional staining techniques (Bell *et al.,* 1974).

The frontal ganglion (FG) of *Achoea janata* and *Philosamia ricini* consists of conspicuous NSCs which are classified into A - and B-types. The A-type of cells are positive to PF and PARF, the material stains deeply purple or purple-red whereas the B-types of cells are negative to “specific neurosecretory” stains and stain green by the counter stain. In the FG 2 to 3, A-cells are clearly observed which are round or oval in form. Depending upon the age of the insect larvae, the cells either contain a heavy accumulation of NSM or a small amount of this material in them. In each larval instar, just after the moult the cells appear depleted of NSM. But abundant A-material in the larvae just before the moulting have been observed. This indicates clearly that in each instar, a fresh synthesis of NSM begins and the material synthesised in the instar accumulates in the neurosecretory cells and is discharged before the moult. In the next instar again a new cycle of synthesis of NSM starts. It may be added that the B-cells do not appear to discharge their material before moult as is done by the A-cells, because, copious amount of B-material have been observed in the B-cells of moulting instar whereas at the same time the A- cells contain a small amount of their specific secretions (Awasthi, 1980b).

The neurosecretory material in A-cells occur either in the form of discrete granules or colloids or droplets that fuse to form clumps as observed in the heavily loaded cells. The A-cells are very distinct that can be conveniently seen even in the lowest magnification of the microscope. They are almost identical in shape and size in both the insects (measure about 25-27.5 μm in diameter and nucleus measuring 12.5 μm). The axons after their origin seem to bend and run into the recurrent nerve. Occasionally, transport of NSM along the axons have also been observed. The location of the neurosecretory cells within the FG varies from individual to individual. Normally, they are found in the central part of the ganglion, but are always superficial in position. The NSCs may also be located either close to the root of the recurrent nerve or frontal ganglion connective. It may be added that the A-cells, which stand out well with PF and PARF and are fuchsionphilic in nature (positive to paraldehyde fuchsin and resorcin fuchsin) are occasionally very lightly stained by the PAVB technique (Awasthi, 1980b). The neurosecretory cells have also been reported in the optic lobe of several insects e.g. Beattie, (1971), has reported that in the region of the distal optic chiasma of each optic lobe of *Periplaneta americana*, there is a group of about 120 monopolar neurosecretory cells. These cells do not stain with paraldehyde fuchsin but remain acidophilic after oxidation. They stain red or sometimes indigo with the azan technique. Histochemically, the neurosecretory material is positive for protein and the amino acids tryptophan and arginine but negative for 1, 2-glycols and strongly acidic groups. At the ultrastructural level, the cytoplasm of the cells contain many elementary neurosecretory granules 100 to 170 nm in dia. The cells also contain well-developed Golgi bodies and endoplasmic reticulum. The axons from these cells run towards the interior of the optic lobe. In this region, axons containing dense granules (mean diameter 70 nm) and synaptic vesicles synapse onto the axons from the neurosecretory cells. The neurosecretory axons then cross over to the anterior side of the optic lobe and run towards the brain (Beattie, 1971).

TRANSPORT OF THE NEUROSECRETORY MATERIAL

It may be remarked here that all the axons of the medial neurosecretory cells of one brain lobe, converge at one point and form a bundle of medial neurosecretory pathway. The medial neurosecretory pathways formed in such a manner run on the dorsum of the protocerebrum for a shorter or longer distance, depending on the insect species/order and decussate each other. The pathways after the decussation, diverge and run on the dorsum of the protocerebrum of the opposite side. In some insects they emerge from the posterior part of the protocerebrum while in others they traverse the deutocerebrum and emerge from the posterior part of the tritocerebrum as nervi corporis cardiaci-I (NCC-1). In a similar manner lateral neurosecretory pathways are formed but their axons do not cross over each other ie there is no chiasma formation in the lateral neurosecretory pathways. These emerge from the brain as nervi corporis cardiaci-II (NCC-II). In some insects both NCC-I and NCC-II fuse in the hinder part of the brain before emergence and they come out in the form of a common nervi corporis cardiaci (NCC). In certain instances, a third nerve emerges from the brain and is termed as nervi corporis cardiaci-III (NCC-III) which probably transport the neurosecretory material of the tritocerebral neurosecretory cells. The NCC-I, NCC-II and NCC-III or the common NCC, enters the corpora cardiaca of the same side in insects in general, but in Dermaptera and Heteroptera the NCC-I enters the cephalic aorta (AO) and NCC-II enters to the corpora cardiaca (CC).

3

THE NEUROHAEMAL ORGANS

The neurohaemal organs (NHOs) are those which store the neurohormones secreted by the neurosecretory cells of the brain and other ganglia and release them into the haemolymph. The neurosecretory material (NSM) is stored in the neurohaemal organs (NHOs) for a shorter or a longer duration. The neurohormones may be released as such or prior to release, they may be modified in the neurohaemal organ(s) by its own intrinsic secretions. Two principal types, and numerous other types of neurohaemal organs have been recognized in insects (see : Awasthi 1974, 1976e). The principal types of neurohaemal organs are corpora cardiaca and cephalic aorta and others are the neuropilar neurohaemal organs, NCA-II in Gryllidae, perisympathetic organs etc. Account of some of these NHOs is given here.

THE CORPORA CARDIACA

The glandular nature of CC was first recognized by DeLerma (1933). Later, comprehensive studies were made in different insect orders, and their account is given in considerable detail by Awasthi (1974, 1976e).

In most of the insect orders, the corpora cardiaca perform the function of a principal neurohaemal organ. The corpora cardiaca are a pair of oval, spherical, spindle-shaped, elongate or "U" shaped neuroglandular bodies, lying behind the brain on the dorsal part of the foregut and are associated with cephalic aorta. They appear silvery white with a bluish tinge in living condition. In some insects they are found on the lateral part of the foregut in between the dorsal and ventral dilator muscles. They are absent in Collembola. The corpora cardiaca receive the neurosecretory axons of the neurosecretory cells of the brain *via* neurosecretory pathways and nervi corpori cardiaci- I and -II. The corpora cardiaca are fused in some insects and separate in others. In some insects they are differentiated into two distinct regions viz. (1) one region is composed of neurosecretory axons and axon-terminals of the cerebral neurosecretory cells and the other (2) glandular part containing intrinsic neurosecretory cells but in many insects no such distinction is possible. The cells of corpora cardiaca in staining properties are similar to the medial A-cells of the brain, and also contain axons loaded with neurosecretory material. The corpora cardiaca are surrounded by a connective tissue sheath consisting of an acellular stroma. While the surface of the gland may be smooth in some insects, it more often shows deep invaginations. As remarked, the internal structure of the gland is complex and individual structural elements may differ in considerable detail from insect to insect. A part of the corpora cardiaca consists of nerve bundles and the enlarged axon-terminals of the neurosecretory cells of the brain which contain accumulated neurosecretory material. Many of the axon terminals occur near to the outer surface but some are also observed on the stormal part. The glands consist of two types of cells (1) the chromophil cells or intrinsic neurosecretory cells, similar to the neurosecretory A - cells of the brain and (2) the chromophobe cells or glial cells or interstitial cells.

It is believed that the intrinsic neurosecretory cells of the corpora cardiaca secrete a hormone which regulates heart beat (see : P. 54). It will be seen in the figures that the gland is composed of different types of neurosecretory axon-terminals containing granules of different electron density.

Scharrer (1963) has also remarked that the corpora cardiaca of *Leucophaea* contain two classes of intrinsic elements viz. parenchymal and interstitial cells. The parenchymal cells produce a secretory material first visible in the Golgi zone of the perikaryon in the form of distinct electron opaque granules. These undergo changes (gradual loss of electron density, emergence of internal structure) as they accumulate in cellular processes. The parenchymal cells according to her are best classified as neuroglanduar elements since in addition to secretory inclusions they possess characteristic of ganglion cells. The interstitial cells (the chromophobe cells of the light microscope) are non-glandular structures, and are of the nature of connective tissue. They encapsulate the perikarya of parenchymal cells as well as cellular processes. The cytoplasmic processes include: (a) relatively short ones belonging to parenchymal cells (intrinsic cells) and (b) long axons whose cell bodies lie within the central nervous system (extrinsic elements). Many of the latter contain electron opaque granules of the kind found in electron micrographs of typical neurosecretory cells (A- cells). These extrinsic granules represent the second category of secretory products stored within the corpora cardiaca. By comparison with the product of intrinsic gland cells, the neurosecretory granules seem to be fairly stable. Neither type seems to pass through the connective tissue sheath of the corpora cardiaca in the form of descrete granular entities. Scharrer (1963) has further pointed out that the sheath which sends branches into the interior of the corpora cardiaca, has the properties of a basement membrane. It represents a pathway for the exchange of substances between the cellular components of the corpus cardiacum and the surrounding haemocoel. As remarked earlier, Scharrer (1963) has also stated that the dual character of the corpus cardiacum *viz* that of a storage and release centre for extrinsic neurosecretory substances and of an endocrine organ in its own right, is herewith established beyond doubt.

Bowers and Johnson (1966) have similarly reported that the corpus cardiacum of *Myzus* (Aphid) is composed of intrinsic secretory cells, glial or interstitial cells and a complicated region of intertwined granule filled axons. Three types of neuro effector junctions were observed in the latter region.

Secretory granules, presumably from brain or CC neuroglandular cells, are also present in axons of nerves leaving the CC and going to various target organs such as the corpora allata, pericardial cells and visceral and somatic muscles. Granule containing axons are present in each of these tissues and show contacts with the cells which resemble synapsis. It is suggested that in these instances, the secretory material acts as a local hormone. At two sites along the lateral nerve from the CC, glandular cells containing large (200-290 m μ) granules were observed (Bowers and Johnson, 1966).

In the stick insect *Carausius*, the corpus cardiacum contains cells of three main categories (1) axons of neurons located in the brain, (2) intrinsic secretory neurons, and (3) glial cells (Smith and Smith, 1966) (for details see: figures).

The ultrastructure of corpus cardiacum of the house cricket, *Gryllodes sigillatus* has been described by Awasthi (1977), which is similar to other insects. The corpora cardiaca are externally bound by a thick layer (2300-3500Å) of connective tissue sheath, composed of numerous filamentous structures. In the gland three types of granules in the neurosecretory axon terminals and pre-terminals have been distinguished on the basis of size, shape and opacity of the contents. The 1st type of granules are electron dense of mixed nature (500 - 1400Å). The 2nd type of granules having electron lucent space between the matrix and the boundary membrane and are labelled as dense cored vesicles. They are of about 900-1700Å in diameter. The 3rd type of granules are electron dense oval or spherical in outline with indistinct boundary membrane and are approximately 1100 to 2300Å in diameter. The intrinsic cells are of 2 types having a large number of mitochondria and plentiful free ribosomes. In addition the CC also contain glial cells, ordinary axons and tracheoles.

Thus, the ultrastructure of the corpora cardiaca in insects in general has revealed the presence of extrinsic and intrinsic elements. The extrinsic elements are the neurosecretory axons and their terminals with NSM of cerebral origin i.e. of the neurosecretory cells located in the brain. The intrinsic elements of the corpora cardiaca are the chromophilic or intrinsic cells of the gland and the interstitial or glial cells. Since this acts as the neurohaemal organ for both the extrinsic and intrinsic secretions, granules of both the origin are stored in the gland having different size and electron density.

In the flies, NSM is stored in the CC or aorta (AO) or in both (Thomsen, 1952, Normann, 1965, Deoras and Bhaskaran, 1966-1967; Awasthi, 1972c) and in mosquito *Aedes aegypti*, it is stored in the dorsal mass that lies above the anterior part of the hypocerebral ganglion. The dorsal mass, functionally corresponds to the corpora cardiaca of other insects (Burgess and Rempel, 1966). In *Culiseta inornata* also (Burgess, 1971) the NSM is stored in the dorsal mass. Meola *et al.* (1970) have recorded an accumulation of medial neurosecretory material in the corpus cardiacum of *Aedes sollicitans*. In other 4 - mosquitoes viz *A. aegypti, A. taeniorhynchus, A. triseriatus* and *Culex pipiens*, an accumulation of PF - positive material (NSM) was observed in the corpora cardiaca as in *A. sollicitans* (Meola and Lea, 1972). Electron micrographs of the corpus cardiacum of *A. sollicitans* reveal that the ganglion cells and neuropile are not concentrically arranged as in other insect ganglia. Instead, the corpus cardiacum is a linearly arranged ganglion in which the ganglion cells are centrally located between an anterior and posterior neuropile. Thus the corpus cardiacum consists of three regions: an anterior neuropile, a central region of ganglion cells, and a posterior region containing neuropile but mainly composed of neurosecretory terminals of the lacunar neurohaemal site.

The corpus cardiacum is a neurohaemal organ located in the head of mosquitoes in contrast to other Diptera, in which it is found in the neck and thorax. Two neurohaemal sites were found in this corpus cardiacum; an aortic neurohaemal site in which neurosecretory axons extend between the muscle fibres of the aorta and terminate in the lumen, and a lacunar neurohaemal site in the posterior region of the corpus cardiacum. In the lacunar site the neurosecretory axons terminate in extracellular channels which are confluent with the haemocoel. The axons of the lacunar site are associated with a type of glial cell restricted to this region. Five different types of neurosecretory axons were noted in the corpus cardiacum based upon relative size and density of the granules and density of the axoplasm. Two different types of axons were usually found paired in the lacunar release site, encapsulated by a glial cell process (Meola and Lea, 1972). From these observations it appears that the CC perform the function of the neurohaemal organ in the mosquitoes and in the flies also.

In dipterous larvae, the NSM is stored in the basal part of the ring gland (see: for details under Weismann's ring). In adults, the CC and AO, both perform the function of a neurohaemal organ because they contain neurosecretory axons and abundant NSM (Normann, 1965, Thomsen, 1969, King *et al.*, 1966a, Aggarwal and King, 1971, Awasthi, 1972c). King *et al.*, (1966a) observed that the nerve entering the CC gives a beaded appearance due to the accumulation of neurosecretory granules in the localized buldges along the axons. Further, King *et al.* (1966b) have remarked that the corpus cardiacum of *Drosophila* consists of a central bundle of axons surrounded by cortical cells. Many of the axons are thought to originate as projections from specialized, secretory neurons residing in the pars intercerebralis. Neuroscretory spheres are carried posteriorly in these axons by travelling bulges. The cortical cells contain myriads of neurosecretory spheres suspended in their cytoplasm and most of these are believed to have been donated by contiguous axons. Cortical cells are also characterized by an extensive, rough surfaced endoplasmic reticulum consisting of interconnected bloated cisternae which arise from localized, blister-like proliferations of the outer membranes of the nuclei. The cortical cells of *fes* females also swell, and their nuclear blisters reach immense sizes. Presence of intrinsic secretory cells in the CC with secretory products have been reported by Thomsen (1969) in *Calliphora* and Aggarwal and King (1971) in *Drosophila*. Aggarwal and King (1971) add that the intrinsic cells of *Drosophila* CC, contains large number of neurosecretory granules in the perikarya and resemble the neurosecretory cells of the brains of house fly, *Musca domestica* (Ramade, 1966) and *Calliphora erythrocephala* (Bloch *et al.,* 1966).

The corpora cardiaca of earwig, *Euborellia annulipes* as in other insects, clearly show the intrinsic neurosecretory cells, the axon terminals and pre-terminals containing electron dense granules 400-1200 Å in diameter. The processes of these cells contain mitochondria and multivesicular bodies. In addition to these intrinsic elements, there are two types of extrinsic axons (1) ordinary axons and (2) axon terminals of the extrinsic neurosecretory B - cells of the brain. They contain mixed granules (electron dense as well as electron lucid) of 400-900 Å in diameter. The neurosecretory axons are provided with mitochondria of different shapes and sizes and multivesicular bodies along with neurosecretory granules. Thus in the CC three types of axon terminals and pre-terminals are visible (1) axons of intrinsic cells (NS-III) (2) axon terminals and pre terminals of extrinsic B-cells of the brain (NS-II) and (3) ordinary axons. The NS-II granules of CC are comparable to NS-II of aorta. In connection with CC, it may be mentioned that the axon-terminals and pre-terminals containing large electron dense granules (NS-I of A-cells) are totally absent (Awasthi, 1976c).

The organization of the corpora cardiaca and corpora allata of the female alfalfa weevil, *Hypera postica* (Tombes and Smith, 1970), generally parallels that reported for other insects with extrinsic neurosecretory axons, intrinsic neurosecretory cells, and interstitial or glial cells appearing in both glands.

Connecting the brain with the cardiacum are the extrinsic neurosecretory axons containing granules between 600 and 2200 Å. Granules with a diameter of up to 4000 Å are found in the intrinsic secretory cells of the corpus cardiacum.

From the lobed corpus cardiacum, axons extend along a short nerve trunk into the ovoid corpus allatum. A characteristic of this gland is the presence of a layer of axons, ensheathing most of the corpus allatum and situated beneath an external basement membrane. These axons are believed to have their origin in the protocerebral neurosecretory cells. Although the corpus allatum is known to be engaged in the release of hormonally active material, the cytoplasmic organization does not strongly reflect such activity. The glands of the diapausing weevil appear similar to those of the active weevil with a reduction in cytoplasm and a consequent folding of the plasma membrane (Tombes and Smith, 1970).

Thus, the corpora cardiaca perform the functions of a principal NHO in Phasmida, Dictyoptera, Orthoptera, Neuroptera, Coleoptera, Diptera, Isoptera, Homoptera, Hymenoptera, Plecoptera, Ephemeroptera, and many others but not in Dermaptera, Heteroptera and Lepidoptera (for details see : Awasthi, 1974, 1976e).

THE CEPHALIC OR ANTERIOR AORTA

The aorta in insects is a tubular anteriorly extended part of the heart. It is always dorsal to the foregut. The part of aorta (AO) which receives the axons of the nervi corporis cardiaci-I (NCC-I) or common nervi corporis cardiaci (NCC) as the case may be, is much wider than its remaining proximal and distal parts. The lateral aortic walls are thick whereas the dorsal and ventral walls are thin (Unnithan *et al.*, 1971, Awasthi, 1973c). The NSM in the majority of the bugs has been observed to accumulate all along its circumference. The axons proximally extend either up to the level of the CC (Dogra, 1967 b,c) or up to the level of the hypocerebral ganglion (Awasthi, 1973c). Distally the axons extend for about 1/2 to 1/3 of the length of aorta. The NSM in axons appear as a continuous column or broken at regular intervals giving thereby a beaded appearance. Thus, the NSM is stored in the axon-terminals and intercellular spaces. In the heteropteran insects, the wall of aorta is either spongy (Dogra, 1967a,b) or is made up of thick and loose tissues (Awasthi, 1973c) for the accommodation of NSM.

At the level of light microscopy, the aorta appears to be composed of two layers (1) an inner thick layer of loose or spongy tissues and (2) an outer thin layer. The former layer contains neurosecretory axons and NSM and the latter appears to be almost devoid of NSM.

Ultrastructural studies of Unnithan *et al.* (1971) have revealed that the inner and the outer surfaces of aorta are covered with one or more layers of connective tissues, the latter is about 2500 Å thick. The space between the outer and the inner layer is occupied by muscle fibres, nerve fibres with or without granules, glial cells and tracheols. Longitudinal and circular muscle fibres also occur in aorta.

On the basis of morphology of granular material, the axon-terminals and pre-terminals have been classified by Unnithan *et al.* (1971) into three types viz. (i) those with spheroid electron dense granules of 1000-2600 Å in diameter, (ii) with elongate electron dense granules of 2100 x 1200 Å in diameter and (iii) with less electron dense spheroid - oval granules of 900 - 2400 Å in diameter. These three types of granules are also reported to occur in the NCC axons. The axons and their terminals contain neurosecretory granules, mitochondria neurotubules, lysosomes and multivesicular bodies. Unlike other heteropteran insects in *Metochus* the NCC-1 after entering the aorta ends in a bulbous structure. From the latter, on each side arise about 20 or more secondary axons that spread out in the inner aortal wall. The bulbous structure of NCC-I, the neurosecretory axons and the intercellular spaces in the aortal wall, all contain abundant NSM (Awasthi, 1973c). The studies conducted on a very large number of crypto- and gymnoceratan Heteroptera, has clearly shown that in this order cephalic aorta functions as the principal NHO; the NSM of A - cells of cerebral origin is not found in the corpora cardiaca. However there are only a few exceptions e.g. *Rhodnius*, where it has been conclusively reported that the CC performs the function of a neurohaemal organ.

Likewise, a mention must be made here about the function of a principal NHO in Dermaptera. Although Gabe (1966) while reviewing insect neurosecretion pointed out that in Dermaptera, (*Forficula auricularia, Labidura riparia* and *Anisolabis maritima*) the NCC-I and NCC-II, formed by the axons of the medial and lateral neurosecretory cells of the brain respectively, joins the rostral part of the corpora cardiaca. Thus, the corpora cardiaca, according to him functions as neurohaemal organ in Dermaptera, but the observations of Awasthi (1975c, 1976a, b) on *Labidura* and *Euborellia* have clearly revealed that in these earwigs the CC are totally devoid of PF, PARF and PAVB positive NSM. In these earwigs, the medial neurosecretions are stored only in the cephalic aorta or the anterior aortal wall, and not in the corpora cardiaca, The function of aorta as a principal NHO in Dermaptera has been further studied by Awasthi (1976c) with electron microscope.

The ultrastructural studies of aorta of *Euborellia* have shown that it is covered on both inner and outer surfaces by an extremely thin stromal lining. The most notable feature of this organ is that almost the entire space between the inner and outer stroma is occupied by two types of terminals and pre-terminals of neurosecretory axons whose granules differ in size and electron density. The first type of axon (NS-I) most abundant in the entire aorta contains large electron dense granules of about 1000-2200 Å in diameter. The second type of axons (NS-II) contain small, mixed granules (electron dense as well as electron lucent) 400 - 900 Å in diameter. It has been suggested that the large electron dense granules belong to A-cells of the brain. Likewise a few axon terminal and pre-terminals found among A-cell axons appear to belong to the medial B-cells; and the small less electron dense granules (NS-II) seem to represent B-cell secretion of the medial groups. Glial cells and muscles are poorly developed in the cephalic aorta. The large electron dense granulas are found only in the cephalic aorta and B-fibres occur in abundance in the corpora cardiaca but less so in aorta. The electron microscopic study has confirmed that in Dermaptera, the aorta serves as a principal NHO for the products of the medial neurosecretory cells of the brain (Awasthi, 1976c).

Em studies on *Euborellia* have confirmed the role of the aorta as a neurohaemal organ for the medial neurosecretory cells and the corpora cardiaca for the lateral neurosecretory cells of the brain. The corpora cardiaca also act as a reservoir for the intrinsic secretions (Awasthi, 1976c). Thus light and electron microscope studies have clearly demonstrated that in Heteroptera and Dermaptera, aorta performs the function of a principal neurohaemal organ. The corpora cardiaca in these two orders act as the NHO for the lateral neurosecretions as well as intrinsic secretions of the gland (for details see: Dogra, 1967a,b, 1969; Awasthi, 1969, 1972a, 1973bc, 1975c, 1976bcd, 1980a; Srivastava, 1969, 1970; Srivastava and Dogra, 1969; Srivastava and Tiwari, 1973; Srivastava *et al.,* 1974; Tiwari and Srivastava, 1975; Tiwari *et al.,* 1975).

OTHER NEUROHAEMAL ORGANS

Corpus allatum as a neurohaemal organ

It is an established fact that the corpus allatum hormone (CAH) maintains larval characters in young stages and in adult insects CAH regulates a number of metabolic processes (see : Doane, 1973; Slama *et al.*, 1974). However, these epithelioid non-nervous glandular bodies act as NHO for NSM of cerebral origin in certain insects. Kobayashi and

Yamashita (1959) as well as Ichikawa and Nishiitsutsuji - Uwo (1959) are of the view that the CA of *Bombyx mori* and *Philosamia cynthia ricini* store and release NSM, elaborated in the neurosecretory cells of the brain. Singh and Awasthi (1981b) and Awasthi and Singh (1986b) have shown that the corpora allata in *Amsacta collaris* and *Achoea janata* act as NHO. Ultrastructural studies of Nishiitsutsuji-Uwo (1961) have shown that NSM of cerebral origin is stored in the CA and thus the latter act as NHO in Lepidoptera.

Perisympathetic Neurohaemal Organs

It is an established fact that the ganglia of the ventral nerve cord in insects are provided with neurosecretory cells. The NSM produced in them is either stored in the neuropile mass or the terminals of the neurosecretory axons. Due to the accumulation of a large amount of NSM, these small areas appear as "ganglia like" swellings. The peripheral disposition of these "ganglia like" swellings which are designated as perisympathetic neurohaemal organs (PNOs) or perisympathetic organs (POs), close to the circulating haemolymph, facilitate the liberation of neurohormones. Raabe (1965, 1966) probably for the first time suggested that the medial nervous system in insects contain many neurohaemal organs. Working on the central nervous system of stick insects viz. - *Clitumnus extradentatus* and *Carausius morosus* Raabe (1965, 1966) observed ganglia like swellings along the central nervous system. These swellings were not positive to CHP and AF but were positive with Azan and were very similar to the corpus cardiacum. Chalaye (1966, 1967) also observed such swellings along the medial nervous system of *Locusta migratoria*. Raabe (1965, 1966) on the basis of histological observations suggested that the NSM of NSCs of the ganglia of ventral nerve cord is released from segmentally arranged NHOs which appears as swellings on the medial or transverse visceral nerves and are called perisympathetic neurohaemal organs (PNOs).

A considerable detailed information is available on the subject which has been reviewed by Awasthi (1976e). Ultrastructural studies conducted by several workers on these swellings and their role as NHO has been confirmed. Provansal-Baudez and Baudry-Partiaoglou (1983) have given a comprehensive account on the POs in three coleopteran insects. Ultrastructural comparison of different types of perisympathetic organs (POs) in three species of Coleoptera (*Chrysocarabus auronitens, Oryctes rhinoceros*, and *Tenebrio molitor*) showed that the structure of these organs was not related to their morphological types but to their topography. Two kinds of PO structure may be distinguished : compact median and diffuse lateral. They were similar in that both were surrounded by thin neural lamellae and exhibited numerous glial cells originating in the perineurium (type I perineurial cells) as well as abundant neurosecretory endings. They were different in as much as in median POs, the neurosecretory endings were generally surrounded by perineurial processes but in transverse POs, these endings were sheathless. Only one type of neurosecretory axon was distinguished in the median organs but three or four in the transverse. The nature of the processes by which neurosecretory granules are released may depend on the type of neurosecretory axon. For instance, exocytosis always occurred for dense spherical granules, and granule fragmentation was visualized for granules of smaller size. (Provansal-Baudez and Baudry- Partiaoglou, 1983). Raabe *et al.* (1974) have also given an account of the work done on POs in insects.

Brady and Maddrell (1967) have studied the ultrastructure of the neurohaemal organs in the medial nervous system of many insects and found segmentally arranged swellings in *Carausius morosus, Schistocerca gregaria* and *Periplaneta americana*. These swellings show that they are not ganglia but contain numerous neurosecretory axon branches and suggests that they serve to release neurosecretory material into the haemolymph. In *Carausius* and *Schistocerca* the material occurs as electron-transparent vesicles respectively 1100-2500 Å and 400-800 Å in diameter but in *Periplaneta* as a mixture of electron transparent vesicles 800-1500 Å in diameter and electron-dense granules 550-1000 Å in diameter. In addition, a small number of neurosecretory axons containing classical electron dense granules have been observed. Thus, these organs contain at least two types of neurosecretions. Ultrastructural evidence suggests that the electron transparent vesicles in *Carausius* and *Schistocerca* probably comprise

"C-type" neurosecretory material, which stains with Azan but not with paraldehyde fuchsin (Brady and Maddrell, 1967).

Further, a comparative ultrastructural study of insects was made by Baudry-Partiaoglou (1986), whose primitive perisympathetic organs consisted either of a single median neurohaemal formation per segment (*Periplaneta americana*) or of two transverse formations (*Carausius morosus*), or again, of three- one median and two transverse - as in *Locusta migratoria*. The results showed that the perisympathetic organs always comprised two types of structure (α and β), even when a single formation exists. In *Periplaneta* and *Carausius*, both types were seen to coexist in the same neurohaemal formation. In *Locusta*, however, they were separated, the α structure constituting the median organ and the β structure, the two transverse organs. Each of the two structures have special features; the β structure displays the usual characteristics of a loose neurohaemal organ, i.e., it is penetrated by sinuses and has neurosecretory endings devoid of a glial coat; the α structure, on the contrary, forms dense compact organs whose endings are covered with a continuous glial layer. During the evolution, this last structure gradually separated from the other neurohaemal formations. It was observed to correspond to a single type of neurosecretory cell, distinct from all other cell types by the particular mode of its release. These results for primitive perisympathetic organs confirm earlier findings for advanced organs. They indicate that both the α and β structures are generally present in insects, thus showing their importance in the physiology of these animals (Baudry-Partiaoglou, 1986).

The Nervi Corporis Allati-II in Gryllidae

In gryllids, from each side of the CC, a short nerve arise viz nervus corporis allati-I (NCA-1). These nerves, after innervating the corpus allatum of the ipselateral side emerge as nervus corporis allati-II (NCA II). The latter finally joins the suboesophageal ganglion. In the house cricket *Acheta* (Gaude and Weber, 1966, Weber and Gaude, 1971) and *Gryllodes* (Awasthi, 1968, 1969a) the swollen part of NCA-II acts as NHO. The content of the swollen or bulbous part of NCA-II is similar to that found in the corpora cardiaca. It has been reported that most of the fibre of NCC terminate within the CC but a number of them emerge from the CC and form NCA-I, supply CA and then emerge to form a swollen structure or bulbous structure along the mid way of NCA-II (Awasthi 1968, 1969a). Ultrastructural studies of Weber and Gaude (1971) on *Acheta* have confirmed the role of NCA-II as NHO in Gryllids and their contents are released in the same manner as from the other NHOs. Thus, in *Acheta, Gryllodes* and other gryllids NCA-II performs the function of a NHO (for details see : Awasthi 1974, 1976e).

In addition to above mentioned NHOs, there are other neurohaemal organs like lateral cardiac nerves in *Periplaneta* (Johnson, 1966a), intraganglionic NHO in *Hydrophilus* (Gundavia and Ramamurty, 1972) and hypocerebral ganglion in *Epilampra* (Awasthi, 1975b).

NEUROHAEMAL ORGANS OF APTERYGOTA

In Collembola the NCC-I and NCC-II after emerging from the brain fuse to form a common nerve (NCC) which lies ventral to the oesophagus. The terminals of the common NCC in each side swells up to form a bulbous structure. This swollen structure functions as NHO, some times the NHO is associated with the cephalic blood sinus (Cassagnau and Juberthie, 1967).

In Thysanura the CC performs the function of a neurohaemal organ (see : Gabe, 1966). In Campodidae (Diplura) the CC are absent (Hanstrom, 1940, Cazal 1948) but paired corpora cardiaca occur in Japygidae. The neurosecretory pathways in Japygidae resembles to that of pterygote insects, the two pathways composed of axons of neurosecretory cells of the brain, decussate in the pars-intercerebralis. Thereafter, traversing the brain emerge from the posterior part of the brain as NCC-I. The NCC-I of each side joins the CC of the same side. A copious amount of NSM is present in the CC which clearly suggests their role as a neurohaemal organ.

4

THE CORPORA ALLATA

The secretory nature of the corpora allata (CA) was first suggested by Nebart (1913). Their endocrine function was indicated by Ito (1918) and the function was elucidated by Wigglesworth (1935). Cazal (1948) classified them into four types (see below). Wigglesworth (1954) clearly suggested that the CA is the source of juvenile hormone (JH) in *Rhodnius*. It is well known that they are the site of origin of JH in all insects and their significance in controlling various vital events is well documented (see : Engelmann, 1968, 1970; Novak, 1975, Doane, 1973, Raabe, 1982, 1989 and others).

STRUCTURE OF THE CORPORA ALLATA

The corpora allata are non nervous, epithelioid, oval, spherical, elongate or leaf like glandular bodies which lie on either side of the corpora cardiaca. They are paired structures in most of the insects but in some insects eg. - Dermaptera, Gymnoceratan Heteroptera and Diptera they are unpaired bodies. Each corpus allatum is connected with the corpus cardiacum of the same side by nervi corporis allati-1. (NCA-I) which carries fibres from the neurosecretory cells of the brain. The glands consist of parenchymal cells and are innervated by the neurosecretory axons of cerebral origin. On the basis of their structure four types of corpora allata have been recognized viz.

(I) The Vesicular type - In Thysanura and Phasmida the corpora allata are hollow balls and the parenchymal cells are distributed in the peripheral zone.

(II) **The Pseudolymphoid type** - These are formed by small parenchymal cells with reduced deeply staining cytoplasm resembling the lymphoid tissue of vertebrates. They are found in Odonata and Ephemeroptera.

(III) **The Microcellular type** - The corpora allata of this type are most common and are found in Orthoptera, Dictyoptera, Coleoptera Hemiptera etc. They are composed of numerous small parenchymal cells.

(IV) **The Macrocellular type** - As the name indicates, they are composed of large parenchymal cells and their number is small. They have large lobulate nuclei. They are found in Hymenoptera, Lepidoptera, Diptera and Trichoptera.

ULTRASTRUCTURE OF THE CORPORA ALLATA

Ultrastructural studies have shown that the cells of the active corpora allata are characterized by increased cytoplasmic and nucleolar volume as compared to the cells of inactive gland (Slama *et al.*, 1974). Scharrer (1964a) has studied the ultrastructure of CA in *Leucophaea*, Waku and Gilbert (1964) in *Hyalophora*, King *et al.* (1966b) in *Drosophila,* Panov and Bassurmanova (1970) in *Eurigaster*, Joly *et al.* (1969) in *Locusta*, Odhiambo (1966) in *Schistocerca*, Fukuda *et al.* (1966) in *Bombyx*, and Thomsen and Thomsen in *Calliphora* (1969, 1970). Most of the investigators agreed that the spherical, ovoid, irregular nuclei with smooth membranes were either larger or more numerous in the active glands. The cytoplasm of these cells contain numerous mitochondria which are elongate with well developed cristae, suggesting thereby that they are the site of high respiratory metabolism. The cytoplasm also contains free and abundant ribosomes. The Golgi apparatus is usually inconspicuous and do not show intensive secretory activity. The plasma membrane is folded and the cell membranes are interdigitate. The intercellular spaces contain neurosecretory axons and tracheoles.

The stromal sheath is acellular. In most of the insects the cells of CA have well developed agranular or smooth surfaced endoplasmic reticulum. The CA show cycles of secretory activity associated with which are the changes in their size. The active glands have large cells with a lot of dense cytoplasm and scattered nuclei. The active gland also shows a marked increase in cellular and nuclear size as well as a shift in nuclear - cytoplasmic ratio. The inactive gland is small with a poor amount of cytoplasm and aggregated small nuclei. Further, the inactivity of the CA is accompanied by a decrease in number and a change in the appearance of some cytoplasmic organelles. The mitochondria tend to be smaller and the ergastoplasm is reduced to scattered wisps of ribosome - studded membranes (Scharrer, 1964a).

According to Scharrer (1964a) aside from changing outlines, all corpus allatum cells have long, gradually thinning processes. These penetrate deeply into the parenchyma, where they interlock with those of other cells, many processes eventually seem to reach the surface of the gland where the secretory products are released into the haemolymph. These have to pass through an acellular connective tissue layer that shares tinctorial and ultrastructural properties with those of a boundary or basement membrane. This stromal element forms a sheath and branching processes that extend into and permeate (i.e. diffuse through) the parenchyma. It seems to represent a system of channels not only for the release of the secretory and other cellular products but for the entry of nutrients and perhaps chemical messenger substances. She has further remarked as in other cases, the neurosecretory material can be observed in the form of structurally distinctive elements i.e. as electron opaque granular inclusions within axon-terminals that become contiguous with corpus allatum cells (Scharrer, 1964a).

Almost similar observations have also been made on the corpus allatum of other insects so far studied e.g. Waku and Gilbert (1964) have studied the ultrastructure of the corpora allata of *H. cecropia* and observed changes strikingly during pupal-adult transformation. In the diapausing male pupa, which has inactive glands, the cell organelles are generally small and few in number. In the young adult, a stage where JH is actively secreted the cell organelles are much larger and in greater concentrations. The mitochondria become large and numerous with many cristae packed in an electron dense matrix and a complex endoplasmic reticulum fills the cytoplasm. The Golgi apparatus on the other hand does not change markedly and remains inconspicuous. Nuclei of the adult cells contain more numerous chromatin blobe than those of the pupa. Both adult and pupal corpora allata are covered by a thick acellular basement membrane which involutes deeply into the parenchyma particularly in the pupa.

The plasma membranes of the peripheral cells of the adult gland are extremely convoluted. The secretory product of the active gland was not identified with certainty. However, two types of characteristic electron dense bodies which may represent moieties of the secretory substance are seen both on the surface and within the adult cells. They are absent from the inactive pupal cells. Neurosecretory axons penetrate and appear to innervate both the pupal and adult corpora allata. Cell organelles in the senescent moth show marked signs of aging and degeneration (Waku and Gilbert, 1964).

In the slug moth prepupae, *Monema flavescens*, before beginning of diapause, the secretory cells of the corpus allatum increase in size with the enrichment of the cytoplasm in quantity and in number of organelles, while RNA synthesis also starts. As diapause progresses, the secretory cells have many large unstained vacuoles in the cytoplasm, which were demonstrated to contain some substances of lipoidal nature. It is not clear that the substance is juvenile hormone itself or a material related to juvenile hormone. Agranular ER is the most characteristic organelle connected with mitochondria and situated around the vacuoles during diapause. Electron micrographs show that agranular ER and mitochondria have an essential role for the production of juvenile hormone (Takeda, 1977).

Awasthi (1976c, 1979b) has described the ultrastructure of corpus allatum of an earwing, *Euborellia annulipes* and reported that the gland consists of compactly arranged parenchymal cells with distinct cells boundaries. The cells contain a large number of mitochondria of different shapes, oval, spherical, elongate and Y-shaped. Golgi bodies, and endoplasmic reticulum is poorly developed. Electron dense inclusions are found in abundance in the cytoplasm of CA cells which are supposed to be the lysosomes. The gland is innervated by neurosecretory and non- neurosecretory axons. The nuclei of the corpus allatum cells are oval and fairly large occupying most of the cell. The nuclear membrane is double layered with distinct foldings. Occasionally, the cells are separated by glial processes and are associated with tracheoles. The corpus allatum is devoid of axons with large electron dense granules of medial A - cells which are confined to the aorta only and are also absent in corpus cardiacum.

The corpora allata secrete juvenil hormone (JH) which controls metamorphosis and yolk deposition in the eggs. A number of other functions like tanning of the cuticle, green/brown polymorphism in Locusts, mating behaviour in social and other insects are performed by the juvenile hormone. Since in young stages it maintains larval characters and inhibits adult differentiation it is also called as neotenin.

SPECIAL INCLUSION BODIES OF THE CORPUS ALLATUM

Scharrer (1971) presented ultrastructural evidence for the manufacture, storage and release of a distinctive cellular product into CA cells of the insect *Leucophaea*. According to Scharrer (1971) an exclusive feature of the *Leucophaea* corpus allatum are characteristically structured bodies which are found both intra-and extracellularly. They have been referred to as 'C' bodies, a neutral designation used for the time being for lack of a better term. The material giving rise to C-bodies, originates in the Golgi zone and acquires a characteristic, regularly structured appearance of exceptionally high electron density. The C-bodies are of different shapes and sizes.

The differential distribution of this easily identified product in various sites in the organ (CA) permits the re-construction of a sequence of dynamic events involving transport from the intra-to the extra cellular-compartment which affords access to the circulation. A small amount of C-body material becomes incorporated into the multivesicular bodies. Another cytoplasmic compartment into which C-body material may find its way is the smooth surfaced endoplasmic reticulum. However, in the absence of a definite proof of direct communication between Golgi elements and agranular reticulum, no conclusion can be drawn about the possible participation of latter organelle in the maturation and intracellular transport of this material. What can be stated however is that, during their intra-cytoplasmic existence, C-bodies undergo some transformation. Confluence of small, solid elements yields inclusions which as they become larger, acquire the typical "perforated" appearance. Depending on their size, some of the C-bodies may have only one "hole" in the centre, some perhaps 3 or 4 and others many more than that (Scharrer, 1971).

It has been further remarked that the variability in the occurrence, distribution and fate of the C-body material during various periods of the animal's life cycle and under experimentally altered conditions, suggests a relationship with the glandular (i.e. secretory) functions of the corpus allatum. Conspicuous deposits of C-body material, especially in the extra-cellular stroma, parallel situations in which hormonal activity of the organ seems to be low or temporarily suspended. Thus, ultrastructural manifestation of intermittent sluggishness in the system, as observed in *Leucophaea*, provide valuable cues for the existence of a periodically changing production line involving a specific glandular material (for details see : Scharrer, 1971).

5

THE PROTHORACIC OR ECDYSIAL GLANDS

Although the prothoracic glands of *Cossus cossus* were described by Lyonet in 1762 but till 1902, a difinite information was not available on them. Toyama (1902) identified these glands as epidermal invagination of the labial segment of the head. Ke (1930) extensively studied these glands in lepidopteran larvae and named them prothoracic glands on the basis of their location. Later numerous workers have experimentally demonstrated that they secrete a hormone needed for moulting and metamorphosis (Fukuda, 1940, 1941, 1944, Bodenstein, 1936, Williams, 1946, 1948 etc.). Herman and Gilbert (1966) proposed a new name to the prothoracic glands viz. "ecdysial glands" which was later followed by a large number of workers. Now both the terms, prothoracic and ecdysial glands are in common use, one based on the location and the other relates to function. The prothoracic glands or ecdysial glands (causing moulting or ecdysis) have been discovered in the larval stages of all insects (see : Herman, 1967). Extensive work has been done on the above mentioned glands in the past few decads which has been discussed in several reviews (Gilbert, 1964; De Wilde, 1964; Wigglesworth, 1964; Scharrer, 1965; Novak, 1966, 1975; Doane, 1973; Raabe, 1982, 1989; Slama *et al.*, 1974 and many others).

THE PROTHORACIC OR ECDYSIAL GLANDS

The prothoracic or ecdysial glands (PGsor EGs) are paired glandular structures located in the ventro lateral areas of the prothorax extending into the mesothorax in some species and some time partially fused in the median line. They are often clearly associated with the lateral tracheal trunk near to the thoracic spiracle. In same insects eg. *Bombyx* and *Achoea* they appear as compact bodies while in others they may be diffusely distributed in the prothorax and mesothorax as in *Philosamia* and *Hyalophora*. Herman and Gilbert (1966) have classified the insect orders into 2 main groups on the basis of the anatomy of EGs. In the Ist group, the gland cells are few but large and more are less diffusely distributed in the thorax and these glands are termed as "diffused type" In the 2nd group, the glands are composed of compact numerous small cells as in *Bombyx* and *Achoea* and are known as "compact type". In *Amsacta* they are of intermediate type hence named as "compact cum diffused type" (Awasthi and Singh, 1983).

Like corpora allata, they are also non-nervous epithelioid bodies. In the living condition the gland cells appear transparent. In each larval instar there is a distinct cyclicity in the secretory activity. In the EG of insects there occur certain structural changes which may indicate secretory activity or inactivity of the glands. These criteria of activity include for example : changes in the total volume of the gland, cytological changes in the shape and size of the nucleus, changes in the chromatin patterns, occurrence of cytoplasmic vacuoles, and development of other cytoplasmic organelles. Without respect to the type of growth, which can be endomitotic in most Endopterygota and mitotic in Exopterygota, the gland volume increases successively during the period of larval growth in all insects. The growth increments measured in separate larval instars are roughly proportional to growth of the whole body or growth of some other organs. The maximum size of the ecdysial glands is reached in the last larval instar. Within each larval instar there occur considerable variations in the glandular volume, as well as in the size and number of the parenchymal cells. These changes are precisely synchronized with the course of the intermoult period. Maximum growth and probably also maximum secretory activity occurs usually at the first third of a larval instar. This period is characterized by maximum growth of the whole body and by the beginning of the mitotic divisions in the epidermis (Slama *et al.*, 1974; Novak, 1975).

Further, in Apterygota where the EG function during the adult stage there occurred cyclical changes in the gland volume with maximum growth towards the middle of the intermoult period and minimum at the time of ecdysis. The difference between the maximum and minimum volume was 3 to 1 (Watson, 1964). In cockroaches which represent Exopterygote insects, there also appeared morphological and physiological changes in the EG during the larval period. The changes had cyclical character and were related to the moult cycles. For example, there were changes in structure and size of the mitochondria, changes in internal structure of the nucleus, etc. However, the size of the nuclei did not change at the corresponding moments of the successive larval instars. Furthermore, mitoses were most common at the middle of the intermoult period (Scharrer, 1964b). In *Locusta*, mitoses in the EG appeared two days after the ecdysis, e.g. significantly before the appearance of mitoses in the epidermis. The gland volume increased faster than that of the whole body until 4th larval instar. Since then the gland volume still increased but less than that of the body.

In *Drosophila*, which represents Endopterygote insects, the nuclei progressively increased in size during the period of larval growth. They became highly polyploid and the secretory activity of the gland was probably increased. It was expected that the gland released more hormone before puparium formation than during the period of larval growth (King *et al.*, 1966a). Comparisons between the volumetric increase of EG and CA during the larval stage have indicated large differences, especially with respect to the last larval instar. Thus, in young larval instars of *Pieris* the ratio of CA to EG volumes was 1 to 2, while in the last larval instar it was as much as 1 to 29. The EG volume was doubled throughout the last larval instar. It was suggested that maximum secretory activity occurred 2 days after each ecdysis in young larval instars. During metamorphosis of *Tenebrio* the EG may attain maximum activity 3 to 4 days before ecdysis, e.g. 2 days before the gland volume begins to decrease (see : Slama *et al.*, 1974).

At rest the nuclei are small and oval but in the active gland they become enlarged and lobulated, the basophilic cytoplasm is deeply stained. Furthermore, it has been observed that in *Amsacta* and other moth larvae the initiation of the secretory activity takes place from the nucleus. The active gland cells are characterized by having the highly branched nuclei reaching to the cell periphery. They are provided with abundant cytoplasm, numerous vacuoles and well developed brush borders. The glands are totally inactive in diapause pupae which are characterized by a large and compact nucleus and a small amount of cytoplasm devoid of vacuoles (Awasthi and Singh, 1983). The ecdysial glands degenerate in majority of the insects after the final moult but they remain preserved in some cases like *Schistocerca* and *Amsacta*. The glands are innervated by the nerves issuing from the suboesophageal, prothoracic and mesothoracic ganglia. The prothoracic glands in insects in general are innervated at least by 4 pair of nerves viz.: (1) one pair of suboesophageal transverse (2) one pair of prothoracic transverse (3) one pair of mesothoracic transverse and (4) the branches of medial-1. In some cases six pairs of nerves have also been reported (Singh and Awasthi, 1980).

Ultrastructural features of the prothoracic glands of cockroaches viz. *Leucophaea* and *Blaberus* have been studied during larval growth and degeneration by Scharrer (1964b, 1966).

Scharrer (1964b) remarked that the ultrastructural characterization of the prothoracic glands derives interest from the fact that they are the only known source of a steroid hormone (ecdysone) among invertebrates. Scharrer (1964b) studied the nymphal glands of *Leucophaea* and *Blaberus* at graded intervals over a period of several intermoult cycles. According to her among the prominent cytological features of this gland are long and intricately interwoven cellular processes that may be separated from each other by extracellular channels with an average width of 0.5 microns. The plasma membranes of the cells facing these spaces are coated by an external lamina (the boundary membrane) representing extensions of the same extracellular material that provides the connective tissue sheath of this and other insect organs. Micropinocytotic caveolae and vesicles present along this large cellular surface area may facilitate the transport of material between glandular cells and haemolymph or vice versa.

The cytoplasm contains a fairly moderate amount of membranous structures. Golgi units are distinctive but never spectacular, ergastoplasmic elements are sparse in most cells. This sparsity as well as the absence of membrane bounded secretory granules is in line with the known steroidal nature of the secretory product involved. Some cells show a striking array of microtubules. Certain structural attributes of nucleoli, mitochondria and lysosome-like bodies seem to vary in conjunction with distinct phases of the intermoult cycle.

Regarding the degenerating parenchymal (prothoracic gland) cells Scharrer, (1966) has pointed out that the early signs of breakdown appear in groups of nuclei whose substance becomes segregated into patches of contrasting electron density characteristic of pycnosis.

The most conspicuous change in the cytoplasm of parenchymal cells concerns the appearance of large heterogeneous inclusion bodies in which various cellular elements become segregated. These compartments seem to represent autophagic vacuoles within which the gradual degradation of much of their contents takes place, presumably under the influence of lysosomal enzymes. Undigested swirls of membranous character may remain sequestered within these packets for some time.

At advance stages of cellular atrophy, plasma membranes and nuclear envelopes have gradually disappeared and masses of protoplasm undergoing autolysis become invaded by a greater number of haemocytes that are present in the nymphal glands. These phagocytic elements appear to engulf debries of parenchymal cells as well as some degenerating connective tissue elements. After the completion of the regressive process, the axial band of musculature characteristic of the nymphal gland persists on its own. The resorption of the prothoracic gland in the newly emerged insect is the result of physiological autolysis and seems to be aided by the activity of phagocytic haemocytes (Scharrer, 1966).

Similar features are observed in the prothoracic gland-cells of *Drosophila* by King *et al.* (1966a). The secretory products are also thought to be manufactured by the smooth surfaced endoplasmic reticulum. Joly *et al.*, (1969) found extremely well developed Golgi apparatus, a rich ergastoplasm, abundance of free ribosomes and cytoplasmic vesicles in the prothoracic glands of *Locusta*.

THE WEISMANN'S RING*

It is also known as ring gland. It is found in the larvae of cyclorrhaphan Diptera like - house flies, flesh flies and blow flies etc. The gland is formed by the corpus cardiacum-hypocerebral ganglion, corpus allatum and prothoracic glands. The basal part of the ring is formed by fused corpus cardiacum plus hypocerebral ganglion, and is composed of chromophil or intrinsic secretory cells. The sides are formed by the elongated cells of ecdysial or prothoracic glands. The glands are highly innervated. The upper part of the ring is formed by a corpus allatum. The central part of the ring gland is occupied by the cephalic aorta. It is connected to the brain by nerves and is also richly supplied with trachea. In adults, the cells of the prothoracic glands degenerate before the imaginal ecdysis and the corpus allatum is connected to the corpus cardiacum-hypocerebral ganglion by nerves. The ring gland receives neurosecretions via neurosecretory axons of the brain.

Ultrastructure of the ring gland of *Drosophila melanogaster* during second and third larval instars have been described by King *et al.* (1966a). This organ contains tissues of four types : (1) corpus allatum cells, (2) prothoracic gland cells (3) corpus cardiacum cells and (4) tracheal cells. Axons bearing spherical neurosecretory granules pass from the corpus cardiacum to the corpus allatum. Ecdysone and juvenile hormone are thought to be manufactured by smooth surfaced endoplasmic reticulum contained in the prothoracic gland cells and the corpus allatum cells respectively. The morphology of these cells at the two developmental stages studied suggests that the switch over from larval to pupal development is preceded during the third larval stage by an increasing hormone secretion by the prothoracic gland cells and a decreasing secretion of the cells of the corpus allatum.

---

* For the sake of convenience account of the ring gland is given in this section, though only a part of the ecdysial gland is associated with the ring.

6

THE CHEMICAL NATURE OF INSECT HORMONES

A brief mention may be made here of the chemical nature of insect hormones especially brain or activation hormone (BH or AH), ecdysone or moulting hormone (MH) and corpus allatum or juvenile hormone (CAH or JH).

BRAIN HORMONE OR ACTIVATION HORMONE

There is still some controversy regarding the chemical nature of AH or BH. Gersch (1961, 1972), Gersch and Sturzebecher (1970) concluded that the activation hormone, was a protein or peptide, likewise Novak (1975) has remarked polypeptide nature of activation hormone, secreted by the medial neurosecretory cells of the brain. Ichikawa and Ishizaki (1963), and Nishiitsutsuji-Uwo (1972) are also of the view of protein or peptide nature of BH or AH.

On the other hand, Williams (1963, 1968) suggested that the BH (AH) is mucopolysaccharide. Kobayashi and Kirimura (1958), Kobayashi *et al.* (1962) and Gilbert (1964) concluded that BH is steroidal, (cholesterol) in nature. Further studies of Kobayashi and Yamazaki (1966) confirmed the proteinic nature of brain hormone. Thus, most of the workers are of the opinion that the activation hormone or brain hormone is peptide (protein) in nature. But Slama *et al.* (1974) have remarked that so far nothing definite can be said about the chemical nature of AH except that it seems to be a protein with molecular weight between 10,000-50,000. Raabe, (1982) concluded that various investigations have been conducted by different teams in an attempt to isolate brain hormone. It has not yet been accurately identified but is believed to be a protein with a high molecular weight, and for more details see Raabe (1989).

ECDYSONE OR MOULTING HORMONE

Slama *et al.* (1974) have remarked that some ultrastructural patterns, such as the rich perinuclear zone with abundant ribosomes, appearance of the Golgi complex and a well developed ergastoplasm in some species indicated secretion of proteinaceous material but other features such as the well developed smooth surfaced endoplasmic reticulum, vesicles released into inter-cellular spaces, lamellated structures or abundant glycogen in the chondriome as found in some other species, suggested the ultrastructural signs of steroid secretions are also present.

Butenandt and Karlson (1954) extracted crystalline moulting hormone, which they called "ecdysone". Karlson and Hoffmeister (1963) and Karlson *et al.* (1963) on the basis of their extensive work, concluded that the ecdysone is steroidal in nature. A large number of other workers have confirmed the conclusion of Karlson and his co-workers that the ecdysone or moulting hormone is steroidal in nature and cholesterol is the precursor of the ecdysone.

CORPUS ALLATUM HORMONE OR JUVENILE HORMONE

It has been suggested that the corpus allatum hormone (CAH) is synthesized in the tubules of endoplasmic reticulum, transported into vesicles, taken up and dissolved by lipid droplets and finally released in lipid-droplets into haemolymph. This would be consistent with the assumption of the lipid nature of the hormone (Slama *et al.* 1974). It was also proposed that the CAH might possibly be a steroid (Gilbert, 1964), a peptide or protein. The assumption that CAH was a protein was based on certain ultrastructural features of corpus allatum cells and also on morphological and functional analogies between CA and adenohypophysis of vertebrates (Novak, 1975).

Since corpus allatum hormones or juvenile hormones are highly lipophilic substances, which are poorly soluble in water, it was postulated that the hormones would circulate in the insect haemolymph bound to a hydrophilic carrier such as a protein or lipoprotein (Wigglesworth, 1964). Such carriers have now been identified in a number of species. Highnam and Hill (1979) have remarked that the insect JH was a farnesol or some very closely related terpenoid compound, and when Schmialek in 1963 isolated farnesol from the male *Hyalophora cecropia*, the issue seemed to be beyond doubt (see : Highnam and Hill, 1979).

Gilbert (1964), Wigglesworth (1964, 1970), Doane (1973), Slama *et al.* (1974), Novak (1975), Highnam and Hill (1979), Raabe (1982, 1989) and others have discussed the chemical nature of brain hormone or activation hormone, moulting hormone or ecdysone and corpus allatum hormone or juvenile hormone in considerable detail.

7

THE RELEASE SITES OF NEUROHORMONES

Numerous workers like Smith and Smith (1966) in *Carausius morosus*, Johnson (1966) in *Calliphora stygia*, Scharrer (1968) in *Leucophaea maderae, Periplaneta americana, Blaberus cranifer*, Normann (1965, 1974) in *Calliphora erythrocephala*, and others have studied in considerable detail the ultrastructure of the sites and machanisms by means of which the neurosecretory products are released from their sites of origin and reaches the target organs. There is no fixed way for all insects. In different insects the neurosecretory material reaches to the target organs in different manners.

In the following pages a brief account of the mechanisms of release of neurohormones produced by the neurosecretory cells of various ganglia, and how they reach to their target organs is given.

A direct transfer of neurohormones along the axons probably occurs in Aphids (Johnson, 1963). Similarly neurosecretory axons of cerebral origin pass through the CC of *Calliphora, Sarcophaga* and other insects to the pricardial cells, aorta, corpora allata, ecdysial glands, alimentary canal and reproductive organs.

In addition to above manner (i.e. a direct transport of secretory product), majority of the workers have described through ultrastructural studies that the neurosecretory material in insects is released in two main ways viz. (1) by exocytosis and (2) by the fragmentation of granules and diffusion of their contents.

(1) BY EXOCYTOSIS

It is well known fact that the neurosecretory granules are encased in a delicate membrane contributed by the Golgi complex. This membrane fuses with the plasma membrane of the cells, the wall in between dissolves as a result of enzymatic activity and the contents of the droplets or vesicles are released in the extracellular or extra axonal space as the case may be e.g. *Carausius morosus, Calliphora erythrocephala* and many others. Neurosecretion by exocytosis has been given in considerable detail by Normann (1965, 1974, 1976) and Smith and Smith (1966).

According to Normann (1965) the corpus cardiacum neurosecretory cells of the blow fly, *Calliphora erythrocephala* release the contents of their granules by exocytosis; and the presence of omega-like figures is the characteristic ultrastructural indication of this process. Omega figures are formed by fusion of granule membranes (*vide supra*) with axolemma, the contents thus being voided and allowed to dissolve and escape by diffusion. Clusters of small vesicles (300-400 Å) at the axolemma can also be observed in actively secreting axonal endings. These configurations were previously interpreted as indications of a different release mechanism viz. intracellular fragmentation of granules. However, substantial evidences supports the idea that in neurosecretory cells "omega figures" and "vesicles clusters" represent consecutive phases of exocytosis viewed as a dynamic sequence of membrane phenomena (Normann, 1969). The microvesiculation provides for membrane retrieval and prevents additions of excess membrane area to axolemma during secretion (Normann, 1970). Experimental work of Normann (1974) have revealed that exocytosis of neurosecretory granules in the corpus cardiacum of the blow fly, *Calliphora erythrocephala* was induced both by electrical stimulation and by depolarization with K^+. This secretory activity was quantitatively determined by the frequency of Ω figures and vesicle clusters which are the ultrastructural indications of two distinct phases of the exocytotic process. Experimentally elicited exocytosis was reduced when Ca^{2+} were omitted from the bathing medium. This results as Normann (1974) has pointed out supports the idea that the coupling between excitation (depolarization) and exocytosis involves Ca^{2+} influx through the axolemma. Barium effectively substitues for calcium in the secretory process. High levels of magnesium in the bathing medium however, decreased exocytosis, possibly by interfering with calcium influx. Normann (1976) in a review on "Neurosecretion by exocytosis" has given a comprehensive account on this subject which has been outlined briefly above.

A mention may be made here of the corpora cardiaca and the process of release of secretory product in the stick insect, *Carausius morosus* as described by Smith and Smith (1966). As already stated (*vide supra*) the CC contains three main categories viz. axons of neurons located in the brain, intrinsic secretory neurons and glial cells. Within the organ the cellular components form a loosely knit tissue incorporating the ramifying extracellular spaces confluent with haemolymph via the peripheral basement membrane sheath of the organ. The electron microscopic account relates particularly to the mode of release of neurosecretory products into extracellular milieu. This evidence indicates that the CC is the site of release of both extrinsic and intrinsic secretions but does not support the suggestion that intrinsic neurosecretory materials synthesized within the CC are discharged via axon-terminals at sites outside the corpus cardiacum They have suggested that the release of both extrinsic and intrinsic material takes place by fusion of the membrane surrounding the neurosecretory droplet with the plasma membrane of the nerve cell in the region where the glial sheath is absent. The material contained in the granule is released in the extracellular space by means of exocytosis as pointed out by Normann (1974, 1976). Smith and Smith (1966) stressed that in this functionally versatile organ, the electron microscope may reveal the mechanism of release of active neurohumoral material, but is at present unable to differentiate between the rich variety of these pharmacologically active products which affect numerous target organs throughout the body.

(2) BY THE FRAGMENTATION OF GRANULES AND DIFFUSION OF MATERIAL

Ultrastructural studies conducted by Johnson (1966b) on *Calliphora stygia* have shown that the sites of release of NSM comprise clusters of small pale vesicles adjacent to the plasma membrane with electron dense material and neurosecretory granules associated with them. These release sites were found in some axons in which the plasma membrane is adjacent to the basement membrane and in other axons in which the plasma membrane is contiguous with the plasma membrane of the cells of the corpus allatum and with other axons containing neurosecretory granules.

According to Scharrer (1968) the ultrastructural specializations characteristic of sites of release of neurosecretory material (NSM) from axons of several cockroaches has revealed that discharge of NSM may take place within or outside of the neurohaemal organs and is not restricted to the axon terminals. Structurally distinctive areas serving this functions i.e. the sites of release of NSM, occur intermittently and may be more or less transient. Many of these specialized zones face the extra-cellular stroma that forms sheaths and partitions of neurohaemal organs (corpora cardiaca, perisympathetic organs), others contact various cellular elements like nerve fibres with or without neurosecretory granules, glial cells, non-neuroendocrine cells.

Irrespective of the milieu, the release sites are characterized by small electron lucent vesicles clustered near the internal surface of the plasma membrane and by variously shaped accumulation of electron dense material on either side of the plasma membrane. It has been further reported that the small vesicles observed seem to be the result of fragmentation of neurosecretory granules prior to the discharge of their contents. These ultrastructural features are strikingly similar to those of the presynaptic component of conventional interneural junctions. However, the functional implications of this morphological resemblance appear to be limited. In neurosecretory systems, physiological phenomena comparable to "chemical" transmission are out of the question in the absence of postsynaptic cells as clearly remarked by Scharrer (1968). Furthermore, the variously shaped intra and extra-cellular dense material in synaptoid areas seems to represent a neurosecretory product in transit and is therefore not equivalent to dense material customarily found within or on either side of the regular synaptic cleft (for details see : Scharrer, 1968).

According to Maddrell (1966) in *Rhodnius* the diuretic hormone produced in the neurosecretory cells of the ventral ganglion, pass out along the axons and is released into the blood via a large number of small lateral branches terminally swollen (POs?) and ending beneath the nerve sheath. Accordingly, the presence of numerous swollen endings facilitate the rapid discharge of the hormones (by any one of the above mentioned manners).

A large number of self explanatory figures have been given in the book which clearly indicate the manners by means of which the product(s) of the neurosecretory cells (neurohormones) are released in the blood stream.

The other endocrine glands viz. - corpora allata and ecdysial or prothoracic glands are influenced by the neurosecretory cells of the brain and release their secretory products i.e. corpus allatum hormone or juvenile hormone and moulting hormone or ecdysone directly into the circulating haemolymph.

8

SOME HORMONALLY CONTROLLED PHENOMENA

It is an established fact that almost all life processes in insects are controlled by the hormones. In the following pages only a brief account of some of the hormonally controlled phenomena is given, and for details see the bibliography.

HORMONAL CONTROL OF METAMORPHOSIS

Insect metamorphosis is controlled by a variety of hormones of which the most important are (1) brain hormone, (2) ecdysone and (3) juvenile hormone. Most of the work with insect hormones has been carried out by ligaturing the larvae or pupae at various levels between anterior and posterior ends, by making grafts or by the injection of extracts of endocrine glands or synthetic hormones.

BRAIN HORMONE

This is secreted by the neurosecretory cells of the brain which are usually 4-clusters - a pair of medial and a pair of lateral cluster. The hormone secreted by these cells pass down the nerve axons of these cells and accumulates in the corpus cardiacum. Release of the hormone into the blood occurs at the corpus cardiacum level through several ways i.e. either by disintegration of granules and their diffusion into the blood or by means of exocytosis.

Experimental investigations in *Cecropia* silkworm show that both the median and lateral neurosecretory cells are necessary for the formation of this hormone. In fact the brain hormone is a mixture of several hormones secreted by different types of neurosecretory cells present in the brain.

Role of the brain hormone on metamorphosis

This hormone has an indirect effect on metamorphosis. The chief function of this hormone is to control the secretion of the prothoracic glands. Consequently, the removal of the brain has the same result as the removal of the prothoracic glands. In both cases the secretion of ecdysone is stopped and the insect enters into a state of diapause. Thus, diapause is due to the failure of the endocrine activity.

ECDYSONE

This is also known as moulting hormone. Fraenkel (1935) found that when mature fly larvae were ligatured, only the front end was able to form a puparium while the hinder part remain unchanged. Blood transfusion from the front to the hind end even in the ligatured fly larvae immediately caused a change resulting in the pupation of the posterior part. From these observations, Fraenkel suggested that the growth of the larva into the pupa was dependent on a hormone secreted somewhere in the anterior end. Fukuda (1940) clearly demonstrated that it is secreted by the prothoracic glands or ecdysial glands. Finally it was Butenandt and Karlson (1954) who isolated the hormone in a highly purified stage and named it as "Ecdysone". Chemical study of ecdysone has clearly revealed that it is composed of two closely related molecules viz. α-ecdysone and β-ecdysone and it is steroidal in nature.

Role of the ecdysone on metamorphosis

Ecdysone is the only hormone which has a central control on the developmental events at the cellular level in insects. When the prothoracic glands stop the secretion of ecdysone the result is developmental standstill. This is what happens during diapause. It must be mentioned here however that the development of muscles during metamorphosis is not governed by ecdysone alone. There is another factor called neurosecretory factor or a neurosecretory control (hormone) which when produced during the same period when ecdysone is released in the blood causes muscle development. This observation clearly points that there is a close integration of the prothoracic secretory process with the activity of the neurosecretory system. Further integration is through the neurosecretory control of the brain over the prothoracic gland.

JUVENILE HORMONE

This is secreted by corpora allata, a paired or unpaired small cephalic endocrine glands. The hormone is known as juvenile hormone, (JH) corpus allatum hormone or neotenin which prevents the real change of stage in the developing insect. Wigglesworth (1934) was the first to demonstrate this function of the juvenile hormone in his experiments with *Rhodnius prolixus*. However other workers such as Bounhiol (1938) and Fukuda (1944) and several others have confirmed the observations of Wigglesworth by their experiments on caterpillars. The JH is terpenoid in chemical nature.

Role of the juvenile hormone on metamorphosis

Its chief function is to maintain the immature stages i.e. larvae or nymphs of insects without causing pupation or the development of adult stage i.e. the actual metamorphosis is blocked. In a normal metamorphosing insect the corpora allata decline its secretory activity and thus a sudden change from an immature young stage to adult happens. Removal of corpora allata in the juvenile period of growth in insects have clearly shown a precocious metamorphosis in the life history.

In short the process of metamorphosis takes place in the following way -

The brain hormone secreted by the neurosecretory cells stimulates the prothoracic glands to secrete ecdysone or moulting hormone, but moulting occurs only when the juvenile hormone is either absent or present in very small amounts. In the last larval instar of Holometabola the corpora allata becomes almost inactive, hence a low concentration of JH is present causing larva to moult into a pupa. Similarly during later part of the pupa or late pupa, again the corpora allata become inactive and in total absence of JH the pupa moults into an adult insect. In Hemimetabolous insects also the same phenomena is repeated i.e. the corpora allata become inactive in the last instar nymphs and in the complete absence of JH imaginal ecdysis takes place. Thus metamorphosis in insects is controlled by a delicate balance in the timing of secretion and concentration of two hormones viz. JH and MH secreted by the corpora allata and ecdysial or prothoracic glands. The total absence of JH is the most significant factor for imaginal ecdysis in both, Hemimetabolous and Holometabolous insects.

If much JH is present then larva will moult into a larva, if small amount of JH is present then pupation takes place and in the total absence of JH adult emergence occurs. And if the balance of hormones is disturbed then monsters, intermediate between larva and pupa or between pupa and adult are produced and the phenomena is termed as prothetely or metathetely (for details see p. 106).

HORMONAL CONTROL OF METAMORPHOSIS IN OTHER INSECTS

This general principle of the control of metamorphosis by secretion of juvenile hormone throughout the larval stages, and the cessation of secretion causing transformation to the adult, has been found to apply in almost all insects but there are variations in detail that are characteristic of the different groups. In the stick insect *Carausius*, which normally has seven larval instars before the female becomes mature, removal of the corpora allata at an early stage, say in the second-stage or third-stage larva, is followed by two moults and then the insect begins to lay eggs of half the normal size (Wigglesworth, 1954). Conversely the implantation of corpora allata from young insects into *Carausius* in the fifth or sixth stage may cause them to make as many as four extra moults and become giant forms 15 cm long, or double their normal length. A similar response is seen in the Madeira cockroach, *Leucophaea*. There are usually eight larval stages; if the corpora allata are removed in the fifth or sixth stage they make two moults before adultoid characters are developed. This delay presumably results from the persistence of substantial quantities of juvenile hormone in the blood and tissues of the young insects after the corpora allata have been removed. In the cricket *Gryllus*, the grasshopper *Melanoplus* and in *Locusta* extra larval stages can be produced by implantation of corpora allata (Wigglesworth, 1954, 1964). In *Sialis* (Neuroptera) precocious metamorphosis can be induced by decapitation or by extirpation of the corpora allata at any of the larval stages from the first to the ninth (Wigglesworth, 1964).

Further variants are to be found in holometabolous insects. Bounhiol (1938) early succeeded in extirpating the corpora allata in young silk-worm larvae in the fourth, third and even in the second instar. Such larvae then proceeded to spin cocoons and to pupate at the next moult; and the pupae gave rise to diminutive adults. The second-stage larvae produced pupae weighing only 25 mg as compared with the normal pupal weight of 1 g or more (Bounhiol, 1938). The tiny adults to which they give rise are more or less normal in appearance and they lay eggs, not so large as usual but by no means reduced in size in proportion to the body of the mother (Wigglesworth, 1954). As in *Rhodnius*, so in all larvae of *Bombyx* after the first instar (as shown by Fukuda) metamorphosis can be induced simply by decapitation if this is done at the proper time e.g., in third instar larvae deca- pitated between 40 and 60 hours after moulting. Similar results have been obtained in the wax moth *Galleria*; not only are diminutive pupae and imagines produced by removal of corpora allata from the very young larvae, but full-grown larvae receiving multiple implants of cor- pora allata from young larvae will continue larval development with the eventual production of giant pupae and adults (Wigglesworth, 1954).

In *Tenebrio* (Coleoptera) as many as six extra larval moults have been obtained after implantation of corpora allata from young larvae. The corpora allata have a controlling effect on metamorphosis in the honey-bee (Hymenoptera). The major group in which the action of the juvenile hormone is not so clear is that of the Diptera. If the corpus allatum is taken from a young larva of *Calliphora* and implanted into the abdomen of a fifth-stage larva of *Rhodnius* a localized patch of larval cuticle is produced overlying the implanted gland (Wigglesworth, 1954), and experiments on *Drosophila* and *Calliphora* respectively have afforded some evidence of the production of partially larval structures resulting from implantation of corpora allata into pupating larvae of these Diptera but precocious metamorphosis, or giant larvae and adults, have not so far been produced. These are insects in which the imaginal discs are exceptionally important in adult development. Perhaps that is why such experiments have not proved successful. It might be expected that the more generalized group of Diptera, the Nematocera, might show such effects more readily. But the Tipulid *Nephrotoma suturalis* given massive doses of the juvenile hormone of cecropia surface application or by injection into the pupa within a few minutes after the larval skin had been shed, developed into completely normal adults of both sexes (Wigglesworth, 1970).

The other group in which the action of the corpus allatum is exceptional is the Apterygota, at the other end of the evolutionary scale. Here moulting continues in the adult insect in which the thoracic glands persist and the moults alternate with cycles of reproduction. In the silver fish, *Lepisma* there is a constant correlation between the body size and the length of the genital appendages; the adult form is attained gradually. This relation is scarcely upset if the corpora allata are extirpated : it is not possible to produce adults of unduly small size as can be done in other insects (Wigglesworth, 1964).

HORMONAL CONTROL OF HEART BEAT

The activity of the heart in insects is controlled by a variety of factors among which endocrine or hormonal control of the heart beat seems to be one. The corpus cardiacum plays an important role in controlling the beating of insect heart. Cardioacceleratory substances have also been extracted from the ganglia of the ventral nerve cord. Likewise, metameric neurohaemal organs also contain substances which stimulate heart beat in insects. Mordue and Goldsworthy (1969) reported that the corpus cardiacum extract accelerated the heart beat. These workers found that the glandular lobes of the corpora cardiaca of locusts produce 53% and storage part 37% increase in the heart beat rates.

Unger (1957) separated two physiologically different substances from the brain, suboesophageal ganglion, corpora cardiaca as well as from the ventral nerve cord of *Periplaneta americana* and termed them as neurohormone-C and neurohormone-D. The neurohormone-D stimulated the frequency of the heart beat and increased its amplitude but use of high concentration of neurohormone-C increased the frequency of heart beat but decreased its amplitude. At higher concentration neurohormone-C stopped the heart beat causing systolic collapse.

Traina *et al.* (1976) separated four heart acceleratory factors from the corpora cardiaca of *Periplaneta americana* and named them 1a, 1b, 2a, 2b. All the four factors in chemical nature are peptides.

A variety of chemicals have been extracted from the corpora cardiaca of insects. Barton-Browne and others (1961) have suggested that the active substance which increase the heart beat in chemical nature is similar to adrenalin. Cameron (1953) isolated O-diphenol from the intrinsic cells of the corpora cardiaca of *Periplaneta americana* which was similar to adrenalin. Davey (1961, 1962) and Ralph (1962) demonstrated that the substance increasing the heart beat is a peptide, which acts on pericardial cells. The substance extracted from the pericardial cells was indole alkylamine. The indole alkylamine is secreted by the pericardial cells when corpus cardiacum extract is injected. Colhoun (1963) and Garsch *et al.*, (1961) have shown presence of serotonin in the cerebral neurosecretory cells and corpora cardiaca of *Periplaneta americana*.

Thus, O-dephenol, serotonin and an adrenalin like substance, were extracted from the corpora cardiaca and indole alkylamine, from the pericardial cells upon stimulation by the corpus cardiacum extract. In fact which one is responsible for the heart beat control cannot be said with certainty. However, it is clear from experimental evidences in hand, that the corpus cardiacum directly controls the heart beat. The detailed process and the exact chemicals involved in the process are yet not clearly understood in insects in general (for details see. Awasthi, 1983).

HORMONAL CONTROL OF DIAPAUSE

The period of suspended development at any stage of the life cycle, under adverse condition(s) accompanied by greatly decreased metabolism is called diapause. This phenomenon is governed by a variety of external and internal factors. This is an adaptation to tide over unfavourable environmental conditions like cold, draught, extremes of temperatures, humidity, photoperiod, nutritional, vitamin and other deficiencies. The other factor(s) which determine the diapause and are equally responsible are the endocrine glands (see : Doane, 1973; Novak 1975; Raabe 1982, 1989)

The hormonal regulation of diapause was earlier proposed by Wigglesworth (1934) and was confirmed by the experiments of Williams (1946) on *Hyalophora cecropia* which normally enters diapause just after pupation and may be induced to resume development by several months of chilling at 3-5°C. In Saturniids, the pupal diapause is caused by the absence of ecdysones, which results from inactivity of the neurosecretory cells of the pars intercerebralis. Diapause has been broken in a number of species by the injection of α or β-ecdysone, with the onset of adult development. A permanent diapause may be produced even in related non-diapausing species by brain removal. This response will vary from species to species.

Thus, the neurosecretory cells of the brain are the controlling elements in the initiation and termination of hormonally regulated diapause. However, developmental arrest is not necessarily the result of a lack of ecdysone in all cases. Adult diapause may be brought on by corpora allata inactivity under the control of neurosecretory cells of the protocerebrum. In this case also as in the case of ecdysone regulated diapause, reactivation of the neurosecretory cells in the brain is required to terminate the conditions and restore reproductive activity.

Three principal types of diapause can be distinguished on the basis of their hormonal mechanism :

(1) Diapause caused by a lack of brain hormone and moulting hormone. According to Novak (1975), to this category belong all cases of larval and pupal diapause known so far and doubtless also, embryonic diapause.

(2) Diapause caused by lack of brain hormone and juvenile hormone, this mechanism is concerned with imaginal diapause.

(3) Diapause caused by the action of neurosecretory factor produced by the suboesophageal ganglion of the female and affecting the development of the eggs. This category comprises early embryonic diapause.

The diapause caused by a deficiency of the brain hormone and moulting hormone is the most common type of diapause and the type so far most thoroughly investigated. It consists of an arrest of growth and, development due primarily to the absence of moulting hormone itself is a consequence of an inhibition of brain hormone production. It is likely that the majority of both the pupal and the larval diapause results from the failure of the cerebral neurosecretory cells to produce thoracicotropic hormone.

Adult Diapause

The adult diapause is the result of a deficiency in the brain hormone and juvenile hormone i.e. inactivity of the neurosecretory cells of the pars intercerebralis and corpora allata. Morphologically, the adult diapause is only seen in the interruption of growth of the ovaries and in the suppression of the functions of the accessory glands of both the male and female individuals. In the fly *Calliphora*, the implantation of corpora allata can induce the ripening of some of the eggs but complete termination of diapause is possible only by the simultaneous action of juvenile hormone and brain hormone. Apart from this, the presence of BH seems indispensable for the activity of CA. Evidences for CA (JH) regulation of adult diapause comes from a number of other species e.g. *Pyrrhocoris apterus, Hypera postica, Necrophorus fossor, Galeruca tanaceti, Pterostichus nigrita, Leptinotarsa decemlineata, Aedes aegypti, Culex pipiens, Calliphora erythrocephala* and *Gryllodes sigillatus* etc. The experimental evidences clearly suggest that the basic intrinsic factor of imaginal diapause lies in the neurosecretory cells (NSCs) of the brain which appears to be part of the mechanism of action of external stimuli (for details see: Doane, 1973; Novak, 1975; Raabe, 1982, 1989).

Pupal Diapause

The principal cause of pupal diapause is the failure of endocrine activity i.e. activity of the NSCs of the brain and the ecdysial glands. Thus, to break the pupal diapause, presence of active ecdysial glands (EGs) is necessary in addition to BH. Implantation of active EGs also breaks the diapause. It is important to note that implantation of one or more active brains into an isolated abdomen in *Hyalophora* has no effect whereas the implantation of active EGs breaks the diapause and development continues. Injection of ecdysone has the same effect as the implantation of EGs. Pupal diapause can take place at any point during the pupal stage. Its occurrence in a given species is bound to be in the same phase of the pupal period and in most species it is genetically fixed. The common example of pupal diapausing insects are *Amsacta moorei, Amsacta collaris, Achoea janata* and others.

Larval Diapause

It has been found that in the diapausing penultimate instar nymphs of *Gryllus campestris* ecdysial glands show reduced secretory activity and the brain appears to be responsible for this. Both the brain and the EGs have also been shown to control diapause in larvae of *Sialis lutaria*. It has been remarked that the brain controls the activity of the EGs and CA, and is the principal control centre of diapausing larvae.

Early Embryonic Diapause

The early embryonic diapause is associated with the diapausing hormone (DH) secreted by the suboesophageal ganglion. Removal of the suboesophageal ganglion from diapause producing pupae shortly after pupation completely eliminates diapause from the eggs eventually produced; when the operation is delayed, a progressively larger proportion of eggs enter diapause. The incidence of the diapause in the eggs is conditioned by endocrine activity of the neurosecretory cells of the suboesophageal ganglion of females. The extirpation of suboesophageal ganglion at the beginning of the pupal instar causes eggs which were determined for diapause through their mother (eg. female *Bombyx mori*), to develop without diapause. In the same way univoltine females can be caused to lay nondiapausing eggs, a thing which they never do under normal condition, conversely if up to a critical period a female pupa which has been conditioned to lay non-diapausing eggs receives a suboesophageal ganglion from female which would have laid diapausing eggs, it will also lay diapausing eggs. The same effect can be produced by the implantation of the suboesophageal ganglion of either females or males of a univoltine race. That the relevant DH is non-specific for species, genus or family was shown by implantations of the suboesophageal ganglion from other Lepidoptera. But in this case also the neurosecretory cells of the brain play a key role. In diapause producers, the brain stimulates the production and release of DH from the suboesophageal ganglion, in nondiapause producers the brain inhibits hormone production and release from suboesophageal ganglion (see: Doane, 1973; Novak, 1975; Raabe, 1989).

Diapause in the Reproductive Stage

Reproduction like growth, may suffer a periodic arrest and as with growth this arrest may be the direct effect of an adverse environment, or it may be a true 'diapause' which persists even under favourable conditions. In *Dytiscus* and various other beetles, the gonads revert to a resting state after the first reproductive period; they show renewed activity about the same date the following year, sometimes in a third or fourth year also. This seems to be a deep-seated rhythm and not a simple effect of warmth following the winter cold; for in *Carabus coriaceus* and *Leistus* spp. the breeding season does not begin until the late summer or autumn and *Dytiscus marginalis* lays its eggs in March, *D. semisulcatus* in October (Wigglesworth, 1965).

These reproductive cycles are probably controlled by hormones. The full development of the eggs in *Dytiscus marginalis* is dependent on the corpus allatum. Re-absorption of the oocytes takes place in July as soon as they have grown to 0.8 mm in length; in February they reach 2 mm before they break down and in March they attain the normal length of 7 mm. But if corpora allata are implanted, the full development of the eggs can be produced at any season and if the corpora allata are removed at the height of egg production, the ovaries regress. In

Carabids, as in *Dytiscus*, the corpus allatum hormone prevents degeneration of the oocytes and permits ripening of the ovaries (Wigglesworth, 1965).

The hormonal regulation of adult diapause has been studied in great detail in the Colorado potato beetle, *Leptinotarsa*. The induction of diapause is a short-day effect which supervenes in the late summer and autumn. It is associated with clear-cut changes in behaviour and metabolism : the beetles burrow into the soil where they come to rest, the rate of oxygen consumption falls to a low level and ovarian development is arrested. These changes can be reproduced by surgical removal of the corpora allata, and there is little doubt that their immediate cause is the suspension of juvenile hormone secretion (Wigglesworth, 1970).

Under long-day treatment the corpora allata are activated probably by hormonal stimuli from the medial neurosecretory cells. Short-day treatment, on the other hand, not only inactivates the neurosecretory cells and so removes the hormonal stimulation of the corpus allatum but there is also active inhibition of the corpus allatum by way of their nerve connections. Diapause in *Leptinotarsa* can also be induced by feeding an aged potato foliage. This is not a direct effect of poor nutrition but is again due to neuroendocrine arrest brought on by sensory information about the host plant. The corpora allata are inhibited; but reproduction can be restored by the implantation of active glands even though the food remains unchanged (see : Wigglesworth, 1970).

In hibernating adults of *Leptinotarsa* the flight muscles show extreme degeneration, the muscle fibres are reduced in diameter and the mitochondria or sarcosomes are virtually absent. Similar signs of degeneration follow extirpation of the corpora cardiaca and corpora allata. If several of the active complexes are re-implanted there is a very rapid regeneration of muscle fibres and a new formation of mitochondria as the beetles come out of diapause. Under normal conditions the beetle works its way into the superficial layer of the soil where it rests until the capacity for flight is restored. Quantitative assays for juvenile hormone in the blood of adult *Leptinotarsa* show that this falls to a very low level during diapause, but rises again during long-day treatment (Wigglesworth, 1970).

Very similar results have been obtained on the Chrysomelid beetle *Galeruca tanaceti*; here the use of ^{35}S-DL-cystine in autoradiographic studies has confirmed that the rate of turnover of cystine in the neurosecretory cells increases about two-fold when a mature female begins to oviposit and about ten-fold when a female goes from diapause to oviposition. In diapausing females the labelled cystine builds up in the neurosecretory cells; in ovipositing females it accumulates in the corpus cardiacum and is discharged within 24 hours (see: Wigglesworth, 1970).

Mosquitoes become more or less immobile in the autumn and after taking meals of blood they tend to develop an enlarged fat body instead of maturing their ovaries, a condition known as 'gonotrophic dissociation'. In *Culex pipiens*, this change seems to be closely related with the temperature if given blood and kept warm at any time in the winter they will develop eggs. But if *Anopheles maculipennis* race *atroparvus* is fed with blood during the winter and kept at 29°C, although some develop eggs, others lay down reserves in the body and *A. maculipennis* race *messeae* in the autumn shows complete 'gonotrophic dissociation' and fails to produce eggs at any temperature. In this state there is no absorption of developing oocytes as described above and the ovaries show no signs of growth at all (Wigglesworth, 1965).

The cessation of blood-feeding in *Culex pipiens* and the gonotrophic dissociation in *Anopheles maculipennis* during the winter diapause are the result of exposure to a short (12-hour) photoperiod. The changes are doubtless associated with arrest of secretion in the neurosecretory cells and in the corpus allatum but the details have not been described. The same is probably true of the reproductive arrest that occurs in many other insects, such as *Polistes,* where the neurosecretory cells of the pars intercerebralis become laden with neurosecretory material during the winter diapuse and emptied again when diapause comes to an end, *Musca autumnalis*, the locusts *Nomadacris* and *Anacridium* where again the neurosecretory cells are filled with secretion during diapause and the corpora allata are small and inactive and in th alfalfa weevil *Hypera* where topical application of the methyl ester of 10, 11-epoxyfarnesenic acid effectively terminates the summer diapause (see: Wigglesworth, 1970).

An unusual example of an environmental factor affecting hormone secretion is seen in the rabbit flea *Spilopsyllus cuniculi*. The ovaries of fleas kept on male or non-pregnant female rabbits remain immature as in diapause whereas those on a pregnant host are mature by the time the young are born. The factor involved is available only during the final week of pregnancy and not at all in male or non-pregnant female rabbits. The factor disappears from the female after parturition but is present in her nestlings for at least 7 days. The factor does not have solely a trigger-like action in initiating ovary development, the ovaries undergo regression in maturing fleas transferred to hosts that do not supply the factor. It is suggested that the 'yolk-forming hormone' normally secreted by the corpora allata may only be produced by rabbit fleas when they obtain the postulated factor. Whether the steroid hormone levels in the host are involved is still uncertain (Wigglesworth, 1970).

On various aspects of insect diapause a comprehensive account is given by Wigglesworth (1964, 1970) DeWilde (1964), DeWilde and DeLoof (1973). Doane (1973), Novak (1975), Highman and Hill (1979), Raabe (1982, 1989) and many others.

HORMONAL CONTROL OF REPRODUCTION

Wigglesworth (1936) for the first time clearly pointed out that corpora allata secrete a gonadotrophic hormone in *Rhodnius prolixus*. Now it is well established that in many insects the egg maturation is controlled by the corpora allata. In several insects viz. *Pieris brassicae, Galleria mellonella, Leucania separate* and *Carausius morosus* removal of the corpora allata in the larvae reduced the total egg production of the adult (Engelmann, 1968). In the cockroach, *Leucophaea*, the implantation of active corpora allata into normally fed pregnant whose own corpora allata were inactive, caused egg maturation in about 80% of the females but only 45% starved females when received active corpora allata matured eggs. Thus, nutritional factor is also involved in yolk deposition in the maturing eggs.

During the time of egg maturation i.e. yolk deposition in the oocytes, the size of the corpora allata increase markedly in majority of the insects showing enhanced activity (Engelmann, 1970). Cyclic changes in the corpus allatum volume which are correlated with egg maturation cycles have been reported for many insects.

In *Calliphora erythrocephala* the neurosecretory cells are essential in reproduction. Removal of the neurosecretory cells eliminated egg maturation completely. Reimplantation of neurosecretory cells into females previously deprived of neurosecretory cells resulted in resumption of egg maturation in some flies. The mosquito, *Aedes taeniorhynchus* failed to mature eggs in the ovaries after destruction of the pars intercerebralis but after reimplantation of pars intercerebralis, maturation resumes. Unlike *Calliphora*, in *Aedes taeniorhynchus* corpus allatum is indeed essential for egg maturation. Allatectomy is followed by complete cessation of egg maturation (Lea, 1967). In these cases both the neurosecretory cells of the brain and corpus allatum furnish hormones that are needed for egg maturation. In *Sarcophaga bullata* also two hormones, one secreted by the neurosecretory cells of the brain and the other, by corpus allatum is needed for egg maturation. In *Sarcophaga bullata* egg maturation was abolished after the destruction of neurosecretory cells of the brain. In the case of *Sarcophaga bullata* it was suggested that corpus allatum promotes protein synthesis essential for egg maturation and the neurosecretory cells of the brain act as a true gonadotrophic hormone, facilitating the incorporation of yolk in the oocytes.

In *Locusta migratoria* also, both corpora allata and neurosecretory cells of the brain are essential for egg maturation i.e. both the hormones viz. corpus allatum hormone and brain hormone are needed for egg maturation. Removal of the neurosecretory cells of the brain completely prevented egg maturation in several insects e.g., *Schistocerca gregaria, Schistocerca paranensis, Anacridium, Gomphocerus, Aedes aegypti* and *Tenebrio* etc. which clearly indicates that only neurosecretory cells of the brain are essential in these insects for egg maturation.

The enhanced activity of the neurosecretory cells of the brain was observed during egg maturation. In *Schistocerca*, the rate of protein synthesis in the fat body as well as concentration of protein in the haemolymph was reduced in the pars intercerebralis cauterised females. Failure to mature eggs in the pars intercerebralis cauterised/ operated insects was the consequence of low protein metabolism, which did not allow the activation of the corpora allata. It was suggested that the corpora allata facilitate the deposition of available proteinaceous yolk into the growing oocytes but not involved in protein synthesis.

Thus, both the hormones are needed for maturation of eggs. The principal function of the hormone secreted by the corpora allata or neurosecretory cells of the brain or both, as the case may be, is to promote the incorporation of the proteinaceous and lipid yolk in the growing oocytes. It was found that the blood protein level decreased with the approach of complete egg growth and was paralleled by a decreased corpus allatum activity. In addition to hormonal factors, several other factors like nutritional, mating, environmental etc. are essential indirectly with egg maturation process.

It can be said that the neurosecretory cells stimulate and activate the corpora allata and the secretion of corpora allata, facilitate yolk synthesis and yolk deposition in the eggs i.e. maturation of eggs (for details see : DeWilde, 1964, DeWilde and DeLoof, 1973; Engelmann, 1968, 1970; Raabe, 1982, 1989).

HORMONAL CONTROL OF REPRODUCTION IN SOCIAL INSECTS

The main features of reproduction in social insects is the inhibition of reproduction exerted by the queen on the other members of the colony, thus determining the existence of castes. Pheromones and nutrition affect caste determination through the CA, which are inhibited in both primitive social hymenopterans and highly social wasps, bees, and termites.

In *Polistes annularis* topical JH application promoted oocyte maturation in subordinate females and caused changes in their behaviour. Moreover, in summer-diapausing females of *Polistes metricus,* JH caused ovarian development. In *Bombus terrestris* the JH titre is low in the queenright workers and high in the queenless workers. In addition, during the prepupal development of workers and queens, both JH and ecdysteroids peak at intervals of a few hours, but the worker peaks precede the queen peaks by about 20 hr, suggesting that the effects determined by JH may be time-dependent (see, Raabe, 1989).

In bees, the role of JH is still controversial and the CA were even considered not to be indispensable either to vitellogenin synthesis or to queen determination from young worker larvae (Rembold *et al.* 1974). However, JH determination shows that the JH titre is 10 times higher in 3-day-old queens than in workers of the same age. This is precisely the moment at which the ovarian anlagen differentiated into ovarioles in larvae raised in queen cells, but atrophied in worker cell larvae. The idea that JH triggers reproductive development in bees was also supported by the results of experiments involving JH treatment of larvae reared on worker jelly, which differentiated into either intercastes (Rembold, 1976) or perfect queens. In other experiments it was reported that JH might have antagonistic effects, depending on the doses applied (Ramamurty and Engels, 1977).

The data relevant to ants also pointed to the CA as triggering sexual development in *Myrmica rubra* and *Pheidole pallidula* thus inducing the production of eggs determined to develop into reproductives.

In termites, the implantation of corpora allata in pseudergates (a term designating fully grown larvae) resulted in the moulting of the receptor insect into supplementary reproductive, showing that the corpora allata help to trigger sexual processes, although they also intervene in the caste determination of soldiers.

Few data are available concerning the role of ecdysteroids in reproductive determination in social insects. High ecdysteroid levels were reported in ant larvae destined to develop into workers, but the opposite situation prevails in bee nymphs, since the future queen pupae have higher ecdysteroid titres than the worker pupae. No general conclusion regarding the role of ecdysone can therefore be drawn at the present time (see: Engelmann, 1970, Raabe, 1989).

HORMONAL CONTROL OF UNUSUAL MODES OF REPRODUCTION IN APHIDS

Unusual modes of reproduction are not rare in insects, and may occur sporadically or continuously. They include parthenogenesis, viviparity, and paedogenesis, which are frequently combined. The neuroendocrine regulation of these events have been studied only in aphids.

In aphids, the reproductive processes proceed in completely different ways depending on the season and day length. During the long day summer period, several generations of viviparous parthenogenetic females (termed virginoparae) develop and during the short days of autumn give rise to a generation of sexual males and females, the latter being termed oviparae. The structure of the female reproductive system is different in the two morphs : oviparae form large, yolky, haploid eggs, whereas virginoparae develop, during their larval growth, numerous smaller eggs which evolve parthenogenetically into diploid embryos (Raabe, 1982).

Juvenile hormone influences several features of aphid morphs and acts both on the treated insects and on the progeny incubated in their abdomen causing the latter to develop in parthenogenetic viviparous females. Cautery of certain medial neurosecretory cells however, abolished the normal response to day length and induced the appearance of oviparous females and males instead of viviparous parthenogenetic females (also see : P. 99). The existence of a virginopara -promoting substance is therefore probable. This substance might possibly intervene by stimulating CA functioning (see : Raabe, 1989).

HORMONAL CONTROL OF METABOLISM

Studies of a large number of investigators have shown that almost all vital events, including metabolism of protein, lipid, carbohydrate, respiratory and water are controlled by the insect hormones. The work has been reviewed by Awasthi (1979a).

Protein metabolism

It has been demonstrated that in *Calliphora* deprived of their medial neurosecretory cells, the activity of protease was greatly decreased. Reimplantation of the neurosecretory cells caused normal release of proteases. Production of proteoltic enzyme is regulated by eating of meat and presence of medial neurosecretory cells. In

Calliphora, Nauphoeta, Glossina, Rhodnius, Periplaneta, midgut protease synthesis is controlled by medial neurosecretory cells but in *Leucophaea, Dytiscus, Sarcophaga* and *Blatta* secretion of intestinal proteases is independent of medial neurosecretory cells; it may be controlled by neural stimuli or secretogogues.

In the insects which are deprived of their medial neurosecretory cells and corpora allata, an accumulation of protein in the haemolymph was observed. In some insects only corpora allata regulate protein synthesis. In *Schistocerca*, neurosecretory cells control the haemolymph protein concentration i.e. protein availability to the oocytes. There are two views, according to one corpora allata produce a gonadotropin but the protein metabolism associated with egg production is under the control of neurosecretion. The other view is that corpus allatum hormone is involved in the regulation of protein metabolism in *Sarcophaga bullata*. Thus, in some insects protein metabolism during egg maturation depends on either corpora allata or neurosecretory material or on the secretions of the both.

The female specific protein-synthesis is controlled by the corpora allata or corpus allatum hormone. In *Leucophaea* female specific protein or "vitellogenin" which is induced by juvenile hormone analogue, makes up about 90% of the total yolk protein. The female specific protein is essential for egg maturation. In *Drosophila* yolk protein is synthesised in the fat bodies outside the ovaries and is stored in the haemolymph where it appears as female specific protein and from here it is taken up by the growing oocytes (see review : Awasthi, 1979a).

Carbohydrate metabolism

The corpora cardiaca regulate the carbohydrate metabolism in insects in various ways. The action of carpus cardiacum extract depends on the lowering of the amount of glycogen and increasing the haemolymph trehalose concentration. The corpora cardiaca of *Calliphora* and *Periplaneta* contain a hyperglycaemic hormone which when liberated into the blood increases the trehalose contents. In *Periplaneta* two hormones are present in the corpora cardiaca which are responsible for increasing the haemolymph sugar concentration. After injection of corpus cardiacum and brain extracts an increase in the trehalose content in the haemolymph in *Periplaneta* was observed and it was remarked by Ralph and McCarthy (1964) that the active factor of the corpus cardiacum is produced by the neurosecretory cells of the brain. But Steele (1961) is of the view that the hyperglycaemic hormone is secreted by the intrinsic cells of the corpora cardiaca which stimulate the production of blood trehalose from glycogen stored in the fat body cells. Highnam and Goldsworthy (1971) are also of the opinion that the hyperglycaemic factor extractable from the corpus cardiacum is the production of neurosecretory cells of the brain. It appears that the hyperglycaemic factor is synthesised in the brain and stored in the corpus cardiacum from where it is released into the blood.

Lipid metabolism

In the mosquitoes, the neurosecretory system causes a decrease in glycogen synthesis while it stimulate the synthesis of triglyceride. Cauterisation of medial neurosecretory cells from the brain in the female *Calliphora*, soon after the final moult, caused reduction in the quantity of lipid in the fat body. According to Verman *et al.*, (1965) the allatectomized females of *Periplaneta* store 68% more lipid than is done by the unoperated females, the turn over rate of both triglycerides and phospolipids being slowed down in operated females. Gilbert (1967) has pointed out that the uptake of lipids by the ovaries in *Leucophaea* was stimulated by corpus allatum hormone. It is thus clear that the corpora allata are involved in the lipid metabolism and facilitating egg maturation.

Allatectomy of one-day old female desert locusts caused an accumulation of lipid in fat body which was due to the continuation of lipid deposition in the fat body after the expiry period of somatic growth. Lipid accumulation in allatectomised locusts is more likely to be related directly to the lack of juvenile hormone. Implantation of active corpora allata into one day old adult female locusts resulted in a premature development of oocytes and a decrease in fat body lipid accumulation. The juvenile hormone may have a dual effect on the fat body metabolism, that of suppressing lipid synthesis and of stimulating vitellogenic protein synthesis. Increasing synthesis of lipid by the fat body would then account for the accumulation of lipid in the fat body after allatectomy.

Respiratory metabolism

Removal of corpora cardiaca depresses the respiratory metabolism in the adult male cockroach, *Blaberus*. In the cardiacectomised-allatectomised cockroaches the rate of respiratory metabolism was very low. The corpora cardiaca extract produces an effect on respiratory metabolism at the level of mitochondria. Brain hormone has a stimulative effect on oxygen consumption of the whole body or on isolated body parts. The fat body, Malpighain tubules, pericardial cells, intestine and corpora allata show increased respiratory metabolism in response to brain hormone. Removal of neurosecretory cells of the brain led to a 20% reduction of the normal oxygen consumption and considerably reduced nutritional uptake (Keeley, 1967).

The corpora allata also play an important part in respiratory metabolism. Decrease in oxygen consumption was caused by allatectomy in *Calliphora* and *Carausius* whereas an increase in oxygen consumption was observed following implantation of corpora allata. In *Tenebrio* the juvenile hormone analogue influences the respiratory metabolism in the same way as the implantation of corpora allata i.e. an increase in oxygen consumption and protein synthesis was observed following corpus allatum implantation. Similarly an increase in the corpus allatum hormone concentration in the haemolymph of adult insects is followed by an increase in respiratory metabolism. However, in certain insects like *Gryllus bimaculatus, Pyrrhocoris* and *Leucophaea* removal or implantation of corpora allata or corpus allatum hormone have no pronounced effect on the respiratory metabolism (see : Awasthi, 1979a).

Water metabolism

The various parts of the neuroendocrine organs probably contain diuretic and antidiuretic factors or a factor is released by one of the same part of the endocrine system which performs different functions in different species (Raabe, 1989).

In addition to brain and corpora cardiaca, the corpora allata also play an important role in regulating water excretion in insects. Delphin (1965) reported that the cells in the ventral ganglia of *Schistocerca gregaria* produce an antidiuretic factor. In *Rhodnius*, the diuresis is controlled by a hormone produced by small paired neurosecretory cells present in the ventral ganglionic mass. The diuretic hormone is released by the axon terminals of the neurosecretory cells of thoracico-abdominal ganglionic mass in *Rhodnius*. An antidiuretic hormone in the thoracic ganglia and a diuretic hormone in the abdominal ganglia of *Chaoborus* was demonstrated by Gersch (1969). Mordue (1963) demonstrated both the diuretic and antidiuretic factors in *Locusta*. In *Periplaneta* the neurosecretory cells produce an antidiuretic hormone causing water retention. Highnam *et al.* (1965) are of the opinion that the neurosecretory cells of the brain in *Schistocerca* produce a diuretic factor. According to Nicolson (1976) the diuretic hormone is probably produced in the brain and released from the corpora cardiaca. Both diuretic and antidiuretic effects of neuroendocrine system depend on the (1) age, (2) the water availability to the insects for maintaining the water balance and on (3) several internal and external factors.

HORMONAL CONTROL OF PIGMENTATION OF THE CUTICLE

Hormonal control of tanning of the cuticle is now an established fact. Fraenkel and Hsiao (1962) have demonstrated that the pars intercerebralis controls the tanning in *Calliphora*. They are of the view that the neurosecretory cells of the brain contain a melanogenic hormone. Likewise, Girardie (1964) has shown that colouration of the cuticle in *Locusta* is controlled by the cells of the pars intercerebralis. According to Roussel (1966) implantation of a pair of corpora allata of an adult male or female *Gryllus bimaculatus* into nymphs of the last but one or the last instar induces at the subsequent moult an overall clearing of the cuticle due to the dispersion or disappearance of the black pigment.

Extirpation of corpora allata in the last but one or the last instar induces at the subsequent moult a complete black pigmentation in all the operated *Gryllus bimaculatus*. According to Roussel (1967) the corpora allata strictly control pigmentation, their absence always induces black pigmentation. The control by corpora allata is a direct one and acts on the integument (see: Raabe, 1989).

HORMONAL CONTROL OF GREEN/BROWN POLYMORPHISM IN LOCUSTS AND GRASSHOPPERS

Rowell (1967) performed a number of experiments to find out the factor(s) involved in the colour changes in the locusts/grasshoppers. He observed that the nymphs of *Acanthacris ruficornis, Humbe tenuicornis* and *Gastrimargus africanus* were implanted with a single corpus allatum from a young adult donor of the same species and in all the 3-species of the post-operated nymphal instars included significantly more green forms than the controls and in *A. ruficornis* and *H. tenuicornis* significantly more green adults. It is concluded that corpus allatum control green/brown polymorphism in Acridoidea in general. Pener (1967) reached to the same conclusion that corpus allatum directly or indirectly controls colours changes in *Schistocerca gregaria*. Thus it is clear from ablation and implantation of corpus allatum experiments performed by numerous workers that colouration in the locusts/grasshoppers is controlled by endocrine glands, specially the corpora allata (for details see: Raabe, 1982, 1989).

HORMONAL CONTROL OF BEHAVIOUR

Insects show a greater variety of hormonally influenced behaviours than do any other group of invertebrates. This is due at least in part to the fact that the neuroendocrinology of insects has attracted more attention than that of all other groups combined. Consequently, the neuroendocrine relationships of insects are the best known among the invertebrates. Additionally, insects-exhibit the richest and most varied behaviours seen among the invertebrates. This richness of behaviour may be directly related to this group's heavy reliance on hormonal mechanisms. The use of hormones to switch from one behavioural state to another most likely reduces the number of neurons needed to control a particular behaviour and may thereby allow a greater variety of behaviour to be packed into a small nervous system (Truman and Riddiford, 1974). Evidence for hormonal involvement exists for a variety of behaviours shown by a number of insect groups.

EFFECTS OF HORMONES ON BEHAVIOUR

The effects of hormones on insect behaviour are somewhat analogous to the effects of pheromones on behaviour. The latter have been classified by Wilson and Bossert (1963) into two types : releasers and primers. In a hormonal context, a releaser effect denotes a relatively rapid behavioural response which is directly triggered by the hormone. Additional sensory stimuli are not necessary for the behaviour to occur. This action is similar to that of a releaser pheromone.

Modifier effects

In most insects the individual is capable of a variety of behavioural options when faced with a given stimulus. It can attack, flee, attempt to copulate, ignore the stimulus, etc. Consequently, one response must be promoted and the other repressed, depending upon the "mood" of the insect. Hormones are often involved in the determination of "mood" and, indeed, this is the most common behavioural effect of hormones. As humoral agents which have access to all cells, they are well suited to effect tonic alterations in the responsiveness of the nervous system. A classic example of a hormone induced alteration in behaviour is illustrated by the effect of JH on the response of a female *Gomphocerus rufus* to a courting male. After the imaginal ecdysis, the female grasshopper displays primary defence- the approaches of males are met with kicking movements of the hind legs and escape reactions. After exposure to JH, primary defence disappears. Now the female faced with the same stimulus of a courting male stridulates in answer to the male's call and allows him to mount and copulate (see: Truman and Riddiford, 1974).

The eclosion hormone has a similar effect on the mating response of the silkmoth, *Antheraea pernyi*. In these moths the male is attracted to the female by a volatile sex pheromone that is perceived by sensory cells in the sensilla trichodea which cover the antennae of the male moth. These receptors are finely tuned to respond to the emanations of conspecific females. The irrelevant odours of most heterospecific females are, therefore, effectively filtered out at the receptor level. In the case of the Pernyi moth, males invariably mate with virgin females within 15 minutes of being placed with them in a cage under dim illumination (Riddiford, 1970).

But this response to the sex pheromone does not occur if the male has not been exposed to the eclosion hormone. When the pupal cuticle was removed from a pharate moth which had finished adult development but had not yet emerged, the animal showed no behavioural response to the sex pheromone, even under optimal conditions. This behavioural block was not in the ability to detect pheromone, since the antennae from pharate male moths gave normal electroantennogram responses to the female pheromone. Instead, this block must reside centrally. Whatever the nature of the block, it is rapidly removed when adult behaviour is "turned on" by the eclosion hormone (see: Truman and Riddiford, 1974).

Releaser effects

Some hormones serve to trigger directly a specific piece of behaviour. This action can be seen in the capacity of the phallic nerve stimulating hormone to release copulatory movements in cockroaches. Similarly, the hormonal control of "calling" behaviour in silkmoths (see p. 98) is presumably mediated through a releaser effect. The eclosion hormone also acts as a releaser because it directly triggers the performance of the pre-eclosion behaviour (Truman and Riddiford, 1974).

HORMONAL CONTROL OF PRE METAMORPHIC BEHAVIOUR

The hormonally induced peripheral changes do influence behaviour is seen in the premetamorphic behaviour of *Hyalophora cecropia*. In this species the onset of metamorphosis is signalled when the caterpillar ceases to feed, wanders from the food plant, and begins to spin a cocoon. These events are potentiated by a low JH level and perhaps triggered by a neurosecretion from the brain. The construction of the cocoon is accomplished by simple figure-eight movements of the head and by more complex movements of the body. Caterpillars whose spinnerets were blocked, nevertheless, performed the spinning movements. Thus, the actual spinning of the silk thread is not a prerequisite for the behaviour to occur. By contrast, when the silk glands were removed from early last instar larvae, the caterpillars fed and behaved normally until fully grown, but then did not show wandering behaviour. Spinning behaviour was also omitted. From this evidence it appears that low JH titres influence spinning behaviour at least partially through the JH- controlled growth of the silk glands. Most probably the hormone titres also have a direct influence on the CNS of the larvae.

HORMONAL CONTROL OF COPULATORY MOVEMENTS IN MALE COCKROACHES

The first example of a hormone directly releasing a specific behaviour was the hormonal triggering of copulatory movements in cockroaches. In both mantids and cockroaches the motoneurons in the terminal abdominal ganglion that innervate the phallic musculature are normally inhibited by centres located in the head. Decapitation of male cockroaches induced rhythmical coordinated movements of the phallomeres that mimicked the movements seen during mating. Under these conditions recordings from the nerves leading from the terminal ganglion to the phallomeres showed rhythmical motor bursts. This pattern of bursting in the phallic nerves could also be induced by application of corpora cardiaca extracts to the intact insect or to the nerve cord (Milburn *et al.* 1960). Motor bursts first appeared about 10 minutes after application and attained a maximum frequency about 20 minutes later - a time course similar to that seen after decapitation. Apparently, the motor programmes responsible for copulation are under nervous inhibition, presumably by centres in the suboesophageal ganglion. The "phallic nerve stimulating hormone" acts on these centres to block the inhibition and thereby release the proper motor programme.

HORMONAL CONTROL OF PRE-ECLOSION BEHAVIOUR IN MOTHS

One of the best characterized systems in insects involves the hormonal release of the pre-eclosion behaviour in silk moths. At the end of adult development the "pharate" moth initiates a stereotyped motor programme that culminates in the eclosion (emergence) of the animal from the old pupal cuticle. This motor programme, the pre-eclosion behaviour, is triggered by the "eclosion hormone" that is released from the corpora cardiaca into the blood. The hormone is produced in the brain by cells in the median neurosecretory cell-cluster (Truman, 1973b), and it appears to be a peptide.

Studies on the hormonal release of the pre-eclosion behaviour have concentrated on *Hyalophora cecropia*, a species with a relatively complex behaviour (Truman, 1971a, 1973a). The series of abdominal movements shows three distinct phases and last about 1.25 hours. An initial 0.5 -hour period of hyperactivity is followed by a comparable period of relative or complete quiescence, and then a second period of intense activity. During the first phase the movements are typically rotational motions of the abdomen whereas in the last phase they are strong peristaltic contractions that move anteriorly along the abdomen (the eclosion movements). This pre-eclosion behaviour begins within 10 to 15 minutes after injection of the pharate moth with the eclosion hormone.

The complete pre-eclosion behaviour is also seen after the hormone is injected into abdomens that have been isolated from pharate moths (Truman, 1971a). Thus, the neural circuitry necessary to respond to the hormone and to perform the pre-eclosion behaviour is located within the abdominal nervous system. This region of the moth CNS is composed of about 3000 neurons arranged in a chain of 3 simple segmental ganglia and a compound terminal ganglion.

HORMONAL CONTROL OF ECLOSION RHYTHM

Species such as the saturniid moths *Hyalophora cecropia* and *Antheraea pernyi* always emerge at a definite time, which varies for each species. Eclosion regulation is under brain control, and if the brain is removed, eclosion occurs at random. A brain implantation or the injection of brain extracts triggers the characteristic behaviour that precedes eclosion, and this occurs within 1-2 hr, even in isolated abdomens. By making reciprocal implantations in both species, the eclosion rhythms obtained correspond to the donor and not to the host (Truman and Riddiford, 1970; Truman, 1971a), but the characteristic behavioural programmes of each species are retained. Hence the implantation of a *Hyalophora* brain in *Antheraea* induces the programmes of the latter, but at the time of *Hyalophora*.

The brain thus plays a humoral role. The active factor, termed eclosion hormone, is secreted in the pars intercerebralis (Truman, 1973b) and stored in the corpora cardiaca (*vide supra*). It triggers eclosion behaviour, which includes several successive steps in *Hyalophora* : a phase of rapid rotational abdominal movements, a rest period, a new phase of violent activity consisting of peristaltic movements, and finally eclosion. This is followed by spreading of the wings, caused by the pressure of the blood that fills them, which is pulsed by contraction of the abdominal longitudinal muscles. Once this process is complete, the degeneration of the abdominal longitudinal muscles begins under the control of eclosion hormone (Truman, 1973a). Although no degeneration takes place in isolated abdomens, it does take place if the hormone is injected.

Eclosion hormone acts on the central nervous system (CNS), where it triggers the initiation of the species-specific programme. This was demonstrated in isolated abdomens, where the action potential of the nerves controlling the muscles involved in preecdysial contractions was recorded. The results differ widely according to whether the eclosion hormone is present or absent. If it is present, many similar bursts of spikes are first observed in the successive ganglia, but reverse from one side to the other, corresponding to rotational movements. In the next phase of peristaltic contractions, the impulses are identical on both sides, but are phase-shifted from back to front. When isolated central nervous systems are subjected to the eclosion hormone, they respond by the generation of a programme of motor activity that mimics the one expected during the preeclosion and eclosion behaviours. The initiation of these behaviours is elicited by only a few minutes exposure of the central nervous system to the hormone (see : Raabe, 1989, Awasthi, 2001).

The tanning processes begin as soon as the young moth has spread its wings. The question arises whether bursicon, the tanning hormone, is different from eclosion hormone. By using different organs of the tobacco horn-worm *Manduca sexta,* Truman tested eclosion hormone activity on *Antheraea* and bursicon activity on *Manduca* (1973c). These experiments revealed that eclosion hormone is located in the brain and in the corpora cardiaca, whereas bursicon is found in the abdominal ganglia and in the perisympathetic organs. Experiments in which different parts of the brain were removed indicate that eclosion hormone is located in the pars intercerebralis and not in the lateral protocerebral neurosecretory cells.

Eclosion hormone is not constantly present in the brain-corpora cardiaca complex. It appears in *Antheraea* at the onset of adult development, namely on the 7th day of pupal life, increases progressively to reach a plateau on the 10th day in the brain and on the 13th day in the corpora cardiaca. It then diminishes considerably in the corpora cardiaca on the 19th day, namely, just before eclosion, while its level in the haemolymph rises sharply (Truman, 1973b). The time of appearance of eclosion hormone in the haemolymph of *Manduca* was determined by isolated wing bioassay. It was demonstrated that this hormone is released in the blood only at a restricted time that is determined by a circadian clock once development has been completed (Reynolds *et al.,* 1979)

The studies concerning eclosion hormone were extended to *Manduca* (Truman, 1973c). It was demonstrated that in *Manduca,* as in silk moths, an eclosion hormone was present causing early eclosion when injected into pharate adults and inducing the release of a stereotyped behaviour that allows shedding of the old skin. The target sites were shown to be scattered throughout the central nervous system, and a coordinated response was dependent on all the segmentary sites (Truman, 1979).

Investigations in *Manduca* embryos, larvae, and pupae revealed that not only adult eclosion, but also embryonic larval and pupal ecdysis were under the endocrine control exerted by eclosion hormone (Truman *et al.* 1980, 1981), which causes premature ecdysis in larvae and pupae. The amounts of eclosion hormone in the ventral nerve cord were determined during all larval instars. In final instar larvae, eclosion hormone increases during feeding period, reaches a plateau, and drops sharply at ecdysis. In the pharate pupae, eclosion hormone appears in the haemolymph about 30 minutes before the onset of ecdysis (Truman *et al.,* 1980), and its titre rises sharply and declines steadily during a period of 5 hr (Reynolds *et al.,* 1979).

The role of eclosion hormone is well-known in the adult *Manduca* (*vide supra*). However, the origin of eclosion hormone is not the same throughout postembryonic development; in the larvae and pupae it is produced in the ventral nerve cord, whereas in the adult it is produced in the brain and released in the corpora cardiaca. Eclosion hormone activity is found in several insects of the main orders (Truman *et al.,* 1980; Truman *et al.*, 1981).

THE HORMONAL INHIBITION OF FIREFLY FLASHING

The flashing of the firefly beetle, *Luciola* is modulated by a hormonal mechanism which acts peripherally. In fireflies an exposure to light will inhibit the flashing of the beetle. Magni (1967) showed that in *Luciola italica* this photically induced flash inhibition was mediated by two mechanisms : a central one which acted at the level of the neural pacemaker and a peripheral mechanism which acted on the lantern. The involvement of a hormone in this peripheral inhibition was shown by connecting the haemocoels of two beetles (*Luciola lusitanica*) and then illuminating the eyes of one. This treatment markedly depressed the intensity of spontaneous flashing in the unilluminated partner. When the testes were removed from the stimulated insect, the inhibition seen in the attached partner was abolished (Brunelli *et al.*, 1968 a,b).

When the light organ was stimulated directly, the intensity of the elicited flash was unaffected by either decapitation or denervation of the lantern. But if the eyes of the denervated firefly were illuminated, the intensity of the electrically induced flash was markedly reduced. Inhibition was again abolished by removal of the testes.

Innervation of the testis is essential in this inhibition. Transection of the nerve cord anterior to the ganglion which innervates the testis destroyed the inhibitory response to illumination of the eyes.

Consistent with this finding was the result that electrical stimulation of the nerve leading to the testis inhibited the flash elicited by electrical stimulation of the lantern (Bagnoli *et al.*, 1970).

The material responsible for flash inhibition is apparently stored in granules found in the cortex of the testis (Brunelli *et al.,* 1970). Histological studies showed a marked depletion of these cortical granules after the insect was exposed to light. Moreover, these granules, when isolated from testes homogenates by centrifugation in a sucrose density gradient, inhibited the induced flashing of the lantern when injected into males.

The flash-inhibiting substance was identified as noradrenalin (Bagnoli *et al.*, 1972). This identification was based on the low doses (6×10^{-12}g) of noradrenalin needed to inhibit flashing, the fluorimetric measurements of high levels (75×10^{-12} g per male) of noradrenalin and of the absence of adrenalin in the gonads, and autoradiographic evidence of the uptake of ^{3}H- dopamine into the cortical granules.

The action of noradrenalin must be peripheral since its application inhibits flashing but does not affect the size or frequency of efferent volleys coming to the lantern. Since the neurotransmitter released from the neuroeffector organ within the lantern also appears to be adrenergic in nature, there are several possible sites where noradrenalin could exert its inhibitory effect. Bagnoli *et al.* (1972) present two alternative hypotheses to explain its inhibitory action : (1) noradrenalin may act to block transmitter release at the neuroeffector junction; or (2) it may act somewhere between transmitter release and light production.

Also, noradrenalin interferes with the response of the lantern to direct electrical stimulation (Bagnoli *et al.*, 1972).

HORMONAL CONTROL OF LARVAL BEHAVIOUR AT THE OUTSET OF METAMORPHOSIS

In holometabolous larvae the endocrine activity of the corpora allata begins to decline midway through the last instar and reaches a low level by the time of pupation (Williams, 1961). Accordingly, towards the end of the last instar, the blood titre of JH decreases. During this period of endocrine changes, there are also major behavioural alterations. In Lepidoptera, the mature last stage larva typically ceases feeding and leaves the food plant either to burrow into the ground or to begin cocoon construction. In the linden moth, *Mimas tiliae*, Piepho *et al.* (1960) showed that the mature larva becomes positively geotactic and descends from its tree to the ground. The positive geotaxis could be reversed by the implantation of active corpora allata. Therefore, the sign of larva geotaxis is influenced by the JH titre in the blood.

The declining titre of JH apparently facilitates the performance of the premetamorphic behaviour but does not trigger its occurrence. Thus, in another sphinx moth, *Manduca sexta*, the cessation of feeding is followed by increased locomotor activity, voiding of the gut, appearance of a pink pigment in the dorsal epidermis, and exposure of the heart. All of these events occur at approximately the same time. Experiments involving neck ligation of mature *Manduca* larvae showed that at least the last two events occur in response to a factor released from the head. Presumably, the hormonal control of red pigmentation in *Manduca* is similar to the control reported for *Cerura vinula*, in which the pigmentation change is mediated through the brain- prothoracic gland system. The behavioural events associated with the appearance of pink pigment in *Manduca* (i.e. gut emptying and increased locomotion) are probably also triggered by the same endocrine signal. One might suspect that some of the premetamorphic behaviour of larvae is potentiated by the low JH titre and then released in response to the prothoracicotropic hormone or to ecdysone (see : Truman and Riddiford, 1974).

HORMONAL CONTROL OF COCOON CONSTRUCTION IN MOTHS

The role of JH in cocoon construction was first indicated by the classic experiments of Bounhiol (1938) using *Bombyx mori*. Allatectomy early in the penultimate larval instar caused precocious pupation and, as prelude to this, the precocious spinning of a miniature cocoon. By contrast, implantation of active corpora allata into last stage larvae caused a supernumerary larval moult and the delay of cocoon construction.

In the greater was moth, *Galleria mellonella*, a silken structure is spun at the end of each instar but the shape changes depending upon the instar. The penultimate ie. 4th stage larva spins a thin-walled moulting tube. The mature larva constructs a tough-walled cocoon. When active corpora allata from fourth instar larvae were implanted into a last stage larva, the latter spun a structure which was intermediate between a cocoon and a moulting tube.

In some other lepidopterans the untimely appearance of JH prevents spinning. The application of JH mimics to mature Cecropia larvae delayed the onset of spinning for 1 to 3 days but did not alter the type of cocoon which was eventually spun (Riddiford, 1972). Application of JH mimics to last instar *Cerura* larvae also disrupted spinning behaviour but the data indicate that the disruption may simply be due to the abnormal development of the silkgland.

Neurosecretion may also play a role in the type of cocoon constructed. In the silkworm, *Samia cynthia*, a hormone from the brain reportedly directs the construction of either a summer or a winter cocoon. The type of cocoon spun is correlated with whether or not the insect will diapause. Awasthi (1985a) has pointed out that the *Philosamia ricini* larvae treated with 25 μl of JHA once only on day 3, constructed on days 8 and 9 extremely thin cocoon through which the prepupae were clearly seen and the site of JHA application in the pupae remained untanned. The cocoons were provided with two wide-valves, one anterior and the other posterior. The anterior valve was more than 1.5 cm in diameter.

The larvae treated with 12.5 μl JHA on days 3, 4 and 6 indicated prolongation of larval life and inhibition of cocoon formation. The results indicated that out of the treated larvae (1) some of the larvae pupated without constructing cocoons (naked pupae); (ii) others constructed very thin and light cocoons with two big valves. Thus profound influence of JHA was observed in cocoon construction.

HORMONAL CONTROL OF ACTIVATION OF ADULT BEHAVIOUR IN MOTHS

In the giant silk moths, a "pharate" moth just prior to the normal time of adult emergence shows marked deficits in adult behaviour. This can be readily seen in the response of male *Antheraea pernyi* to virgin females. When placed in a cage with virgins under dim illumination, males become excited, undertake a search for the female, and invariably copulate within 15 minutes. Pharate male moths, from which the pupal cuticle has been peeled away, remain completely passive when presented with the same stimuli. This behavioural block seen in pharate males is not due to the inability to detect the sex pheromone because the antennae of pharate moths give a normal electroantennogram response when presented with this odour. The blockade is undoubtedly central and is removed about 2-3 hours after injection of the peeled moths with the eclosion hormone. Besides, quite complex behaviour, simpler behaviours such as the righting reflex also become "turned on" by the eclosion hormone.

The appearance of adult behaviours at eclosion likely occurs through the removal of inhibition. In species of the moth *Automeris*, the adult rocking display behaviour can be elicited during the late stages of adult development. Similarly, in *A. pernyi, A. polyphemus*, and the sphinx moth *Manduca sexta* the flight and "warm-up" motor programmes are also seen in developing animals. Removal of the pupal cuticle from *A. pernyi* about four-fifths of the way through adult development releases distinct eclosion movements. Yet the ability of these animals to show these behaviours declines as development nears its end and on the day of adult emergence they cannot be elicited. In *A. pernyi* the disappearance of spontaneous eclosion movements coincides with the onset of sensitivity to the eclosion hormone. Thus it appears that the neural circuits controlling at least some aspects of adult behaviour are functional prior to adult eclosion, but that they are repressed by inhibitory neurons. The eclosion hormone presumably acts on these inhibitory neurons to release this inhibition (see : Truman and Riddiford, 1974, 1977).

In adult *Manduca sexta*, the "shut-off" motoneurons degenerate along with their muscles. Certain groups of small neurons also rapidly degenerate after eclosion. These latter cells may have had a role in generation of certain pupal behaviour patterns and in the inhibition of adult behaviour in the pharate adult. This neuron death is apparently caused by the eclosion hormone, but it does not seem to be the mechanism by which the hormone triggers muscle degeneration or the release of adult behaviour patterns. This conclusion is based on the fact that the closely related silk moths show the same behavioural and muscular changes at eclosion as does *Manduca*, but their neurons do not then degenerate. Consequently, neuron death is not the necessary outcome of the eclosion hormone "shutting-off" neurons, but rather a mechanism that *Manduca* uses to rid itself of circuitry that is no longer useful. However, this phenomenon has served to indicate which neurons in the moth CNS are "turned-off" by the eclosion hormone (Truman and Riddiford, 1977).

HORMONAL CONTROL OF REPRODUCTIVE BEHAVIOUR

The influence of hormones on insect behaviour has been most thoroughly studied in connection with sexual maturation and reproduction. During its life the adult insect often goes through several transitions from one behavioural state to another - i.e. a non-receptive virgin becomes receptive. Since all aspects of reproductive behaviour from mate attraction to the deposition of fertilized eggs must be smoothly coordinated with both the internal physiology and the external environment, it is not surprising that hormones play a large role in this coordination.

Male sexual behaviour

Compared to the female, the male of most insect species shows rather simplified behaviour. Mate-finding, copulation and, perhaps, feeding are his main pursuits. In accordance with this reduced behavioural repertoire, one also finds a paucity of cases of hormonal involvement in either the "maturation" of male behaviour or its actual expression. In most instances male behaviour matures in concert with the metamorphosis of the nervous system. Thus, at adult ecdysis or shortly thereafter, suitable stimuli will evoke the full copulatory behaviour. This behavioural state then persists until death.

Grasshoppers and locusts

The neuroendocrine control of male behaviour has been most widely studied in the grasshoppers and locusts. In the desert locust, *Schistocerca gregaria*, and the red locust, *Nomadacris septemfasciata*, the corpora allata are required for the maturation of sexual behaviour in males. In both *Schistocerca* and *Nomadacris* allatectomy of the male soon after ecdysis completely prevented the onset of sexual behaviour which normally occurred about 10 days post-ecdysis. The implantation of several pairs of active corpora allata into an allatectomized male promoted the reappearance of this behaviour (Loher, 1961, 1967).

The regulation of JH secretion in *Schistocerca* is apparently controlled by the brain via the nervus corporis allati. Sectioning of the nerve in freshly emerged males severely inhibited the appearance of male behaviour. But, eventually about 10 per cent of the insects displayed some sexual activity. This low level of sexual behaviour was presumably due to a small amount of JH which is secreted from the denervated gland. Brain control of the activation of the corpus allatum was also suggested by the observation that flight activity accelerated sexual maturation in these males have shown that similar activity in *Schistocerca* or crowded *Locusta* females caused accelerated oocyte growth through the increased release of the corpora allata-activating neurosecretion (see : Engelmann, 1968b, 1970; Raabe, 1989).

As suggested by Odhiambo (1966), it is very likely that JH acts directly on the central nervous system of *Schistocerca* to control the development or unmasking of the neural circuitry involved in courtship and mating. This unmasking of the proper circuits appears to require the continuous presence of JH since allatectomy of a mature male will lead to the disappearance of sexual behaviour (Loher, 1961; Pener, 1967).

In the male migratory locust, *Locusta migratoria migratorioides*, JH has a similar role but its control is not as complete as that seen in *Schistocerca*.

In *Locusta* the C-type neurosecretory cells of the pars intercerebralis appear to be responsible for the activation of the corpora allata which in turn leads to the onset of sexual behaviour. Cauterization of the pars intercerebralis completely prevented the appearance of sexual behaviour. Importantly, when 6 pars intercerebralis were implanted into cauterized males, both sexual behaviour and the mature yellow pigmentation were restored (Truman and Riddiford, 1974).

Although the action of the C-cells of the brain is partially through their stimulation of JH secretion, they also have direct behavioural effects. This additional role was indicated by the fact that allatectomy only partially reduced male sexual behaviour, whereas cauterization of the pars intercerebralis completely abolished it. Moreover, the behavioural effects of cauterization could be partially reversed by pars intercerebralis implants but not by CA implants. Accordingly, Pener *et al.* (1972) concluded that the C-type neurosecretory cells are required for the performance of sexual behaviour and that the CA then regulate the level of this behaviour.

The interaction between the corpora allata and the neurosecretory system of the brain provides a control over the onset of sexual behaviour which is responsive to environmental conditions. In the grasshopper, *Oedipoda miniata*, as well as in some subspecies of *L. migratoria*, the males shows a photoperiodically controlled diapause. In these insects the sexual behaviour of the male was severely inhibited or eliminated under long-day conditions, but it was restored by the implantation of active corpora allata. Presumably, as has been shown in *Leptinotarsa decemlineata*, the photoperiod controls the release of neurosecretion which then activates the corpora allata (De Wilde and De Boer, 1969; DeWilde and DeLoof, 1973).

Although the onset of male behaviour may not be controlled by JH, the neurosecretory hormone discovered by Milburn *et al.* (1960) almost certainly has a role during courtship and mating.

Diptera

In Diptera the only evidence for hormonal involvement in male behaviour is in the yellow dung fly, *Scatophaga stercoraria* (Foster, 1967). When the fly was deprived of food after emergence, both sexual behaviour and the accessory glands remained undeveloped. This same syndrome was also seen in fed flies which were allatectomized immediately upon emergence. The effects of prey deprivation or of allatectomy could be reversed simply by implanting active corpora allata. Thus, in males of this fly, food intake apparently serves to activate the corpora allata which, in turn, promote the onset of sexual behaviour.

HORMONAL CONTROL OF THE DEVELOPMENT OF RECEPTIVITY IN VIRGIN FEMALES

The role of the corpora allata

Whereas the corpora allata seem not to be necessary for male sexual activity in most insects, the reverse is true in the case of the female. But one must carefully distinguish between direct effects of JH on the behaviour of the female towards the courting male and indirect effects of JH in mediating sex pheromone production and thereby attraction of the male.

The classic example of the distinction between JH effects upon behaviour and upon pheromone production is seen in the cockroach, *Byrsotria fumigata*. When allatectomized females were placed with males, no mating occurred. This effect was not due to lack of receptivity on the part of the female, but rather to her failure to produce the volatile sex pheromone. When such females were coated with extracts containing pheromone, they were vigorously courted and copulated readily. Thus, the CA of *Byrsotria* are not involved in the onset of sexual receptivity and, indeed, the JH mediated onset of pheromone production occurs two days prior to normal appearance of sexual behaviour (Barth and Bell, 1970).

At the other end of the spectrum one finds insects in which the CA completely control the onset of female receptivity. The best known example is that of *Gomphocerus* (Loher, 1962, 1966).

Loher and his co-workers have shown that allatectomy of last instar or newly emerged females completely prevented the onset of sexual receptivity. Allatectomized females remained in a permanent state of "primary defence" and responded to a courting male by kicking and escape. The implantation of active CA into such females then led to a state of "copulatory readiness" - courting males were allowed to mount and copulate. The application of JH mimics such as farnesyl methyl ether also served to promote receptivity in allatectomized females. Copulatory readiness was not only initiated but also maintained through the action of the CA. Allatectomy of sexually mature females led to the loss of receptivity and the reappearance of primary defence behaviour within six days (Loher, 1966).

As in the case of some male acridids, the neurosecretory cells of the pars intercerebralis of the female *Gomphocerus* has a role in turning on the CA. They thereby indirectly regulate both oocyte maturation and the appearance of sexual receptivity (see : Engelmann, 1968, 1970).

Cockroaches provided the first evidence that hormones were involved in the onset of sexual behaviour in insects. Engelmann (1968) demonstrated that allatectomy of newly emerged *Leucophaea maderae* females led to a marked reduction in the number of females which became receptive and mated (30 per cent). When CA were implanted into such females, normal sexual receptivity ensued.

Besides the role of the CA, there is evidence that the neurosecretory cells of the brain may also be needed for the onset of receptivity in cockroaches. In *Nauphoeta cinerea*, Roth and Barth (1964) hypothesized that the delayed onset of receptivity which was observed in allatectomized females may be due to a neurosecretion from the brain. In *Byrsotria* a histological examination of the brain revealed a change in neurosecretory activity at the time of parturition (a time when females again become receptive). However, in this species one finds no comparable neurosecretory changes during the initial maturation of sexual receptivity.

Barth (1968) suggested that the prolonged courtship necessary to induce receptivity in allatectomized *Leucophaea* as well as the delayed onset of receptivity in allatectomized *Nauphoeta* indicated that neural centres controlling this behaviour may be turned on by environmental influences, presumably working through the neuroendocrine system. Similarly, he suggested that the retardation of sexual receptivity in *Blattella germanica, B. vaga* and *Leucophaea* which can be induced by starvation may be due to the prevention of release of the proper neuroendocrine agent (see : Roth and Barth, 1964).

The only direct evidence for the involvement of a brain neurosecretion has been reported for *Leucophaea* (see : Engelmann, 1968, 1970). Cauterization of the pars intercerebralis of the female completely abolished mating behaviour. The proper behaviour could not be restored by implantation of active CA. Therefore, the absence of receptive behaviour was not simply due to the lack of a tropic factor which is necessary for JH release. Rather, a neuroendocrine agent was probably directly involved in the appearance of the behaviour. But, as discussed above for female *Gomphocerus*, without additional implantation data this evidence cannot be taken as unambiguous support for a neuroendocrine involvement.

In Diptera the role of the CA in promoting female receptivity was first demonstrated in *Drosophila melanogaster*. The female of this insect becomes receptive to courting males during the second day of adult life. When active corpora allata-corpora cardiaca complexes were implanted into pharate females 17 to 19 hrs. before eclosion, the females became precociously receptive by 24 hrs. after eclosion.

Sexual activity in *Musca domestica* is likewise influenced by the CA. Allatectomy of freshly emerged females did not prevent sex pheromone production and consequently, attraction of the male occurred normally. But operated females did not usually respond to the mounting male by protrusion of the ovipositor. Thus, successful copulation was prevented. JH application to allatectomized females caused the onset of receptive behaviour.

In Diptera the most complete studies of the role of the CA in promoting female sexual behaviour have been carried out on mosquitoes, especially *Aedes aegypti*. The newly emerged females is sexually unreceptive and although males may attempt to copulate with her, successful insemination does not occur. By one to two days after emergence, depending upon the particular strain, females accept males and are inseminated (Truman and Riddiford, 1974, 1977).

The onset of receptivity could be substantially delayed by removal of the CA from newly emerged females. But this condition was not permanent, and such females showed a considerable percentage of inseminations by 12 days after eclosion. The eventual appearance of receptivity might again be due to the secretion of a small amount of JH prior to CA removal. This possibility is likely since Lea (1968) has shown that the CA of female mosquitoes are active within 5 minutes of emergence.

As with other insects, implantation of CA completely restored receptivity to allatectomized females (Lea, 1968). Similarly, application of a JH mimic to female *A. aegypti* within 30 minutes after emergence caused the precocious onset of sexual receptivity (Gwadz *et al.* 1971b).

HORMONAL CONTROL OF OVIPOSITION BEHAVIOUR

The stimulation of oviposition and the cessation of receptivity are closely related behaviours which most probably represent two manifestations of the same process, i.e. a switchover from the "virgin mode" to the "mated mode". Thus, changes in oviposition rate can provide a quantitative measure to the extent of this behavioural switch.

In many insects there seems to be a release of neurosecretory material at the time of oviposition. On the basis of histological studies, this relationship has been claimed for *Schistocerca gregaria*, the cockroach, *Trichoblatta sericia*, the pyrrhocorid bug, *Iphita limbata* the beetles, *Leptinotarsa decemlineata* and *Galeruca tanaceti* and the silkmoth, *Bombyx mori* (see: Truman and Riddiford, 1977).

In *Iphita*, pairs of bugs copulate for 3 weeks during which time the eggs of the female mature. When blood was taken from females during the period around the time of oviposition and injected into partially gravid mating females, mating was terminated and oviposition induced (Nayar, 1958). These injections also served to deplete the neurosecretory cells of the brain. Since a similar depletion was noted after injection of extracts of mature ovaries, it was concluded that the factor in the blood was derived from the ovary. This material then acted on the neurosecretory cells of the brain to cause the release of an oviposition-stimulating hormone. Further evidence for this hypothesis was obtained by the implantation of two sets of neurosecretory cells from a mating female into a nearly gravid female. This treatment rapidly stimulated oviposition. When neurosecretory cells were implanted into young females which lacked eggs, quivering movements of the genital plates (a characteristic behaviour which accompanies oviposition) were triggered (Nayar, 1958).

In *Schistocerca* the injections of blood or corpora cardiaca homogenates from ovipositing females as well as the implantation of the corpora cardiaca caused oviposition movements in virgin females. Electrical stimulation of the central nervous system or enforced hyperactivity (both of which cause release of neurosecretory material) also induced oviposition. Besides the neurosecretory cells of the pars intercerebralis those in the glandular lobe of the corpora cardiaca and in the third thoracic ganglion also seem to be involved (see: Delphin, 1965; Truman and Riddiford, 1974).

The median neurosecretory cells of the mated female of *Rhodnius prolixus* similarly release a substance which influences oviposition. In the walking sticks, *Carausius morosus* and *Clitumnus extradentatus*, the control over oviposition is more complex. The neurosecretory cells of the pars intercerebralis have a role (Dupont-Raabe, 1952a,b), but so also do those in the suboesophageal, thoracic, and first four abdominal ganglia.

The switch in oviposition rate observed in mated *H. cecropia* females is mediated by a hormone which is apparently released from the intrinsic cells of the corpora cardiaca (Truman and Riddiford, 1971). An increased oviposition rate after mating was seen in both intact and allatectomized Cecropia females. But removal of both the CA and CC abolished the response to mating. Thus it appears that the intrinsic cells of the CC also mediate the change in oviposition behaviour (see: DeWilde and DeLoof, 1973).

In *Bombyx mori*, injections of blood from mated females caused increased oviposition. Similarly, the blood from mated Cecropia females stimulated egg-laying when injected into virgin females. The blood injections did not elicit an immediate oviposition response, but rather caused a marked increase in the number of eggs deposited per night for the next one to two days. The effects of the blood were similar to those of the bursa implants except that they were transient. Most probably, the active agent in the blood was the factor which is secreted by the bursa of a mated female. This factor then presumably acts through the neuroendocrine system of the female to cause the release of the oviposition-stimulating hormone from the corpora cardiaca (Truman and Riddford, 1974).

HORMONAL CONTROL OF MIGRATION AND ORIENTATION

Certain environmental conditions provoke migration of some insects from their old habitat to a new one. This move may be of single individuals over small distances, such as the movement of aestivating insects from their host plant into the soil. Or it may consist of the mass movement of insects over hundreds of kms. as is seen for locusts. Irrespective of the extent of the migration, the endocrine state of the insect appears to be of great importance.

Gregarious and migratory behaviours in locusts

Most studies of hormonal influences on migratory behaviour have been focussed on the migratory locusts, *Schistocerca gregaria* and *Locusta migratoria migratorioides*. When reared in isolation, the hoppers of these species show a moderate level of activity; the adults tend to be solitary and usually do not perform long sustained flights. Crowding of young locusts results in highly active, gregarious hoppers which display marching behaviour and which show rapid oriented locomotion in response to certain stimuli. The adults are gregarious and undertake long sustained flight (Johnson, 1969).

An indication of an endocrine basis for the differences in locomotor activity observed in the two phases was provided by the observation that solitary hoppers have larger prothoracic glands than do gregarious hoppers. Moreover, in the adult the prothoracic glands persist in the solitary form but rapidly disappear in gregarious locusts.

When prothoracic gland homogenates were injected into gregarious hoppers, marching was decreased thus supporting the hypothesis of a possible role for these glands in the control of activity. Similarly, implantation of prothoracic glands from solitary adults into gregarious adults also decreased the duration of sustained flight of the recipient (Michel, 1972). But in neither study was the decrease striking (50 per cent at most) nor was it comparable to the marked difference in locomotor activity exhibited by the two phases. Since activity is a labile behaviour which can be influenced by many factors, including injury due to injection or implantation of foreign tissues, one cannot be confident that the above effects were specifically due to ecdysone.

The corpora allata have also been implicated in the control of locust activity. Cassier (1963) first showed that CA implants decreased the latency of the phototropic response shown by gregarious *Locusta*, and led to an increase in the speed of movement. In *Schistocerca*, males, Odhiambo (1965) then demonstrated that allatectomy severely decreased spontaneous locomotor activity. In addition, these allatectomized insects showed abnormally large deposits of glycogen and lipids in the fat body. Odhiambo (1965) hypothesized that JH enhanced locomotor activity, as well as sexual activity, by a direct action on the nervous system and that the increased fat and glycogen reserves resulted from the persistent inactivity of the operated males.

This hypothesis was partially confirmed by the demonstration that allatectomized *Locusta* which were forced to exercise 2 hrs. each day deposited less fat than did unexercised, allatectomized controls (Strong, 1968) (also see: P. 105).

Migration and Behaviour

Johnson (1969) put forth a model in which adult migration was triggered when ecdysone was absent and the JH titre was low. The rise in JH titre then caused both the cessation of migratory behaviour and the onset of oogenesis. This model was tested in the milkweed bug, *Oncopeltus fasciatus*. In this bug the migratory behaviour of the adult is determined primarily by the photoperiod to which it was exposed during larval life. Long-day reared *Oncopeltus* show little or no tendency to make sustained flights, and oogenesis begins in the female relatively soon after adult ecdysis. By contrast, short-day bugs typically undertake long sustained flights, and thus the onset of reproduction is delayed.

Rankin (1974) examined the role of the CA in controlling migratory behaviour in *Oncopeltus*. The migratory status of the bugs was determined by the percentage of animals which performed sustained, tethered flight. When a JH mimic was applied to either long-day or short-day males, and the insects tested three days later, a greater proportion showed sustained flight than did controls of comparable age. In a parallel experiment, the implantation of three CA significantly increased the number of males which made sustained flights.

JH initially appeared to have different effects in the female (Rankin, 1974). When applied to 5-day-old females, the JH mimic did not stimulate flight behaviour. When it was administered to very young (2-day-old) females, a significant but transient increase in flight activity was observed. Since the applied JH caused the rapid onset of oogenesis in the day 5 females, these results were not surprising because migratory behaviour in *Oncopeltus* is terminated with the onset of oogenesis.

Thus, in *Oncopeltus* the onset of migratory behaviour is apparently caused by high titre of JH. In the females, migratory behaviour is later shut off in response to the JH-induced maturation of the ovaries. The nature of this effect of the ovaries is unknown.

Unlike *Oncopeltus*, the ovaries do not influence the behaviour of *Leptinotarsa*. Long-day females which were ovariectomized nevertheless continued to show "soil-negative behaviour". Also, allatectomized and ovariectomized females switched from "soil-positive" to "soil-negative" behaviour after implantation of active CA, suggesting that the action of JH is not mediated through its effects on the ovaries (DeWilde and DeBoer, 1969).

Effects of JH on orientation have also been reported in other insects. As described above, positive geotaxis is induced in larvae of *Mimas tiliae* by low JH titre (Piepho *et al.* 1960). Also, in the scarab beetle, *Melolontha melolontha*, implantation of active female CA into migratory males reverses the compass direction of flight. Both JH and ecdysone influence the sign of phototaxis in the caterpillar of *Smerinthus ocellata*.

HORMONAL CONTROL OF CIRCADIAN RHYTHMS

As in most other organisms, insects possess internal clocks which can be synchronized and entrained by daily environmental periodicities. In the absence of external cues these clocks continue to run, but with a periodicity which is slightly different from 24 hrs. Thus, they are "circadian" clocks.

These clocks have been widely implicated in the control of many diverse types of behaviour. Generalized behaviour such as locomotor activity is influenced by a complex array of factors. In this instance the behaviour has numerous potential functions and therefore, is used in various environmental contexts (e.g. escape, food-gathering, mate-finding, etc.). Consequently, the daily stimulus provided by the clock must interact with a network of diverse external and internal stimuli which also influence locomotion. A simpler type of control is often seen with certain specialized behaviours. These are stereotyped behaviour patterns which are displayed under restricted environmental or developmental contexts. The performance of the behaviour is often directly coupled to the output of the clock with little or no influence from other stimuli. In many instances the coupling between the driving circadian clock and the behaviour appears to be hormonal (see : Truman and Riddiford 1974).

Further circadian rhythms, which are often photoperiodically dependent, control several types of processes including various reproductive behaviours, locomotion, ecdysis, and changes in the activity level.

In the crickets *Acheta* and *Teleogryllus,* removal of the pars-intercerebralis made the insect arrhythmic and parabioses caused the activity rhythm to reappear, while in insects deprived of the pars-intercerebralis implantation of a brain from a rhythmic donor failed to restore the normal situation, although it periodically reduced locomotor activity.

In the stick insect, *Carausius* which displays both circadian colour change and circadian activity rhythm, it was shown that the mechanisms triggering these two processes are different, i.e., neuroendocrine control in the former and nervous control in the latter. Removal of the CC-CA and pars-intercerebralis had no effect on behaviour. However, removal of the brain or severance of the circumoesophageal connectives deleted the circadian rhythm and increased activity, showing that control of this rhythm is exerted by the brain. Removal of the suboesophageal ganglion or cutting of the connectives behind this organ permanently immobilized the insects. Studies conducted on cockroaches, suggest that the suboesophageal ganglion stimulates activity while the brain inhibits it. The severance of the ventral nerve cord behind the suboesophageal ganglion allowed circadian activity to continue in the anterior part of the body, while the posterior part remained motionless (Dupont-Raabe, 1957, Raabe, 1982).

Besides experimental investigations, histological, histochemical, and ultrastructural studies were carried out in order to demonstrate putative circadian changes in the neurosecretory cell load. In crickets and cockroaches, the activity of the neurosecretory cells in the pars-intercerebralis displayed a circadian rhythmic activity which is also found in the A-type neurosecretory cells of the ventral nerve cord of the cockroach. RNA and protein synthesis also follow a temporal rhythm, observed in studies of labelled cystein incorporation and acetylcholinesterase level which drops with the rise in nocturnal activity. Serotonin determinations in the brain and haemolymph also revealed cyclic modifications, but no correlation was found between the serotonin content of the haemolymph and the circadian rhythm. However, a close correlation was shown between this rhythm and the serotonin content of the central body, known to be an important connection centre of the brain (see : Raabe, 1989).

In *Drosophila*, neuroendocrine control of the circadian rhythm was suggested for mutant flies whose locomotor rhythm altered. Brain implantation indeed changed the activity pattern of the implanted receiver. Experiments on a cockroach identified the location of the circadian clock as being in the optic lobes whose reimplantation in arrhythmic insects re-established a circadian activity rhythm (see : Raabe, 1989; Awasthi, 2001).

Circadian rhythms of locomotion

The most extensive studies of the hormonal involvement in insect activity rhythms have utilized the cockroach as an experimental subject. The earliest work was that of Harker (1960) who examined the role of the suboesophageal ganglion in the control of this rhythm. According to Harker, decapitation rendered cockroaches arrhythmic, but the implantation of a suboesophageal ganglion into the headless insect restored rhythmicity. Importantly, the phase of the new rhythm was similar to the prior rhythm of the donor of the ganglion, but not to the rhythm that had previously characterized the host. Thus, the suboesophageal ganglion was responsible for the intrinsic timing of activity and did so through a humoral pathway (Harker, 1960). But many authors are not in favour of the view of Harker (see: Raabe, 1982, 1989).

In the noctuid moth, *Noctua pronula*, microcautery of the median neurosecretory cells abolished the nocturnal circadian component of the adult flight rhythm (Hinks, 1967). Injections of serotonin extended the duration of flight activity in normal individuals. Furthermore, high amounts of tryptophan-positive material were found to be present in the A-cells of the median neurosecretory cluster.

Circadian control of specialized behaviours

As mentioned above, specialized behaviours are performed under restricted environmental or developmental conditions. Some of these specialized behaviours such as pheromone release (Riddiford, 1974) may occur repeatedly at a specific time on successive days. Other behaviours such as eclosion (Pittendrigh, 1966) or cocoon spinning occur but once in the lifetime of the individual, and always during a restricted period of the day.

Some specialized behaviours such as oviposition, and eclosion are apparently triggered by neurosecretory hormones and have been considered in detail above. The circadian organization of these behaviours presumably arises from the circadian release of the corresponding hormones. This has been shown in the case of silkmoth eclosion.

As described above, silkmoths emerge during a species-specific time of day (Truman and Riddiford, 1970). Removal of the brain from silkmoth pupae yielded moths which showed abnormal eclosions that were randomly distributed throughout the day and night. When a brain was implanted into the abdomen of each debrained moth, the proper timing of the behaviour was restored. Moreover, when brains were interchanged between pupae of *Hyalophora cecropia* and *Antheraea pernyi* (two species which had very different eclosion times), the resulting moths emerged at the time of day which was characteristic of the species which donated the brain (Truman and Riddiford, 1970). Thus, the brain controls the time of release of the eclosion hormone which in turn activates the proper behaviour.

HORMONAL CONTROL OF SOCIAL BEHAVIOURS

In the social insects, caste determination is under endocrine control, but some of the adult behaviours necessary to the social system have also been shown to be influenced by hormones. In honey bee workers the behavioural change from a nurse bee to a forager bee is induced by juvenile hormone. Also in workers of the wasp *Polistes*, juvenile hormone promotes dominance behaviour that is normally shown only by the queen (see : Truman and Riddiford, 1977).

SEXUAL RECEPTIVITY IN MOSQUITOES

Mosquitoes provide an interesting set of modifier effects that can be related to the action of hormones on a specific region of the nervous system e.g. the terminal abdominal ganglion of the female. Newly emerged *Aedes aegypti* females are refractory to the approaches of males but within 2 to 3 days juvenile hormone transforms the individual into a receptive virgin. With mating the female receives a substance, "matrone", in the accessory gland fluid from the male. Matrone acts to shut off receptivity permanently and is fully effective even when injected into the females body cavity.

The difference in behaviour between receptive and refractory females is confined primarily to changes in the posture and movement of the genitalia. In response to genital contact with the male, a receptive female extends her cerci, thereby ensuring the male a firm grasp and copulation ensues. A refractory female retracts her cerci and thereby avoids copulation. These important responses to contact by the male appear to be mediated by the terminal abdominal ganglion. Surgical isolation of this ganglion from the rest of the CNS caused little change in receptive or refractory behaviour of the appropriate females, but removal of the ganglion, thereby denervating the genitalia, abolished the differences between virgin and mated individuals. Consequently, the difference between the receptive and refractory behavioural programmes must be due to changes occurring within the terminal ganglion, and juvenile hormone and matrone presumably act directly on this ganglion to promote these changes (Gwadz *et al.*, 1971a, b).

HORMONAL CONTROL OF CALLING BEHAVIOUR IN SILK MOTHS

Unlike many other insects, the female wild silk moth emerges with mature ovaries and is sexually receptive. In response to specific environmental cues, virgin females assume a "calling" posture. This behaviour involves extension of the last two abdominal segments which exposes the pheromone glands and allows the release of the sex pheromone. Calling behaviour in virgin females of *Antheraea polyphemus* and *Hyalophora cecropia* can be essentially abolished by removal of the corpora cardiaca.

This behavioural deficiency is most probably due to the lack of the intrinsic cells of the CC because under the conditions of the above experiments the neurohaemal portion of the CC routinely regenerates. Transection of the small nerves between the brain and the CC was as effective as corpora cardiacectomy in preventing calling behaviour. Apparently, the release of the "calling hormone" is controlling neurally by the brain.

The "calling hormone" has been demonstrated in the blood of calling females. When injected into virgin *A. polyphemus* females, blood from calling moths-routinely provoked calling in 1.5 to 2 hours; the blood of non calling virgins did not have this effect (Riddiford, 1974, 1977, Riddiford and Moore, 1977).

The target for the calling hormone appears to be the abdominal nervous system. Abdomens isolated from virgin *H. cecropia* females respond to injection of calling blood or corpora cardiaca extracts within 5-10 minutes by an extension of the terminal abdominal segments in the typical calling position. This behavioural response is turned off in mated females, apparently by a hormone from the mated bursa copulatrix, as implantation of bursae from mated females into virgin females will effectively eliminate the calling response to corpora cardiaca extracts (Riddiford and Moore, 1977).

HORMONAL CONTROL OF POLYMORPHISM

Polymorphism is widespread in insects and has many determining factors and aspects including the morphology and pigmentation of the insect, its reproductive mode, and its behavioural characters. On the whole, polymorphism allows specialization which facilitates the dispersion of the species and above all the sharing of work in social insects.

The corpora allata are known to play a major part in determining polymorphism and phase. In locusts, their role has been established in the triggering of green pigmentation, wing development, ovarian maturation rate, certain behavioural characteristics of the solitary phase and the phase determinism of their offspring.

In aphids, the corpora allata also appear to be involved in morph determination, which in response to crowding and depending on the photoperiod gives rise to viviparous, apterous, parthenogenetic females or alate, oviparous sexuparae. Experiments with topically applied JH or JHA seemed to divert alate development towards the apterous condition, an effect similar to that obtained by long-day photoperiod treatment. Moreover, JH acts on the morph determination of the paedogenetic embryos developing in the abdomen of the treated larvae, which switch from oviparae to parthenogenetic viviparae (Raabe, 1982).

Caste determination in social insects depends on many factors including pheromones and nutritional factors. Juvenile hormone was shown to play an important part in caste regulation in *Polistes, Bombus*, bees, ants and termites (Raabe, 1989).

The involvement of a neurohormone has been demonstrated in the aphid *Megoura viciae* by microcautery of different groups of brain neurosecretory cells in insects reared under various conditions. The morph of the offspring can be distinguished in young larvae by the presence of growing eggs in oviparae and by chains of embryos located in the ovarioles in the virginoparae. The operated insects behave normally when reared under short-day conditions, but under long-day conditions oviparae appear in the offspring; the response to long days in insects previously exposed to short days, i.e., production of virginoparae instead of oviparae, is prevented by microcautery of brain median neurosecretory cells. These neurosecretory cells therefore, appear to be the site of origin of a factor controlling virginoparae production in response to increased day length. Since they send one of their processes to the ventral nerve cord, their secretory products might be delivered directly to the abdomen and perhaps to the reproductive system itself. It is not yet known if the action of the brain in connection with this process is direct or if it is exerted through the corpora allata (Raabe, 1982, 1989).

In the lepidopterans *Polygonia c-aureum, Papilio xuthus*, and *Lycaena phlaeas,* which may develop in distinct spring or summer morphs, a hormonal role of the brain has been clearly shown (see : Wigglesworth, 1970; Raabe, 1989).

HORMONAL CONTROL OF PUPARIATION IN FLIES

Pupariation, which includes different processes preceding sclerotization differs greatly from tanning which occurs after ecdysis because it does not involve tanning of an insect which has just moulted but that of one which has reached the end of the larval stage without yet having undergone apolysis.

Pupariation includes several distinct processes which were precisely defined in *Sarcophaga* using a new technique for measuring haemolymph pressure inside the body. By revealing the alterations occurring in muscle activity, this technique allowed separation of the different events. Pupariation includes the following steps : first, an immobilization stage characterized by pressure fluctuations separated by pauses, then the retraction of the anterior segments accompanied by slight regular fluctuations in pressure, a contraction stage characterized by sharp regular peaks of high magnitude and short duration, followed by the start of the shrinkage stage, during which muscle activity gradually disappears, leaving only the small pulses necessary for the smoothing of the cuticle (Raabe, 1989).

These distinct processes are controlled by neurohormones. When *Calliphora* or *Sarcophaga* larvae are ligatured a few hours before pupariation, the anterior part pupariates before the hind part and the pupariation of the hind part, which is devoid of nerve ganglia, can be accelerated by the injection of haemolymph or nervous tissue homogenate. The active factors are present in the brain, in the ring gland, and also in the fused thoracoabdominal ganglionic mass. In the brain, they are located in the pars-intercerebralis, and in the ring gland, they are restricted to the NH area. Outside the central nervous system, active factors appear to be stored at peripheral sites.

Injections of active haemolymph or nervous system extracts which accelerate tanning also accelerate anterior retraction, but these two activities do not occur concomitantly, either in the haemolymph or in the nervous system. This shows that they correspond to distinct factors, i.e., the anterior retraction factor (ARF), and the puparium-tanning factor (PTF). Two other factors are involved in the behaviour of prepuparial larvae, i.e., pupariation-immobilizing factor (PIF), and puparium-stimulating factor (PSF) (see: Raabe, 1989).

The presence of PIF in haemolymph samples injected into wandering stage larvae causes immobilization within 35 min, but this effect disappears within about an hour. Pupariation immobilizing factor occurs in the haemolymph of the pre-red spiracle stage as well as in contracting larvae; it reaches its highest titre 2 hr after the onset of tanning and vanishes when the latter is completed. The insect's sensitivity of PIF changes during pupariation. It is nil when they have just left the food but increases during the 2 hrs. that precede the onset of the contraction phase. The lateness of the PIF peak suggests that this factor is not only involved in the immobilization process, but might play a part in subsequent processes that also involve the muscle activity which gradually ceases once the shaping of the pupa has been completed (Raabe, 1989).

The effect of PSF is to stimulate the contractions of the somatic muscles of the wandering larvae. This factor can only be detected in the haemolymph during the very short period of strong contractions of the longitudinal muscles. In postfeeding larvae, injection of blood from contracting larvae determines changes in internal pressure (Raabe, 1982).

The role of ecdysone in triggering puparial tanning processes in flies was known long before the role of neurohormones was discovered (Fraenkel, 1935) and ecdysone was considered to regulate dopa decarboxylase synthesis. To obtain a precise idea of the role of ecdysone, it is helpful to examine haemolymph ecdysteroid titres during larval and pupal development. In both *Sarcophaga* and *Calliphora*, pupariation is preceded by a low ecdysteroid titre and a high narrow peak occurs a few hours after pupariation. Therefore, high ecdysone titres do not trigger pupariation itself but the processes following it, i.e., apolysis and pupal cuticle synthesis. Therefore the initial experimental results could not be explained by high ecdysteroid levels but rather by the low concentration of ecdysone. This recalls the mode of action of ecdysone during eclosion. As far as this mode of action is concerned experimental results indicate that it might act through the central nervous system; nevertheless, it is possible that ecdysone has direct effects on tanning, especially the pre-ecdysial tanning of restricted areas (see : Raabe, 1982, 1989).

HORMONAL CONTROL OF PLASTICIZATION

Plasticization in flies, which occurs during air pumping by the ptilinum was shown to be under hormonal control. Recent investigations have demonstrated that several distinct processes take place in the fly during the tanning period, i.e., plasticization and muscle contractions causing pulsing and air pumping. Injection into flies of blood samples from other flies of differents ages shows that pumping is only stimulated by 15-min blood. which suggests that the factor involved is distinct from both eclosion hormone and bursicon which are not present in the haemolymph at this time. Hence the existence of several distinct factors is proposed, including eclosion hormone, a pulsing and pumping factor, a plasticization factor, and bursicon.

In *Manduca*, the plasticization process that precedes ecdysis is dependent on eclosion hormone. A second plasticization occurs after eclosion. This is quite an important event because it allows the expansion of the new cuticle in the wings.

The plasticization which occurs in *Rhodnius* after the blood meal must be mentioned here. The active factor in this case is not bursicon. It is present in the brain and is released at undetermined peripheral sites (for details see: Wigglesworth, 1970; Reynolds, 1974; Raabe, 1982, 1989).

HORMONAL CONTROL OF HARDENING AND DARKENING OF THE CUTICLE

It has long seemed probable that the hardening and darkening of the cuticle which takes place simultaneously throughout the body shortly after moulting is under hormonal control. It was suggested that in *Rhodnius* the neurosecretory cells in the thoracic ganglia might be the source of the hormone in question; but in unpublished experiments in which the abdomen was separated from the entire central nervous system, almost before the insect had escaped from the cast skin by means of a transverse bar firmly pressed down just behind the metathorax, it was found that no delay took place in the hardening and pigmentation of the isolated abdomen (Wigglesworth, 1970. Clearly any hormonal factor responsible had already been released. Delayed hardening is not a normal feature in *Rhodnius*. Cottrell (1964) had observed that Lycaenid butterflies emerging from ants' nests in the soil often showed a prolonged delay between emergence and hardening; but since this was not convenient material for experiment it was decided that he should investigate the problem in Muscid flies. Fraenkel in

1935 had shown that if *Calliphora* was compelled to continue burrowing through loose soil, the hardening and darkening process could be delayed for as long as seven hours. Inflation, hardening and darkening occurred as usual when the insects became free.

These chemical processes are evidently initiated by some nervous mechanism. Cottrell (1964) found that the brain is concerned in the release to the blood of a peptide hormone which acts on the epidermal cells and leads to the normal darkening and hardening of the cuticle. It was shown by Fraenkel and Hsiao (1965) that removal of the medial neurosecretory cells from the dorsum of the brain has the same effect as decapitation suggesting that these are the source of the hormone but they also noted that extracts from the compound ganglion of the thorax were even more active than extracts from the brain in inducing hardening in ligatured flies. The active substance begins to appear in the blood 2-3 minutes after emergence, reaches a maximum at 30-60 minutes and then falls and is quite inactive 10 hours later. These authors named the active substance 'bursicon', a name that refers to the 'tanning' of the cuticle after expansion. According to Wigglesworth (1970) this name conveys the suggestion that the active substance tans the cuticle. That, of course, is not the case. The hormone is a messenger which stimulates the epidermis to carry out all the complex processes involved in hardening and darkening. In *Periplaneta*, 'bursicon' occurs in most parts of the central nervous system, but it seems to be released only from the terminal abdominal ganglion. It appears in the blood soon after ecdysis and builds up to a maximum 1-2 hours later; it is no longer detectable at 8 hours. Section of the nerve cord stops release from the terminal ganglion and release can be induced by electrical stimulation (see: Wigglesworth, 1970).

Another process which takes place at the time of moulting, and at eclosion from the egg, is the active absorption of fluid from the new tracheae and tracheoles so that the system becomes filled with air. This likewise is a process for which the activity of the central nervous system is required. The newly hatched mosquito larvae do not fill their tracheal system while lightly anaesthetized but do so soon after recovery. Here again a hormone is probably involved (Wigglesworth, 1964) indeed, it might be 'bursicon' itself.

HORMONAL CONTROL OF FLIGHT

Experiments performed on locusts and some other insects show the necessity of certain neurohormones for flight to occur. Removal of the neurosecretory cells of the pars intercerebralis causes the disappearance of nocturnal flight (Hinks, 1967) in the noctuid moths *Noctua pronula* and *Agrotis ipsilon* and in the blowfly *Calliphora erythrocephala*, the corpora cardiaca are indispensable to the achievement of flight (Vejbjerg and Normann, 1974).

In the locust *Schistocerca gregaria*, the neurohaemal part of the corpora cardiaca performs an important role in sustained flight, which is prevented by the disconnection or removal of the corpora cardiaca and is only restored after their implantation or the implantation of their neurohaemal part, the glandular part being ineffective (Michel, 1972). The electrocoagulation of the pars intercerebralis also inhibits flight, but its implantation fails to restore it. Thus, according to Michel, flight in *Schistocerca* is stimulated by a substance produced in the pars intercerebralis and activated in the corpora cardiaca.

In *Locusta migratoria*, removal of the pars intercerebralis disturbs flight only in young insects, and it is suggested that its action is exerted in part by the corpora allata, for the injection of juvenile hormone analogues diminishes the effects caused by destruction of the pars intercerebralis (Goldsworthy *et al.,* 1973, 1977). Nevertheless, the glandular part of the corpora cardiaca is indispensable to flight, which is depressed by their removal. The ultrastructural study of the corpora cardiaca in *Locusta* has confirmed the secretory activity of the glandular cells during flight. The rough endoplasmic reticulum of these cells displays sharp flight-related variations in another locust, *Schistocerca*.

By comparing the results obtained on the two species of locust *Schistocerca* and *Locusta*, it may be observed that the pars intercerebralis is more or less essential, and that the corpora cardiaca, always necessary, act either through their neurohaemal region or through their glandular part. The locusts are known to use two distinct metabolites for flight. Trehalose is used for short flights or at the onset of migratory flight, and diglycerides are consumed during the migratory flight. The release of these two metabolites from the fat body is subject to different controls. The release of trehalose is triggered by the hypertrehalosemic hormone from the neurohaemal part of the corpora cardiaca, while the release of diglycerides depends on the adipokinetic hormone produced in the glandular lobe. Thus, it may well be that the discrepancies between the results are due to differences in the flight metabolites used by the locusts in the experimental conditions and to differences in the neurohormones involved in the release of these metabolites. In fact, the experiments were conducted differently in each species. The assays performed by Michel on *Schistocerca* concerned the spontaneous tendency to sustained flight, while Goldsworthy and co-workers, working on *Locusta*, evaluated the flight obtained after the insects were stimulated. Thus two distinct metabolic neurohormones are involved in flight regulation. It remains to be seen if another "behavioural" neurohormone stimulating the activity might be also involved.

Following flight cessation, degeneration processes often take place in the muscles, which depend on the JH titre as well. These processes were extensively studied in *Leptinotarsa*, *Dysdercus* and *Acheta* (see : Pener, 1985; Raabe, 1989).

HORMONAL CONTROL OF PROTHETELY AND METATHETELY

Wigglesworth (1970) remarked that the moulting hormone and the juvenile hormone are sometimes described as being antagonistic. In fact, they do two quite different things. Ecdysone brings about the activation of the epidermal cells that is necessary for growth and moulting. Once growth has been set going in this way the production of larval or of adult characters is determined by the presence or absence of the juvenile hormone. On the other hand, if too much ecdysone is present cuticle deposition may go forward so rapidly that insufficient time is available for the full development of adult outgrowths or for the breakdown of cells producing larval structures hence intermediate forms may be produced. Conversely if the juvenile hormone is secreted in insufficient quantity for the production of the larval form, there may be a partial development of adult characters, again intermediate forms may appear. The lateral neurosecretory cells of the protocerebrum in *Locusta* normally excite the activity of the corpus allatum in secreting juvenile hormone. Hence, if these cells are destroyed by electric cautery in the fourth-stage larva, dwarf adults are produced. But the neurosecretory cells of the medial region of the pars intercerebralis have the converse effect of inhibiting secretion by the corpus allatum and if excess of these cells are implanted, dwarf adults may again be produced. The control of corpus allatum activity is effected by the balance of these two effects (Bloch *et al.*, 1966).

Long before the hormonal regulation of metamorphosis was recgonized, abnormal insects had been described in which larvae had developed partially pupal characters - such as silkworms developing enlarged antennae, or mealworm larvae developing wing pads. This precocious appearance of imaginal characters was termed 'prothetely'. Other examples were reported in which adult insects had retained some pupal characters, or pupae had retained some larval characters. This persistence of juvenile characters was termed 'metathetely', it is equivalent to 'neoteny' (Wigglesworth, 1970).

When the hormonal regulation of metamorphosis was defined, it was pointed out that these anomalies are most readily explained as resulting from an upset in the balance or the timing of hormone secretions (Wigglesworth, 1934, 1936). This upset may be induced by exposure of cultures of *Tenebrio* or *Tribolium* to high temperatures or to low temperatures. They sometimes occur in larvae of *Simulium* invaded by the nematode *Mermis*. In Lepidoptera they are particularly liable to occur in species hybrids - again presumably because the hormone balance has been upset. Partial retention of larval characters in Saturniid pupae and in *Tribolium* maintained in culture has been traced to infection by *Nosema* and is attributed to the liberation by the micro-organism of substances with juvenile hormone activity (Wigglesworth, 1970, Raabe, 1989).

HORMONAL CONTROL OF GENE ACTIVITY IN THE CELL

As has been emphasized already, the epidermis of the insect is particularly favourable material for the study of differentiative changes during growth because it expresses its juvenile or adult character by the type of cuticle it lays down. It may, therefore be possible to judge whether the activity of a single cell is larval, adult or intermediate.

The cuticle has a complex structure, the product of a complex system of enzymes within the cell, its characters determined by the nature of this enzyme system. Since a given cell may lay down soft pale cuticle in the larva and hard dark cuticle in the adult, there must be large changes in the enzymes that are acting in the two stages. But, as have been already seen (Wigglesworth, 1970), the differences are equally great as between one region of the body and another; a hormone which is leading to the production of soft pale cuticle in place of hard black cuticle in one part of the body, and hard black cuticle in place of soft pale cuticle in another part, must be acting in some very general way. And since single enzymes are believed to be products of single 'genes', it is concluded that the absence of the juvenile hormone is commonly concerned in switching the gene system throughout the body from producing larval enzymes to producing adult enzymes (Wigglesworth, 1970).

If the juvenile hormone is switching gene activity, it might be expected that confining our attention to a single type of cell, it would be possible to observe intermediate stages in which certain activities, that is, certain enzymes or groups of enzymes had been changed while others had not; so that the cuticle produced might be adult in some respects but still larval in others.

The trichogen cells which form the cuticular hairs persist from one instar to the next. If fourth-stage larvae of *Rhodnius* are decapitated around the critical period it is possible to select, from among those larvae that moult a graded series with cuticular hairs showing all degrees of intermediate characters between larval hairs and adult hairs (Wigglesworth, 1940).

Conversely it is possible to observe a similar step-like process of switching if fifth stage larvae of *Rhodnius* are allowed to develop normally towards the adult form for a known number of days after feeding and are then switched back to larval development by flooding the system with juvenile hormone. Larvae switched at one particular phase of development have lost the capacity to produce smooth unpigmented larval plaques on the abdomen, but they show spots of wrinkled unpigmented cuticle marking the places where plaques would have appeared if the switch back had been made a little earlier. Thus, the ability to form smooth weakly sclerotized cuticle is lost by the cells below the plaques before the ability to form unpigmented cuticle. Larvae which are switched back a little later show uniformly grey wrinkled cuticle, the capacity to form unpigmented spots has now been lost (Wigglesworth, 1970).

In the larva of *Oncopeltus* in which as in many insects, the area of cuticle laid down by each epidermal cell has clearly visible boundaries, it has been possible to show very elegantly the progressive changes in the type of cuticle laid down by single cells as this progress of switching goes forward.

Wigglesworth (1970) has further remarked that there is another aspect of the control of metamorphosis which is relevant to this discussion. The cells seem to acquire a certain 'inertia', such that once they have begun to form pupal or adult structures they resist the action of the juvenile hormone in causing them to revert to the larval state. Presumably the cell has been 're-programmed' and those components of the gene system (or those segments of the DNA chain in the chromosomes) that are responsible for the production of the adult enzymes, have been brought into action, and this process is not easily reversed.

It was shown long ago by Piepho (see: Wigglesworth, 1954), that this inertia was soon eliminated in the regenerating cells responsible for wound repair. Cell division favours the changing over of the gene programme, presumably through the synthesis of new DNA. On the other hand, there are undoubted examples, as outlined above, of single cells which undergo neither cell division nor endomitosis, in which the programme of enzyme synthesis can be changed. The question arises (I) whether there exist in the cytoplasm of such cells performed templates (sometimes referred to as persistent or long-lived messenger-RNA) which can be brought into action for the synthesis of a changed type of cuticle by a change in the hormones - that is, by the presence or absence of the juvenile hormone, or (II) whether the juvenile hormone acts directly on the DNA of the nucleus and evokes a change in the activity of the gene system (Wigglesworth, 1970).

Whatever may be the actual site of action of the juvenile hormone within the cells, the fact that the cells can be caused to undergo metamorphosis and then to revert to laying down larval cuticle, fits in well with the idea that the programming of the gene system can be switched in either direction. The cells are certainly far more resistant to switching their activity from the adult state to the larval state than they are to switching from larva to adult; but perphaps that is not surprising since the pressure of selection will naturally favour the normal progress of metamorphosis (Wigglesworth, 1970).

SELECTED BIBLIOGRAPHY

Aggarwal, S.K. and King, R.C. (1971). An electron microscope study of adult *Drosophila melanogaster* and its afferent nerves. *J. Morphol.,* 134 : 437-446.

Arvy. L. and Gabe, M. (1953). Donnees histophysiologiques sur la neurosecretion chez quelques Ephemeropteres. *Cellule,* 55 : 201-222.

Awasthi, V.B. (1968). The functional significance of the nervi corporis allati-I and nervi corporis allati-II in *Gryllodes sigillatus. J. Insect Physiol.,* 14 : 301-304.

Awasthi, V.B. (1969a). Studies on the neurosecretory system of active and hibernating adults of the house cricket, *Gryllodes sigillatus* (Walk.). *Anat. Anz.,* 125 : 256 -272.

Awasthi, V.B. (1969b). Effect of temperature on the neurosecretory activity in *Nezara viridula* Linn. (Heteroptera : Pentatomidae). *Experientia,* 25 : 1164-1166.

Awasthi V.B. (1972a). Studies on the neurosecretory system and retrocerebral endocrine glands of *Nezera viridula* Linn. (Heteroptera : Pentatomidae). *J. Morphol.,* 136 : 337-352.

Awasthi, V.B. (1972b). Studies on the neurosecretory cells of the brain of normal and starved red cotton bug, *Dysdercus koenigii* (Fabr.) (Heteroptera : Pyrrhocoridae). *Aust. Zool.,* 17 : 34-35.

Awasthi, V.B. (1972c). Neurosecretory system of the adult flesh fly, *Sarcophaga ruficornis* F. (Diptera : Sarcophagidae). *Int. J. Insect Morphol. Embryol.,* 1 : 133-140.

Awasthi, V.B. (1973a). An analytical study of the neuroendocrine system of *Anacanthotermes macrocephalus* (Isoptera : Hodotermitidae). *Zool. Beitr.,* 19 : 75-81.

Awasthi, V.B. (1973b). Studies *in situ* on the neurosecretory system of *Sphaerodema rusticum* Fabr. (Heteroptera : Belostomatidae). *Zool. Beitr.,* 19 : 151-156.

Awasthi, V.B. (1973c). Neurosecretory system and retrocerebral endocrine glands of *Metochus uniguttatus* Thunb. (Heteroptera : Lygaeidae). *Int. J. Insect Morphol. Embryol.*, 2 : 1-12.

Awasthi, V.B. (1974). Neurohaemal organs in Insects. *Anat. Anz.*, 136 : 244-267.

Awasthi, V.B. (1975a). Response of neurosecretory systems of *Gryllodes sigillatus* (Walk.) and *Sphaerodema rusticum* Fabr. to salt. solution. *Zool. Jb. Anat.*, 94 : 67-73.

Awasthi, V.B. (1975b). The retrocerebral endocrine glands and associated neurohaemal organs of *Epilampra* sp. with reference to cysteine or cystine rich NSM. *Zool. Jb. Anat.,* 94 : 149-157.

Awasthi, V.B. (1975c). Neurosecretory system of the earwig, *Labidura riparia* and the role of the aorta as a neurohaemal organ. *J. Insect Physiol.,* 21 : 1713-1719.

Awasthi, V.B. (1976a). Effect of nutritional factors on the appearance of neurosecretory cells and ovary development in the fleshfly, *Sarcophaga ruficornis* F. (Diptera : Sarcophagidae). *Zool. Jb. Anat.,* 95 : 1-8.

Awasthi, V.B. (1976b). The functional significance of cephalic aorta as a neurohaemal organ in *Labidura riparia* (Pallas) and other Dermaptera. *Zool. Jb. Anat.,* 95 : 109-113.

Awasthi, V.B. (1976c). Ultrastructure of retrocerebral complex of the earwig, *Euborellia annulipes* and role of aorta as a neurohaemal organ. *J. Insect Physiol.,* 22 : 1181-1186.

Awasthi V.B. (1976d). Neurosecretory cells and aorta as a neurohaemal organ in the earwig, *Euborellia annulipes* Lucas (Deramptera : Labiduridae). *Int. J. Insect Morphol. Embryol.,* 5 : 253-260.

Awasthi, V.B. (1976e). Contribution to the neurohaemal organs in Pterygote insects. *Zeit. Mikrosk. Anat. Forsch.*, 90 : 48-78.

Awasthi, V.B. (1977). Ultrastructure of corpus cardiacum of the house cricket, *Gryllodes sigillatus* (Walk.) Orthoptera : Gryllidae). *Zool. Jb. Anat.*, 98 : 476-490.

Awasthi, V.B. (1978). Studies on the neurosecretory system and retrocerebral endocrine glands of *Epilampra* sp. (Dictyoptera : Blattidae). *Zool. Beitr.*, 24 : 447-464.

Awasthi, V.B. (1979a). Hormonal control of metabolism in insects. *Indian J. Zootomy,* 20 : 1-38.

Awasthi, V.B. (1979b). The Ultrastructure of corpus allatum of the earwig, *Euborellia annulipes* Lucas. (Dermaptera : Labiduridae). *Zeit. Mikrosk. Anat. Forsch.* 93 : 982-991.

Awasthi, V.B. (1980a). Neurosecretion in the milkweed bug, *Lygaeus pandurus* Scop. (Heteroptera : Lygaeidae). *J. Hirnforsch.,* 21 : 169-181.

Awasthi, V.B. (1980b). Neurosecretion in the frontal ganglion of *Achoea janata* Linn. and *Philosamia ricini* (Hutt.) (Lepidoptera). *Zeit. Mikrosk. Anat. Forsch.*, 94 : 825-832.

Awasthi, V.B. (1982). The neurosecretory cells of the brain in Lepidoptera. *J. Adv. Zool.*, 3 : 67-76.

Awasthi, V.B. (1983). Cardiac regulators in insects. *J. Adv. Zool.*, 4 : 64-70.

Awasthi, V.B. (1985a). Influence of juvenile hormone analogue on the metamorphosis of *Philosamia ricini* Hutt. (Lepidoptera : Saturniidae). *Entomon.*, 10 : 165-169.

Awasthi, V.B. (1985b). The present status of insect neuroendocrinology in India. *Ind. Rev. Life Sci.*, 5 : 59- 91.

Awasthi, V.B. (2001). *Principles of Insect Behaviour*, Scientific Publishers (India) Jodhpur.

Awasthi, V.B. (2002). *Introduction to General and Applied Entomology*. 2nd edition, Scientific Publishers (India) Jodhpur.

Awasthi, V.B. and Singh, U.V. (1981a) Dynamics of the neurosecretory cells of the brain of *Amsacta collaris* Hampson (Lepidoptera : Arctiidae) during postembryonic development. *J. Hirnforsch.*, 22 : 161-177.

Awasthi, V.B. and Singh, U.V. (1981b). An analytical study of the neurosecretory cells of the brain of *Philosamia ricini* (Hutt.) (Lepidoptera : Saturniidae) during post-embryonic development. *Experientia*, 37 : 272-375.

Awasthi, V.B. and Singh, U.V. (1982a). An analytical study of the neurosecretory cells of the brain of mature larvae and diapause pupae of *Amsacta collaris* Hamp. (Lepidoptera : Arctiidae) under different temperature regimes. *Gegenbaur's Morph.*, 128 : 100-106.

Awasthi, V.B. and Singh, U.V. (1982b). Studies on the neurosecretory cells of the brain of *Achoea janata* Linn. (Lepidoptera Noctuidae) during post embryonic development. *Entomon*, 7 : 285-295.

Awasthi, V.B. and Singh, U.V. (1983). The anatomy and histology of the ecdysial glands of *Amsacta collaris* Hamp. (Lepidoptera : Arctiidae) during post embryonic development. *Zool. Jb. Anat.*, 109 : 467-480.

Awasthi, V.B. and Singh, U.V. (1985). The neurosecretory cells of the brain in the moth, *Utetheisa pulchella* Linn. (Lepidoptera : Arctiidae) during postembryonic development. In: *Recent Advances in Developmental Biology of Animals* (Goel, S.C. and Srivastava, C.B.L., eds.). Rastogi and Co. Meerut, pp. 77-87.

Awasthi, V.B. and Singh, U.V. (1986a). The retrocerebral endocrine glands of *Achoea janata* Linn. (Noctuidae : Lepidoptera) during post embryonic development. *Zool. Jb. Anat.*, 114 : 269-275.

Awasthi, V.B. and Singh, U.V. (1986b). Histochemical analysis of the neurosecretory cells of the brain in Erisilkworm, *Philosamia ricini* Hutt.

(Lepidoptera : Saturniidae) In : *Recent advances in insect Physiology, Morphology and Ecology* (Pathak, S.C. and Sahai, Y.N. eds.). Today and Tomorrow, Printers and Publishers, New Delhi. pp. 25-38.

Awasthi, V.B. and Singh, U.V. (1994). Studies on the ecdysial glands of *Utetheisa pulchella* Linn. (Lepidoptera : Arctiidae) during post embryonic development. In : *Perspectives in Entomological Research* (O.P. Agrawal, ed.) Scientific Publishers (India) Jodhpur.

Awasthi, V.B. and Singh, U.V. (2003). Studies on the retrocerebral endocrine glands of *Utetheisa pulchella* Linn. (Arctiidae : Lepidoptera) during post embryonic development (in press).

Baehr, J.C. (1968). Etude histologique de la neurosecretion du cerveau et du ganglion sous - oesophagien de *Rhodnius prolixus* Stal. (Hemiptere). C.R. Acad. Sci., 267: 2364-2367.

Bagnoli, P., Brunelli, M.D., Ajello, V. and Magni, F. (1970). Further evidence for peripheral inhibition of flashing and for role of the male gonads in *Luciola lusitanica* (Charp). *Arch. Ital. Biol.*, 108 : 180-206.

Bagnoli, P., Brunelli, M., Magni, F. and Viola, M. (1972). The identification of a flash inhibiting substance from male gonads of *Luciola lusitanica* (Charp). *Arch. Ital. Biol.*, 110 : 16-35.

Bargmann, W. and Scharrer, B. (1970). *Aspects of Neuroendocrinology.* Springer Verlag. Berlin.

Bart, A. (1963). Donnees histologiques et experimentales sur le systeme neurosecreteur de l'lnsecte Apterygote *Petrobius maritimus* (Leach). *Gen. Comp. Endocrinol.*, 3 : 398-411.

Barth, R.H. (1968). The comparative physiology of reproductive processes in cockroaches. Part I. Mating behaviour and its endocrine control. *Adv. Reprod. Physiol.*, 3 : 167 - 201.

Barth, R.H. and Bell, W.J. (1970). Physiology of the reproductive cycle in the cockroach, *Byrsotria fumigata* (Guerin). *Biol. Bull. woods Hole*, 139 : 447-460.

Barton-Browne, L., Dodson, L.F., Hodgson, E.S. and Kiraly, J.K. (1961). Adrenergic properties of the cockroach corpus cardiacum. *Gen. Comp. Endocrinol.*, 1 : 232-236.

Bassalingappa, S., Mugali, R.N., Tharabai, P. and Awasthi, V.B. (1984). NSM-containing cysteine and or cystine in brain and other neuroendocrine glands, and its possible influence on fecundity of the termite, *Odontotermes obesus* Holmgren (Termitidae : Isoptera) an *in situ* study. *Indian Zoologist*, 8 : 135-138.

Bassurmanova, G.K. and Panov, A.A. (1967). Structure of the neurosecretory system in Lepidoptera, Light and electron microscopy of type A - neurosecretory cells in the brain of normal and starved larvae of the silkworm, *Bombyx mori. Gen. Comp. Endocrinol.*, 9 : 245-262.

Baudry - Partiaoglou, N. (1983). Ultrastructure of two opposite types of abdominal perisympathetic organs in the Heteropteran, *Roscius elongatus* Stal. (Phyrrhocoridae) : Abundant release of neruosecretory product by intraganglionic organs. *Gen. Comp. Endocrinol.*, 50 : 407-422.

Baudry-Partiaoglou, N. (1986). Diversity of neurohormonal release sites in insects : Coexistence of two different (α and β) types of neurohaemal structure with perisympathetic organas. *Gen. Comp. Endocrinol.*, 62 : 254-267.

Beattie, T.M. (1971). Histology, histochemistry and ultrastructure of neurosecretory cells in the opitc lobe of the cockroach, *Periplaneta americana. J. Insect Physiol.*, 17 : 1843-1855.

Bell, R.A., Borg, T.K. and Ittycheriah, P.I. (1974). Neurosecretory cells in the frontal ganglion of the tobacco hornworm, *Manduca sexta*. *J. Insect Physiol.*, 20 : 669-670.

Bloch, B. Thomsen, E. and Thomsen, M. (1966). The neurosecretory system of the adult *Calliphora erythrocephala*. III. Electron microscopy of the medial neurosecretory cells in the brain and some adjacent cells. *Z. Zellforsch.,* 70 : 185-208.

Bodenstein, D. (1936). Das determination geschehen bei insekten mit. Ausschluss der fruhembryonader Determination. *Ergeb. Biol.*, 13 : 174-234.

Borg, T.K. and Bell. R.A. (1977). Ultrastructure of the neurosecretory cells in the brain of diapausing paupae of the tobacco hornworm, *Manduca sexta* (L.). *Tissue and Cell.*, 9 : 567-574.

Borg, T.K. and Marks, E.P. (1973).Ultrastructure of the median neurosecretory cells of *Manduca sexta in vivo* and *in vitro*. *J. Insect Physiol*. 19: 1913-1920.

Bounhiol, J.J. (1938). Recherches experimentales sur le determinisme de la metamorphose chez les Lepidopteres. *Bull. Biol. Fr. Belg.*, 24 : 1-199.

Bowers, B. and Johnson, B. (1966). An electron microscope study of the corpora cardiaca and secretory neurons in the aphid, *Myzus persicae. Gen. Comp. Endocrinol.*, 6 : 213-230.

Brady, J. and Maddrell, S.H.P. (1967). Neurohaemal organs in the medial nervous system of insects. *Z. Zellforsch.*, 76 : 389- 404.

Brunelli, M., Buonamici, M. and Magni, F. (1968a). Mechanisms for photic inhibition of flashing in fireflies. *Arch. Ital. Biol.*, 106 : 85-99.

Brunelli, M., Buonamici, M. and Magni, F. (1968b). Effects of castration on the inhibition of flashing in fireflies. *Arch. Ital. Biol.*, 106,:100-112.

Brunelli, M., Buonamici, M., Magni, F. and Viola, M. (1970). The role of the male gonads in the peripheral inhibition of flashing in *Luciola lusitanica. Arch. Ital. Biol.* 108: 1-20.

Burgess, L. (1971). Neurosecretory cells and their axon pathways in *Culiseta inronata* (Williston) (Diptera: Culicidae). *Canad. J. Zool.*, 49 : 889-901.

Burgess, L. and Rempel, J.G., (1966). The stomodaeal nervous system, the neurosecretory system and the gland complex in *Aedes aegypti* (L.) (Diptera

: Culicidae). *Canad. J. Zool.*, 44 : 731-765.

Butenandt, A. and Karlson, P. (1954). Uber die Isolierung eines Metamorphose-Hormons der Insekten in Kristallisierter Form. *Z. Naturforsch.*, 96 : 389-391.

Cameron, M.L. (1953). Secretion of O-diphenol in the corpus cardiacum of insects. *Nature*, London, 172 : 349-350.

Cassier, P. (1963). Action des implantations de corps allates sur la reactivite phototropique de *Locusta migratoria migratorioides* (R&F), phase *gregaria* (Insecte, Orthopteroide). *C.R. Acad. Sci., Paris*, 257 : 4048-4049.

Colhoun, E.H. (1963). Synthesis of 5-hydroxytryptamine in the American Cockroach. *Experientia*, 19 : 9-10.

Cassagnau, P. and Juberthie, C. (1967). Structures nerveuses neruosecretion et organes endocrines chez Collemboles, Le complexe cerebrale des Entomobryo-morphes. *Gen. Comp. Endocrinol.*, 8 : 489-502.

Cazal, P. (1948). Les glandes endocrines retrocerebrales des insectes, (Etude Morphologique). *Bull. Biol. Fr. Belg.*, (Suppl.), 32 : 1-227.

Chalaye, D. (1966). Recherches Sur la destination des produits de neurosecretion de la chaine nerveuse ventrale du criquet migrateur, *Locusta migratoria. C.R. Acad. Sci.* (Paris), 262 : 161-164.

Chalaya, D. (1967). Neurosecretion au niveau de la chaine nerveuse ventrale de *Locusta migratoria migratorioides* R. et F. (Orthoptere : acridien). *Bull. Soc. Zool.* France, 92 : 87-108.

Cottrell, C.B. (1964). Insect ecdysis with particular emphasis on cuticular hardening and darkening. *Adv. Insect Physiol.*, 2: 175-218.

Davey, K.G. (1961a). The mode of action of the heart accelerating factor from the corpus cardiacum of Insects. *Gen. Comp. Endocrinol.*, 1 :24.

Davey, K.G. (1961b). Substances controlling the role of beating of the heart of *Periplaneta. Nature*, London, 92 : 284.

Davey, K.G. (1962). Changes in the pericardial cells of *Periplaneta americana* induced by exposur to homogenate of the corpus cardiacum. *Quart. J. Micr. Sci.*, 103 : 349- 253.

Delepine, Y. (1965). Recherches sur la neurosecretion dans l'ensemble du systeme nerveux central dun lepidoptera, *Galleria mallonalla. Bull. Soc. Zool. France*, 90 : 525-540.

Delphin, F. (1965). The histology and possible functions of neurosecretory cells in the ventral ganglia of *Schistocerca gregaria* (Forsk.) (Orthoptera : Acrididae). *Trans. R. Int. Soc. London*, 117 : 167-214.

De Lerma, B. (1933). Corpe faringei de ortotteri. *Rendie Accad. Lincei.*, 17: 1105-1108.

De Lerma, B. (1954). Osservazioni sulla neurosecrezione in *Hydrous piceus* (Coleoptere). *Pubbl. Staz. Zool. Napoli.* (Suppl.), 24 : 56-58.

DeWilde, J. (1964). Reproduction - Endocrine control. In: *The Physiology of Insecta* (M. Rockstien, ed.). 1: 59-90. Academic Press, New York and London.

De Wilde, J. and De Boer, J.A. (1969). Humoral and nervous pathways in photoperiodic induction of diapause in *Leptinotarsa decemlineata. J. Insect Physiol.*, 15 : 661- 675.

DeWilde, J. and DeLoof, A. (1973). Reproduction: Endocrine control. In: *The Physiology of Insecta* (M. Rockstein, ed.) 1: 97-157, second edition, Academic Press, New York and London.

Deoras, P.J. and Bhaskaran, G. (1966-1967) Studies on the neuroendocrine system in the Indian housefly, *musca nebulo*. Fabr. III Adult Organs. *J. Univ. Bombay*, 35 : 59-72.

Dogra, G.S. (1967a). Studies on the neurosecretory system and the functional significance of NSM in the aortal wall of the bug, *Dysdercus koenigii. J. Insect Physiol.*, 13 : 1895-1906.

Dogra, G.S. (1967b). Studies on the neurosecretory system of *Ranatra elongata* Fabricius (Hemiptera : Nepidae) with reference to the distal fate of NCCI and II. *J. Morphol.*, 121 : 223-240.

Dogra, G.S. (1969). Studies *in situ* on the neuroendocrine system of the giant water bug, *Belostoma indica* (Lep. and Serv.). *Acta Anat.*, 72 : 429-445.

Dogra, G.S. and Tandon, B.K. (1966). An *in situ* study of cystine and cystiene - rich neurosecretory cells in the brain of *Sarcophaga ruficornis* (Fabricius - 1794). (Diptera : Cyclorrhapha). *Proc. R. Ent. Soc., London*, (A) : 41 - 57-66.

Dogra, G.S. and Ewen, A.B. (1970). Histology of the neurosecretory system and the retrocerebral endocrine glands of the adult migratory grasshopper, *Melanoplus sanguinipes.* (Fabr.) (Orthoptera : Acrididae). *J. Morphol.*, 130 : 451-465.

Doane, W.W. (1973). Role of hormones in insect development. In : *Developmental Systems : Insects*, 2 : 291-497. S.J. Counce and C.H. Waddington (eds.). Academic Press, New York.

Dupont Raabe, M. (1952a). Substances chromactives, des crustaces et d insectes. Activite reciproque, repartition difference qualitatives. *Arch. Zool. Exp. Gen.*, 89 : 102-112.

Dupont Raabe, M. (1952b). Contribution a l etude du role endocrine du cerveau et notamment de la pars intercerebralis chez les phasmides. *Arch. Zool. Exp. Gen.*, 89 : 128-138.

Dupont-Raabe, M. (1957). Les mechanisms de l adaptation chromatique chez Les insects. *Arch. Zool. Exp. Gen.*, 94 : 61-292.

Engelmann, F. (1968). Endocrine control of reproduction in insects. *Ann. Rev. Entomol.*, 13 : 1-26.

Engelmann, F. (1970). *The Physiology of Insect Reproduction.* Pergamon Press, Oxford.

Ewen, A.B. (1962). Histophysiology of the neurosecretory system and retrocerebral endocrine glands of the alfalfa plant bug, *Adelphocoris lineolatus* (Goeze)

(Hemiptera : Miridae). *J. Morphol.*, 111 : 253-273.

Flament-Durand, J. (1966). Contribution a l etude de la neurosecretion chez le rat par la methode autoradio-graphiques. Proc. 4th Int. Symp. - Neurosecretion. Cited in : The role of the neurosecretory neuron in neuroendocrine regulatory mechanisms. by Scharrer, B. (1967) : *Amer. Zool.*, 7: 161-169.

Fletcher, B.S. (1969). The diversity of cell types in the neurosecretory system of the beetle, *Blaps mucronata. J. Insect Physiol.*, 15 : 119-134.

Formigoni, A. (1956). Neurosecretion et organes endocrines chez *Apis mellifica* L. *Ann. Sci. Nat. (Zool.),* 18 : 283-291.

Foster, W. (1967). Hormone-mediated nutritional control of sexual behaviour in male dung flies. *Science,* 158: 1596-1597.

Fraenkel, G. (1935). A hormone causing pupation in the fly, *Calliphora erythrocephala, Proc. Roy. Soc. London,* 118 :1- 12.

Fraenkel, G. and Hsiao, C. (1962). Hormonal and nervous control of tanning in the fly. *Science,* 138 : 27-29.

Fraenkel, G. and Hsiao, C. (1965). Bursicon a hormone which mediates tanning of the cuticle in adult fly and other insects. *J. Insect Physiol.,* 11: 513-526.

Fukuda, S. (1940). Induction of pupation in silkworm by transplanting prothoracic glands. *Proc. Imper. Acad.* Tokyo, 16 : 414-416.

Fukuda, S. (1941). Role of the prothoracic glands in differentiation of the imaginal characters in the silkworm pupa. *Annot. Zool. Japan.*, 20 : 9-13.

Fukuda, S. (1944). Hormonal mechanisms of larval moulting and metamorphosis in the silkworm. *J. Fac. Sci. Imp. Univ.* Tokyo, Sect. IV *Zool.* 6 : 477-533.

Fukuda, S., Eguchi, G. and Takeuchi, S. (1966). Histological and electron microscopical studies on sexual differences in structure of the corpora allata of the moth of the silkworm, *Bombyx mori. Embryologia,* 9 : 123-158.

Fuller, H. (1960). Morphologische Und Experimentelle Untersuchungen Uber die neurosekretorischen verhaltnisse im Zentralnervensystem von Blattiden Und Culiciden. *Zool. Jb.*, 69 : 223-250.

Gabe, M. (1966). *Neurosecretion.* Pergamon Press, London.

Gaude, H. and Weber, W. (1966). Untersuchungen zur neurosekretion bei *Acheta domesticus* L. *Experientia,* 22 : 396.

Gersch, M. (1961). Insect metamorphosis and the activation hormone. *Amer. Zool.*, 1 : 53-57.

Gersch, M. (1969). Experimentelle undersuchungen zur endokrinen Regulation des Wasserhaushaltes der Larvae von *Corethra (Chaoborus). Zool. Jb. Physiol.*, 75 : 1-16.

Gersch, M. (1972). Experimentelle Untersuchungen zum Freiset zungsmechanismus von Neurohormone nach elektrischer Reizing der corpora cardiaca. von *Periplaneta americana in vitro. J. Insect Physiol.* 18 : 2425-2439.

Gersch, M. (1975). Prinzipien neurohormonaler Und neruohumoraler steuerung physiologischer Prozesse. Friedrch - Schiller- Universitat, Jena.

Gersch, M., Fisher, F., Unger, H. and Kabitza, W. (1961). Vorkomen von Serotonin in Nervensystem von *Periplaneta americana* (Insecta). *Z. Naturforsch.*, 16 : 351-352.

Gersch, M. and Sturzebecher, J. (1970). Experimentelle stimulierung der Zellularen Activat der prothorakal drusen von *Periplaneta americana* durch den Aktivations faktor. *J. Insect Physiol.*, 16 : 1813-1826.

Girardie, A. (1964). Action de la pars intercerebralis sur le development de *Locusta migratoria* L. *J. Insect Physiol.* 10: 599-609.

Girardie, A. and Girardie, J. (1967). Etude Histologique, Histochemique, et ultrastructurale de la pars intercerebralis chez *Locusta migratoria* L. (Orthoptere). *Z. Zellforsch.*, 78 : 54-75.

Gilbert, L.I. (1964). Physiology of growth and development : Endocrine aspects. In : *The Physiology of Insecta* (M. Rockstein, ed.). Academic Press, New York and London, 1 : 149-225.

Gilbert, L.I. (1967). Changes in lipid content during the reproductive cycle of *Leucophaea maderae* and effects of the juvenile hormone on lipid metabolism *in vitro. Comp. Biochem. Physiol.*, 21 : 237-257.

Gilbert, L.I., and King, D.S. (1973). Physiology of growth and development : Endocrine aspects. In : *The Physiology of Insecta*, (M. Rockstein, ed.). 1 : 249-370. Second edition, Academic Press, New York and London.

Goldsworthy, G.J., Coupland, A.J., and Mordue, W. (1973). The effects of corpora cardiaca on tethered flight in the locust, *J. Comp. Physiol.* 82 : 339-346.

Goldsworthy, G.J., Lee, S.S., and Jutsum, A.R. (1977). Cerebral neurosecretory cells and flight in the locust. *J. Insect Physiol.* 23 : 717-721.

Gundavia, H.S. and Ramamurty, P.S. (1972). On the intraganglionic neurohaemal organ in the ventral nerve cord of *Hydrophilus olivaceus* (Hydrophilidae : Coleoptera). *Experientia* 28 : 401-411.

Gwadz, R.W. Craig, G.B. and Hickey, W.A. (1971a). Female sexual behaviour as the mechanism rendering *Aedes aegypti* refractory to insemination. *Biol. Bull. Woods, Hole*, 140 : 201-214.

Gwadz, R.W., Lounibos, L.P. and Craig, G.B. (1971b). Precocious sexual reciptivity induced by a juvenile hormone analog in females of the yellow fever mosquito, *Aedes aegypti. Gen. Comp. Endocrinol.,* 16 : 47-51.

Harker, J.E. (1960). Endocrine and nervous factors in insect circadian rhythms. *Cold. Spring. Harb. Symp. Quant. Biol.*, 25 : 279-287.

Hanstrom, B. (1940). Inkretorishe Organe, Sinnesorgane und Nerven system des Kopfes einiger niederer Insektordnungen. *Kgl. Svenska Vetensk Akad. Hdl.*, 18 : 1-265.

Herman, W.S. (1967). The ecdysial glands of arthropods. *Int. Rev. Cytol.,* 22 : 269-347.

Herman, W.S. and Gilbert, L.I. (1965). Multiplicity of neurosecretory cell-types and groups in the brain of the saturniid moth, *Hyalophora cecropia* (L). *Nature,* London, 205 : 926-927.

Herman, W.S. and Gilbert, L.I. (1966). The neuroendocrine system of *Hyalophora cecropia* (L.) Lepidoptera : Saturniidae) 1. The anatomy and histology of the ecdysial glands. *Gen. Comp. Endocrinol.,* 7: 275-291.

Highnam, K.C. (1961). The histology of the neurosecretory system of the adult female desert locust, *Schistocerca gregaria. Quart. J. Micr. Sci.,* 102 : 27-38.

Highnam, K.C. (1967). Insect Hormones. *J. Endocrinol.,* 39 : 123-150.

Highnam, K.C. (1968). Neurosecretion in insects. Progress in Endocrinology. *Excerpta Medica International Congress. Series.* 184 : 351-355.

Highnam, K.C. (1971). Estimates of neurosecretory activity during maturation in female locusts. *Insect Endocrines.* Academia Prah., 81-90.

Highnam, K.C. and Goldsworthy, G.J. (1971). Regenerated corpora cardiaca and hyperglycaemic factors in *Locusta migratoria. Gen. Comp. Endocrinol.,* 12 : 83-88.

Highnam, K.C., Lusis, O. and Hill, L. (1965). Neurosecretion and water balance in the male desert locust, *Schistocerca gregaria. J. Zool.,* London, 147 : 201-215.

Highnam, K.C. and Hill. L. (1979). *The Comparative Endocrinology of Invertebrates.* ELBS 2nd Edition, Edward Arnold Publishers, Ltd.

Hinks, C.F. (1967). Relationship between serotonin and the circadian rhythm in some nocturnal moths. *Nature,* London, 214 : 386-387.

Ichikawa, M. and Ishizaki, H. (1963). Protein nature of the brain hormone of insects. *Nature,* London 198: 308-309.

Ichikawa, M. and Nishiitsutsuji-Uwo, J. (1959). Studies on the role of the corpus allatum in the erisilkworm, *Philosamia cynthia ricini. Biol. Bull. Woods Hole,* 117 : 88-94.

Ito, H. (1918). Glandular nature of the corpora allata of Lepidoptera. *Bull. Imp. Tokyo Seric. Coll.,* 1 : 68-103.

Johansson, A.S. (1957). The nervous system of the milkweed bug, *Oncopeltus fasciatus* (Dallas) (Heteroptera, Lygaeidae). *Trans. Amer. Ent. Soc.,* 83 : 119-183.

Johansson, A.S. (1958) Relation of nutrition to endocrine reproductive functions in the milkweed bug, *Oncopeltus fasciatus* (Dallas) (Heteroptera : Lygaeidae). *Nytt. Mag. Zool.,* 7 : 1-132.

Johnson, B. (1963). A histological study of neurosecretion in aphids. *J. Insect Physiol.,* 9 : 727-739.

Johnson, B. (1966a). Fine structure of the lateral cardiac nerves of the cockroach, *Periplaneta americana* (L.). *J. Insect Physiol.,* 12: 645-653.

Johnson, B. (1966b). Ultrastructure of probable sites of release of neurosecretory materials in an insect, *Calliphora stygia* Fabr. (Diptera). *Gen. Comp. Endocrinol.*, 6 : 99-108.

Johnson, C.G. (1969). *Migration and Dispersal of Insects by Flight*. Methuen, London.

Joly, L., Joly, P., Porte, A. and Girardie, A. (1969). Analyse ultrastructurle de l activite des corpora allata des *Locusta migratoria* L. et ses consequences sur la structure que l on doit attributer au mechanisme humoral controlant la metamorphose. *Arch. Zool. Exptl. Gen.*, 110 : 617-628.

Karlson, P. (1964). Biochemistry and mode of action of ecdysone. In : *Experta medica*. International Congress, Series No. 83 : 516-519.

Karlson, P. (1966). Ecdysone des Hautung shormon der insketen. *Naturwissenschaften*, 53 : 445.

Karlson, P. and Hoffmeister, H. (1963). Zur Biogenese des Ecdysons. I Umwan d lung von colesterin in Ecdyson. *Z. Physiol. Chem.* 331 : 298-300.

Karlson, P., Hoffmeister, H., Hopp, W. and Huber, R. (1963). Zur Chemie des Ecdysons. *Justus Leibigs Annl. of Chem.* 662 : 1-20.

Ke, O. (1930) Morphological variation of prothoracic glands in domestic and wild silkworm. *Bull. Sci. Fac. TerKult. Kjusu Imp. Univ.*, 4 : 12-21.

Keeley, L.L. (1967). The corpus cardiacum as a metabolic regulator in *Blaberus discoidalis* Serville (Blattidae). *Distt. Abstr.*, 27 : 3937-3980.

Kerkut, G.A. and Gilbert, L.I. (1985). *Comprehensive Insect Physiology, Biochemistry and Pharmacology*, Vols. 3, 4, 5, 7 and 11, Pergamon Press, Oxford.

King, R.C. Aggarwal, S.K. and Bodenstein, D. (1966a). The comparative submicroscopic cytology of the corpus allatum-corpus cardiacum complex of wild type and *fes* adult female, *Drosophila melanogaster. J. Exp. Zool.*, 161 : 151-176.

King, R.C., Aggarwal, S.K. and Bodenstein, D. (1966b). The comparative submicroscopic morphology of the ring gland of *Drosophila melanogaster* during the second and third larval instars. *Z. Zellforsch.*, 73 : 272-285.

Kobayashi, M. (1957). Studies on the neurosecretion in the silkworm, *Bombyx mori* L. *Bull. Seric. Exp. Sta.*, Tokyo, 15: 181-273.

Kobayashi, M. and Kirimura, J. (1958). The brain hormone in the silkworm, *Bombyx mori* L. *Nature*, London, 181 : 1217.

Kobayashi, M., Kirimura, J. and Saito, M. (1962). The brain hormone of an insect, *Bombyx mori* L. *Mushi,* 36 : 85-92.

Kobayashi, M. and Yamazaki, M. (1966). The proteinic brain hormone in an insect *Bombyx mori* L. (Lepidoptera : Bombycidae). *Appl. Entomol. Zool.*, 1 : 53-60.

Kobayashi, M. and Yamashita, Y. (1959). A function of corpus allatum in neurosecretory system in the silkworm, *Bombyx mori. J. Sericult. Japan*, 28 : 335-339.

Kopec, S. (1912). Uber die Funktionen des Nervensystem der- schmetterlinge ahrend dir successiven studien der Metamorphose. *Zool. Anz.*, 40 : 353-360.

Kopec, S. (1917). Experiments on metamorphosis in insects. *Bull. Int. Acad. Sci.* Cracovie (B), 57-60.

Kopec, S. (1922). Studies on the necessity of the brain for the inception of insect metamorphosis. *Biol. Bull. Woods Hole*, 42 : 323-342.

Kloot, W.G. Van der (1960). Neurosecretion in insects. *Ann. Rev. Entomol.* 5 : 15-59.

Kono, Y. (1973). Light and electron microscopic studies on the neurosecretory controle of diapause incidence in *Pieris rapae crucivora. J. Insect Physiol.*, 19 : 255-272.

Kono, Y. (1975). Daily changes of neurosecretory type-II cell structure of *Pieris* larva entrained by short and long days. *J. Insect Physiol.*, 21 : 249-264.

Kono, Y. (1977). Ultrastructural changes of neurosecretory cells in the pars intercerebralis during diapause development in *Pieris rapae*. *J. Insect Physiol.* 23 : 1461-1473.

Krogh, I.M. and Normann, T.C. (1977). The corpus cardiacum neurosecretory cells of *Schistocerca gregaria*. Electron microscopy of resting and secreting cells. *Acta Zool.*, 58 : 69-78.

Lea, A.O. (1967). The medial neurosecretory cells and egg maturation in mosquitoes. *J. Insect. Physiol.* 13: 419-429.

Lea, A.O. (1968). Mating without insemination in virgin *Aedes aegypti. J. Insect. Physiol.*, 14 : 303-308.

Longley, P.A. (1965). The neuroendocrine system and the stomatogastric nervous system of the adult tsetse fly, *Glossina morsitans. Proc. Zool. Soc. London*, 144, 415-423.

Loher, W. (1961). The chemical acceleration of the maturation process and its hormonal control in the male of the desert locust. *Proc. R. Soc.* B 153: 380-397.

Loher, W. (1962). Die Kontrolle des weibchengesanges von *Gomphocerus rufus* L. (Acridinae) durch die corpora allata. *Naturwiss*. 49: 406.

Loher, W. (1966). Die steuering sexuller verhaltensweisen und der Oocytentwicklung bei, *Gomphocerus rufus* L. *Z. Vergl. Physiol.*, 53: 277-316.

Loher, W. (1967). Die Kontrolle des Weibchengesanges von *Gomphocerus rufus* L. (Acridinae) durch die Corpora allata. *Naturwissenschaften,* 49: 406.

Lyonet, P. (1762). Treite anatomique de la chenille qui ronge la bois du saule. *La Haye*.

Maddrell, S.H.P. (1966). The site of release of the diuretic hormone in *Rhodnius* a new neurohaemal system in insects. *J. Exp. Biol.*, 45 : 499-508.

Magni, F. (1967). Central and peripheral mechanisms in the modulation of flashing in the firefly *Luciola italica. Arch. Ital. Biol.*, 105 : 339-360.

Marks, E.P., Holman, G.M. and Borg, T.K. (1973). Synthesis and storage of a neurohormone in insect brain *in vitro. J. Insect Physiol,* 19 : 471-477.

Michel, R. (1972). Etude experimentale de l influence des glandes prothoraciques Sur l activite de vol due Criquet Peclrin., *Schistocerca gregaria. Gen. Comp. Endocrinol.*, 19 : 96- 101.

Milburn, N.S., Weiant, E.A. and Roeder, K.D. (1960). The release of efferent nerve activity in the roach. *Periplaneta americana* by extract of the corpus cardiacum. *Biol. Bull. Woods. Hole*, 118 : 111-119.

Mordue, W. (1963). Hormonal control of Malpighian tube and rectal function in the desert locust, *Schistocerca gregaria, J. Insect Physiol.*, 15 : 275-285.

Mordue, W. and Goldsworthy, G.J. (1969). The physiological effects of corpus cardiacum extracts in Locusts. *Gen. Comp. Endocrinol.*, 12 :360-369.

Meola, S.M. and Lea, A.O. (1972). The Ultrastructure of the corpus cardiacum of *Aedes sollicitans* and the histology of the cerebral neurosecretory system of mosquitoes. *Gen. Comp. Endocrinol.*, 18 : 210-234.

Meola, S.M., Lea, A.O. and Meola, R. (1970). Corpus cardiacum induced fluctuations in paraldehyde fuschsin material in *Aedes sollicitans*. *Trans. Amer. Micr. Soc.*, 89 : 418-423.

Mc Danial, C.N., Johnson, E., Saum T. and Berry, S.J. (1976). Ultrastructure of active and inhibited prothoracic glands. *J. Insect Physiol.*, 22 : 473-481.

Mitsuhashi, J. (1963). Histological studies on the neurosecretory cells of the brain and on the corpus allatum during diapause in some lepidopteran insects. *Bull. Inst. Agric. Sci.*, 16 : 69-134.

Morris. G.P. and Steel, C.G.H. (1975). Ultrastructure of neurosecretory cells in the pars intercerebralis of *Rhodnius prolixus* (Hemiptera). *Tissue and Cell.*, 7 : 73-90.

Nayar, K.K. (1955). Studies on the neurosecretory system of *Iphita limbata* I. Distribution and structure of nerve ring. *Biol. Bull. Woods Hole*, 108 : 296-307.

Nayar, K.K. (1958). Studies on the neurosecretory system of *Iphita limbata* Stal. V. Probable endocrine basis of oviposition in the female insect. *Proc. Indian Acad. Sci.* B 47: 233-251.

Nayar, K.K. (1973). Elements of Insect endocrinology, P. 1-56. Prentic-Hall of India Private Limited, New Delhi.

Nebart, A. (1913), Die corpora allata der insekten. *Z. Wiss. Zool.*, 104 : 181-358.

Nicolson, S.W. (1976). The hormonal control of diuresis in the cabbage white butterfly. *Pieris brassicae. J. Exp. Biol.*, 65 : 565-575.

Nishiitsutsuj-Uwo, J. (1961). Fine structure of the neurosecretory system in Lepidoptera. *Z. Zellforsch.*, 54 : 613- 630.

Nishiitsutsuji-Uwo, J. (1972). Purification and some properties of insect brain hormone extracted from silkworm heads. *Botyu-Kagaku*, 37 : 93-102.

Normann, T.C. (1965). The neurosecretory system of the adult *Calliphora erythrocephala*. I. The fine structure of the corpus cardiacum with some observations on adjacent organs. *Z. Zellforsch.*, 67 : 461-501.

Normann, T.C. (1969). Experimentally induced exocytosis of neurosecretory granules. *Exptl. Cell. Res.*, 55 : 285-287.

Normann, T.C. (1970). The mechanism of hormone release from axon endings in the insect, *Calliphora erythrocephala*. In: *Aspects of Neuroendocrinology.* (Bargmann, W. and Scharrer, B. eds.)

Normann, T.C. (1974). Calcium dependence of neurosecretion by exocytosis. *J. Exptl. Biol.*, 61 : 401-410.

Normann, T.C. (1976). Neurosecretion by exocytosis. *Int. Rev. Cytol.* 46 : 1-77.

Novak, V.J.A. (1966). *Insect Hormones*. Methuen & Co. Ltd., London.

Novak V.J.A. (1975). *Insect Hormones*. 2nd English Edition. Chapman and Hall, London.

Odhiambo, T.R. (1965). Metabolic effects of the corpus allatum hormone in the desert locust, *Schistocerca gregaria, Nature*, London, 207 : 1314-1315.

Odhiambo, T.R. (1966). The fine strucutre of the corpus allatum of the sexually mature male of the desert locust. *J. Insect Physiol.* 12 : 819-828.

Ozeki, K. (1958). Neurosecretion of the earwig, *Anisolabis maritima. Sci. Pap. Coll. Gen. Ed. Univ. Tokyo.*, 8 : 85-92.

Panov, A.A. and Bassurmanova, O.K. (1970). Fine structure of the gland cells in inactive and active corpus allatum of the bug, *Eurygaster integriceps*. *J. Insect Physiol.*, 16 : 1265- 1281.

Panov, A.A. and Kind, T.V. (1963). The neurosecretory cell system in the lepidopteran brain (Lepidoptera : Insecta). *Dokl. Akad. Nauk. SSSR* (Biol.) 153 : 1186-1189.

Park, K.E. and Yoshitake, N. (1971). Fine structure of the neurosecretory cell in the suboesophageal ganglian of the silkworm, *Bombyx mori*. *J. Insect Physiol.*, 17: 1305-1313.

Pener, M.P. (1967). Effect of the allatectomy and sectioning of the nerves of the corpora allata on oocyte growth, male sexual behaviour and colour change in adults of the *Schistocerca gregaria*. *J. Insect Physiol.*, 13 : 665-634.

Pener, M.P. (1985). Hormonal effects on flight and migration. In : *Comprehensive Insect Physiology, Biochemistry and Pharmacology* (G.A. Kerkut and L.I. Gilbert, eds). 8 : 491-550.

Pener, M.P., Girardie, A. and Joly, P. (1972). Neurosecretory and corpus allatum controlled effects on mating behaviour and colour change in adult *Locusta migratoria migrotorioides* Males. *Gen. Comp. Endocrinol.*, 19 : 494-508.

Piepho, H., Boden, E. and Holz, I. (1960). Uper die Hormonabhangigkeit des Verhaltens Von Schwarmeraupen vor den Hautungen. *Z. Tierpsychol.*, 17 : 261-269.

Pipa, R.L. (1962). A cytochemical study of neurosecretory and other neoplasmic inclusions in *Periplaneta americana. Gen. Comp. Endocrinol.*, 2 : 44-52.

Pittendrigh, C.S. (1966). The circadian oscillation in *Drosophila pseudo-obscura* pupae : a model for the photoperiodic clock. *Z. Pflanzenphysiol.*, 54 : 275-307.

Prabhu, V.K.K. (1981). Recent advances in insect endocrinology in India. In: *Recent Advances in Entomology in India.* (T.N. Anantkrishnan, ed.) Viswanathan, Madras.

Provansal-Baudez, A. and Baudry-Partiaoglou, N. (1983). Ultrastructural comparison of the perisympathetic organs in three Coleoptera : *Chrysocarabus auronitens* F., *Oryctes rhinoceros* L. and *Tenebrio molitor* L. *Gen. Comp. Endocrinol.*, 49 : 383-403.

Raabe, M. (1965). Etude des phenomenes de neurosecretion au niveaue de la chaine nerveuse ventrales des Insectes. *Bull. Soc. Zool.,* France, 90 : 631-654.

Raabe, M. 1966). Recherches sur la neurosecretion dans la chaine nerveuse ventrale du Phasme, *Carausius morosus*. Liasion entre l activite des cellules. Bl et la pigmentation. *C.R. Acad. Sci.* (Paris), 263 : 408-411.

Raabe, M. (1982). *Insect Neurohormones*. Plenum Press, New York.

Raabe, M. (1989). *Recent Developments in Insect Neurohormones*. Plenum Press, New York.

Raabe, M., Baudry, N., Grillot, J.P. and Provansal, A. (1974). The perisympathetic organs of Insects. In : "Neurosecretion: The final common neuroendocrine pathway" (F. Knowles and L. Vollrath, eds.). pp. 59-71. VI. *Int. Symp. Neurosecretion*, London. (1973). Springer Verlag. Heidelberg & New York.

Ralph, C.L. (1962). Heart accelerators and decelerators in the nervous system of *Periplaneta americana. J. Insect Physiol.*, 8 : 431-439.

Ralph, C.L. and McCarthy, G. (1964). Effect of brain and corpus cardiacum extracts on haemolymph trehalose of the cockroach, *Pèriplaneta americana. Nature* London, 203 : 1195.

Ramade, F. (1966). Sur l' ultrastructure de la pars intercerebralis chez *Musca domestica L.C. R. Acad. Sci.* Paris, 263 : 271-274.

Ramamurty, P.S. and Engels, W. (1977). Allatektomie und juvenilhormon-wirkungen auf synthese und Enlagerung von Vitellogenin bei der Bienenkonigin (*Apis mellifica*). *Zool. Jb. Physiol.*, 81 : 165-176.

Rankin, M.A. (1974). The hormonal control of flight in the milkweed bug, *Oncopeltus fasciatus*. In. *Experimental Analysis* of Insect Behaviour (L. Barton-Browne, ed.). Springer Verlag, Berlin and New York.

Rembold, H. (1976). The role of determination in caste formation in the honeybee, In :· *Phase and Caste Determination in Insects*. (M Luscher, ed.), Pergamon Press, New York.

Rembold, H., Czoppelt, C., and Rao, P.J. (1974). Effect of juvenile hormone treatment- on caste differentiation in the honey bee, *Apis mellifera*. *J. Insect Physiol.*, 20 : 1193-1202.

Reynolds, S.E. (1974). Pharmacological induction of plasticization in the abdominal cuticle of *Rhodinius*. *J. Exp. Biol.*, 61: 705-718.

Reynolds, S.E., Taghert, P.H. and Truman, J.W. (1979). Eclosion hormone titres and bursicon and the onset of hormonal responsiveness during the last day of adult development in *Manduca sexta* (L.). *J. Exp. Biol.*, 77 -86.

Rhem, M. (1951). Die Zeitliche Folge der Tatigkeitsrhythmen Inkretorischer organe von *Ephestia kuhniella* Wahrend der metamorphose und des Imaginallebens. *Arch. Ent. Mech. Org.*, 145 : 205-208.

Riddiford, L.M. (1970). Antennal proteins of saturniid moths : their possible role in olfaction. *J. Insect. Physiol.*, 16: 653-660.

Riddiford, L.M. (1972). Juvenile hormone in relation to the larvel-pupal transformation in the cecropia silkworm. *Biol. Bull. Woods Hole*, 142 : 310-325.

Riddiford, L.M. (1974). The role of hormones in the reproductive behaviour of female wild silkmoths; In : *Experimental Analysis* of *Insect Behaviour*. (L. Barton-Browne, ed.). Springer Verlag, Berlin and New York.

Riddiford, L.M. (1977). *Insect Chemoreception* (L. Skirkevicius, ed.) Lithuanian Acad. Sciences.

Riddiford, L.M. and Moore, J.K. (1977). Cited in : Invertebrate system for the study of hormonal effects on behaviour. *Vitamins and Hormones* 35 : 283-315.

Roth, L.M. and Barth, R.H. (1964). The control of sexual receptivity in female cockroaches. *J. Insect Physiol.*, 10 : 965- 975.

Roussel, J.P. (1966). Role des corpora allata et controle de la pigmentation chez *Gryllus bimaculatus* De Geer. *J. Insect Physiol.*, 12 : 1085-1092.

Roussel, J.P. (1967). Fonction des corpora allata et control de la pigmentation chez *Gryllus. bimaculatus* De Geer. *J. Insect Physiol.* 13 : 113-130.

Rowell, C.H.F. (1967). Corpus allatum implantation and green/brown polymorphism in three African grasshoppers. *J. Insect Physiol.*, 13 : 1401-1412.

Rowell, H.F. (1976). The cells of the insect neurosecretory system : constancy, variability and the concept of the unique identifiable neuron. *Adv. Insect Physiol.*, 12 : 63-123.

Rowell, H.F. and Mason, C.A. (1976). Cited in: The cells of the insect neurosecretory system, constancy, variability and the concept of the unique identifiable neuron by H.F. Rowell (1976). *Adv. Insect Physiol.* 12: 63-123.

Saini, R.S. (1966). Neuroendocrine control of oocyte development in the beetle,

Aulacophora faveicollis Luc. *J. Insect Physiol.*, 12 : 1003-1008.

Saini, R.S. (1971). Neuroendocrine control of oocyte development in *Poecilocera picta* (Insecta : Orthoptera). *J. Zool. Lond., 165 : 275-283.*

Schmialek, P. (1963). Uber die Bildung von juvenilhormonen in Wildscidensp-innern. *Z. Naturforsch.,* 186 : 462-465

Scharrer, B. (1963). Neurosecretion. XIII. The ultrastructure of the corpus cardiacum of the insect *Leucophaea maderae. Z. Zellforsch.,* 60 : 761-796.

Scharrer, B. (1964a). Histological studies on the corpus allatum of *Leucophaea maderae* IV. Ultrastructure during normal activity cycle. *Z. Zellforsch.,* 62 : 125-148.

Scharrer, B. (1964b). The fine structure of Blattarian prothoracic glands. *Z. Zellforsch.*, 64 : 301-326.

Scharrer, B. (1965). Recent progress in the study of neuroendocrine mechanism in insects. *Arch. Anatom. Microsc. Morphol. Exp.*, 54 : 331-342.

Scharrer, B. (1966). Ultrastructural study of the regressing prothoracic glands of Blattarian Insects. *Z. Zellforsch.*, 69 : 1-21.

Scharrer, B. (1967). The neurosecretory neuron in neuroendocrine regulatory mechanisms. *Amer. Zool.*, 7 : 161-169.

Scharrer, B. (1968). XIV Ultrastructural study of sites of release of neurosecretory material in Blattarian insects. *Z. Zellforsch.*, 89, 1-16.

Scharrer, B. (1969). Neurohumors and neurohormones. Definitions and terminology. *J. Neuro. Vis. Relat.* (Suppl.). 9 : 1-20.

Scharrer, B. (1970). General principles of neuroendocrine communication. In : *The neurosciences*. Pp : 519-529 (Schmitt, F.O. ed.). The Rockfeller University Press, New York.

Scharrer, B. (1971). Histophysiological studies on the corpus allatum. of *Leucophaea maderae*. V. Ultrastructure of sites of origin and release of a distinctive cellular product. *Z. Zellforsch.*, 120 : 1-16.

Scharrer, B. and Weitzman, M. (1970). Neurosecretion in invertebrates : Current problems in invertebrate neurosecretion. In: *Aspects of Neuroendocrinology*. pp. 1-23. Bargmann, W. and Scharrer, B. (eds.). Springer Verlag, Berlin, Heidelberg - New York.

Scharrer, E. (1928). Die lichtempfindlichkeit blinder Elritzen 1. Untersuchungen. Uber das Zwischenhirn der Fische. *Z. Vergl. Physiol.*, 7 : 1-38.

Scharrer, E. and Scharrer, B. (1937). Uber Drusen Nervenzellen und Neurosek retorische Organe bei Wirbellosen und Wirbeltieren. *Biol. Rev.,* 12 : 185-216.

Scharrer, E. and Scharrer, B. (1963). *Neuroendocrinology.* 289 pp. Columbia University Press, New York.

Schooneveld, H. (1970). Structural aspects of neurosecretory and corpus allatum activity in the adult colorado beetle, *Leptinotarsa decemlineata* Say, as a function of day length. *Netherlands J. Zool.*, 20 : 151-237.

Singh, U.V. and Awasthi, V.B. (1980). The anatomy and histology of ecdysial glands of *Philosamia ricini* Hutt. (Lepidoptera : Saturniidae) during post embryonic development. *Zeit. Mikrosk. Anat. Forsch.,* 94 : 850-864.

Singh, U.V. and Awasthi, V.B. (1981a). The anatomy and histophysiology of the ecdysial glands of *Achoea janata* Linn. (Lepidoptera : Noctuidae) during post embryonic development. *Entomon,* 6 : 183-195.

Singh, U.V. and Awasthi, V.B. (1981b). The neurosecretory cells and retrocerebral endocrine glands of *Amsacta collaris* Hamp. (Lepidoptera : Arctiidae) during post embryonic development. *J. Hirnforsch.*, 22 : 265-278.

Singh, U.V. and Awasthi, V.B. (1982). Studies on the retrocerebral endocrine glands of the Erisilkworm, *Philosamia ricini* Hutt. (Lepidoptera : Saturniidae) during post-embryonic development. *Folia Morphologica,* 30 : 346-354.

Singh, Y.N. and Narain, R. (1980a). Neuroendocrine system and storage of neurosecretory material in the dorsal aorta in a bug *Chrysocoris stollii* Wolf (Heteroptera: Pentatomidae). *J. Hirnforsch,* 21: 443-448.

Singh, Y.N. and Narain, R. (1980b). Neurosecretory system and storage of paraldehyde fuchsin positive neurosecretory material in the dorsal aorta in an earwig, *Anisolabis annulipes* Lucas (Dermaptera: Labiduridae). *Experientia,* 36: 834-835.

Slama, K., Romanuk, M. and Sorm, F. (1974). *Insect Hormones and Bioanalogues.* Springer Verlag, Wien, New York.

Smith, U. and Smith, D.S. (1966). Observations on the secretory processes in the corpus cardiacum of the stick insect, *Carausius morosus. J. Cell. Sci.*, 1 : 59-66.

Srivastava, R.C. (1969). Observations on the neurosecretory axons in the aortal wall of *Halys dentatus* F. (Heteroptera: Pentatomidae). *Experientia*. 25: 534.

Srivastava, R.C. (1970). Morphology of the neurosecretory system and the retrocerebral endocrine glands of adult *Halys dentatus* (Heteroptera: Pentatomidae). *Ann. Ent. Soc. Amer.*, 63: 1372-1376.

Srivastava, R.C. and Dogra, G.S. (1969). Studies on the neurosecretory system of the bug, *Scutelleria nobilis* Fabr. with reference to material in the aortal wall. *Anat. Anz.*, 25: 525-534.

Srivastava, K.P. and Tiwari, R.K. (1973). Secretory factors in the retrocerebral endocrine-aortal complex of the water bug, *Sphaeordema annulatum* (Heteroptera: Belostomatidae). *Gen. Comp. Endocrinol.*, 21: 398-401.

Srivastava, K.P., Tiwari, R.K. and Singh, H.H. (1974). Histological study on the retrocerebral endocrine-aortal complex of the water bug, *Sphaerodema annulatum* (Heteroptera: Balostomatidae) with reference to secretory inclusions. *Z. Mikrosk. Anat. Forsch.*, 88: 585-595.

Sterba, G. and Hoeisel, H. (1964). Darstellung des Neurosekreterischen systems der insekten mit Psuedoisocyanin. *Z. Zellforsch.*, 72 : 31-48.

Stratton, C.J. and Booth, G.M. (1975). Ultrastructure of synaptic vesicles and neurosecretory droplets in insects ganglia after controlled aldehyde fixation. *J. Insect Physiol.*, 21 : 71- 80.

Strong, L. (1965). The relationship between the brain, corpora allata and oocyte growth in the central American locust, *Schistocerca* sp. 1. The cerebral neurosecretory system, the corpora allata and oocyte growth. *J. Insect Physiol.*, 11 : 135- 146.

Strong, L. (1968). Locomotor activity, sexual behaviour and corpus allatum hormone in males of *Locusta. J. Insect Physiol.*, 14 : 1685-1692.

Steele, J.E. (1961). Occurrence of hyperglycaemic factor in the corpus cardiacum of an insect. *Nature*, London, 192 : 680- 681.

Takeda, N. (1977). Histophysiological studies of the corpus allatum during pupal diapause in *Monema flavescens* (Lepidoptera). *J. Morphol.*, 153 : 245-262.

Tewari, H.B. and Awasthi, V.B. (1968). The identical topography of the distribution of alkaline phosphatase and neurosecretory material in the neuroendocrine organs of *Gryllodes sigillatus* (Walk.) (Orthoptera : Gryllidae). *Gen. Comp. Endocrinol.*, 10 : 330-338.

Tiwari, R.K. and Srivastava, K.P. (1975). Studies on the neurosecretory system and retrocerebral endocrine organs in the red cotton bug, *Dysdercus koenigii* Fabr. (Heteroptera: Pyrrhocoridae). *Z. Morph. Tiere*, 81: 355-364.

Tiwari, R.K., Srivastava, K.P. and Singh, H.H. (1975). Studies on the neurosecretory system of the paddy bug, *Leptocorisa varicornis* Fabricius (Heteroptera: Coreidae). *Zool. Beitr.* 21: 105-112.

Thomsen, E. (1952). Functional significance of the neurosecretory brain cells and the corpus cardiacum in the female blowfly, *Calliphora erythrocephala* Meig. *J. Exp. Biol.*, 29 : 137-172.

Thomsen, E. and Thomsen, M. (1969). Fine structure of the corpus allatum of the female *Calliphora* with special reference to hormone formation. *Gen. Comp. Endocrinol.*, 13 : 534.

Thomsen, E. and Thomsen, M. (1970). Fine structure of the corpus allatum of the female blowfly, *Calliphora erythrocephala. Z. Zellforsch.*, 110: 40-60.

Thomsen, M. (1965). The neurosecretory system of the adult *Calliphora erythrocephala*. II Histology of the neurosecretory cells of the brain and some related structures. *Z. Zellforsch.*, 67: 693-717.

Thomsen, M. (1969). The neurosecretory system of the adult *Calliphora erythrocephala*. IV. A histological study of the corpus cardiacum and its connections with the nervous system. *Z. Zellforsch.*, 94 : 205-219.

Tombes, A.S. and Smith, D.S. (1970). Ultrastructural studies on the corpora cardiaca-allata complex of the adult alfalfa weevil, *Hypera postica. J. Morphol.*, 132 : 137-148.

Toyoma, K. (1902). Contribution to the study of silkworm, 1. On the embryology of the silkworm. *Bull. Coll. Agric.* Tokyo, 5 : 73-118.

Traina, M.E., Bellino, M. Serpietri, L., Massa, A. and Foontali, N. (1976). Heart accelerating peptides from cockroach corpora cardiaca. *J. Insect Physiol.,* 22 : 323-329.

Truman, J.W. (1971a). Physiology of insect ecdysis. I. The eclosion behaviour of Saturniid moths and its hormonal release. *J. Exptl. Biol.*, 54 : 805-814.

Truman, J.W. (1971b). Circadian rhythm and physiology with special reference to neuroendocrine processes in insects. In : *Proceedings of the International Symposium on Circadian Rhythmicity*, pp. 11-135. Pudoc. Press. Wageningen, Netherlands.

Truman, J.W. (1973a). How moths "turn on". A study of the action of hormones on the nervous system. *Am. Sci.* 61 : 700-706.

Truman, J.W. (1973b). Physiology of insect ecdysis. II. The assay and occurrence of the eclosion hormone in the Chinese oak silkmoth, *Antheraea pernyi. Biol. Bull. Woods. Hole*, 144 : 200- 211.

Truman, J.W. (1973c). Physiology of insect ecdysis. III. Relationship between the hormonal control of eclosion and of tanning in the tobacco hornworm, *Manduca sexta. J. Exp. Biol.* 58 : 821-829.

Truman, J.W. (1978). Hormonal release of stereotyped motor programmes from the isolated nervous system of the Cecropia silkmoth. *J. Exp. Biol.*, 74 : 151-173.

Truman, J.W. (1979). Interaction between abdominal ganglia during the performance of hormonally triggered behavioural programmes in moths. *J. Exp. Biol.* 83 : 239-253.

Truman, J.W., Taghert, P.H., and Reynolds, S.E. (1980). Physiology of pupal ecdysis in the tobacco hornworm, *Manduca sexta*. I. Evidence for control by eclosion hormone, *J. Exp. Biol.* 88 : 327-337.

Truman, J.W., Taghert, P.H., Copenhaver, P.F., Tublitz, N.J., and Schwartz, L.M. (1981). Eclosion hormone may control all ecdyses in insects. *Nature* London, 291: 70-71.

Truman, J.W., Rountree, D.B., Reiss, S.E., and Schwartz, L.M. (1983). Ecdysteroids regulate the release and action of eclosion hormone in the tobacco hornworm, *Manduca Sexta* (L.), *J. Insect Physiol.* 29 : 895-900.

Truman, J.W. and Riddiford, L.M. (1970). Neuroendocrine control of ecdysis in Silkmoth. *Science*, 167 : 1624-1626.

Truman, J.W. and Riddiford, L.M. (1971). Role of the corpora cardiaca in the behaviour of Saturniid moths. II. Oviposition. *Biol. Bull. Woods Hole*, 140 : 8-14.

Truman, J.W. and Riddiford, L.M. (1974). Hormonal mechanisms underlying insect behaviour. *Adv. Insect Physiol.*, 10 : 297-351.

Truman, J.W. and Riddiford, L.M. (1977). Invertebrate system for the study of Hormonal effects on behaviour. *Vitamins and Hormones*, 35 : 283-315. Academic Press, New York.

Unger, H. (1957). Untersuchungen Zur Neurohormonalen Steuerung der Herztatigkeit bi Schaben *Periplaneta orientalis, P. americana, Phylodromia germanica Biol. Zbl.*, 76: 204-225.

Unnithan, G.C., Bern, H.A. and Nayar, K.K. (1971). Ultrastructural analysis of the neuroendocrine apparatus of *Oncopeltus fasciatus* (Heteroptera). *Acta Zoologica,* 52 : 117-143.

Unnithan, G.C. and Nair, K.K. (1977). Fine structure of the A - cells of the pars intercerebralis of normal and Gamma-irradiated female milkweed bug, *Oncopeltus fasciatus*, (Heteroptera : Lygaeidae). *J. Morphol.*, 154 : 58-82.

Vejberg, K. and Normann, T.C. (1974). Secretion of hyproglycaemic hormone from the corpus cardiacum of flying blowflies, *Calliphora erythrocephala. J. Insect Physiol.* 20 : 1189-1192.

Verman, H.E., Kaplanis, J.M. and Robbins, W.E. (1965). Effect of allatectomy on Lipid biosynthesis and turnover in the female American cockroach, *Periplanta americana* L. *J. Insect Physiol.,* 13 : 513-527.

Waku, Y. and Gilbert, L.I. (1964). The corpora allata of the silkworm *Hyalophora cecropia* : An ultrastructural study. *J. Morphol.*, 115 : 69-96.

Watson, J.A.L. (1964). Moulting and reproduction in the adult firebrat, *Thermobia domestica* (Packard) (Thysanura : Lepismatidae). I. The moulting cycle and its control. *J. Insect Physiol.*, 10 : 305-317.

Weber, W. and Gaude, H. (1971). Ultrastructure des neurohaemalorgans in Nervus corporis allati II von *Acheta domesticus. Z. Zellforsch.*, 121 : 561-572.

Weyer, F. (1935). Uber drusenartige Nervenzellen im Gehrin der Honigbiene, *Apis mellifica* L. *Zool. Anz.,* 112 : 137- 141.

Wigglesworth, V.B., (1934). The physiology of ecdysis in *Rhodnius prolixus.* II. Factors controlling moulting and metamorphosis. *Quart. J. Micr. Sci.*, 77 : 191-222.

Wigglesworth, V.B. (1935). Function of the corpus allatum in Insects. *Nature*, London, 136 : 338.

Wigglesworth, V.B. (1936). The functions of the corpus allatum in the growth and reproduction of *Rhodnius prolixus* (Hemitera). *Quart : J. Micr. Sci.*, 79 : 91-121.

Wigglesworth, V.B. (1940). The determination of characters at metamorphosis in *Rhodnius prolixus. J. Exp. Biol.*, 17 : 201-222.

Wigglesworth, V.B. (1954). *The Physiology of Insect Metamorphosis.* Cambridge University Press, Cambridge.

Wigglesworth, V.B. (1955). The role of the haemocytes in the growth and moulting of an insect, *Rhodnius prolixus* (Hemiptera). *J. Exp. Biol.* 32: 659-663.

Wgglesworth, V.B. (1961). Some observations on the juvenile hormone effect of farnesol in *Rhodnius prolixus* Stal. (Hemiptera). *J. Insect Physiol.,* 7 : 73-78.

Wigglesworth, V.B. (1963). The action of moulting hormone and juvenile hormone at the cellular level in *Rhodnius prolixus. J. Exp. Biol.*, 40 : 231-245.

Wigglesworth, V.B. (1964). The hormonal regulation of growth and reproduction in insects. *Adv. Insect. Physiol.* 2 : 247- 336.

Wigglesworth, V.B. (1965). *The Principles of Insect Physiology*. ELBS and Metheun and coy Ltd. edition.

Wigglesworth, V.B. (1970). *Insect Hormones*, W.H. Freeman and Company, San Francisco.

Wigglesworth, V.B. (1974). *Insect Physiology*. 7th Edn. Chapman & Hall, London.

Williams, C.M. (1946). Physiology of insect diapause. The role of the brain in the production and termination of pupal dormancy in the giant silkworm, *Platysamia cecropia. Biol. Bull. Woods Hole*, 90 : 234-243.

Williams, C.M. (1947). Physiology of insect diapause. II. Interaction between the pupal brain and prothoracic glands in the metamorphosis of the giant silkworm, *Platysamia cecropia. Biol. Bull. Woods Hole.* 93 : 89-98.

Williams, C.M. (1948). Physiology of insect diapause III. The prothoracic glands in the *Cecropia* silkworm, with special reference to their significance in embryonic and post embryonic development. *Biol. Bull. Woods Hole,* 94 : 60-65.

Williams, C.M. (1961). The juvenile hormone II. Its role in the endocrine control of moulting, pupation and adult development in the Cecropia silkworm. *Biol. Bull. Woods, Hole,* 121 : 572- 585.

Williams, C.M. (1963). The juvenile hormone - III. Its accumulation and storage in the abdomens of certain male moths. *Biol. Bull. Woods Hole,* 124 : 355-367.

Williams, C.M. (1968) Ecdysone and ecdysone- analogues : Their assay and action on diapausing pupae of the cynthia silkworm. *Biol. Bull. Woods Hole*, 134 : 344-355.

Wilson, E.O. and Bossert, W.H. (1963). Chemical communication among animals. *Rec. Prog. Horm. Res.*, 19 : 673-716.

PLATES

EXPLANATION OF FIGURES

Note : Brady and Maddrell (1967) and Baudry-Partiaoglou (1986) in their papers have given the account of the neurohoaemal organs etc. of 3-insect orders viz. - Phasmida, Orthoptera and Dictyoptera (Blattaria), which for the sake of convenience have been included with Dictyoptera (Blattaria).

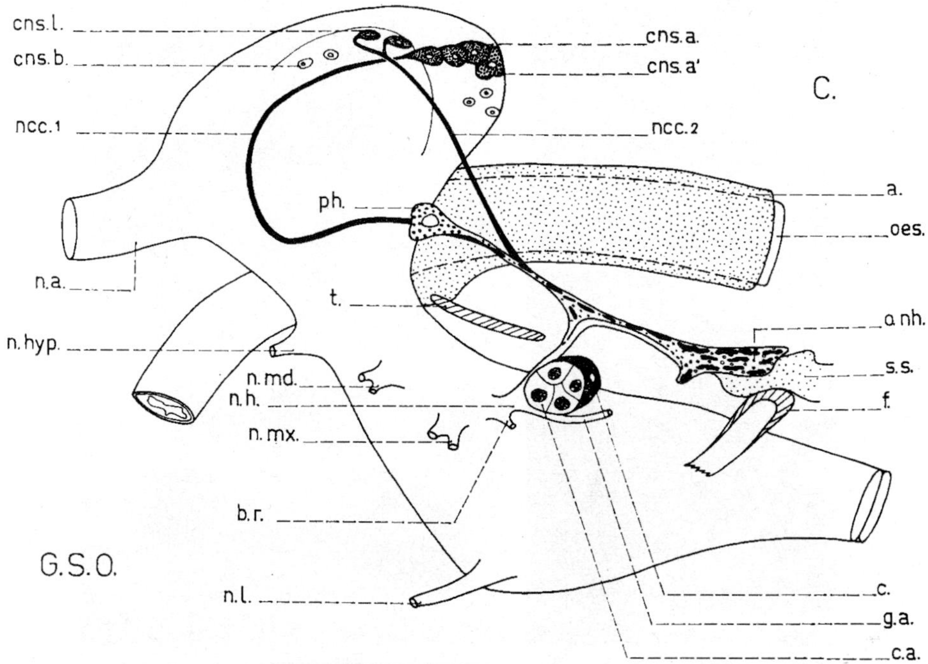

Plate 1. SCHEMA 1. Systeme neuroendocrine d'*Orchesella kervillei,* vue de profil. a. = aorte; b.r. = branche recurrente du nerf d'Hoffmann; C. = cerveau; c. = cellule coiffe; c.a. = corpus allatum; cns a., a', b., l. = cellules neurosecretrices a, a', b, laterales; f. = fulcre; g.a. = branche du nerf d'Hoffmann se rendant a la glande acineuse; G.S.O. = ganglion sous-oesophagien; n.a. = nerf antennaire; ncc.1, ncc.2 = nerfs cardiaques internes, I et externes II ; n.h. = nerf d'Hoffmann ; n.hyp. = nerf hypo-pharyngien; n.l. = nrf labial; n.md. = nerf mandibulaire ; n.m.x. = nerf maxilaire; oes. = oesophage; o.n.h. = organe neurohaemal gauche; ph. = cellule phloxinophile a la sortie du ncc.1; s.s. = sinus sanguin; t = tentorium. (From Cassagnau and Juberthie, 1967)

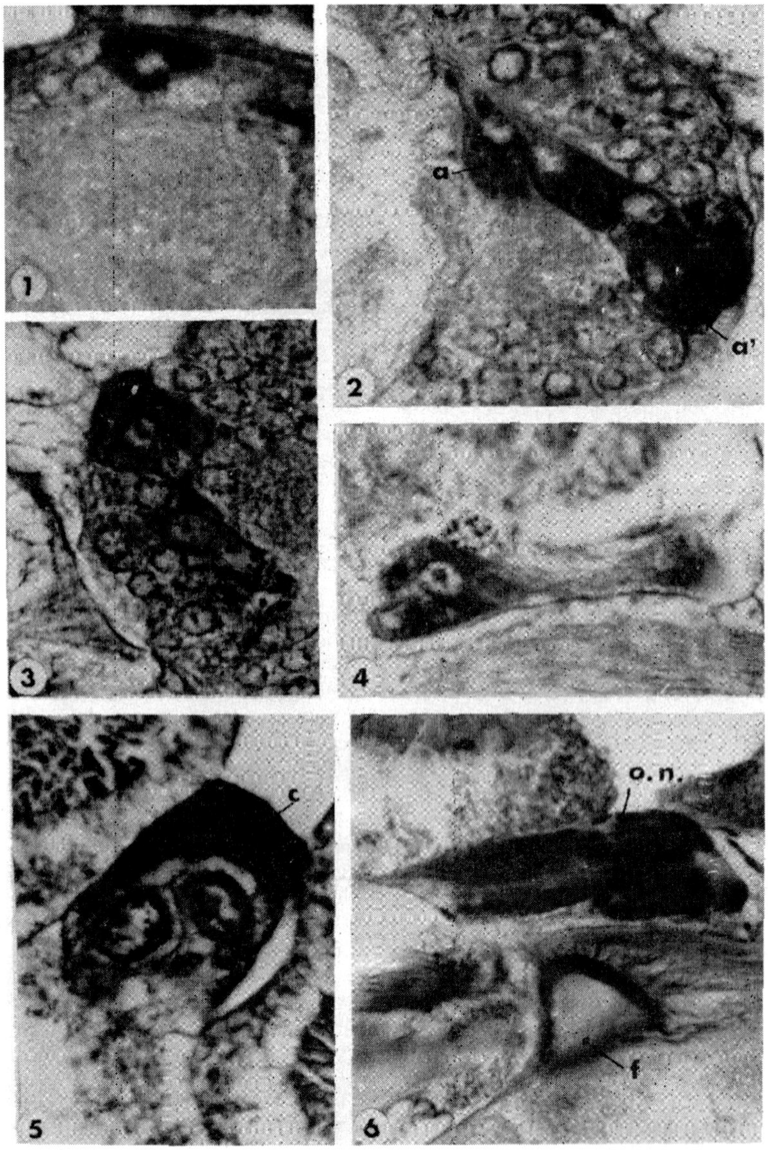

Plate 2. Figs. 1,2,4 *O. kervillei,* Figs. 3,5,6, *O. villosa.* Fig. 1. Cellule neurosecretrice laterale du cerveau en coupe sagittale. X 1600. Fig. 2. Cellules neurosecrtrices nucales a' et a en coupe sagittale. X 1000. Fig. 3. Cellule neurosecretrice nucale a'. X 1000. Fig. 4. Gross cellule phloxinophile au point de sortie du nerf cardiaque. I. X 700. Fig. 5. Corpus allatum avec sa cellule coiffe (c) coloree par le bleu alcian coupe transversale. X 1000. Fig. 6. Organe neurohaemal droit (o.n.) en coupe sagittale; f = fulcre. Bouin, bleu alcian-hemalun-phloxine. X 1000. (From Cassagnau and Juberthie, 1967)

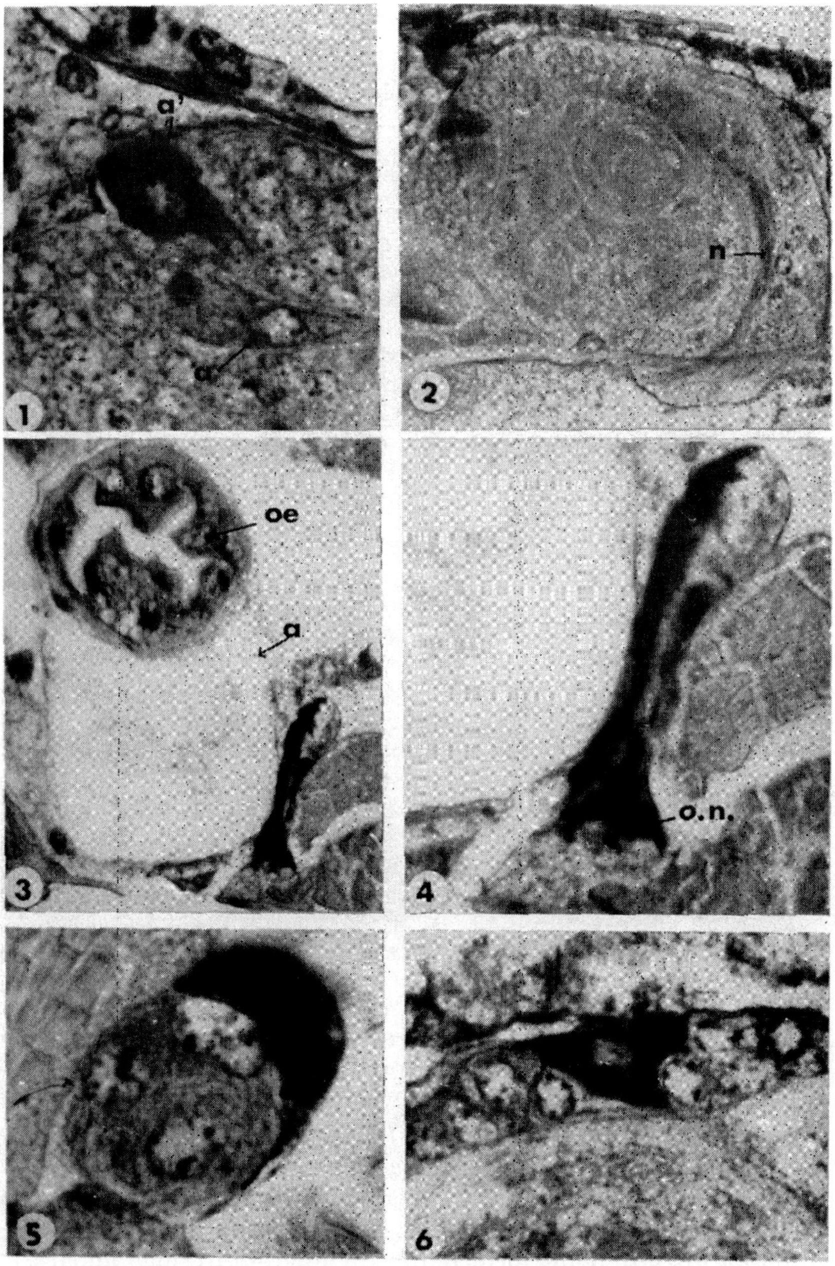

Plate 3. Fig. 1. *Pogonognathellus flavescens.* Cellules neurosecretrices nucales *a'* et a. X 1000. Fig. 2. *Tomocerus minor.* Trajet d'un des nerfs cardiaques internes n dans le cerveau. Coupe sagittale. X 500. Figs 3 et 4. *T. minor.* Situation et structure des organes neurohemaux en coupe transversale; oe, oesophage; a, aorte; o.n., organe neurohaemal. X 600 et X 1500. Fig. 5. *T. minor.* Corpus allatum en coupe transversale, surmonte de sa cellule coiffe (c). X 1500. Fig. 6. *T. minor.* Cellule neurosecretrice laterale en coupe sagittale. Bouin, bleu alcian-hemalun-phloxine. X 1200. (From Cassagnau and Juberthie, (1967)

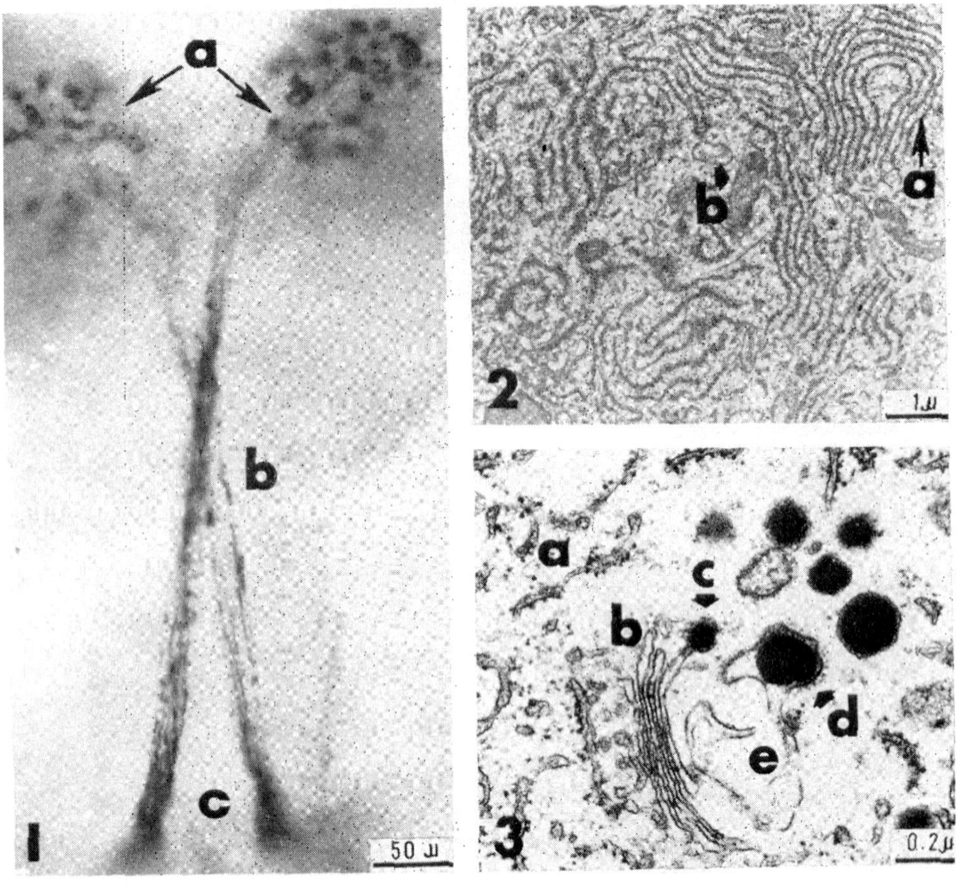

Plate 4. Fig. 1. Brain of *Leucophaea maderae* after 4 days *in vitro*. Victoria blue stainable nsm has accumulated in the perikarya of the mnc (a) and in the fibre tracts (b). The distal flow of the material is evident from the heavy accumulation (c) where the nervi corporis cardiaci were transected. Fig. 2. Stacked arrays (a) of rer are apparent in an incubated brain after 3 days *in vitro*. Structural integrity of mitochondria (b) indicates active metabolism. Fig. 3. After 3 days *in vitro*, neurosecretory granules are still being formed. Rer (a) adjacent to Golgi apparatus (b) presumably provides material for the formation of nsg. An immature nsg (c) is visible within a Golgi tubule and a mature nsg (d) is receiving its bounding membrane as it leaves the distal cisterna (e). (From Marks *et al.*, 1973)

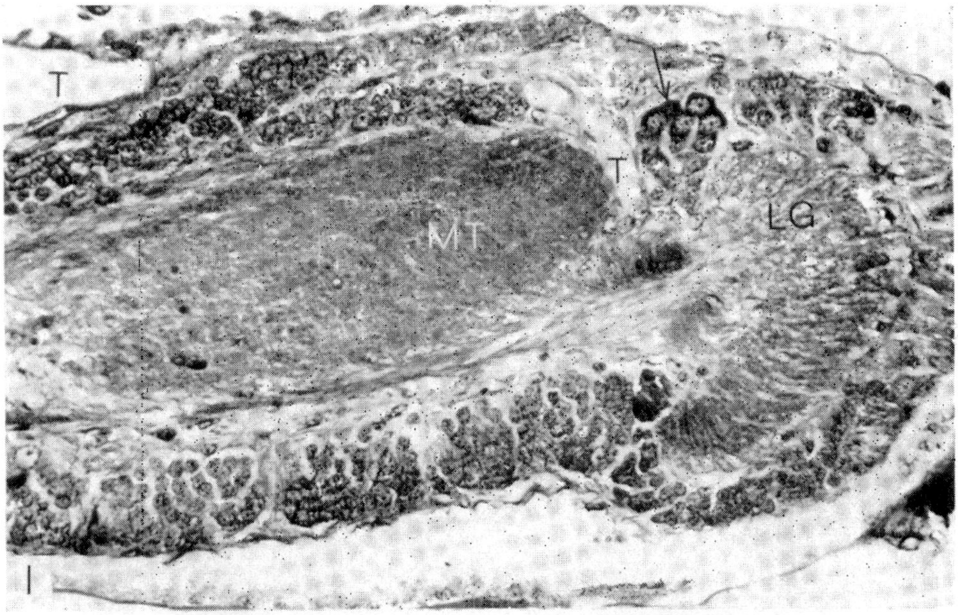

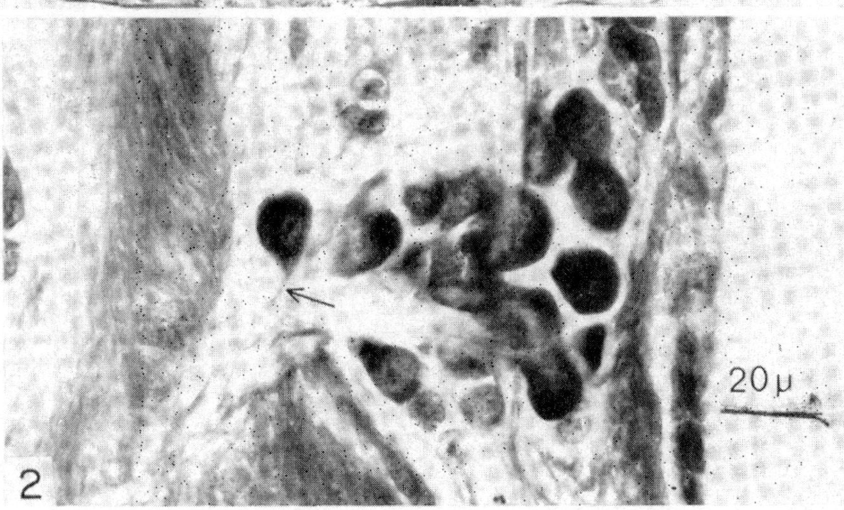

Plate 5. Fig. 1. Horizontal section of the optic lobe of *Periplaneta* showing the neurosecretory cells (arrow) in relation to the medulla terminalis (MT), lamina ganglionaris (LG), and the main trachea (T). Azan technique. Fig. 2. Neurosecretory cells in the optic lobe of *Periplaneta*. An axon (arrow) can be seen running towards the interior of the optic lobe. Azan technique (From Beattie, 1971).

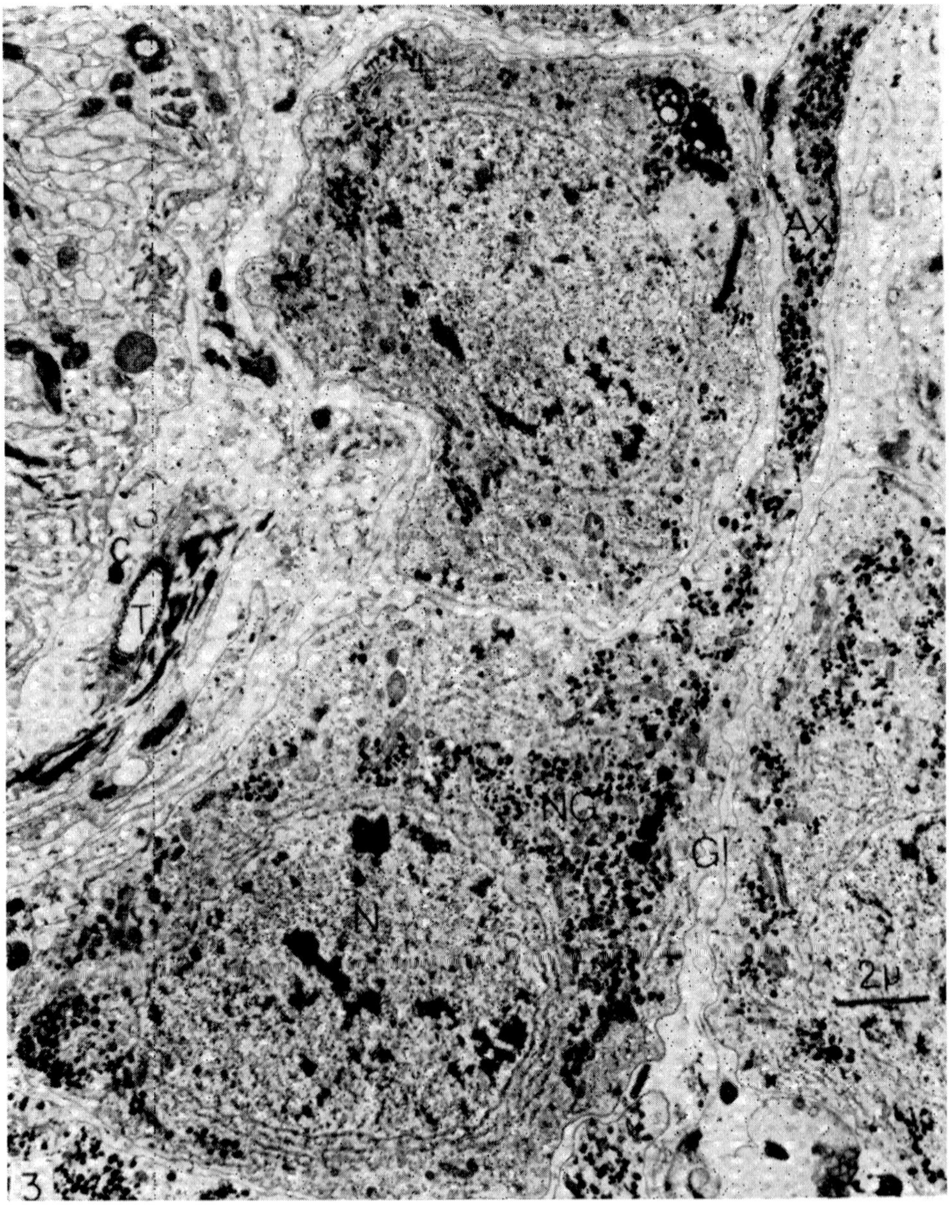

Plate 6. Fig. 3. Electron micrograph (of *Periplaneta*) showing numerous neurosecretory granules (NG) in the cell bodies. The axon (Ax) of one of these cells can be seen running between the other cells. Each nerve cell is surrounded by glial tissue (GI). Trachea (T) are usually present close to the neurosecretory cells (From Beattie, 1971)

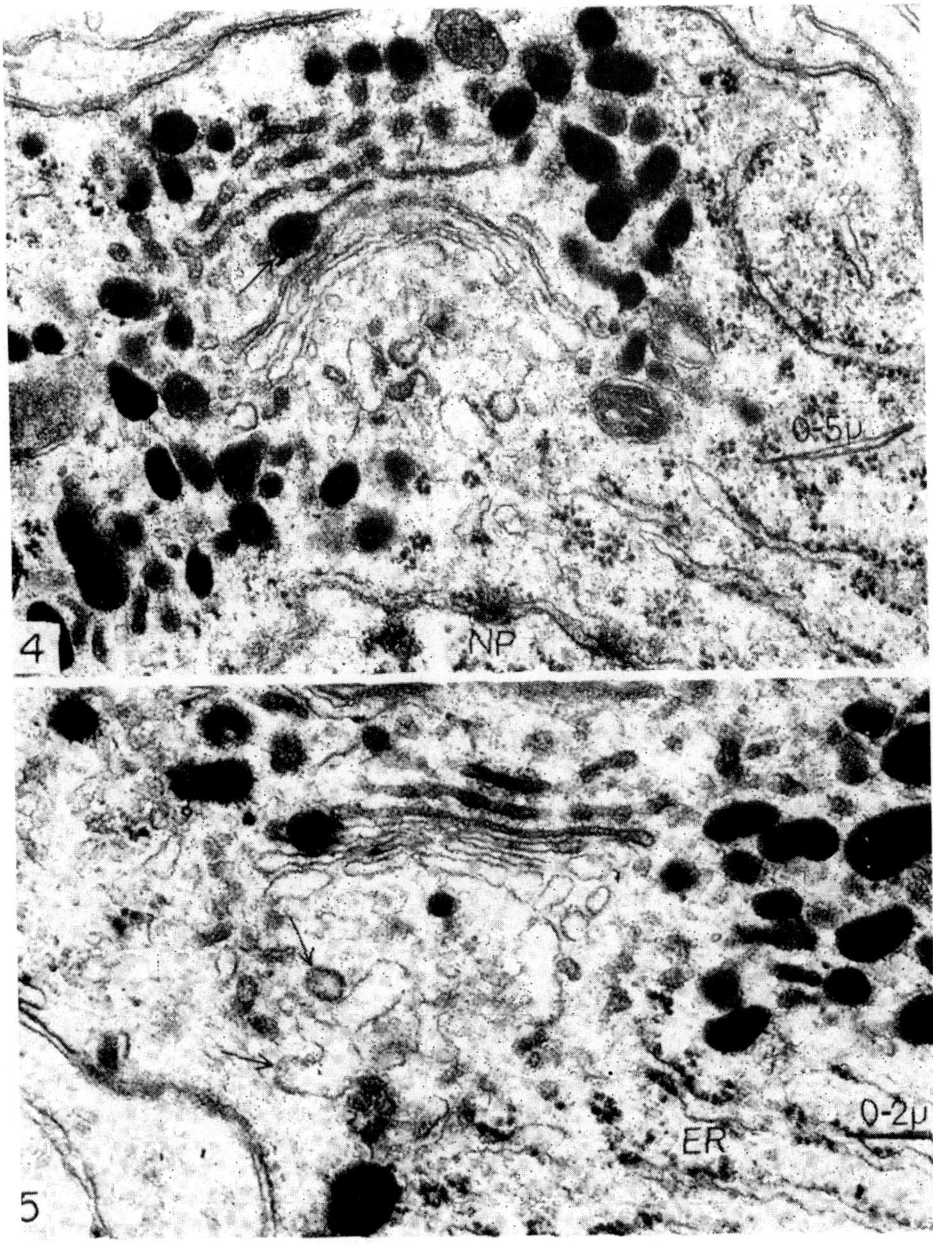

Plate 7. Fig. 4. The Golgi bodies show a marked polarization in their structure. A neurosecretory granule (arrow) can be seen forming at the periphery of the Golgi body. A nuclear pore (NP) is present in the nuclear membrane. (*Periplaneta*) Fig. 5. Another Golgi body showing the apparent continuity between it and the endoplasmic reticulum (ER). Finger-like projections (arrows) from irregular-shaped profiles show an accumulation of dense material. (*Periplaneta*) (From Beattie, 1971)

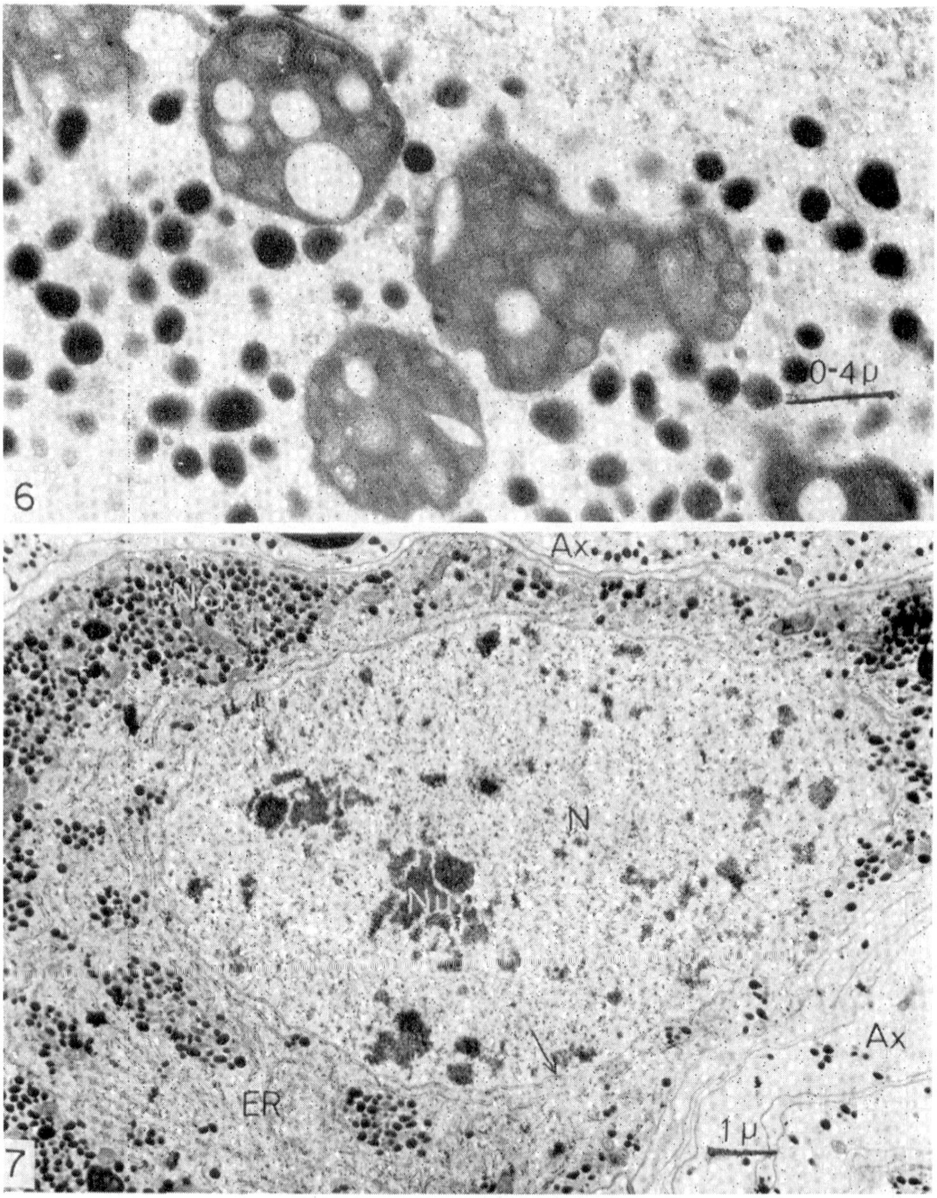

Plate 8. Fig. 6 A group of dense bodies which are possibly autophagic lysosomes. Each body is made up of electron-translucent units and multilaminated units. (*Periplaneta*) Fig. 7. A neurosecretory cell body showing nucleolar material (Nu) within the nucleus (N). A nuclear pore is indicated by the arrow. The endoplasmic reticulum (ER) is fairly well developed and numerous neurosecretory granules (NG) are present. Several axons (Ax) from other neurosecretory cells can be seen passing by this cell. (*Periplaneta*) (From Beattie, 1971)

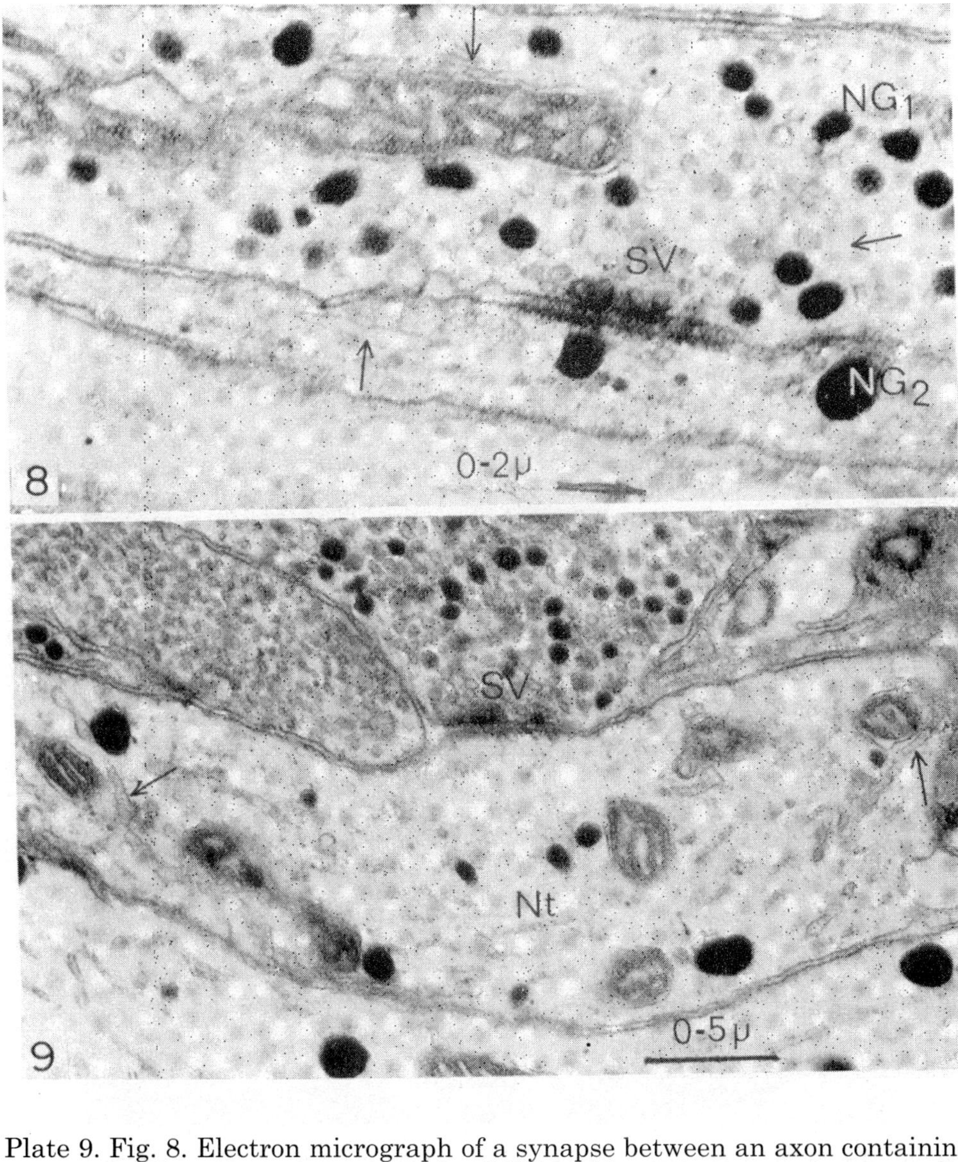

Plate 9. Fig. 8. Electron micrograph of a synapse between an axon containing small neurosecretory granules (NG_1) and large nuerosecretory granules (NG_2). Synaptic vesicles (SV) are present on the pre-synaptic side of the synapse. Tubular endoplasmic reticulum (arrows) is present in both pre-and post-synaptic axons and is often associated with mitochondria (M). (*Periplaneta*) Fig. 9. Similar to Fig. 8, Neurotubules (Nt) and branching tubular endoplasmic reticulum (arrows) are present in the post-synaptic axon. SV, synaptic vesicles. (*Periplaneta*) (From Beattie, 1971).

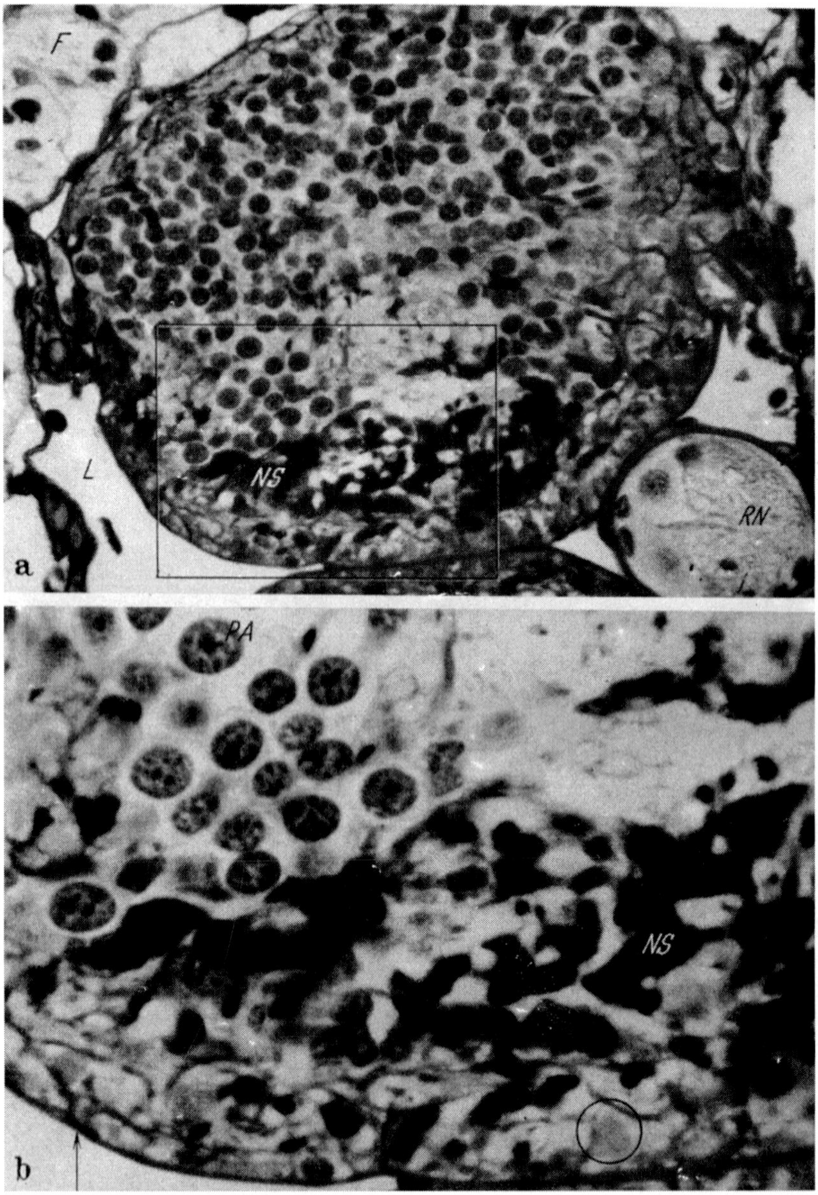

Plate 10. Fig. 1a and b. Photomicrographs of corpus cardiacum of *Leucophaea maderae*, BOUIN, GOMORI's chrome haematoxylin-phloxin. a. cross section of right corpus cardiacum (dorsal surface on left side of figure). Cell bodies mostly in lateral portion of organ; accumulations of extrinsic neurosecretory material in medial portion; intrinsic secretory material mainly near ventral and medial surfaces; nuclei of intrinsic gland cells ovoid, those of interstitial (``glial'') elements irregular in shape. X 400. b. Area indicated in a at higher magnification, showing intrinsic cells and two types of secretory material. X 1000. (*F*) cells of fat body; (*L*) lumen of dorsal vessel; (*NS*) extrinsic neurosecretory material; (*PA*) parenchymal (intrinsic neuroglandular) cell; (*RN*) recurrent nerve; circle- secretory material of intrinsic gland cell; arrow sheath of corpus cardiacum with processes entering corpus cardiacum (From Scharrer, 1963)

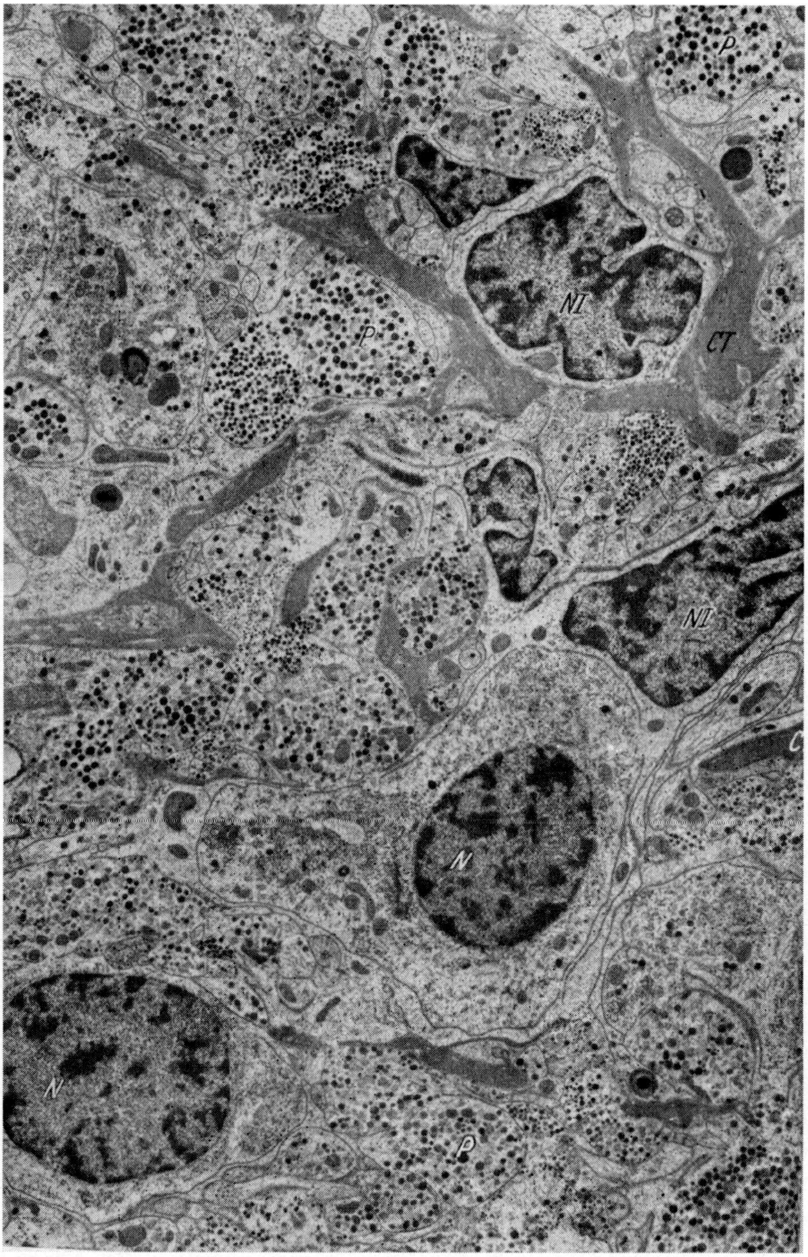

Plate 11. Fig. 2. Low power electron micrograph of the corpus cardiacum of an old normal female of *Leucophaea* showing two types of cell bodies; numerous profiles of cellular processes (*P*) with various types and amounts of granular inclusions; and acellular connective tissue partitions (*CT*). The nuclei (*N*) of the parenchymal (intrinsic neuroglandular) cells are ovoid, those of the glia-like interstitial cells (*NI*) are polymorphic (T) tracheole. X 5,000 (From Scharrer, 1963).

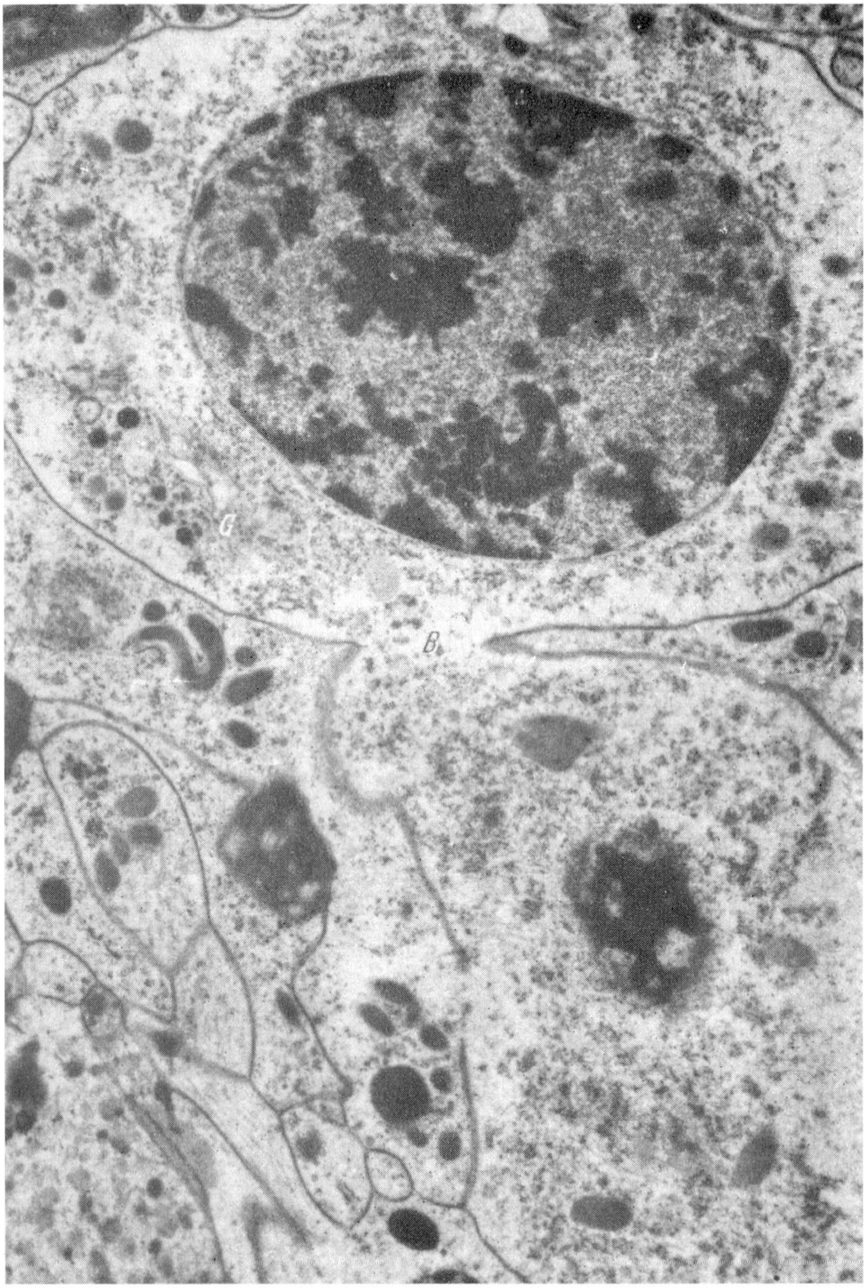

Plate 12. Fig. 3. Parts of two adjacent parenchymal cells connected by cytoplasmic bridge, (*B*) of *Leucophaea* The tangentially cut nucleus of cell at bottom shows nuclear pores. The perikaryon of cell at top contains electron-opaque secretory granules in the Golgi zone (*G*), numerous ergastoplasmic elements close to the nuclear envelope, and some mitochondria. X 11,200 (From Scharrer, 1963)

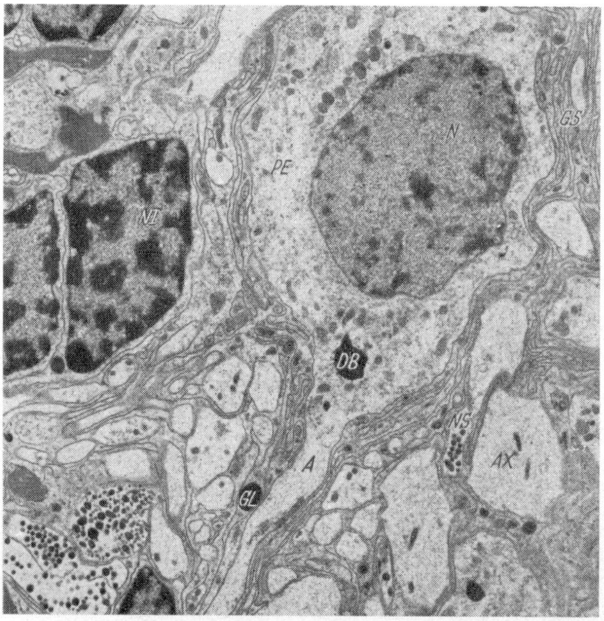

Plate 13. Fig. 4. Parenchymal cell (not actively secreting) from posterior region of corpus cardiacum of freshly moulted seventh instar female nymph of *Leucophaea*. Note interdigitating layers of ``glial sheath'' (*GS*) around perikaryon (*PE*) and axon-like cell process (*A*). The latter contains ``neurotubules'' and an elongate mitochondrion. (*AX*) axon in cross section; (*DB*) dense body; (*GL*) ``gliosome''; (*NS*) small neurosecretory fibre; (*N*) nucleus of parenchymal cell; (*NI*) chromatin-rich nucleus of interstitial (glia-like) cell. X 6,250 (From Scharrer, 1963).

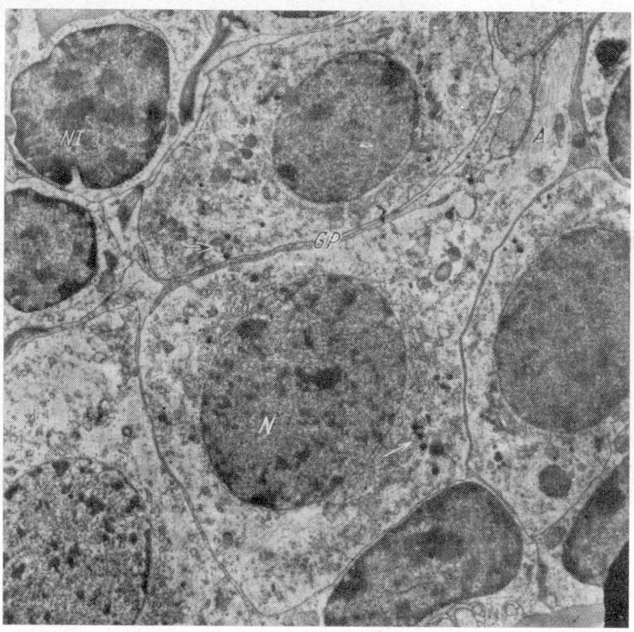

Plate 14. Fig. 5. Group of parenchymal cells from corpus cardiacum of a normal adult (preovulatory) female, *Leucophaea*. Perikaryon and axon-like process (*A*) contain ``neurotubules''. Electron-opaque secretory granules indicated by arrows. (*GP*) tenuous layer of ``glioplasm'' separating adjacent parenchymal cells. (*N*) nucleus of parenchymal cell; (*NI*) nucleus of interstitial cell surrounded by thin coat of cytoplasm. X 4,700 (From Scharrer, 1963)

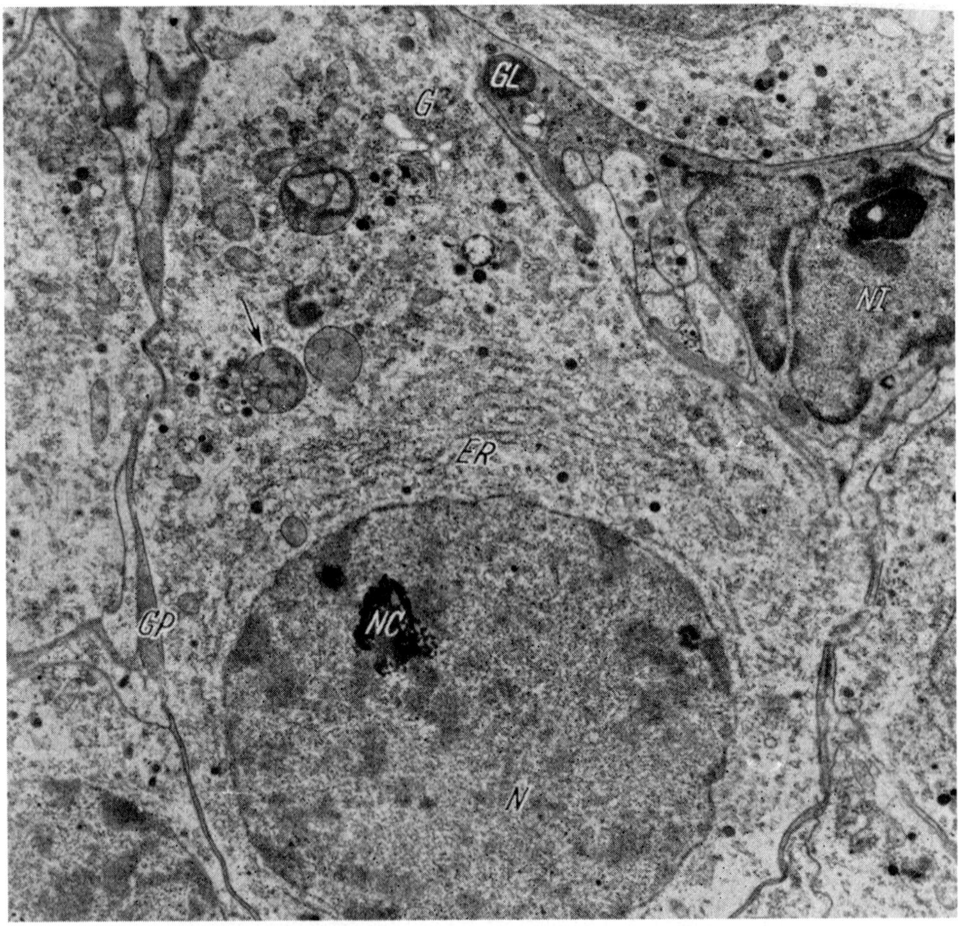

Plate 15. Fig. 6. Pear-shaped parenchymal cell of corpus cardiacum of female *Leucophaea* with cystic ovaries (adult age over 16 months). Note supranuclear ergastoplasm (*ER*), Golgi zone (*G*) with electron-opaque granules, large cytoplasmic inclusion bodies (arrow) seen in more detail in Fig. 12. (*GL*) ``gliosome''; (*GP*) ``glioplasm''; (*N*) nucleus of parenchymal cell; (*NC*) nucleolus; (*NI*) lobated nucleus of interstitial cell with characteristic nucleolus. X 6,800 (From Scharrer, 1963)

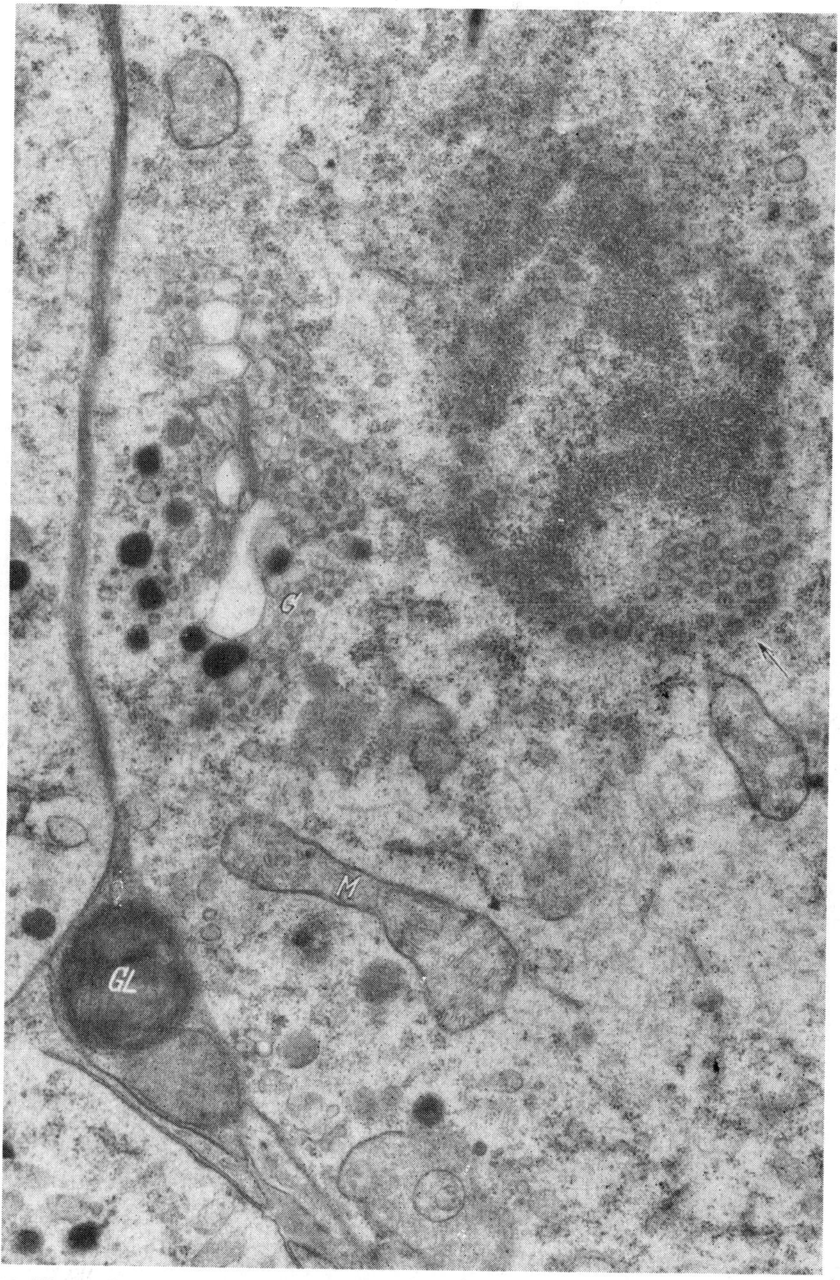

Plate 16. Fig. 7. Part of parenchymal cell of normal female *Leucophaea* (over one year of adult age). The tangentially cut nucleus clearly shows nuclear pores (arrow). Electron-opaqe secretory granules lie within the Golgi zone (*G*), other areas contain scattered elements of ergastoplasm and free ribosomes, ``neurotubules", and mitochondria (*M*), (*GL*) ``gliosome", occupying triangular space between three adjacent parenchymal cells. X 23,600 (From Scharrer, 1963).

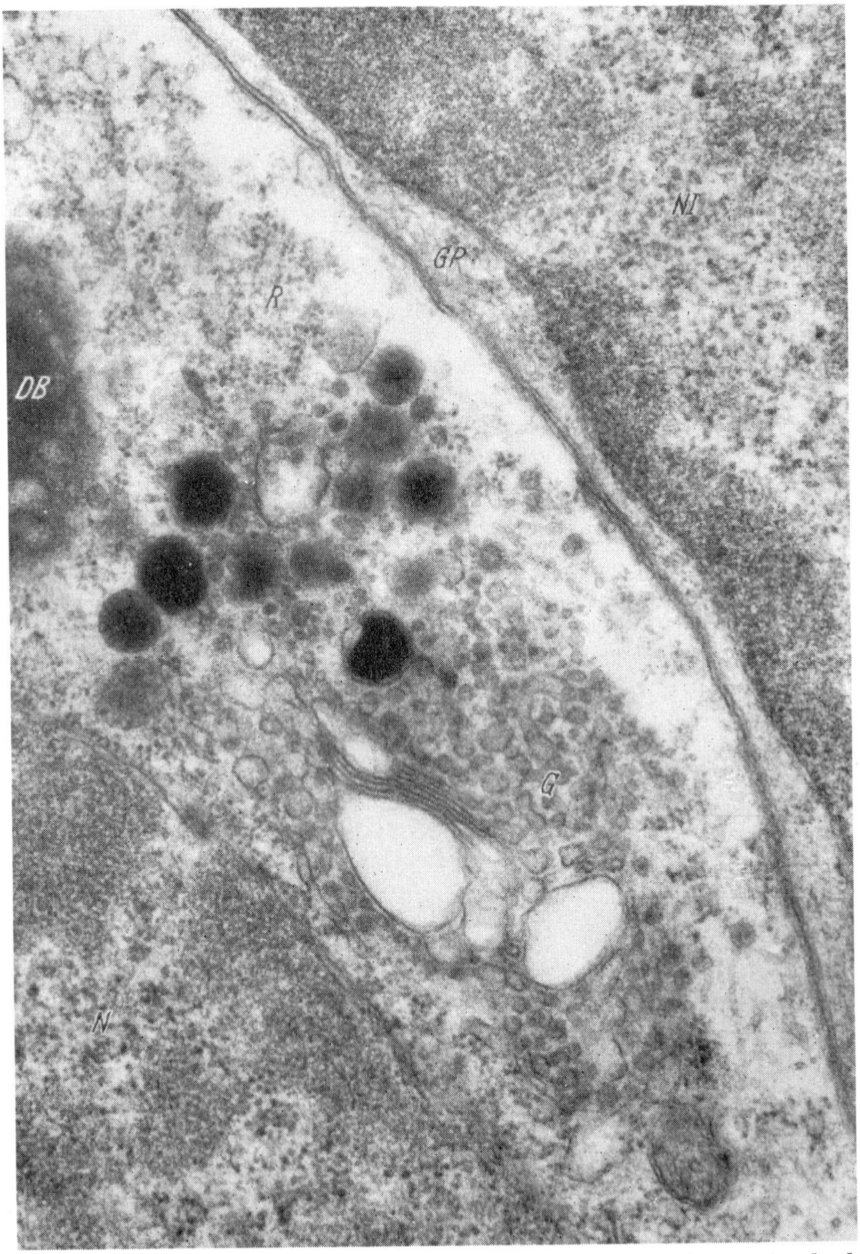

Plate 17. Fig. 8. Golgi zone (*G*) of parenchymal cell of old ovariectomized adult of *Leucophaea*. It consists of small vesicles plus stacks of smooth-surface cisternae and vacuoles. Note intimate spatial relationship with electron-opaque secretory granules some of which show distinct limiting membranes. (*DB*) dense body; (*GP*) thin layer of perinuclear "glioplasm" facing parenchymal cell; (*N*) nucleus of parenchymal cell; (*NI*) nucleus of interstitial cell; (R) free ribosomes. X 42,400 (From Scharrer, 1963)

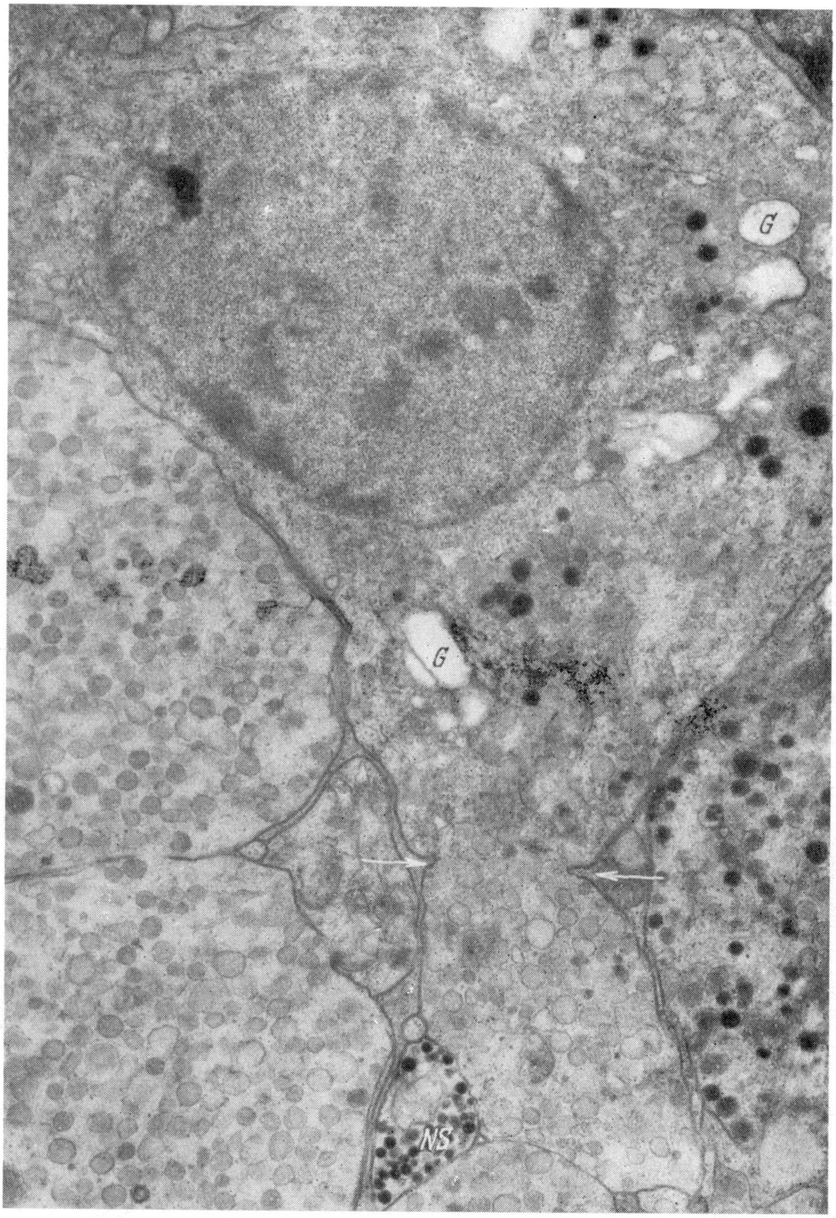

Plate 18. Fig. 9. Parenchymal cell from old pregnant female *Leucophaea*. The plump cellular process, set off from the perikaryon by a constriction (arrows), contains numerous pale granules with membranes. The same type of granule also occurs in the perikaryon intermingled with electron-opaque elements. Large Golgi vacuoles (*G*) are conspicuous. Parts of adjacent cells contain many of the same pale granules along with some darker elements. Together these represent the intrinsic secretory material of the corpus cardiacum. *(NS)* small neurosecretory fibre. X 13,200 (From Scharrer, 1963)

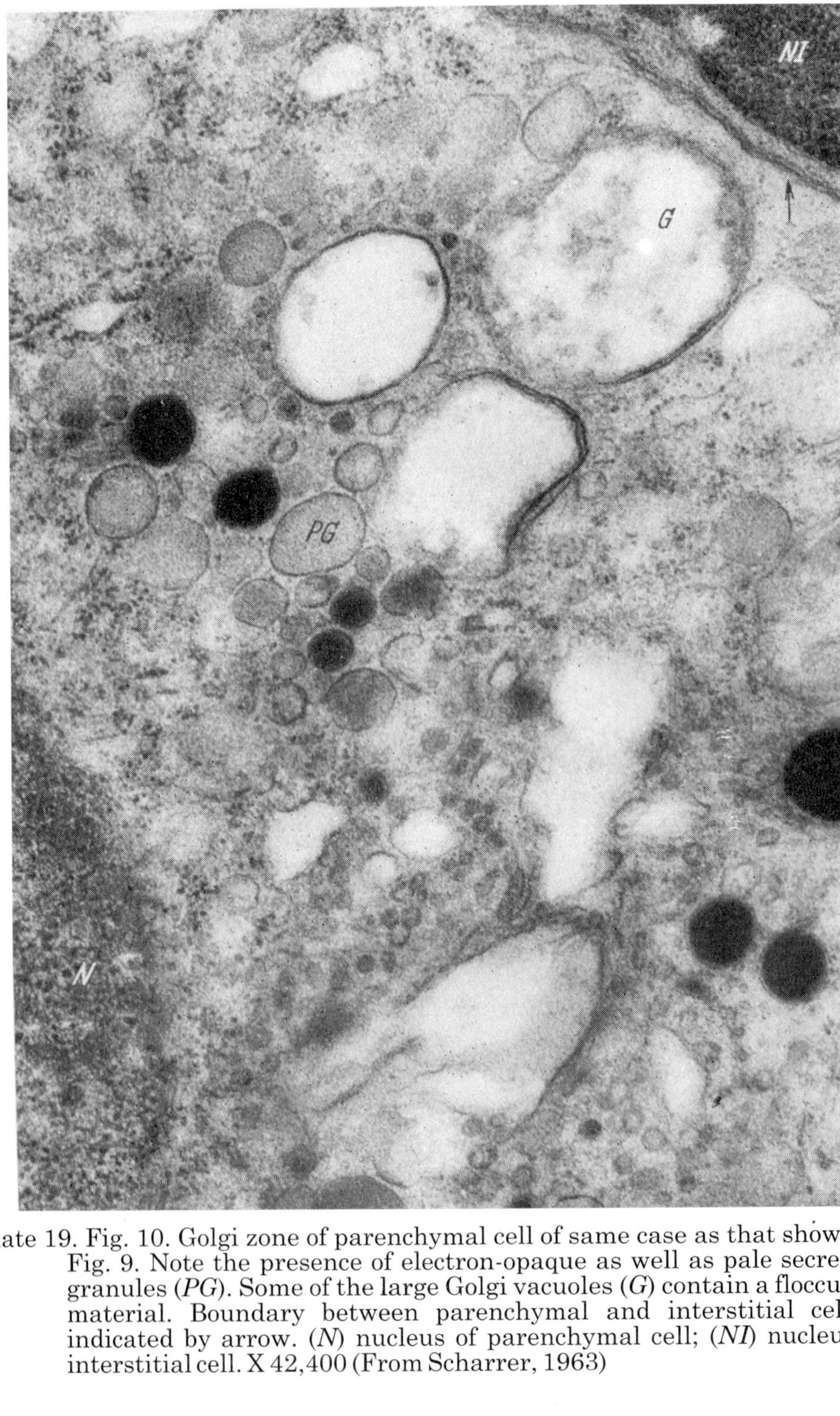

Plate 19. Fig. 10. Golgi zone of parenchymal cell of same case as that shown in Fig. 9. Note the presence of electron-opaque as well as pale secretory granules (*PG*). Some of the large Golgi vacuoles (*G*) contain a flocculent material. Boundary between parenchymal and interstitial cell is indicated by arrow. (*N*) nucleus of parenchymal cell; (*NI*) nucleus of interstitial cell. X 42,400 (From Scharrer, 1963)

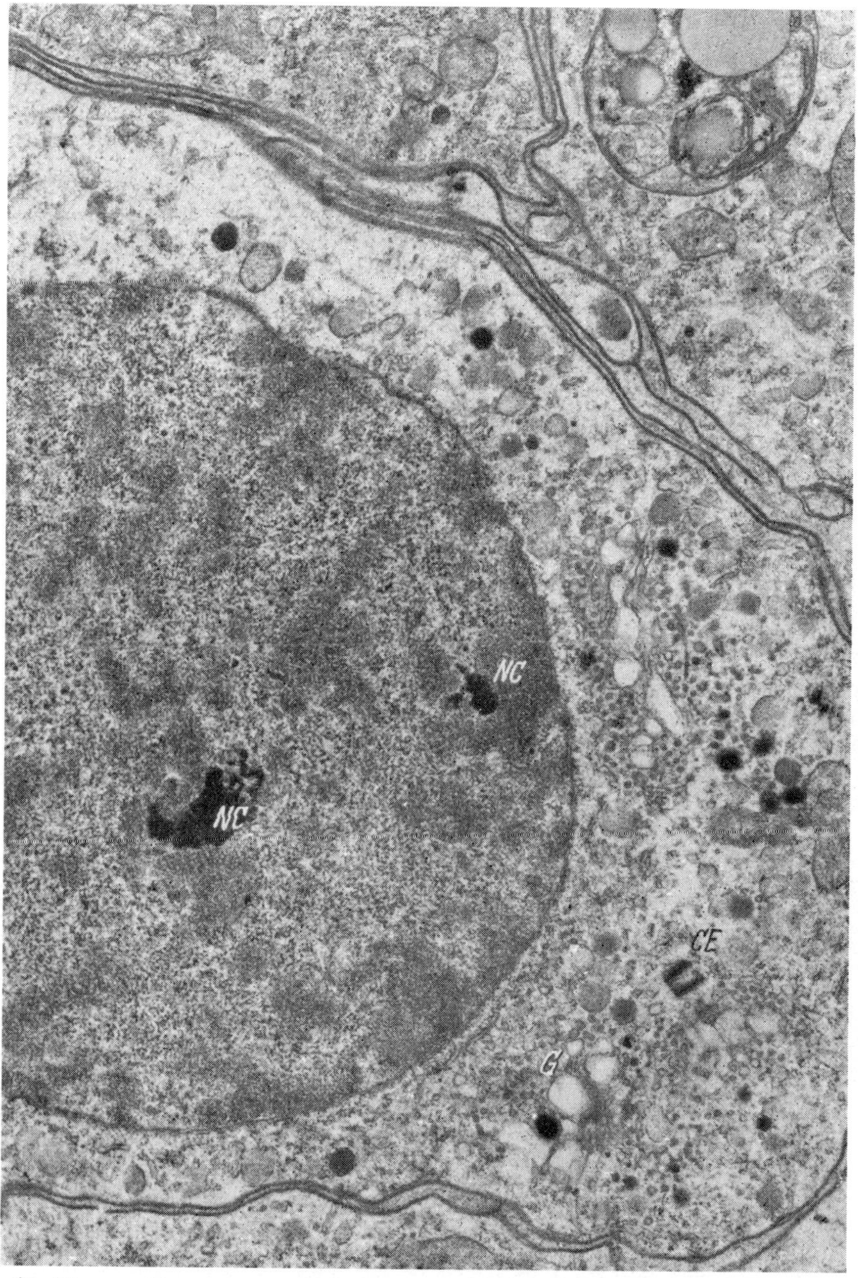

Plate 20. Fig. 11. Parenchymal cell of *Leucophaea* showing centriole (*CE*), Golgi zone (*G*) with secretory granules of varying electron-opacity, some rough-surfaced cytomembranes, and mitochondria. Parts of two nucleoli (*NC*) are visible. The cell is separated from adjacent parenchymal cells by thin layers of "glioplasm" Another cell at upper right contains an inclusion body shown in detail in Fig. 12. X 17,200 (From Scharrer, 1963)

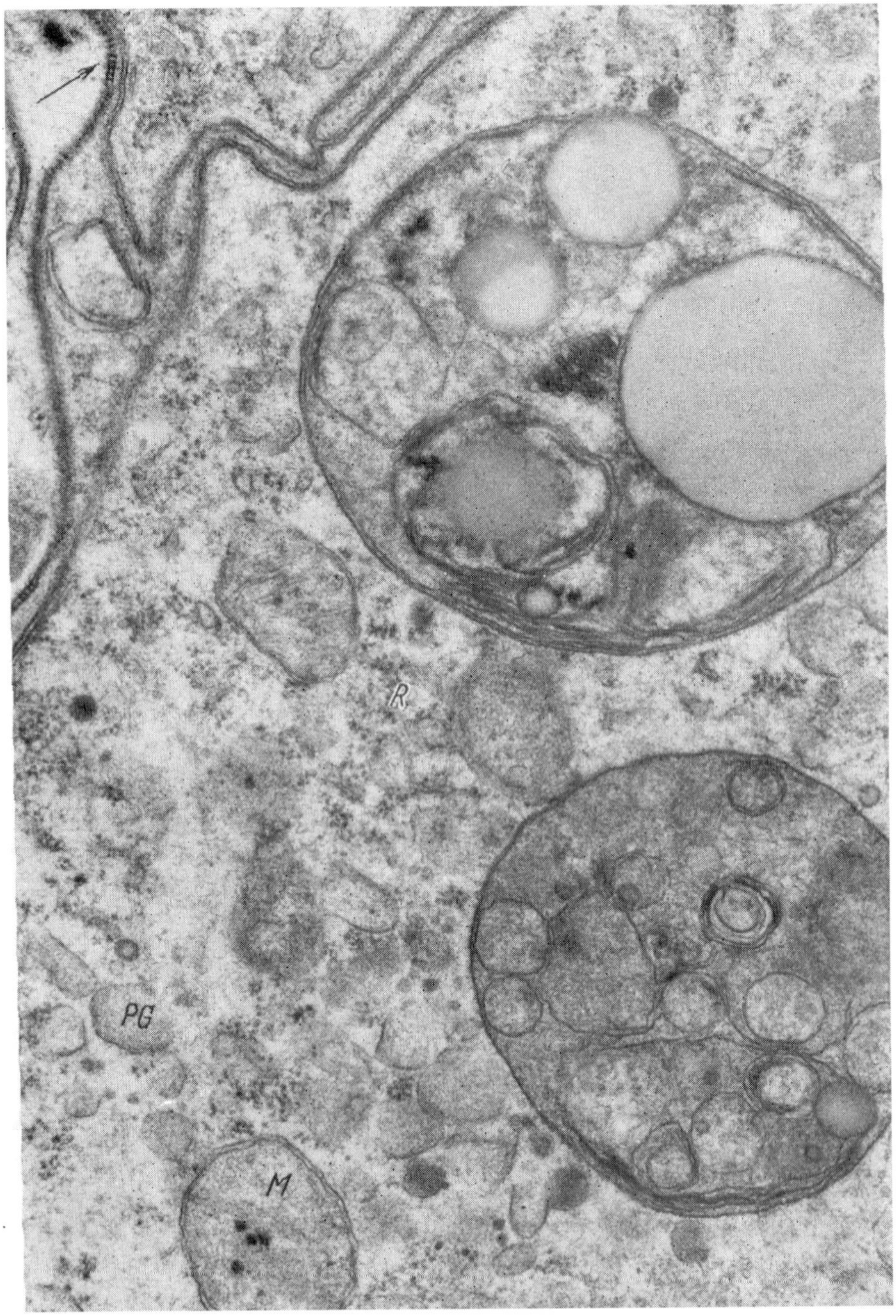

Plate 21. Fig. 12. Part of a parenchymal cell, seen also in Fig. 11. Note two large inclusion bodies part of whose contents seems derived from mitochondria. (*M*) mitochondrion with dark granules in matrix; (*PG*) pale secretory granule; (R) free ribosomes; arrow - septate demosome. X 42,400 (From Scharrer, 1963).

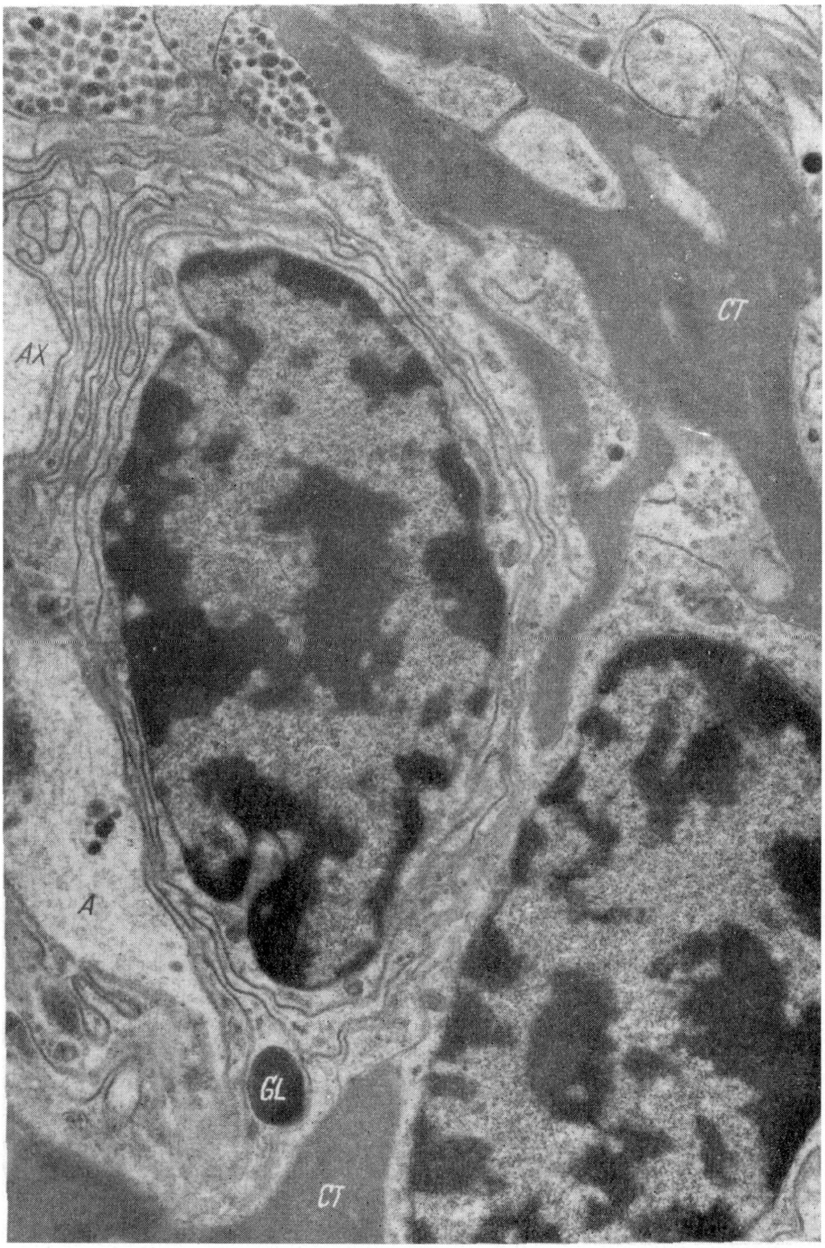

Plate 22. Fig. 13. Two interstitial ``glial'' cells of *Leucophaea* whose perinuclear cytoplasm is characteristically scanty. Next to the cell on the left are several interdigitating ``glial'' membranes. The nuclei are fairly regular in shape and rich in chromatin. *(A)* axon with few secretory granules, and a cluster of small vesicles; *(AX)* axon in cross section with ``neurotubules'' cut perpendicularly to main axis; *(CT)* stroma consisting of extracellular connective tissue projections from the corpus cardiacum sheath; *(GL)* ``gliosome''. X 13,200 (From Scharrer, 1963).

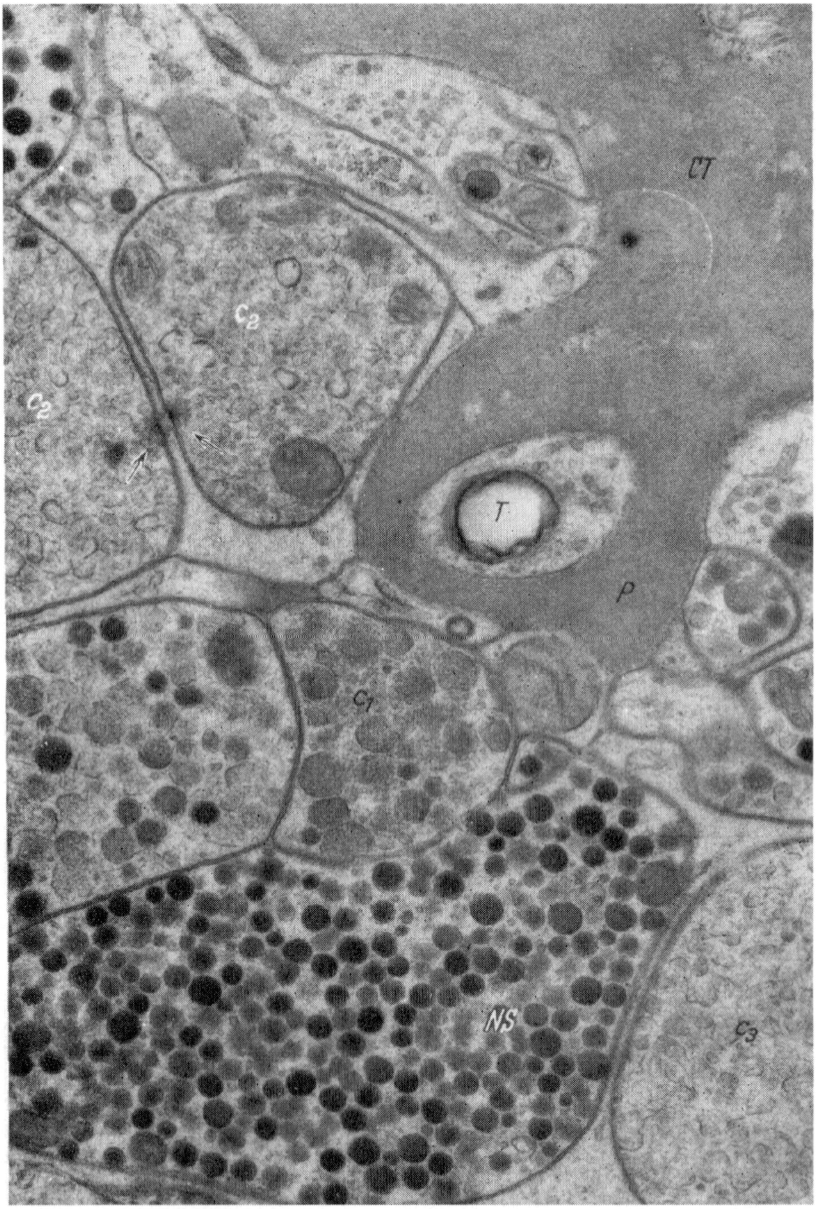

Plate 23. Fig. 14. Area close to the surface of the corpus cardiacum of an old castrated female, *Leucophaea*, showing various types of cellular processes. A neurosecretory fibre (*NS*) with predominantly electron-opaque granules may be contiguous with processes of intrinsic glandular (parenchymal) cells. Among the latter, some contain fairly intact pale granules with internal structure (C_1), others are filled with broken elements whose contents are irregularly spread out (C_2, C_3). The two processes designated as C_2 show a pair of symmetrical ``synapses" (arrows) facing a thin cytoplasmic (probably axonal) partition. (*CT*) connective tissue sheath with short process (*P*) a lacuna of which harbours a tracheole (*T*) X 23,600 (From Scharrer, 1963)

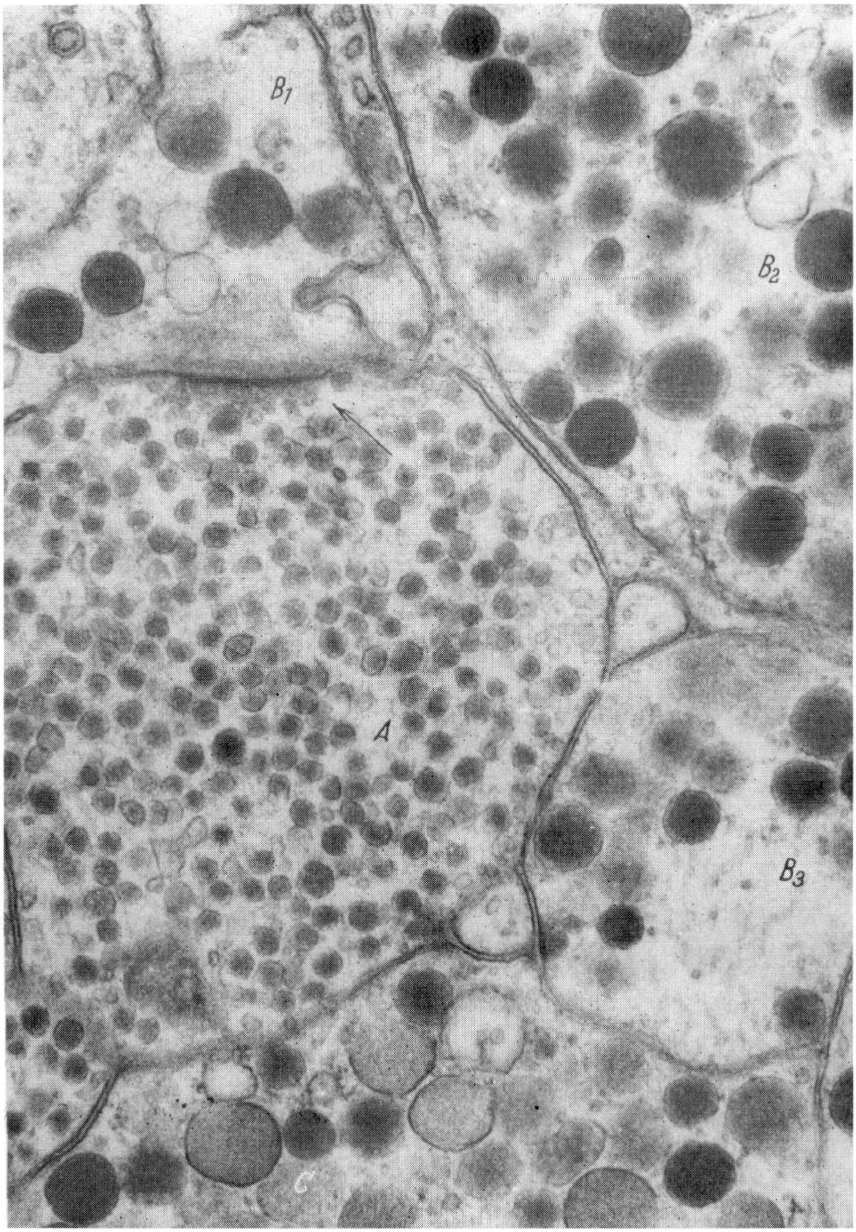

Plate 24. Fig. 15. Different area from corpus cardiacum of same animal as in Fig. 14. An axon terminal *(A)* filled with electron-opaque granules of small average diameter seems to be in synaptic contact (arrow) with another neurosecretory fibre (B_1) that contains dark granules of greater diameter as well as small and large empty vesicles. Other fibres of this second type also contain small vesicles (B_2) and ``neurotubules" (B_3). The pale granules with dark membranes in process *(C)* characterize it as one belonging to a parenchymal cell. X 42,400 (From Scharrer, 1963).

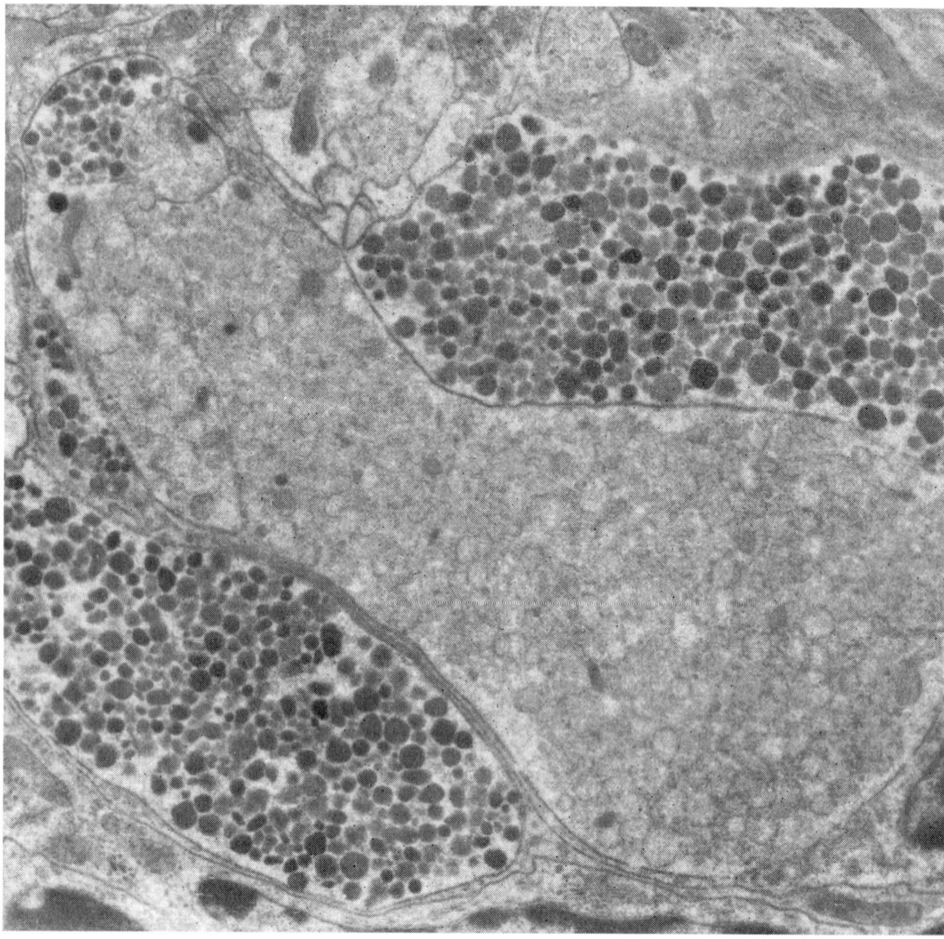

Plate 25. Fig. 16. Two types of adjacent cellular processes from the corpus cardiacum of a freshly moulted seventh instar nymph of *Leucophaea*. The fibres with electron-opaque granules seem to be derived from extrinsic sources, the one in the centre has the typical pale and irregular content of an intrinsic glandular element. X 11200 (From Scharrer, 1963).

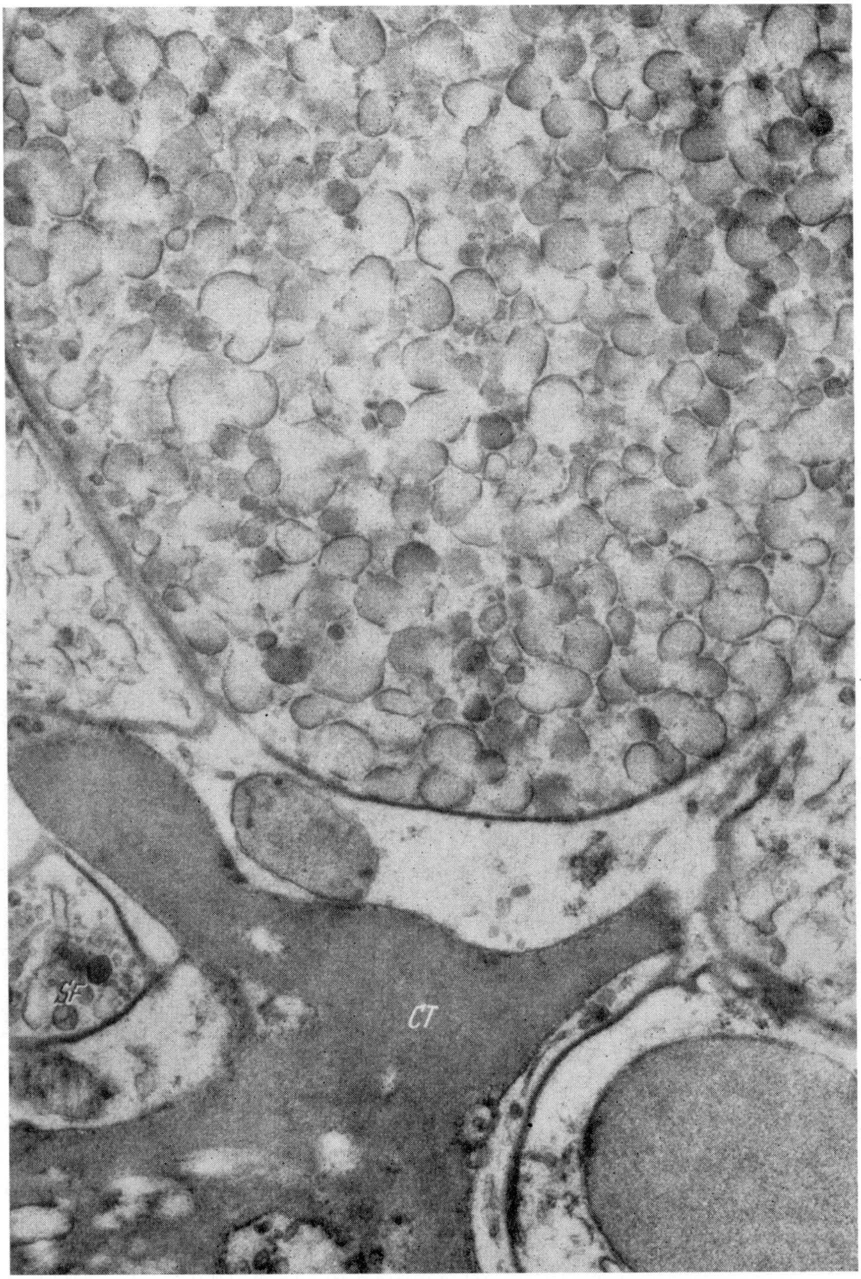

Plate 26. Fig. 17. Bulbous process of a parenchymal cell of *Leucophaea*. The densely packed secretory product consists of pale units many of which are no longer intact. After breaking the distinct dark membranes acquire multiple shapes and the contents of the granules are partially extruded. *(CT)* connective tissue component; *(SF)* small fibre with ``synaptic'' vesicles. X 23,600 (From Scharrer, 1963).

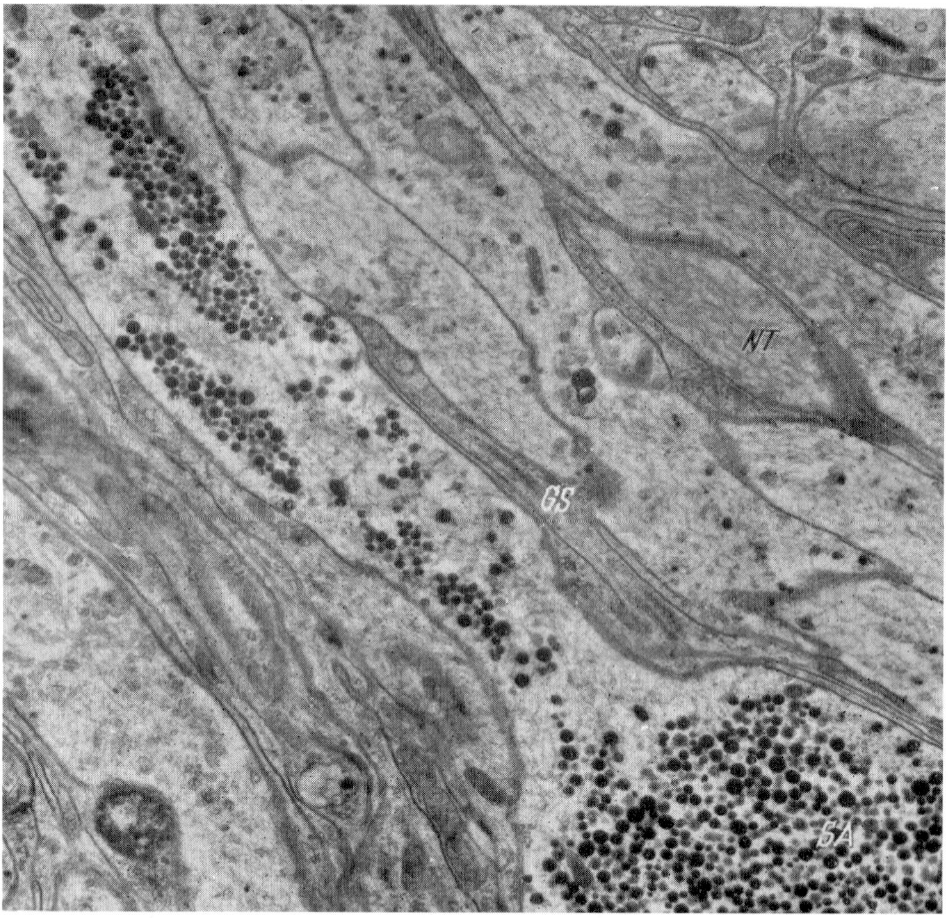

Plate 27. Fig. 18. Longitudinal section through a neurosecretory axon (presumably part of a nervus corporis cardiaci) within the corpus cardiacum of a newly emerged female of *Leucophaea*. Note electron-opaque neurosecretory granules which are particularly abundant in the swollen portion of the axon (*SA*), ``neurotubules'' and occasional mitochondria. Adjacent axons carry almost no neurosecretory material but are rich in ``neurotubules'' (*NT*). Not all parts of the fibres are separated from each other by ``glial'' membranes (*GS*). X 9,000 (From Scharrer, 1963).

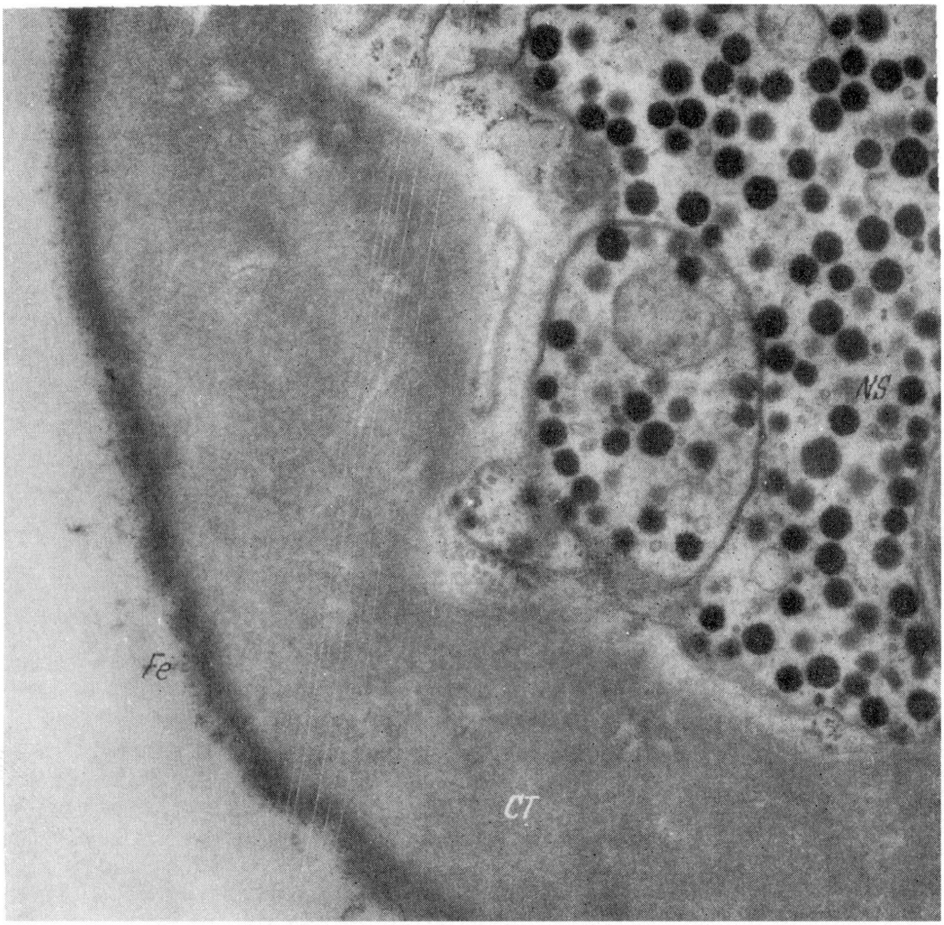

Plate 28. Fig. 19. Surface view of corpus cardiacum of *Leucophaea* with terminals of extrinsic neurosecretory neurons (*NS*). Numerous small and an occasional large empty vesicle lie among typical electron-opaque secretory granules. The latter apparently do not pass through the connective tissue sheath (*CT*) in intact form. Faintly visible collagenous filaments are embedded in the matrix of the sheath material. The distal layer of small black particles (*Fe*) is extraneous; the animal had been injected with saccharated iron oxide before fixation. X 24,400 (From Scharrer, 1963)

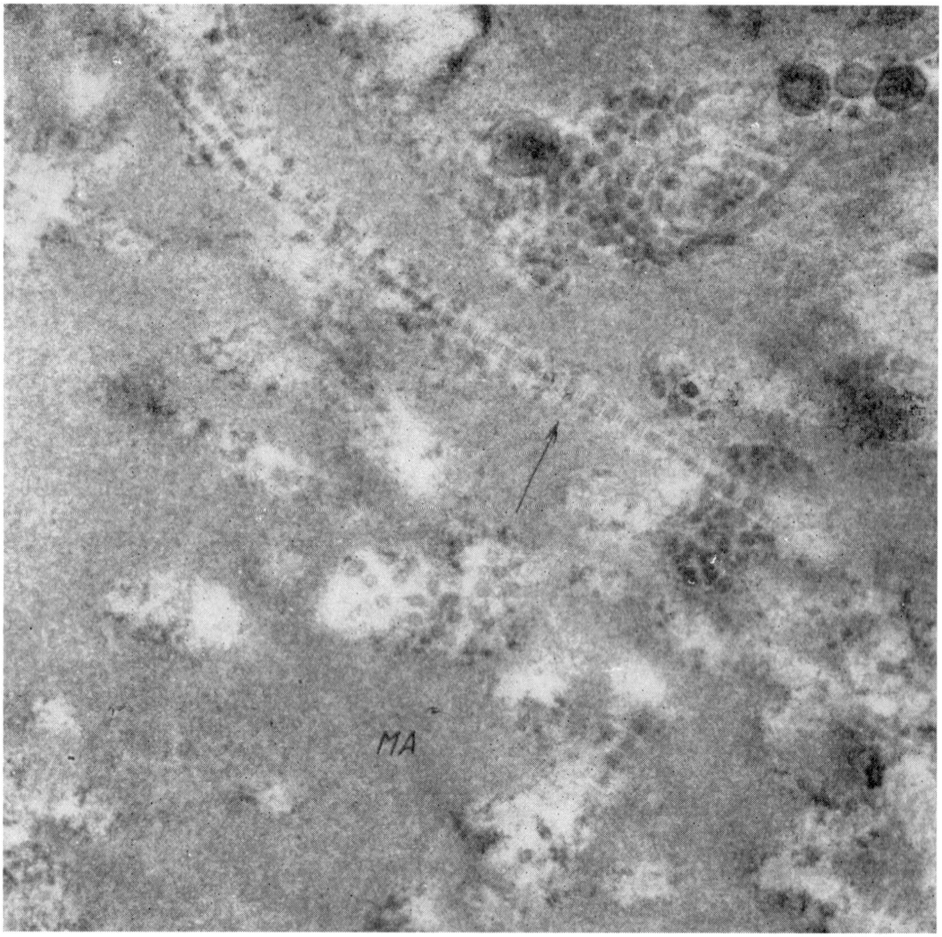

Plate 29. Fig. 20. Part of connective tissue sheath of corpus cardiacum of old female castrate *Leucophaea* showing ``collagenous'' component (arrow) and perforations of materix (*MA*) that contain tubular and/or vesicular elements. X 42,400 (From Scharrer, 1963)

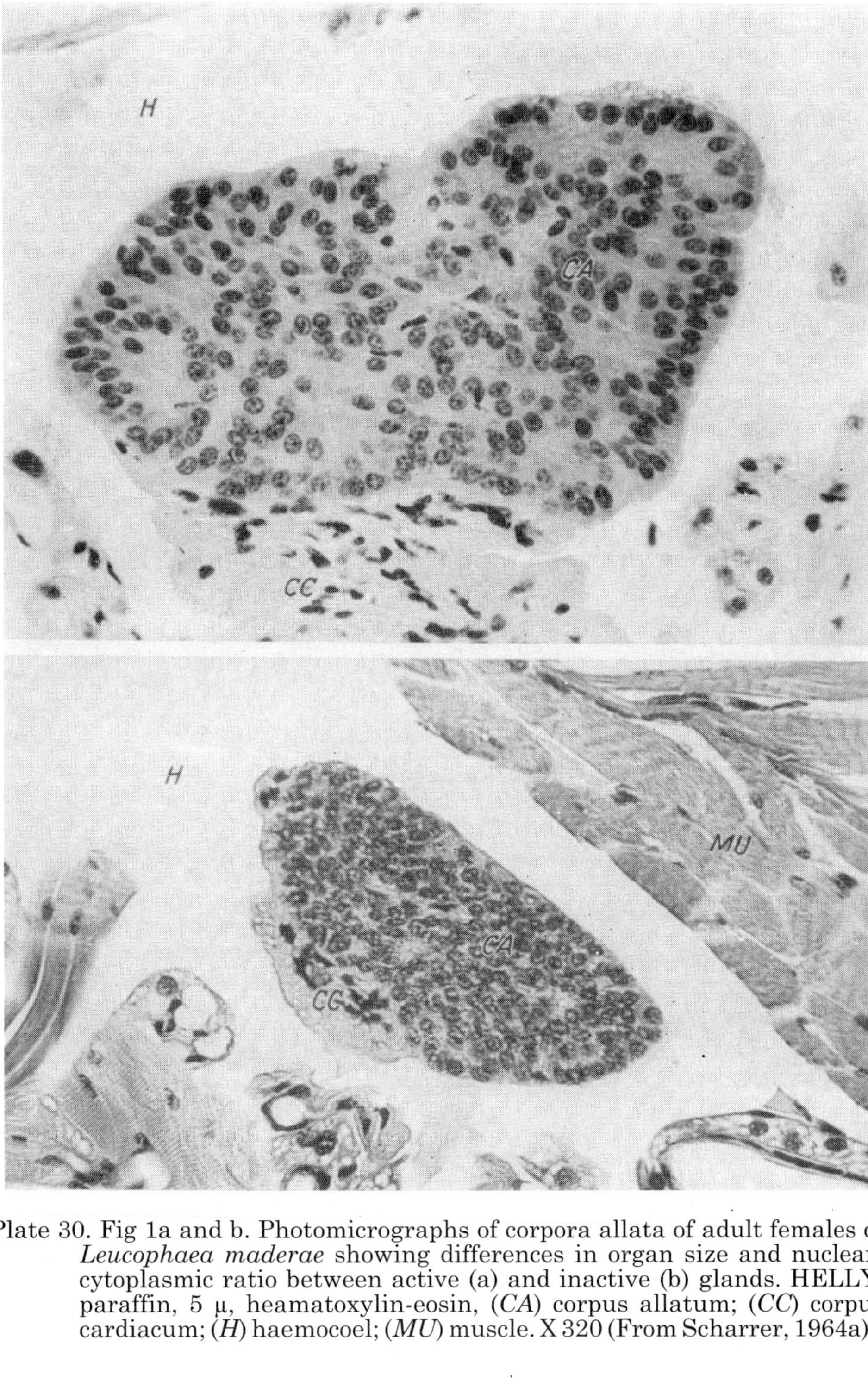

Plate 30. Fig 1a and b. Photomicrographs of corpora allata of adult females of *Leucophaea maderae* showing differences in organ size and nuclear-cytoplasmic ratio between active (a) and inactive (b) glands. HELLY, paraffin, 5 μ, heamatoxylin-eosin, (*CA*) corpus allatum; (*CC*) corpus cardiacum; (*H*) haemocoel; (*MU*) muscle. X 320 (From Scharrer, 1964a).

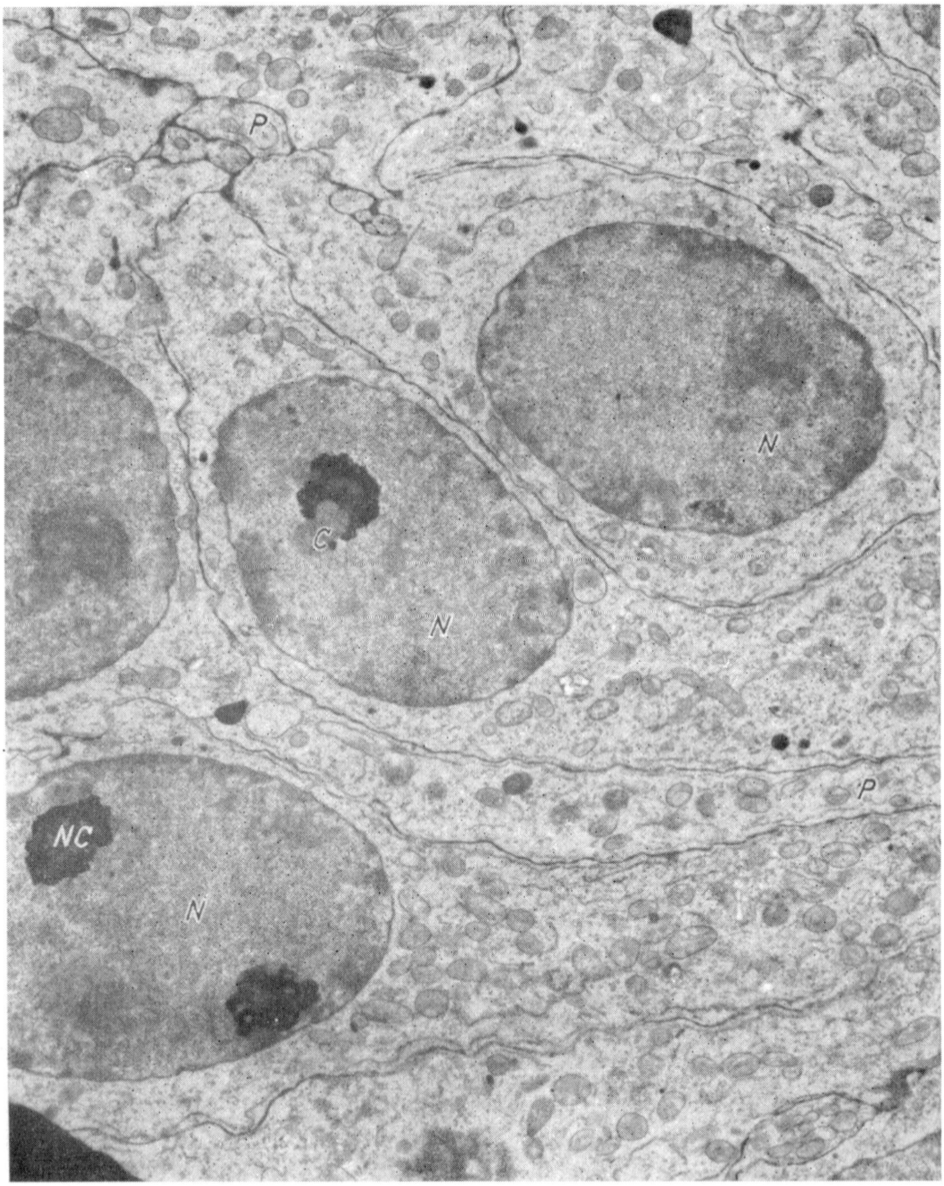

Plate 31. Fig.2 Electron micrograph of active corpus allatum cells of an adult female *Leucophaea* whose oocytes measured 4 mm. One of the ovoid nuclei (*N*) contains two nucleoli (*NC*), another a single nucleolus in close spatial relation to chromatin material (C). Profiles of cellular processes (*P*) in longitudinal and cross section are present. Parallel to the distinct cell boundaries are long ergastoplasmic profiles. In addition to endoplasmic reticulum and free ribosomes, the cytoplasm contains mitochondria, Golgi elements, and electron-dense inclusions that are considered to be lysosomes. X 7,000 (From Scharrer, 1964a)

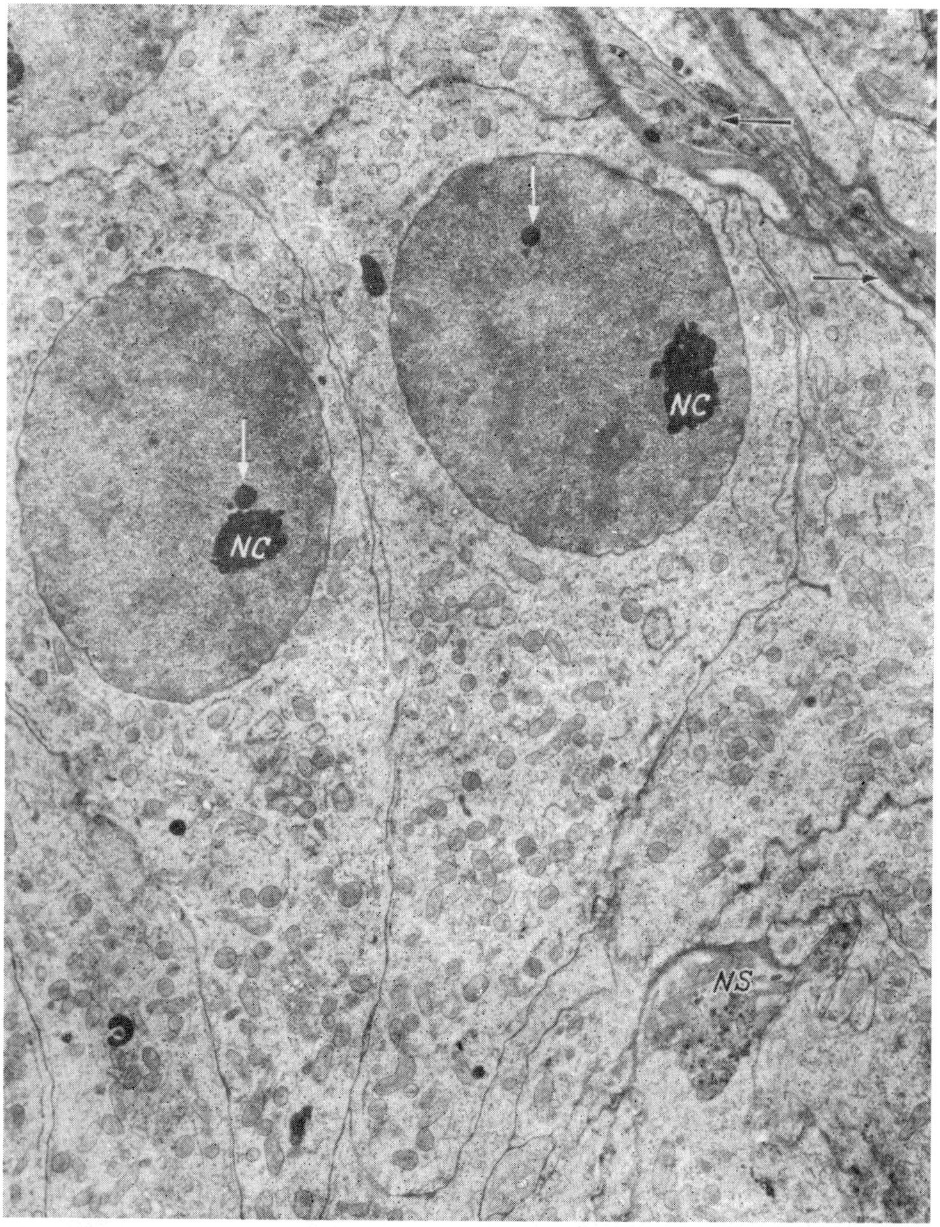

Plate 32. Fig. 3. Two adjacent cells of an active corpus allatum of *Leucophaea* showing ample amounts of cytoplasm and organelles comparable to those in Fig. 2. Each nucleus contains one nucleolus (*NC*) and one spherical dense body (white arrow) of unknown nature. A small neurosecretory fibre (*NS*) lies wedged between corpus allatum cells, separated from them merely by plasma membranes; several others are accompanied by `stroma" (black arrows). X 5, 500 (From Scharrer, 1964a)

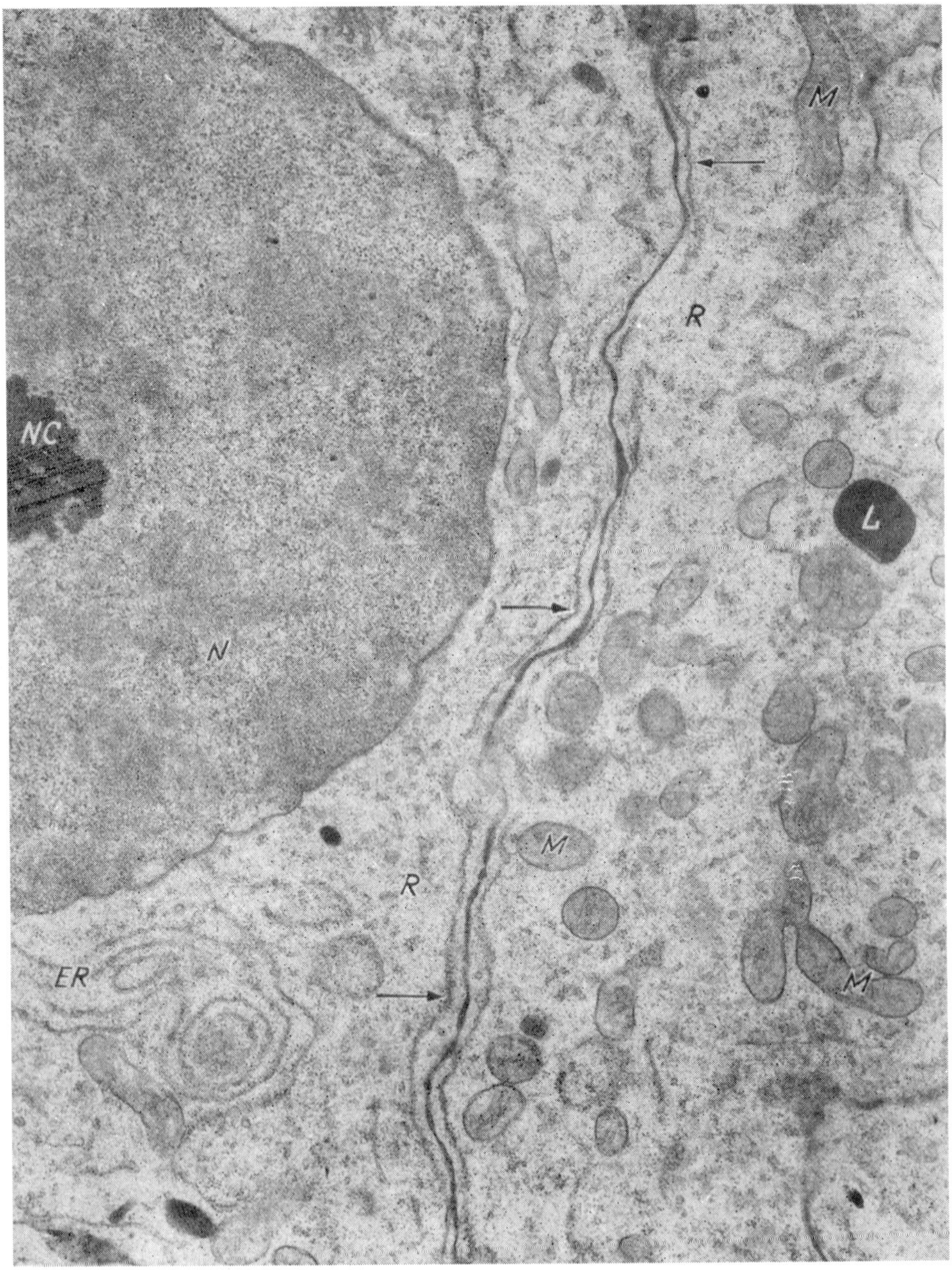

Plate 33. Fig. 4. Parts of adjacent cells of active corpus allatum of *Leucophaea*. (*ER*) small double whorl of ergastoplasm; (*L*) lysosome (?); (*M*) mitochondria (one of them Y-shaped); (*N*) nucleus; (*NC*) nucleolus; (R) free ribosomes. Arrows : long profiles of ergastoplasm running parallel to cell boundaries. X 15, 600 (From Scharrer, 1964a)

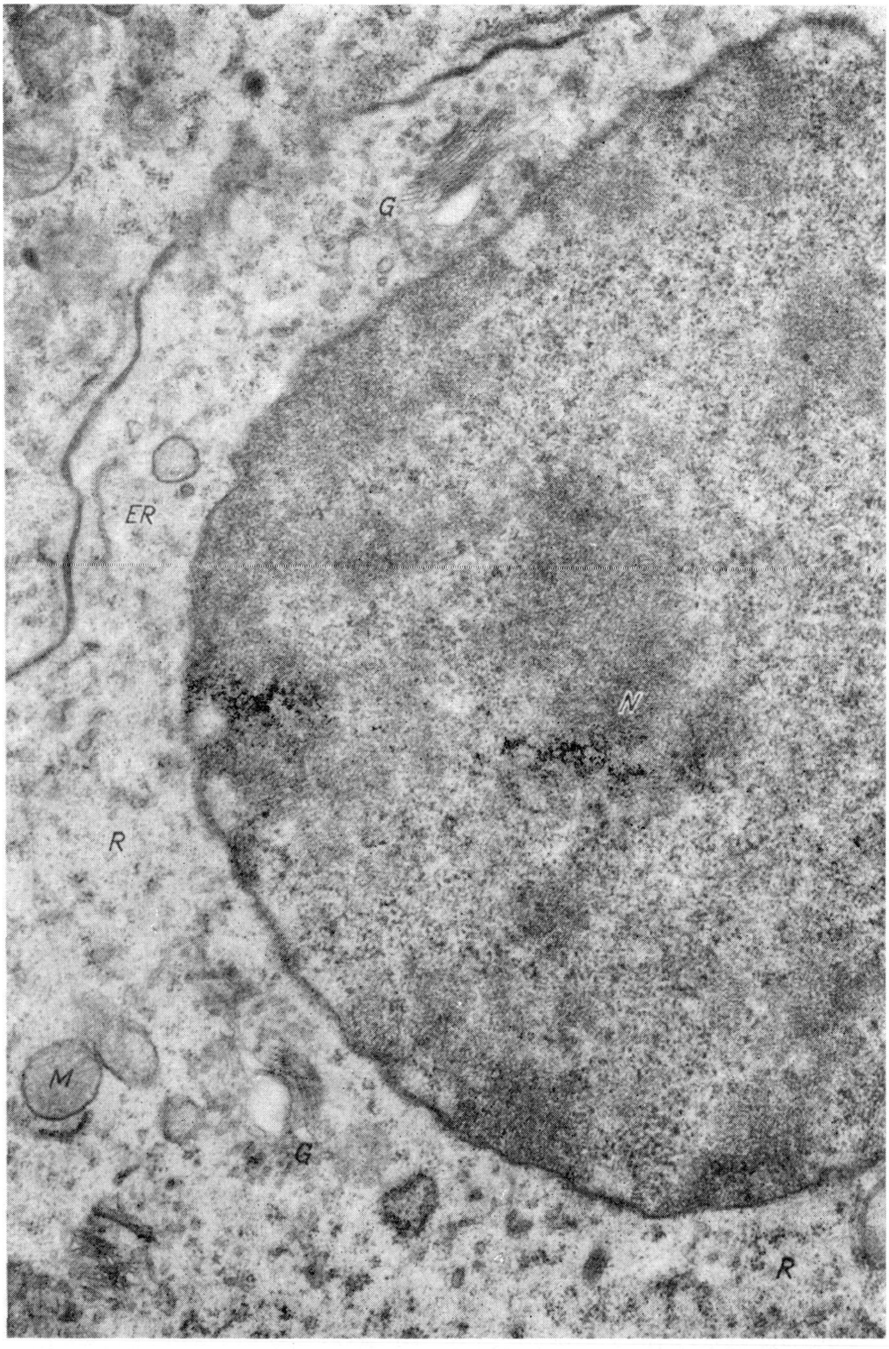

Plate 34. Fig. 5. Part of active corpus allatum cell of *Leucophaea* showing nucleus (N) and several Golgi elements (*G*). There are few short ergastoplasmic units (*ER*), free ribosomes (R), and mitochondria (*M*). X 27,000 (From Scharrer, 1964a).

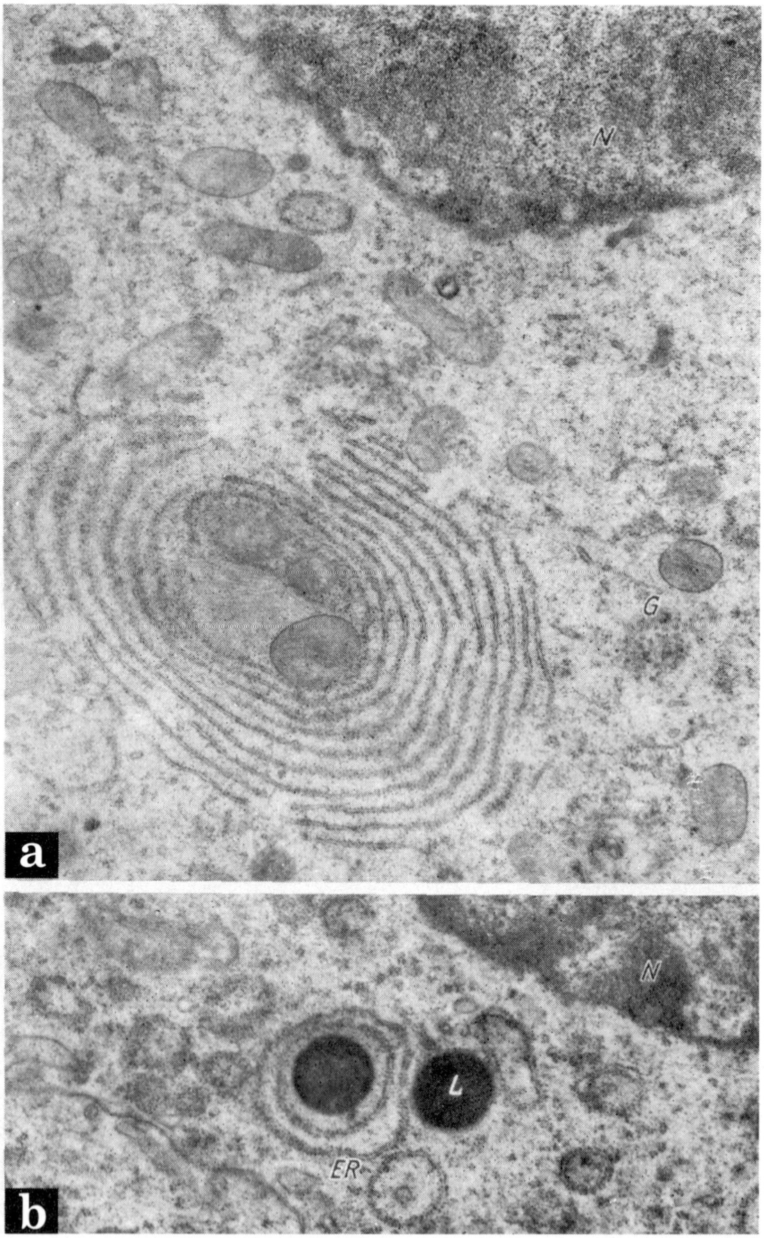

Plate 35. Fig 6a and b. Ergastoplasmic whorls in active corpus allatum cells of *Leucophaea*. a. The peripheral profiles of this large whorl are fragmented. The centre contains remnants of mitochondria. X 15,600. b Two electron dense cytoplasmic inclusions, presumably lysosomes (*L*), one of which is enclosed in two concentric rings of rough-surfaced endoplasmic reticulum (*ER*). Other circular units of ergastoplasm are nearby. (*G*) Golgi element; (*N*) nucleus of corpus allatum cell. X 18,000 (From Scharrer, 1964a).

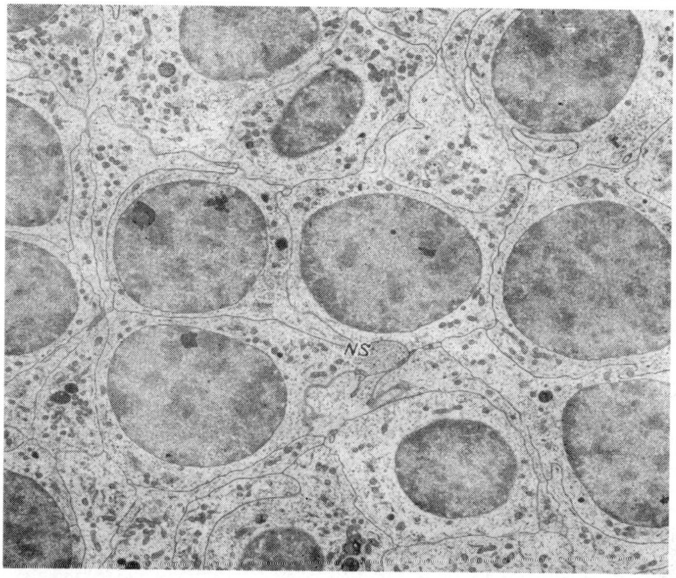

Plate 36. Fig. 7. Group of inactive corpus allatum cells from newly emerged adult female of *Leucophaea* Note relative paucity of cytoplasm and meandering cell boundaries. (*NS*) small neurosecretory fibre wedged between cells X 4400 (From Scharrer, 1964a).

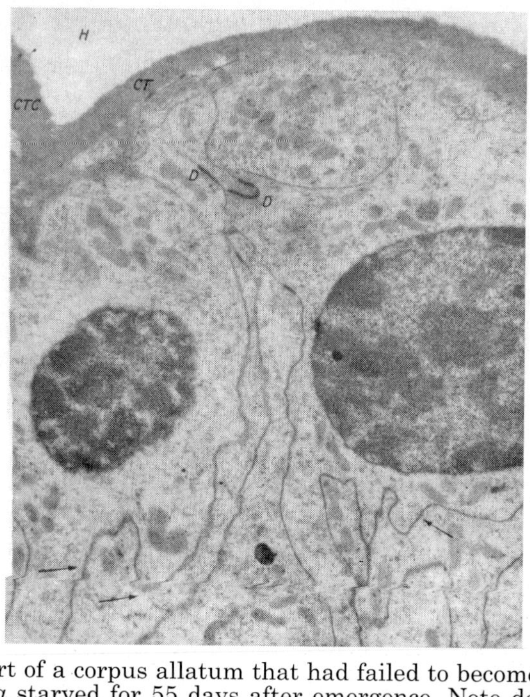

Plate 37. Fig. 8. Part of a corpus allatum that had failed to become active in a female *Leucophaea* starved for 55 days after emergence. Note deep folds in plasma membranes (arrows). The connective tissue sheath (*CT*) which separates the gland from the haemocoel (*H*) continues as the sheath of the adjacent corpus cardiacum (*CTC*). *(D)* desmosomes. X 9,200 (From Scharrer, 1964a).

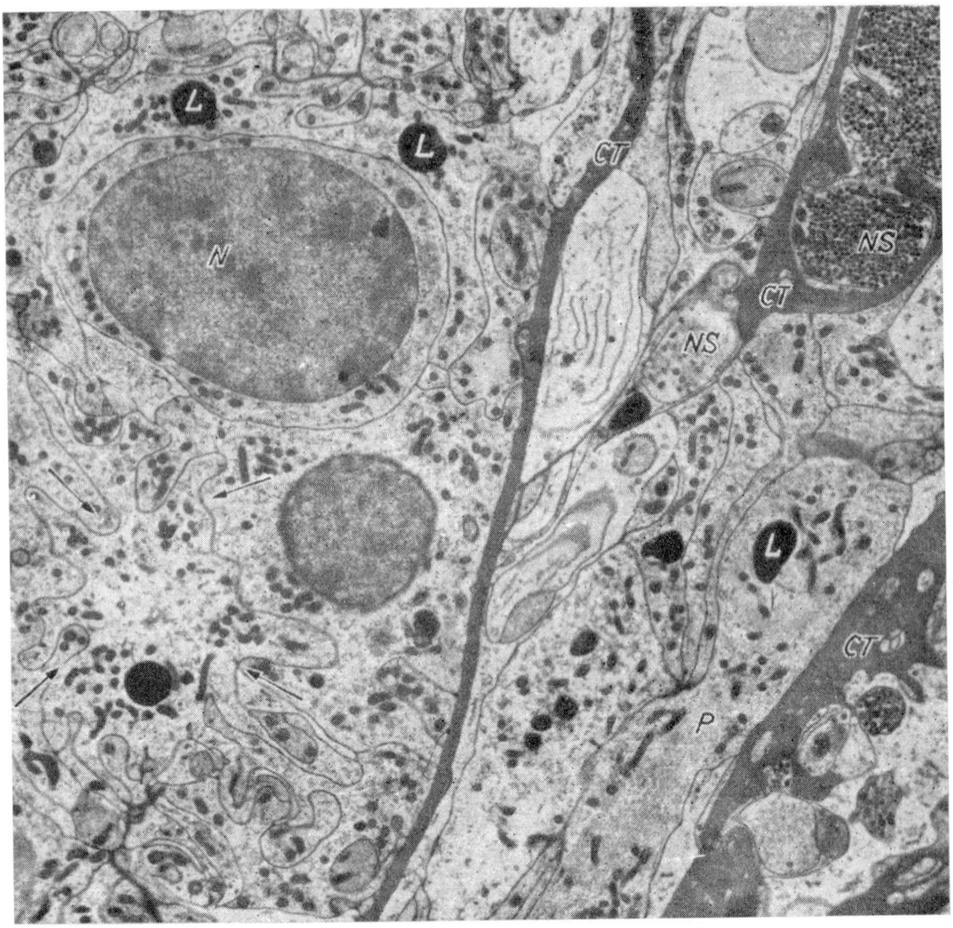

Plate 38. Fig. 9. Corpus allatum-corpus cardiacum junction of a 15 months old female *Leucophaea* fixed shortly after the onset of a reproductive cycle (length of ovarian oocytes 1.5 mm). Arrows indicate extensive folding of plasma membranes in corpus allatum. (*L*) lysosomes; (*N*) nucleus of corpus allatum cell; (*NS*) neurosecretory fibres, surrounded by connective tissue (*CT*) that separates corpus cardiacum from corpus allatum. (*P*) long process of corpus allatum cell with numerous small vesicles, penetrating parts of the corpus cardiacum. X 5,000 (From Scharrer, 1964a).

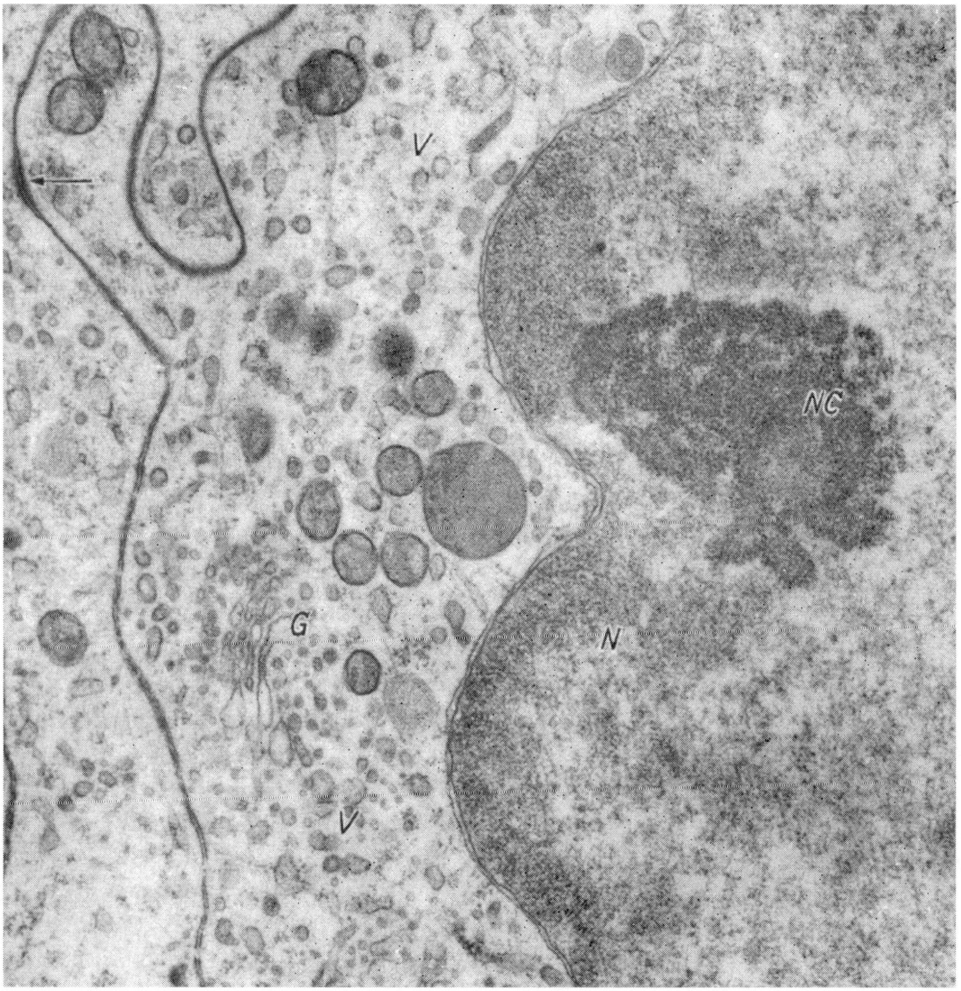

Plate 39. Fig. 10. Detail of corpus allatum cell of newly emerged female of *Leucophaea*. Note deep fold in plasma membrane (upper left hand corner), Golgi element (*G*), and numerous cytoplasmic vesicles (*V*). The nucleus (*N*) is indented; the nucleolus (*NC*) is prominent; the two membranes forming the nuclear envelope are clearly seen. An arrow indicates desmosome. X 23,400 (From Scharrer, 1964a).

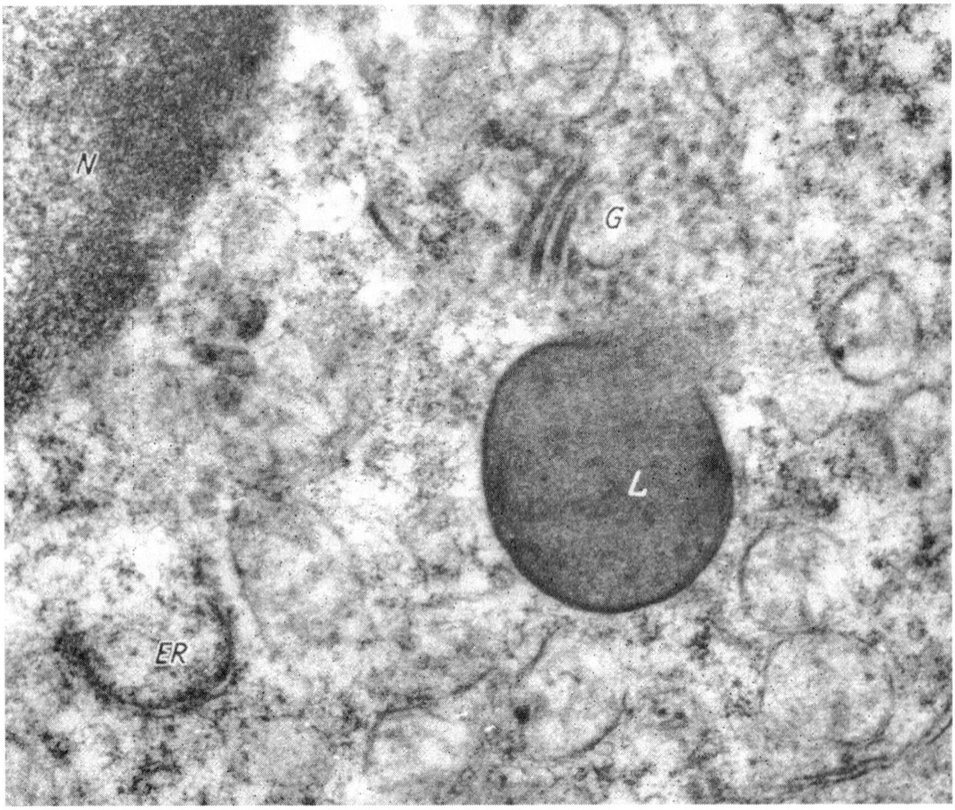

Plate 40. Fig. 11. Detail of corpus allatum cell from young adult female *Leucophaea* fixed after the onset of ovarian growth (oocyte length 1.5 mm) showing a lysosome (*L*) presumably spilling its content. (*ER*) short profile of ergastoplasm; (*G*) Golgi element; (*N*) nucleus. X 36,000 (From Scharrer, 1964a).

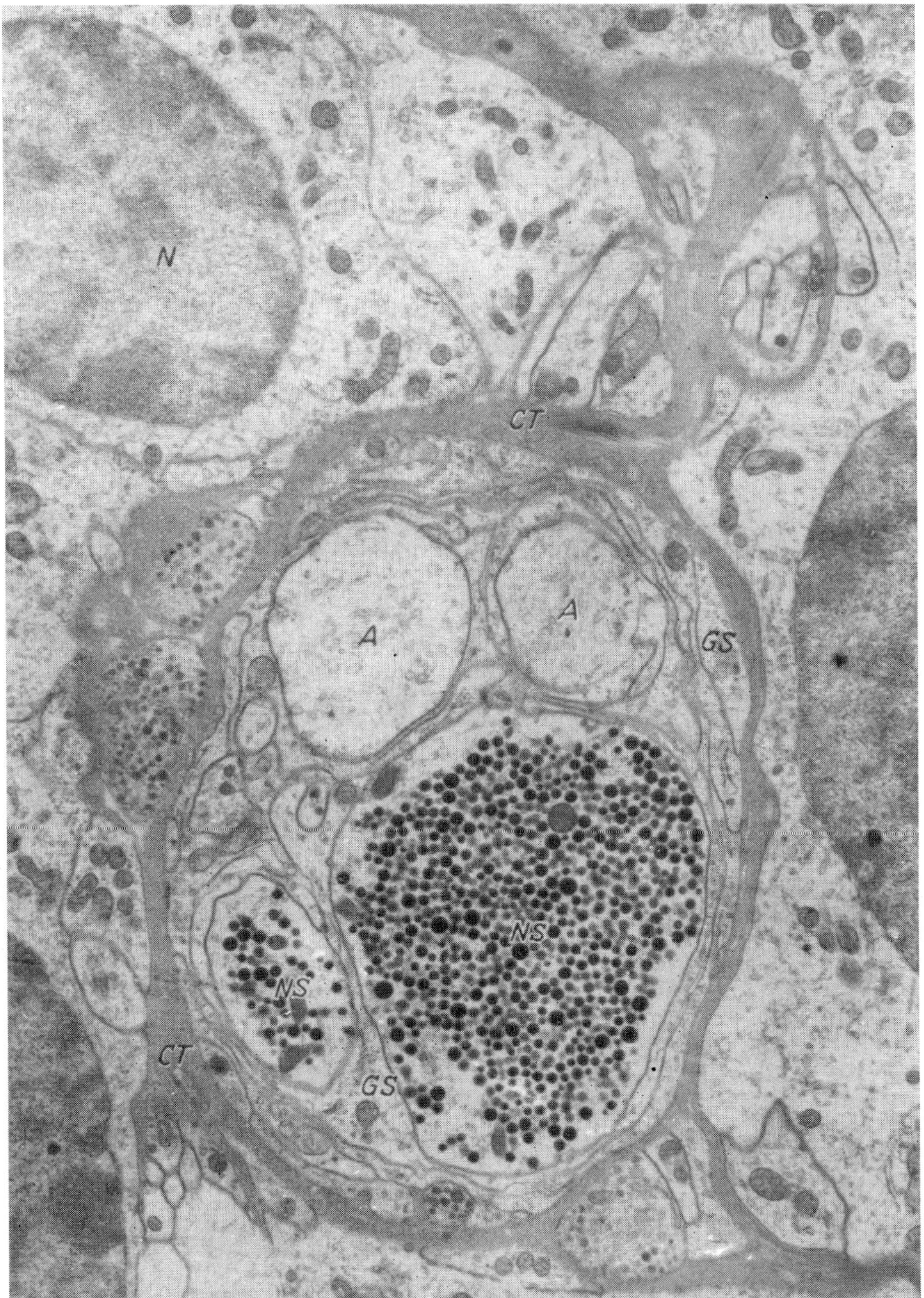

Plate 41. Fig. 12. Corpus cardiacum process within corpus allatum tissue of *Leucophaea*. Profiles of several neurosecretory fibres (*NS*) and ``ordinary'' axons (*A*) are enveloped by sheets of ``glioplasm'' (*GS*) and surrounded by a layer of connective tissue stroma (*CT*). (*N*) nucleus of corpus allatum cell. X 9,600 (From Scharrer, 1964a).

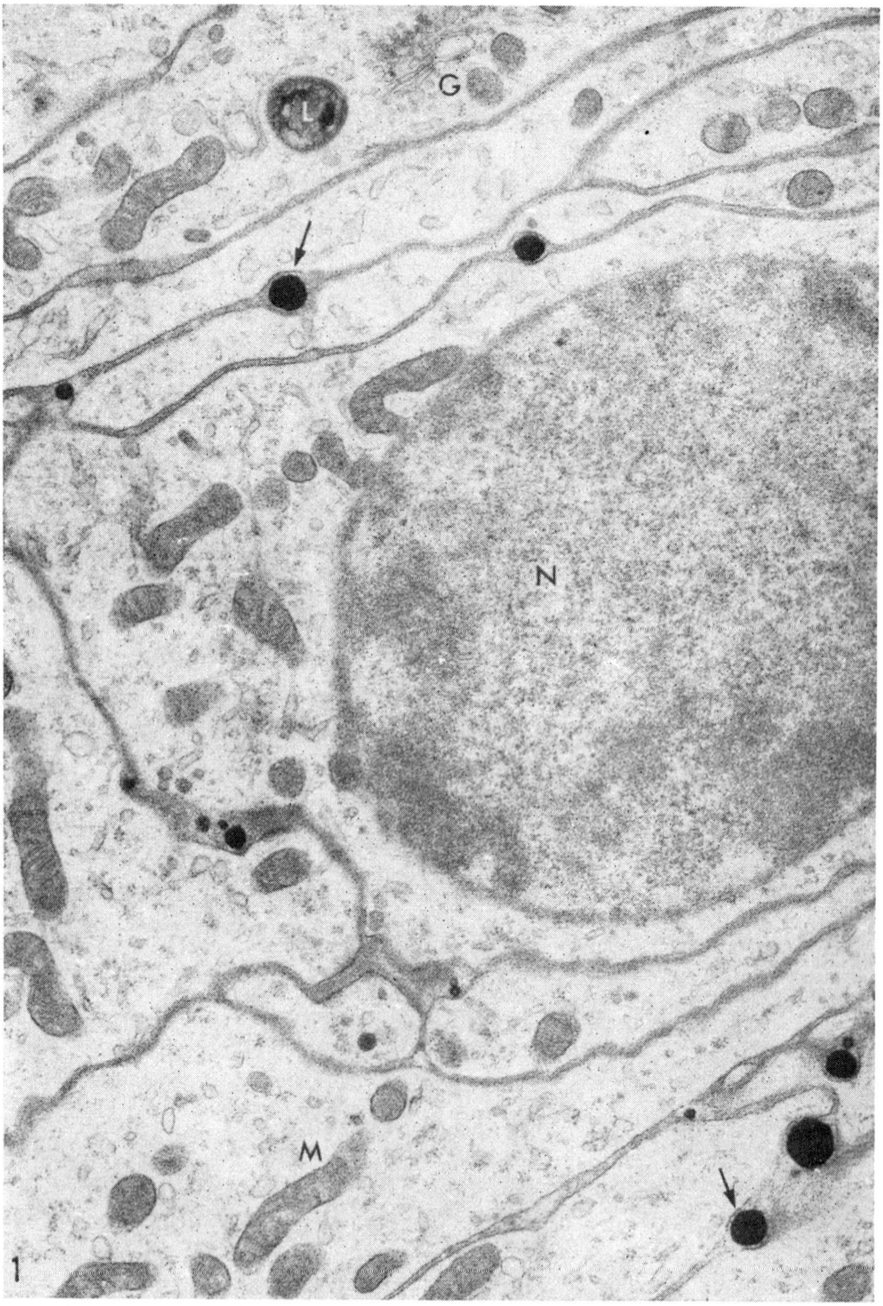

Plate 42. Fig. 1. Parts of corpus allatum cells separated by narrow extracellular channels which contain numerous small C bodies (arrows). N nucleus; M mitochondrion; G. Golgi element; L lysosome. Last instar female nymph of *Leucophaea* fixed 21 days after preceding moult. OsO_4. X 19000 (From scharrer, 1971)

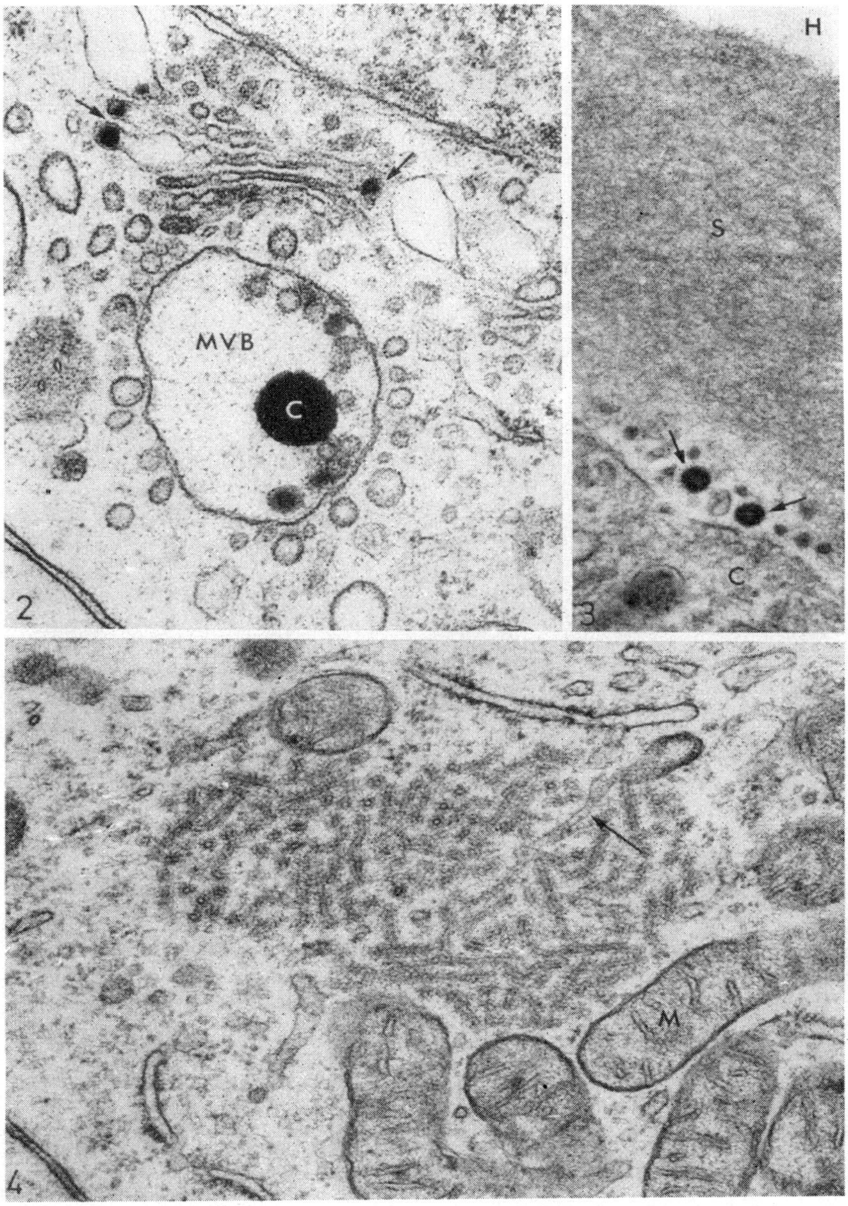

Plate 43. Fig. 2. Golgi zone containing small electron dense granules (arrows) and adjacent MVB containing a cribriform body (C). C. allatum of last instar nymph of *Leucophaea* fixed when emergence to the adult stage was imminent. OsO_4. X 50000 Fig. 3. Surface area of c. allatum of *Leucophaea* showing small extracellular C bodies (arrows) facing stromal sheath (S) that separates gland cell (C) from haemocoel (H). Adult female, pregnant. OsO_4. X 38400 Fig. 4. Group of special microtubular elements, apparently in contact with smooth ER (arrow). Note scattered cisterns of rough ER, parts of which are devoid of ribosomes (upper right). M mitochondrion. Female last instar nymph of *Leucophaea* approaching emergence. OsO_4. X 50000 (From Scharrer, 1971)

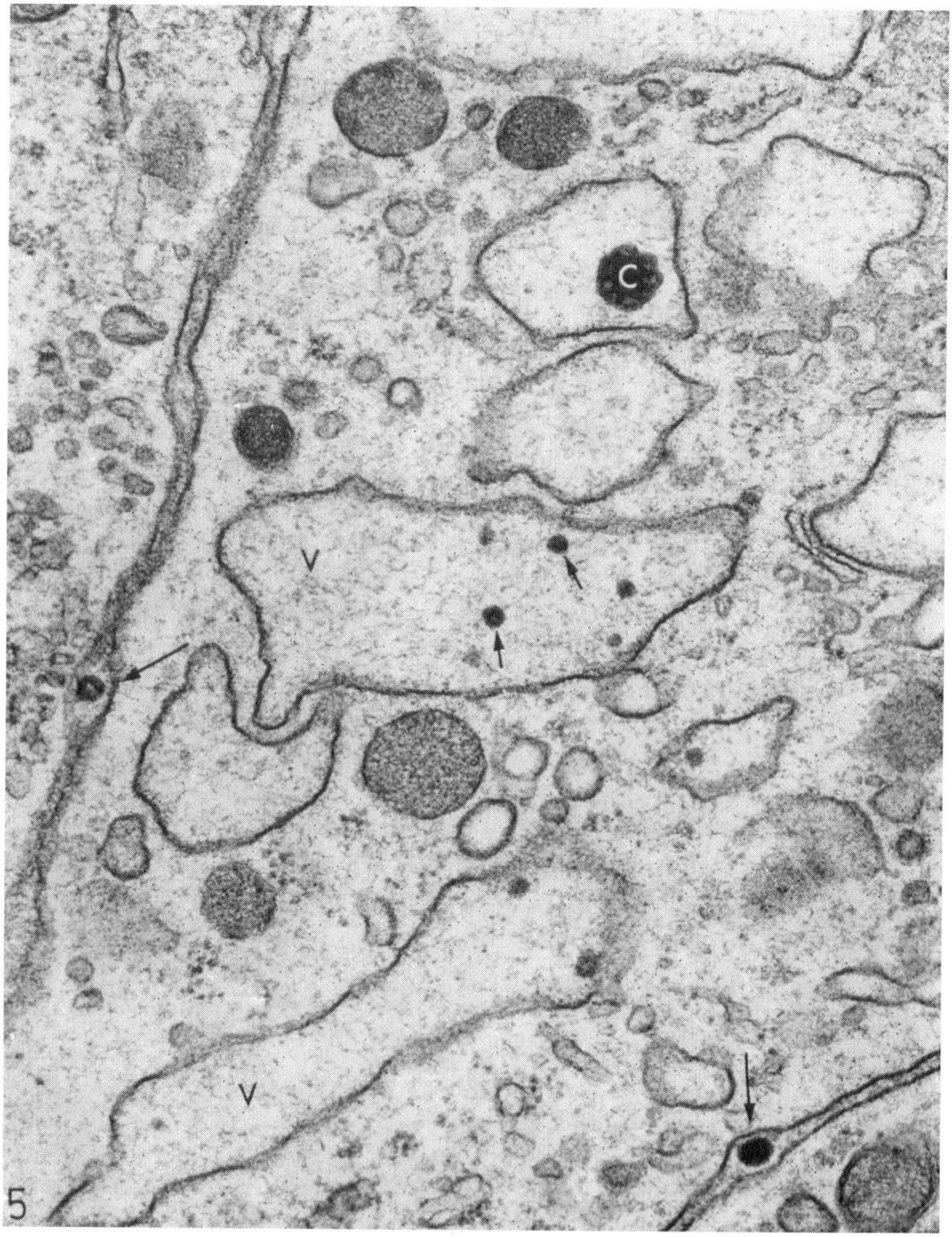

Plate 44. Fig. 5. Large cytoplasmic vacuoles (*V*) considered to be derived from Golgi elements and to be directly related to MVBs with which they share content of flocculent material and C bodies (C) or their precursors (short arrows). Note also two small C bodies wedged between adjacent c. allatum cells (long arrows). Male seventh instar nymph of *Leucophaea*, fixed 20 days. After moult. OsO_4. X 50000 (From Scharrer, 1971)

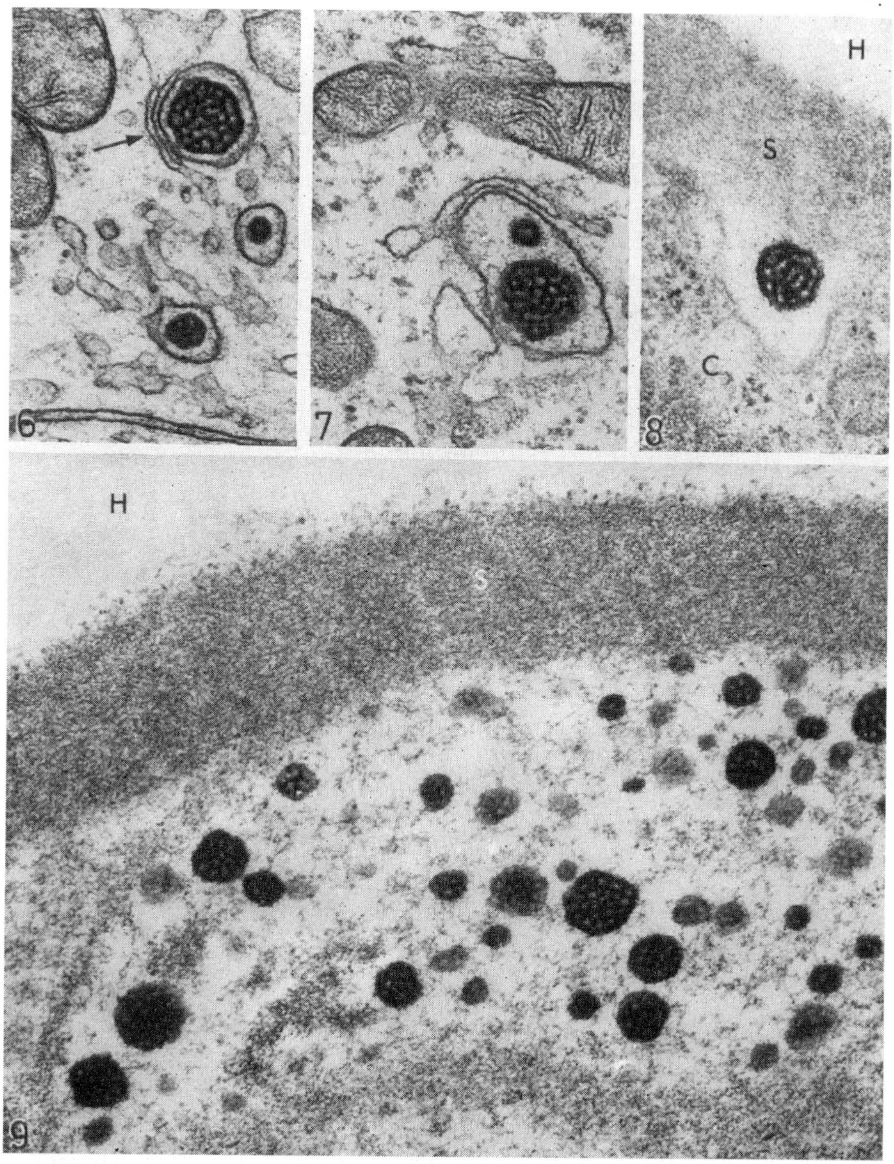

Plate 45. Fig. 6. Three intracellular membrane-enclosed C bodies, the largest of which clearly shows cribriform structure, and accompanying profile of Golgi element (arrow). Male seventh instar nymph of *Leucophaea*. 20 days postmoult. OsO_4. X 50000 Fig. 7. Similar cytoplasmic compartments as in Fig. 6, next to a mitochondrion. Note two differently sized C bodies inside the same spacious vacuole. Last instar female nymph, 21 days postmoult. OsO_4. X 50000 Fig. 8. C body, presumably released from c. allatum cell (C) by exocytosis, facing sheath (S) which provides access to haemocoel (H). Last instar male nymph of *Leucophaea* 13 days postmoult. Glutaraldehyde. X 38400 Fig. 9. Mass of variously sized C bodies in close proximity to c. allatum sheath (S). H haemocoel. Last instar male nymph of *Leucophaea* 13 days postmoult. OsO_4. X 38400 (From Scharrer, 1971)

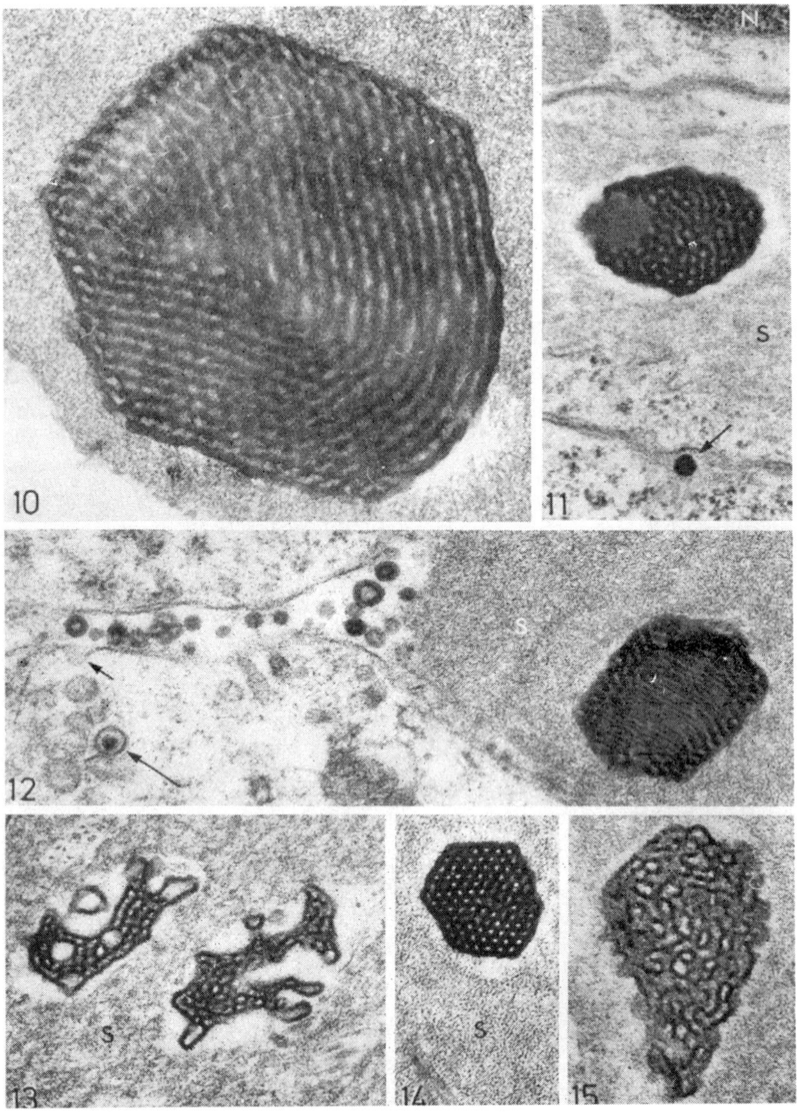

Plate 46. Fig. 10. Large extracellular C body of hexagonal shape. Male adult *Leucophaea* Glutaraldehyde, zinc iodide osmium. X 55200 Fig. 11. Extracellular C body with amorphous component surrounded by halo. Note small intracellular C body (arrow). N nucleus; S. stroma. Female *Leucophaea* adult age 16 1/2 months. OsO_4. X 38400 Fig. 12. Angular C body within stroma (S) and group of small C bodies in electron lucent extracellular space. Also note membrane-enclosed small body (long arrow) within cytoplasm of adjacent cell and indentation of plasmalemma (short arrow). Adult female, pregnant *Leucophaea* OsO_4. X 38400 Fig. 13. Two extracellular C bodies in rarefied areas of stroma (S). Adult female *Leucophaea* starved and dehydrated for 74 days following emergence. OsO_4. X 37600 Fig. 14. Angular cribriform C body in stroma (S) of corpus allatum. Note halo. Female *Leucophaea* adult age 13 days. Glutaraldehyde. X 38400 Fig. 15. Extracellular C body in c. allatum fixed right after emergence of male adult *Leucophaea* OsO_4. X 38400 (From Scharrer, 1971)

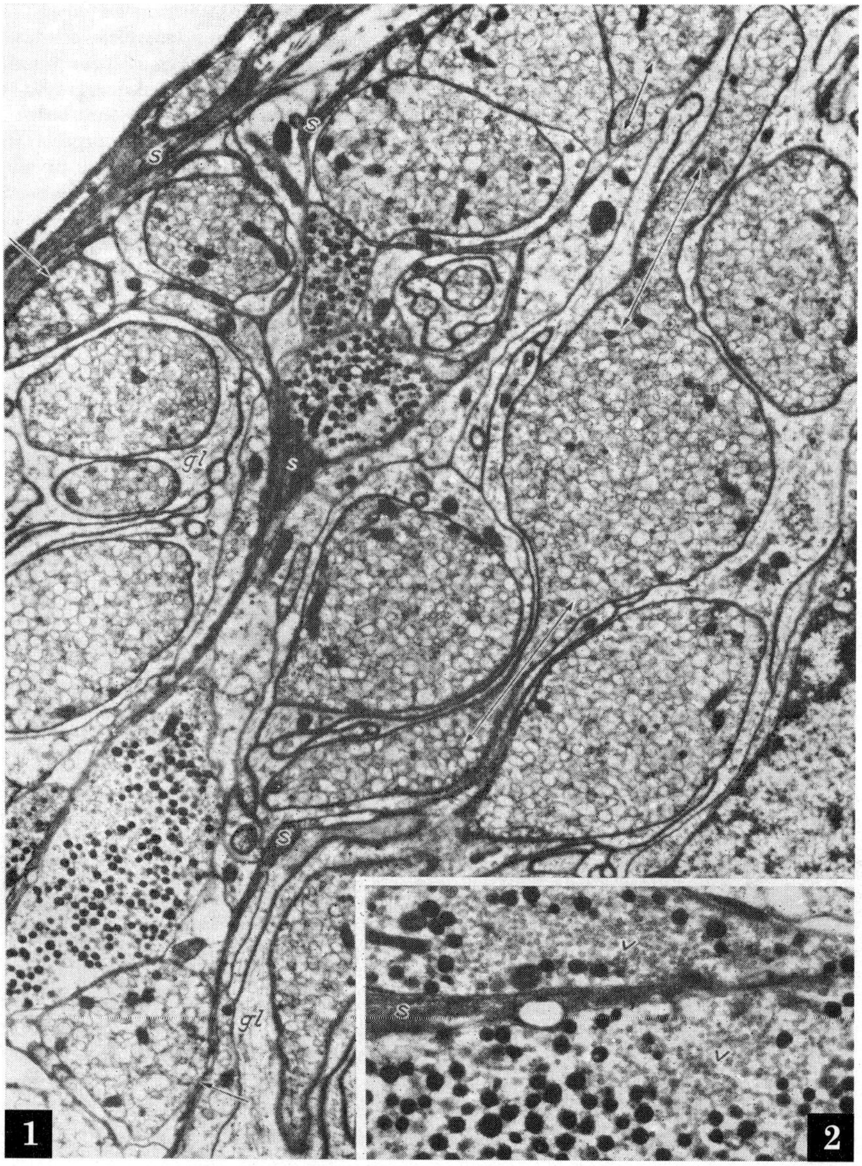

Plate 47. Fig. 1. A field to show the general anatomy of the transverse nerve swelling of *Carausius morosus*. The organ is invested with a sheath, extensions of which (s) penetrate the organ at many places. There are large numbers of axon profiles which contain the electron-transparent vesicles. Some of these appear to be branching (double-headed arrows). Nearly all of them are invested by processes from glial cells (*gl*), but a few make contact with the sheath (arrow). Four axon profiles in the field contain electron-dense granules and these all contact the extensions from the sheath. X 11900 Fig. 2. A micrograph at higher magnification of two axon profiles which contain electron-dense neurosecretory granules to show the large numbers of vesicles (*v*) about 300 Å in diameter which occur in the regions near an extension of the sheath (*s*). X 24500 (From Brady and maddrell, 1967)

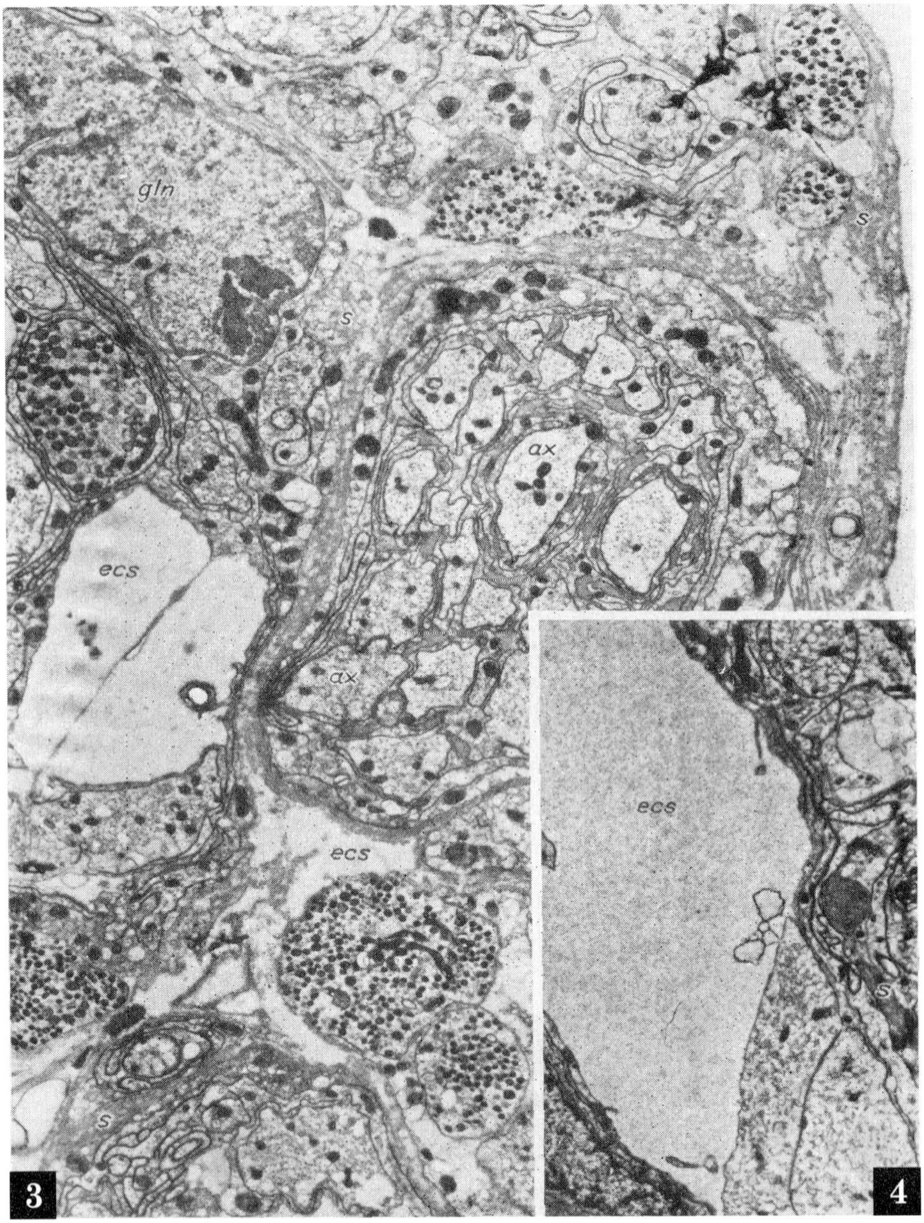

Plate 48. Figs. 3 and 4. Transverse nerve swelling of *Carausius morosus*. Fields to show the large extracellular spaces (*ecs*) which occur in the organ. The field shown in Fig. 3 also includes a glial cell nucleus (*gln*) and a group of several axons (as at *ax*) which have not yet branched and contain neurotubules. The extracellular space shown in Fig. 4 contains a finely flocculent electron-opaque material. Fig. 3. X 8800; Fig. 4 X 7200 (From Brady and Maddrell, 1967)

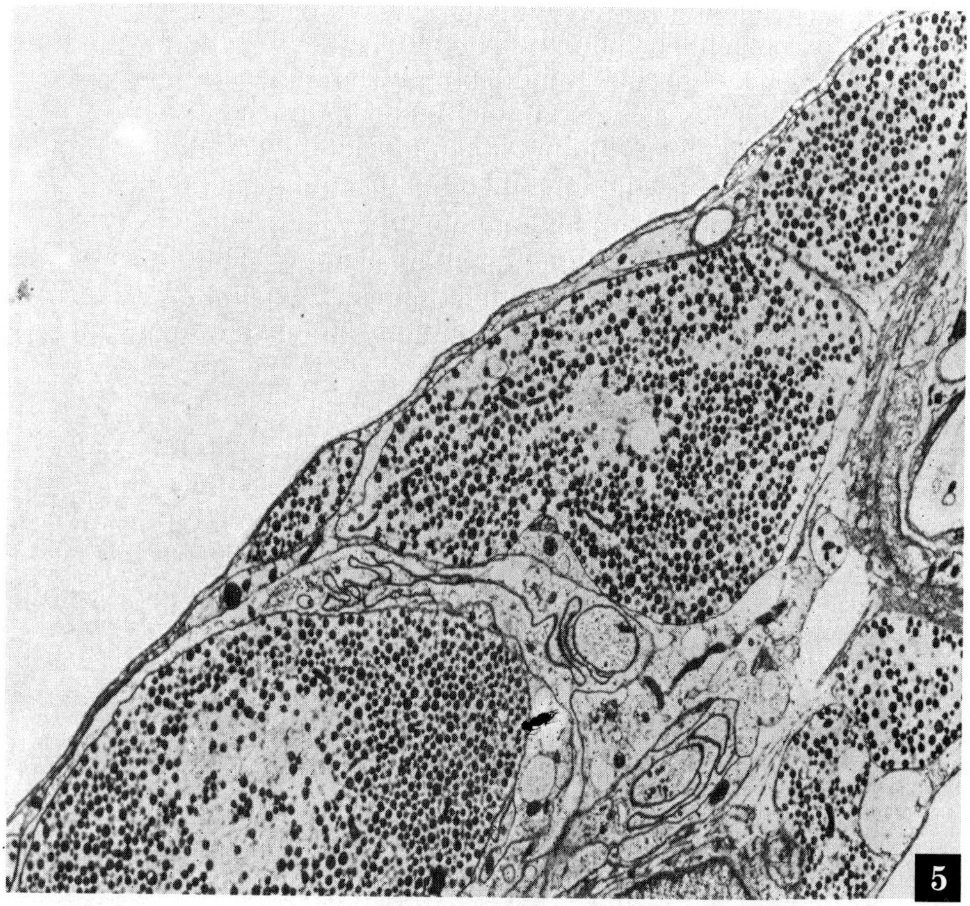

Plate 49. Fig. 5. Transverse nerve swelling of *Carausius morosus*. Part of a section taken near the posterior end of the organ to show the very long axon terminals there which contain electron dense granules X 6000 (From Brady and Maddrell, 1967).

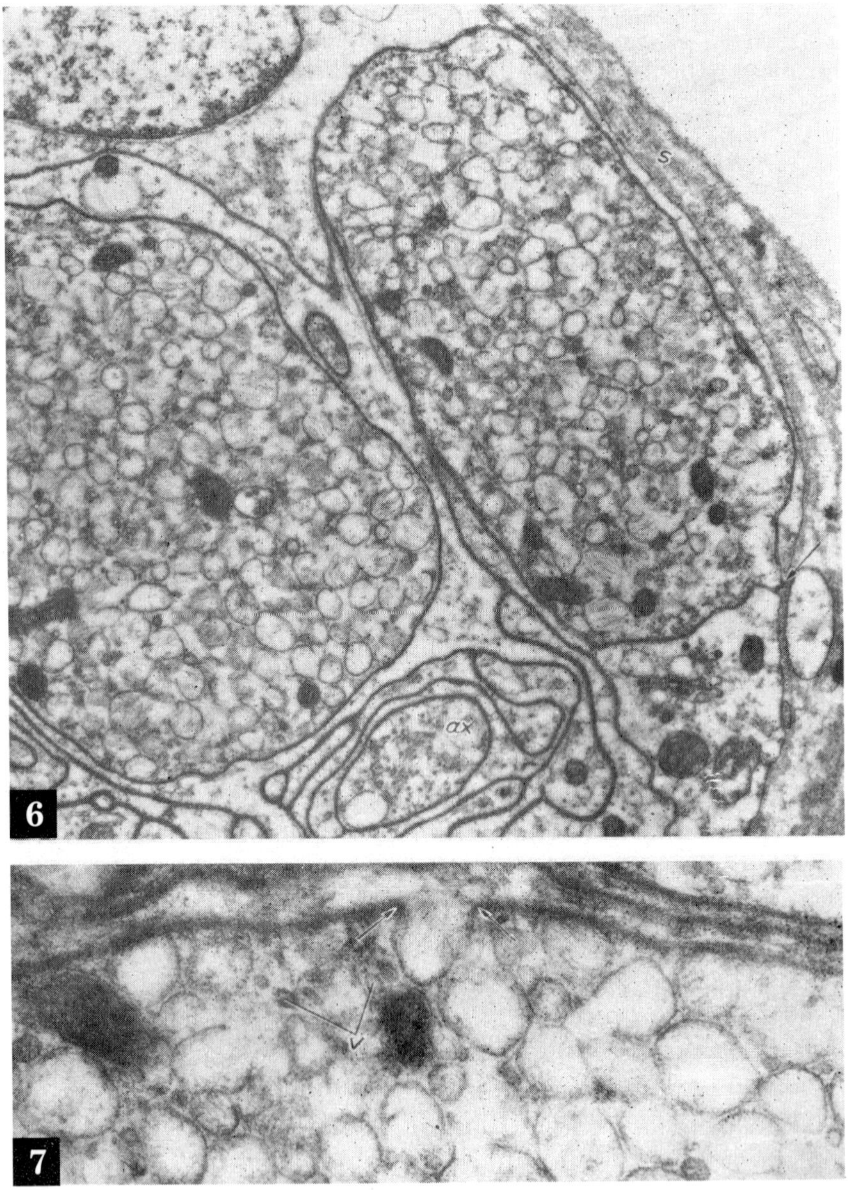

Plate 50. Fig. 6. Transverse nerve swelling of *Carausius morosus*. A micrograph at higher magnification to show in detail the content of those axon terminals which contain electron transparent vesicles. The material in some of the vesicles appears to be faintly straited. The field includes an axon profile (*ax*) which is probably a preterminal portion in that it contains an array of neurotubules. One of the large axon terminals is in direct contact with the sheath (s) at arrow X 24500. Fig. 7. Transverse nerve swelling of *Carausius morosus*. A section through the edge of an axon terminal to show the fusion of the membrane of a neurosecretory vesicle with that of the surface membrane of the axon (arrows). Among the larger vesicles occur several much smaller ones (*v*). X 61500 (From Brady and Maddrell, 1967)

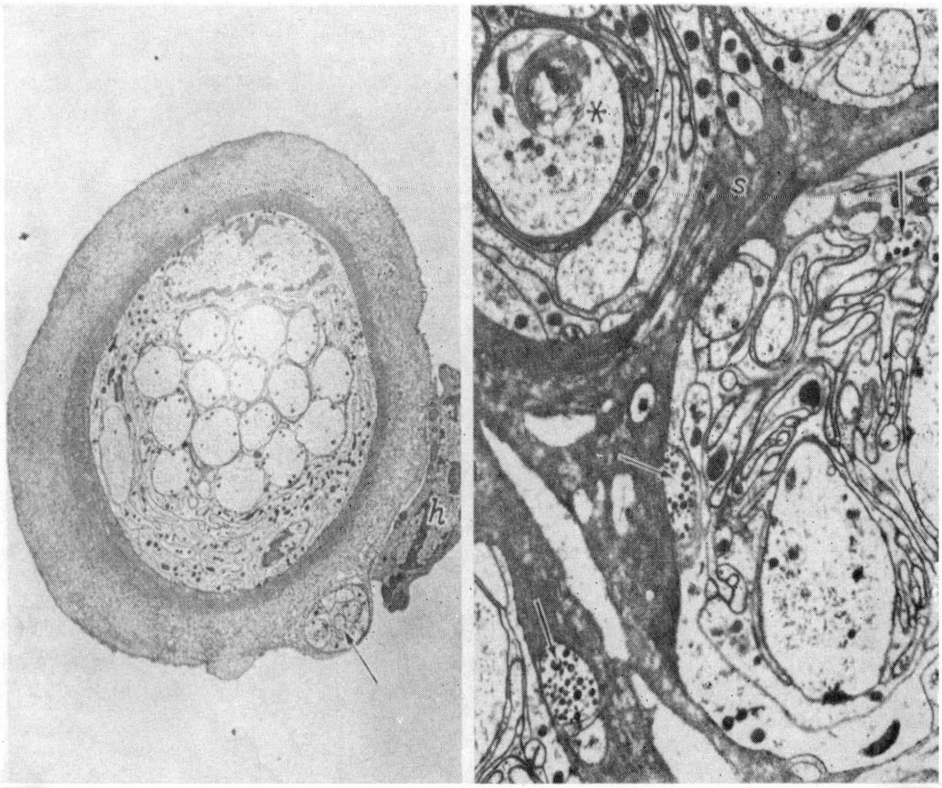

8 9

Plate 51. Fig. 8. The nerve which gives rise to the medial nerve swelling of *Schistocerca gregaria*. The nerve contains relatively few axons and has a very thick sheath. A haemocyte (*h*) lies on the surface of the sheath. Embedded in the sheath is a small bundle of axon branches (*arrow*) which has run back from the main body of the organ. X 2700 Fig. 9. Medial nerve swelling of *Schistocerca gregaria*. A field to show three small axon endings (arrows) which contain electron-dense neurosecretory granules; two of the endings are in direct contact with the inward extension of the sheath (*s*). In one neurosecretory axon ending (asterisk), which contains a few electron-transparent vesicles there is a structure largely made up of concentrically arranged membranes. X 7700 (From Brady and Maddrell, 1967)

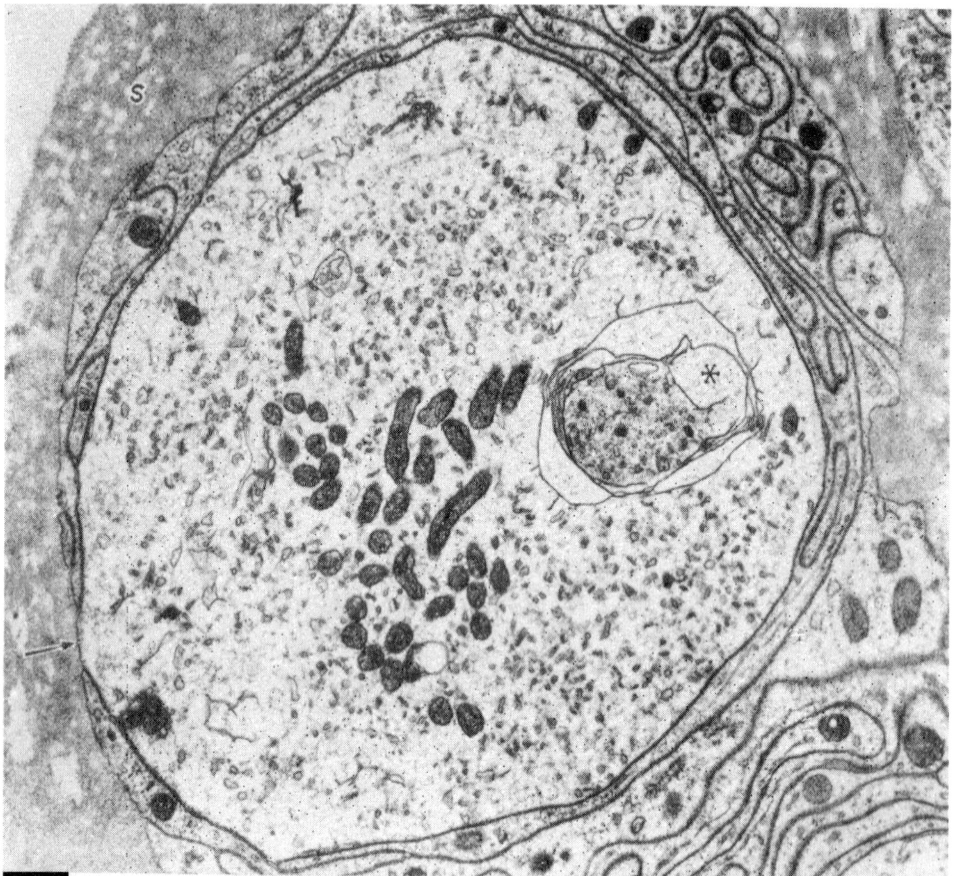

10

Plate 52. Fig. 10. A neurosecretory axon terminal in a medial nerve swelling of *Schistocerca gregaria*. The terminal contains many small electron-transparent vesicles, a group of mitochondria cut in cross-section, and a membrane-bound structure (asterisk) which contains several vesicles. The terminal touches the sheath (*s*) at one point (arrow). X 17100 (From Brady and Maddrell, 1967)

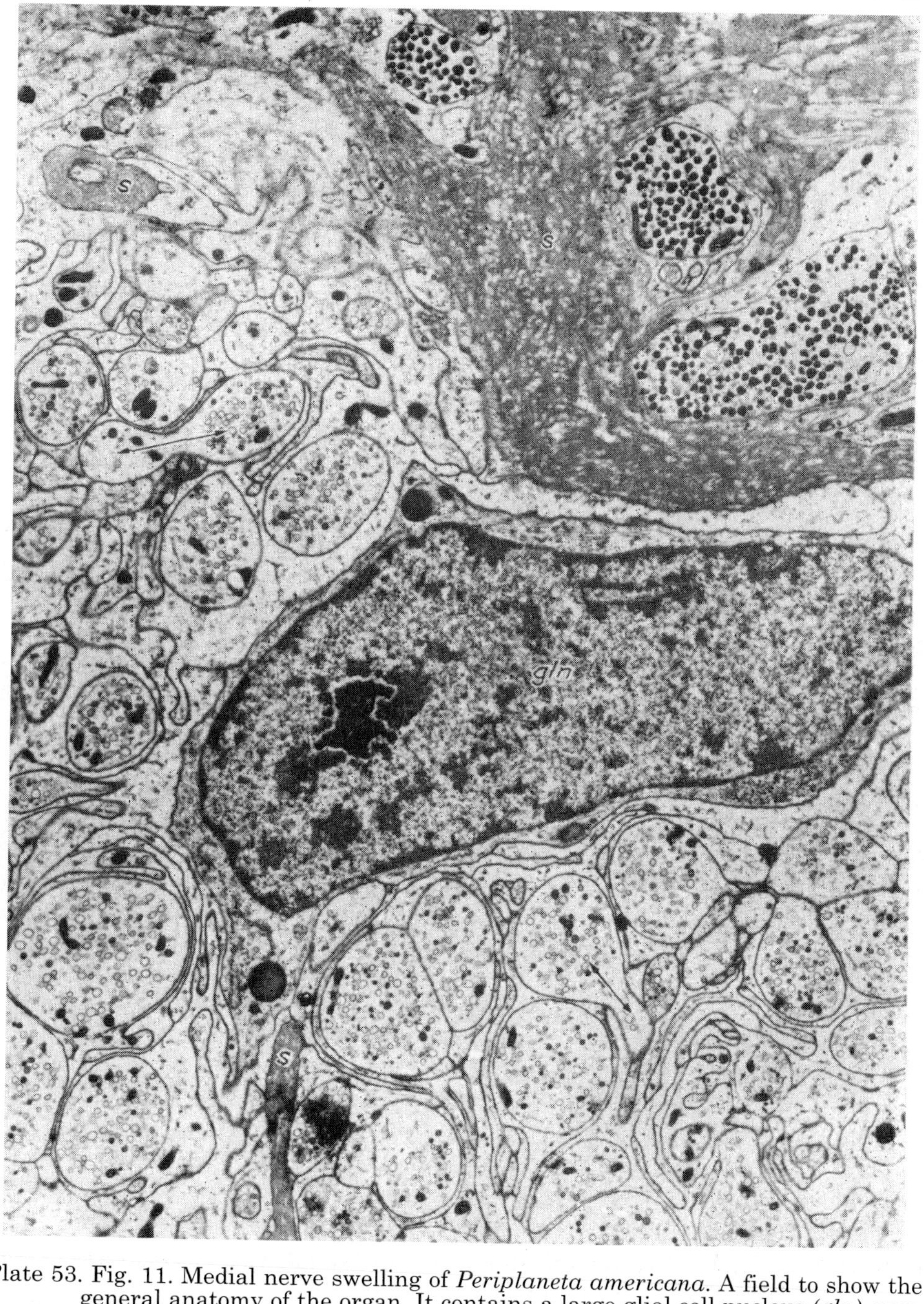

Plate 53. Fig. 11. Medial nerve swelling of *Periplaneta americana*. A field to show the general anatomy of the organ. It contains a large glial cell nucleus (*gln*), very many neurosecretory axons full of a heterogeneous population of neurosecretory vesicles, and three axon endings containing electron-dense granules in contact with the ramifications of the sheath (*s*). Some of the axons containing mixed neurosecretion appear to be branching (double-headed arrows). X 9100 (From Brady and Maddrell, 1967).

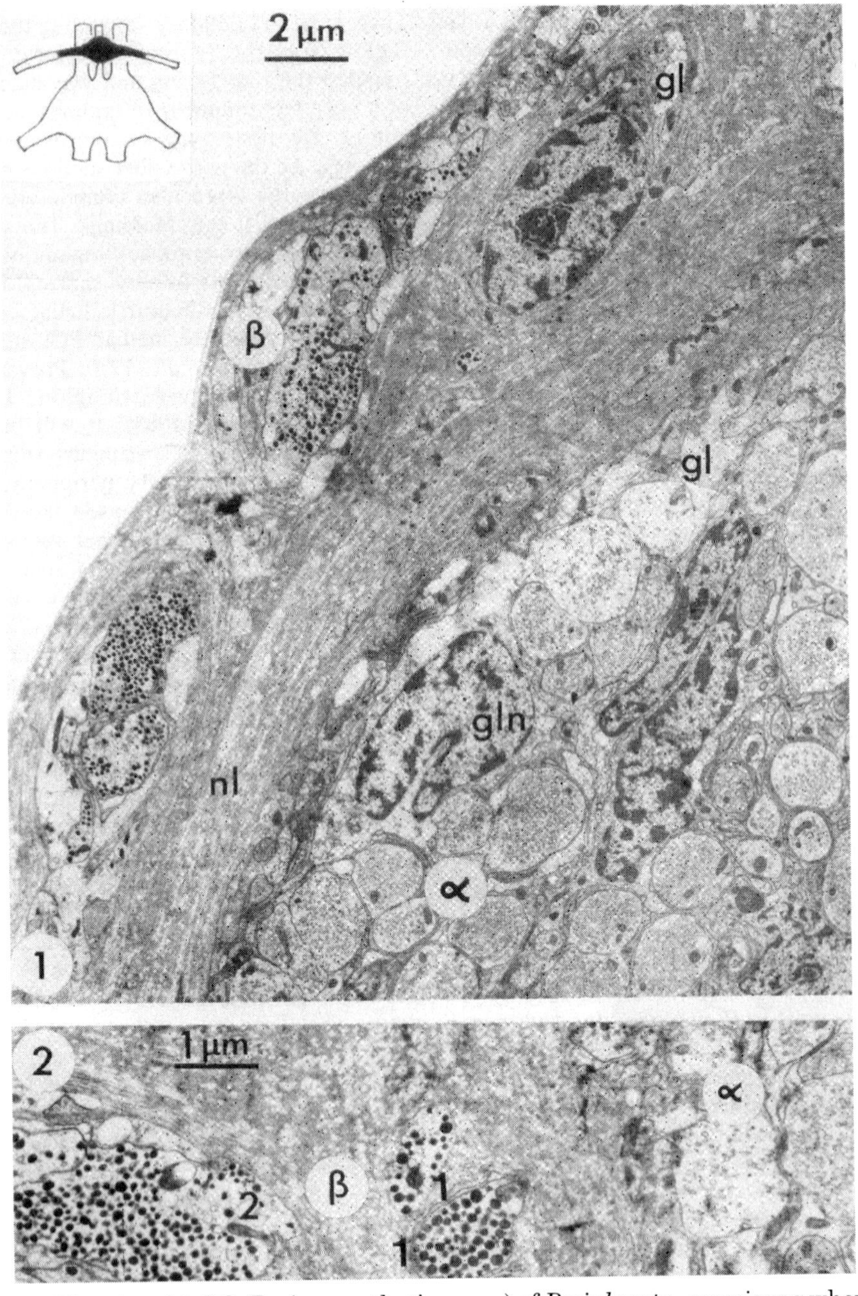

Plate 54. Figs. 1 and 2. PO (Perisympathetic organ) of *Periplaneta americana* where the two types of structure α and β coexist. The central α structure is made up of numerous neurosecretory endings full of clear vesicles. Glial cells ensheath all the neurosecretory endings. The peripheral β structure consists of lobules spread in the neural lamella (nl) where the two types of neurosecretory endings (1 and 2) are always partially naked. Fig. 1. X 5900, Fig. 2. X 12500. (From Baudry-Partiaoglou, 1986)

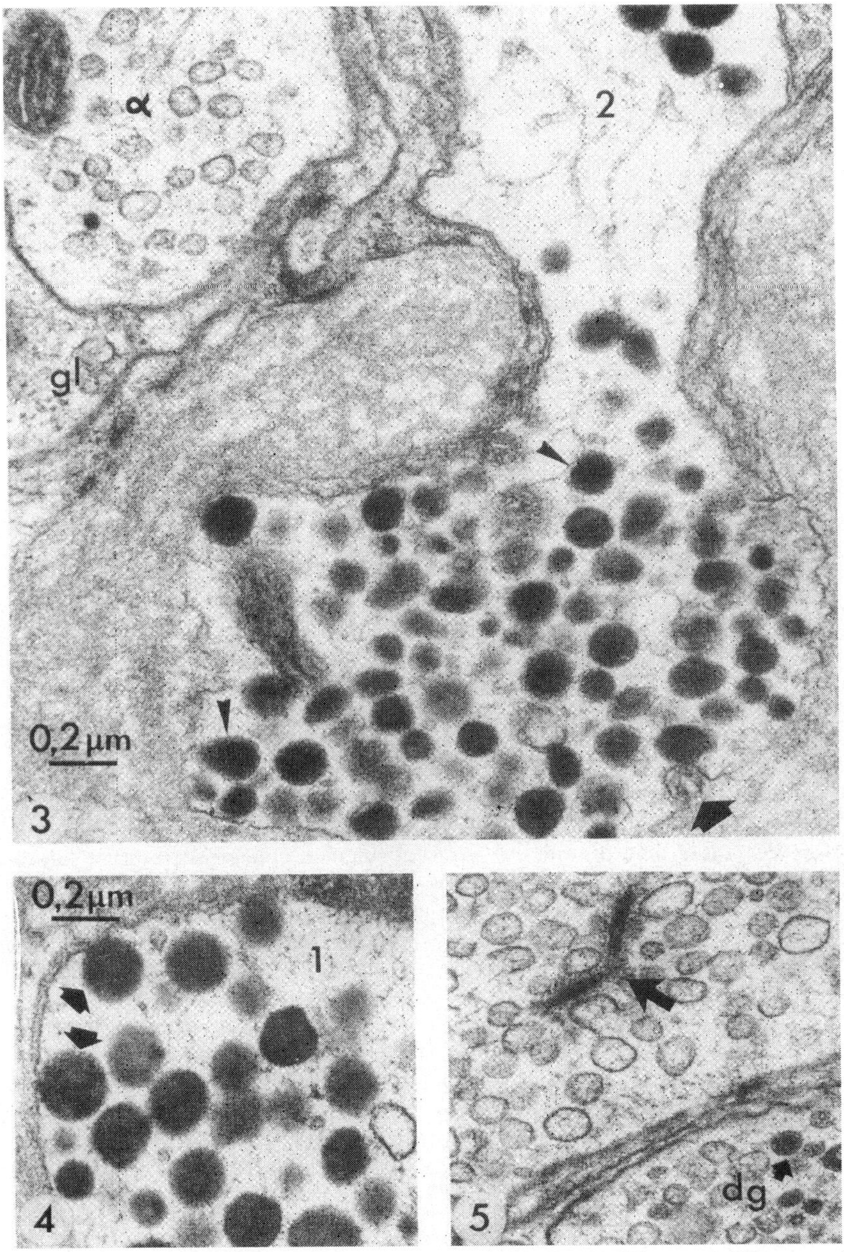

Plate 55. Figs. 3-5. PO of *Periplaneta americana*. X 48,600. Figure 3: The α ending is surrounded by glial cell processes (gl), whereas the type 2 axon is sprouting a neurosecretory ending in direct contact with the neural lamella. The rippléd membrane is obvious around some type 2 granules (arrowhead). Figure 4 : Type 1 ending devoid of glial process in the β structure. Some granules are close to the axolemma (arrow). Figure 5: α endings with clear vesicles. The arrow shows synaptoid areas with accumulation of dense material around synaptic ribbon. Note some dense granules (dg) among clear vesicles. (From Baudry-Partiaoglou, 1986)

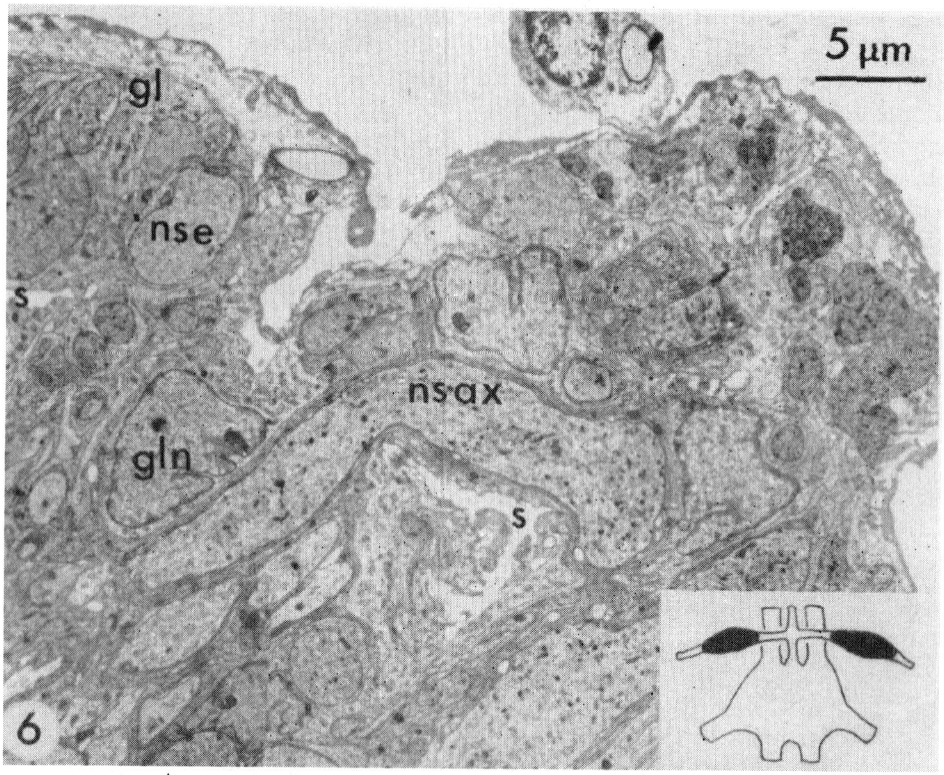

Plate 56. Fig. 6. Part of the α structure in the PO of *Carausius morosus*. Neurosecretory endings contain clear vesicles, whereas neurosecretory axons include an heterogeneous population of vesicles and dense granules. Note the continuous glial coat around axons and endings. X 2900. (From Baudry-Partiaoglou, 1986).

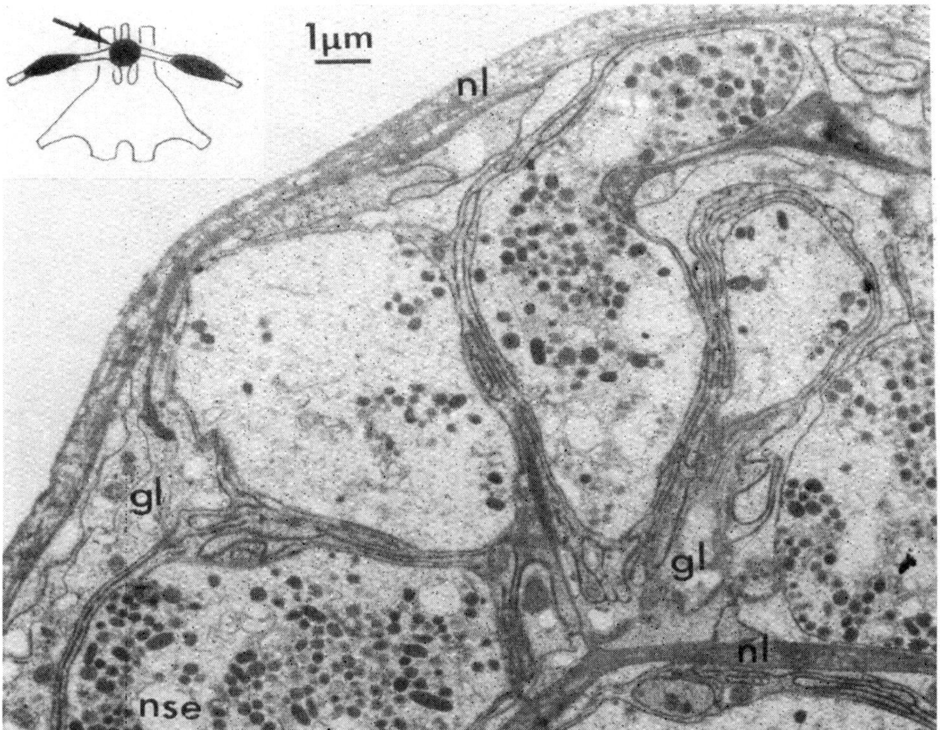

late 57. Figs. 7-9. Median PO of *Locusta migratoria* made up of α structure. Figure 7: Neural lamella invaginations (nl) enter deeply into the organ. All the neurosecretory endings (nse) are surrounded with glial processes (gl). X 7000. Figure 8: High magnification showing the tubulous content of the neurosecretory granules. X 99,000. Figure 9: Synaptoid site with synaptic ribbon (arrow). X 48,600. (From Baudry-Partiaoglou, 1986).

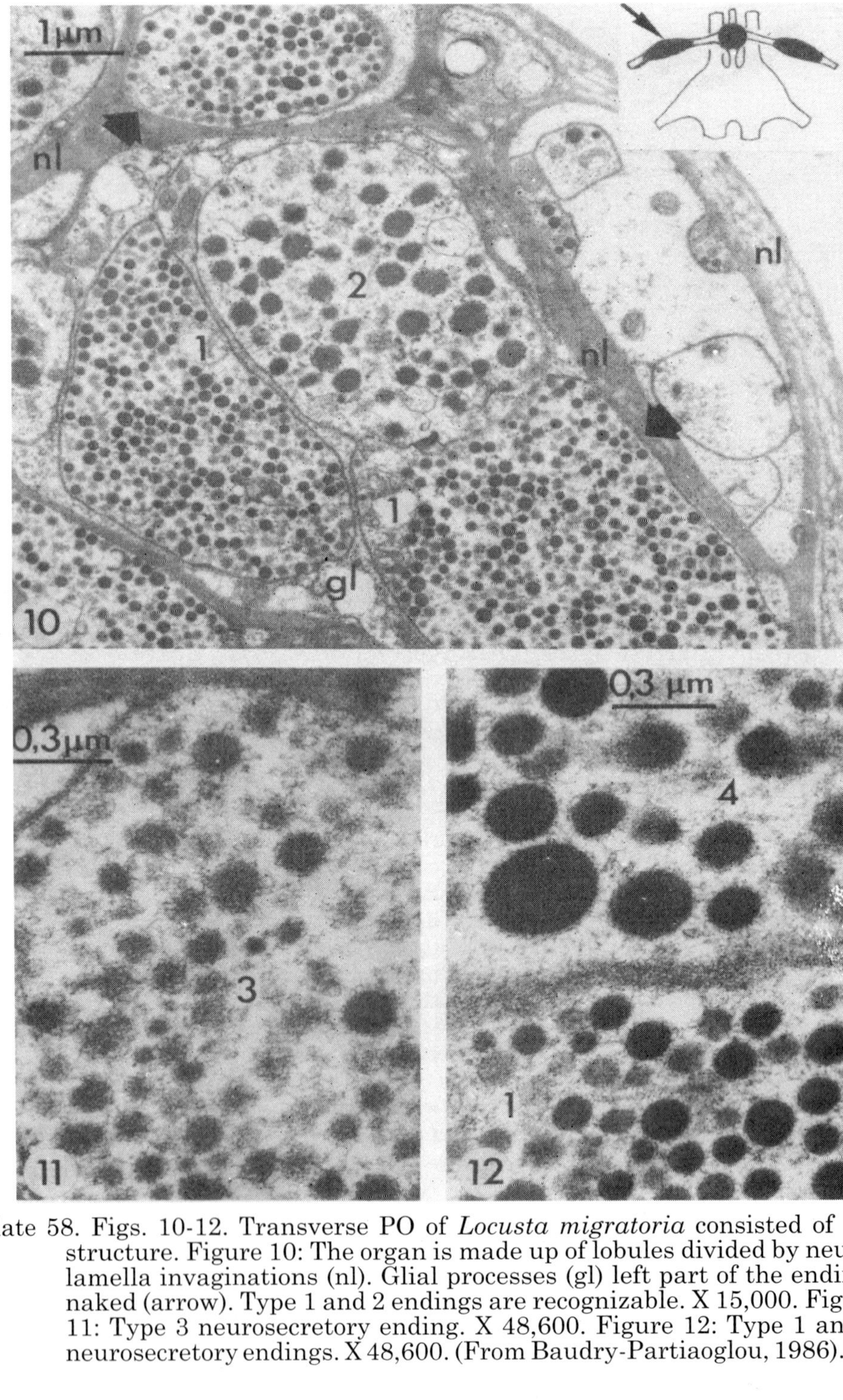

Plate 58. Figs. 10-12. Transverse PO of *Locusta migratoria* consisted of a β structure. Figure 10: The organ is made up of lobules divided by neural lamella invaginations (nl). Glial processes (gl) left part of the endings naked (arrow). Type 1 and 2 endings are recognizable. X 15,000. Figure 11: Type 3 neurosecretory ending. X 48,600. Figure 12: Type 1 and 4 neurosecretory endings. X 48,600. (From Baudry-Partiaoglou, 1986).

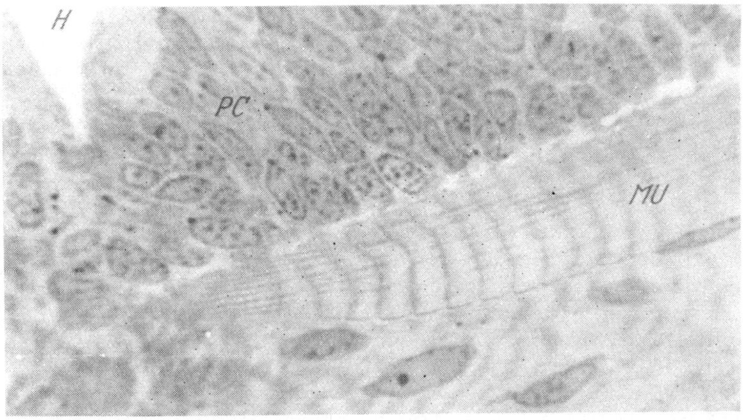

Plate 59. Fig.1 Section of part of prothoracic gland of *Leucophaea* female fixed at time of moult into adult stage. Note layers of parenchymal cells *PC* next to muscle *MU*. A notch in the glandular surface reveals haemocoel *H*. Osmic acid, Epon, about 1 μ toluidine blue. X 720 (From Scharrer, 1964b)

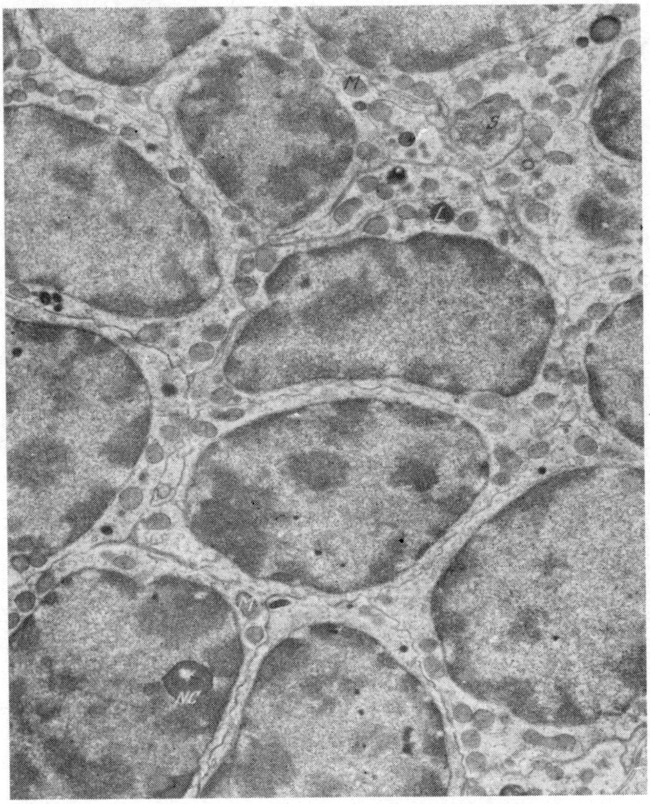

Plate 60. Fig. 2. Low power electron micrograph of prothoracic gland of last instar female nymph of *Leucophaea* fixed several hours after moult. The rather scanty perinuclear cytoplasm contains mitochondria *M* and lysosome-like dense bodies *L*. Note undulating cell boundaries, nucleolus *NC* and intercellular stroma *S*. X 9,000 (From Scharrer, 1964b)

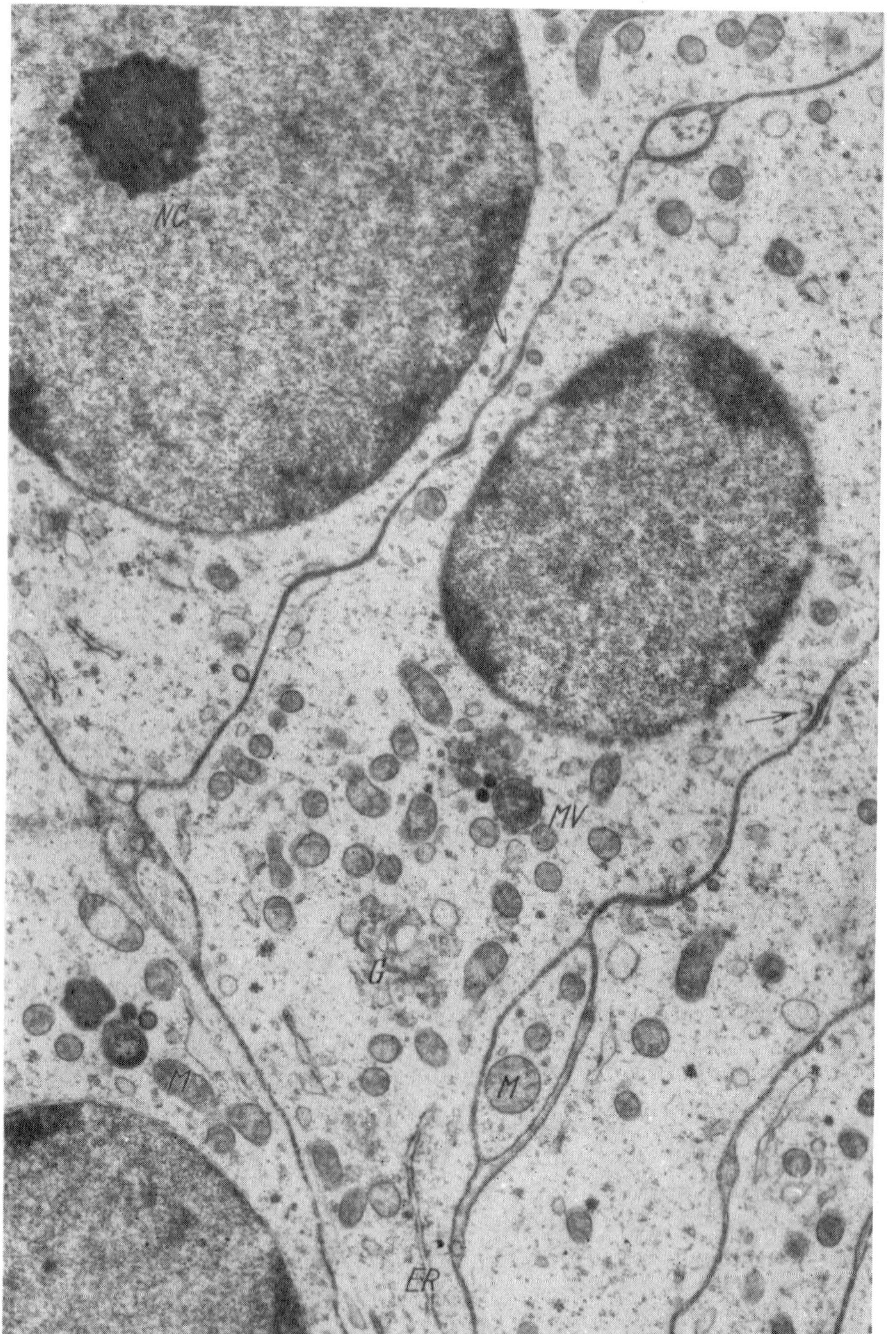

Plate 61. Fig. 3. Contiguous prothoracic gland cells from last instar male nymph of *Blaberus,* 16 days after moult. Note Golgi apparatus G. mitochondria, *M* and sparse endoplasmic reticulum *ER* some of whose elements form subsurface cisterns (*arrows*). A multivesicular body *MV* lies in close proximity to Golgi type vesicles, two of which have electron dense contents, *NC* nucleolus. X 14,000 (From Scharrer, 1964b)

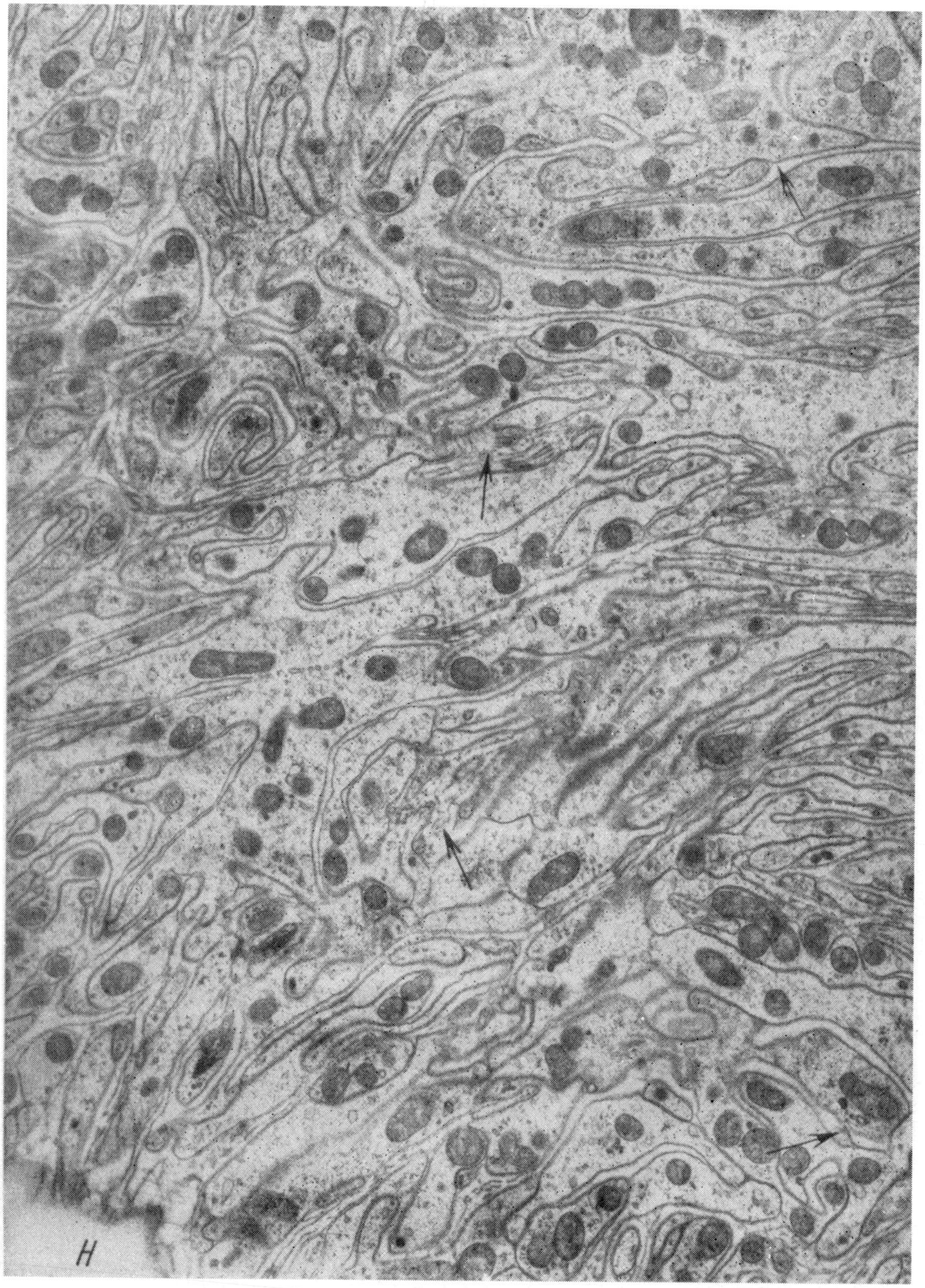

Plate 62. Fig. 4. Surface area of prothoracic gland (*Leucophaea* male, seventh instar nymph, 4 days after moult) showing elaborate interdigitation of processes of parenchymal cells separated by intercullular channels (*arrows*). The connective tissue sheath facing the haemocoel *H* is cut somewhat tangentially. X 7,300 (From Scharrer, 1964b).

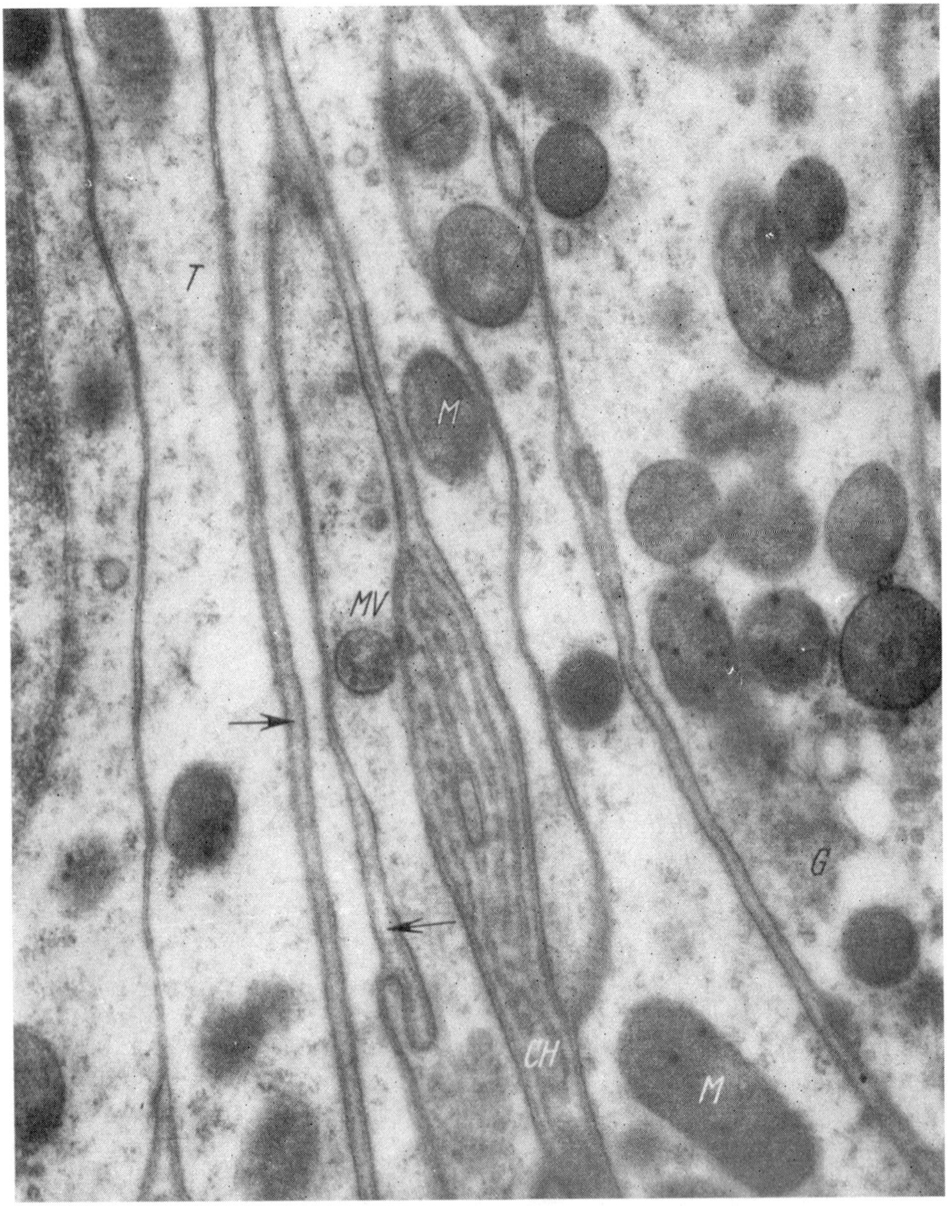

Plate 63. Fig. 5. Detail of extracellular channels (*Leucophaea,* seventh instar female, fixed shortly after moult). In narrow clefts (*arrows)* the phasmalemma is coated with an external lamina (boundary membrane). Multiple layers of the same material appear in wider channels *CH*. *G*. Golgi apparatus, *M* mitochondrion, *MV* multivesicular body, *T*. tonofilament. X 25,000 (From Scharrer, 1964b)

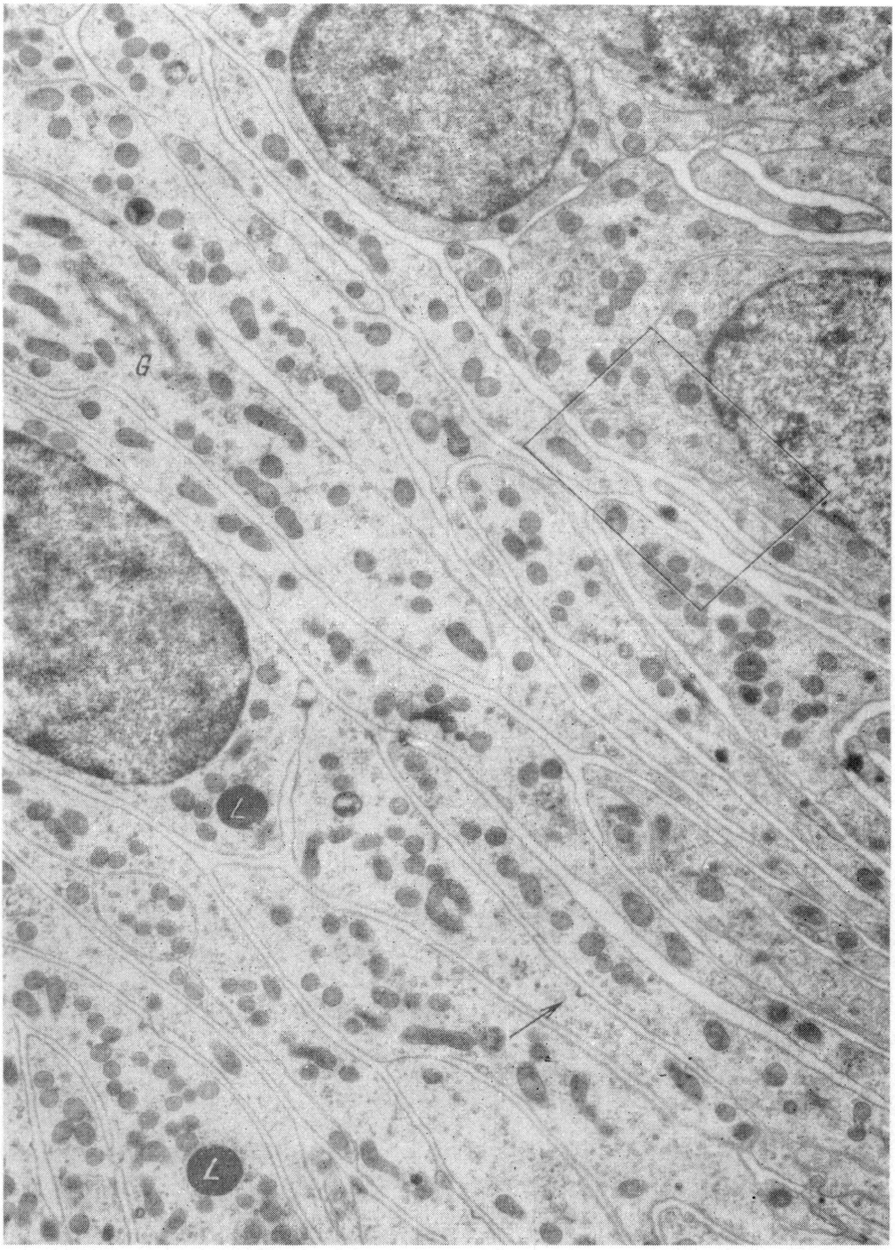

Plate 64. Fig. 6. Peripheral zone of prothoracic gland (*Leucophaea*, last instar female, 8 days after moult). Note relatively dense cytoplasm in cell group at uppe right, and distinct channels separating cellular elements. *G.* Golgi apparatus, L. lysosome-like body, (*arrow*) micropinocytotic pit. Area indicated by rectangle shown at higher power in Fig. 7. X 8,000 (From Scharrer, 1964b)

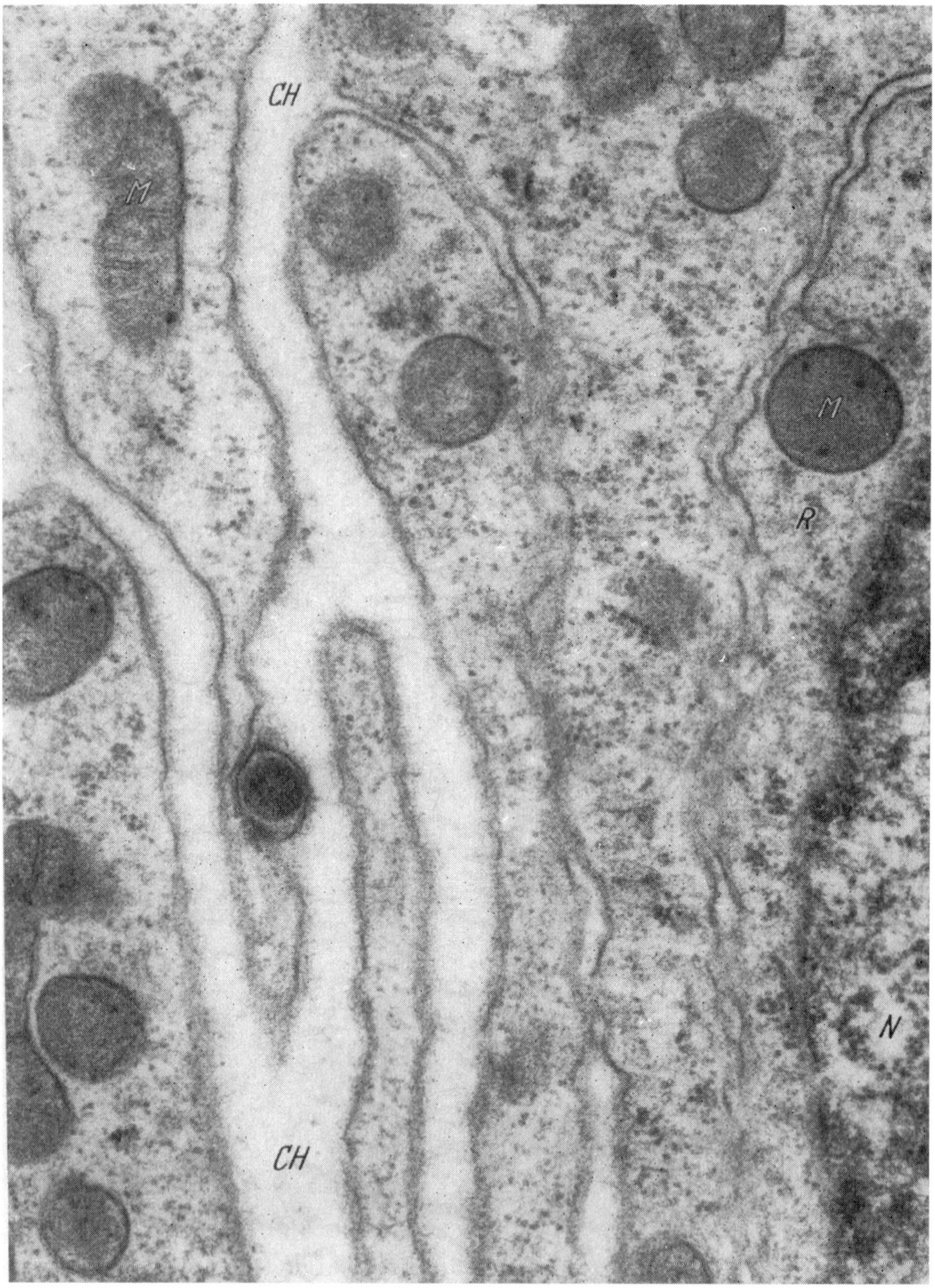

Plate 65. Fig. 7. Detail of area indicated in Fig. 6. Branching extracellular channels *CH* are coated with external laminae of connective tissue matrix. Note abundance of free ribosomes *R*, especially in cell at right where they seem to blend with those of the nucleus *N*. Some mitochondria *M* show dark granules. X 38,400 (From Scharrer, 1964b)

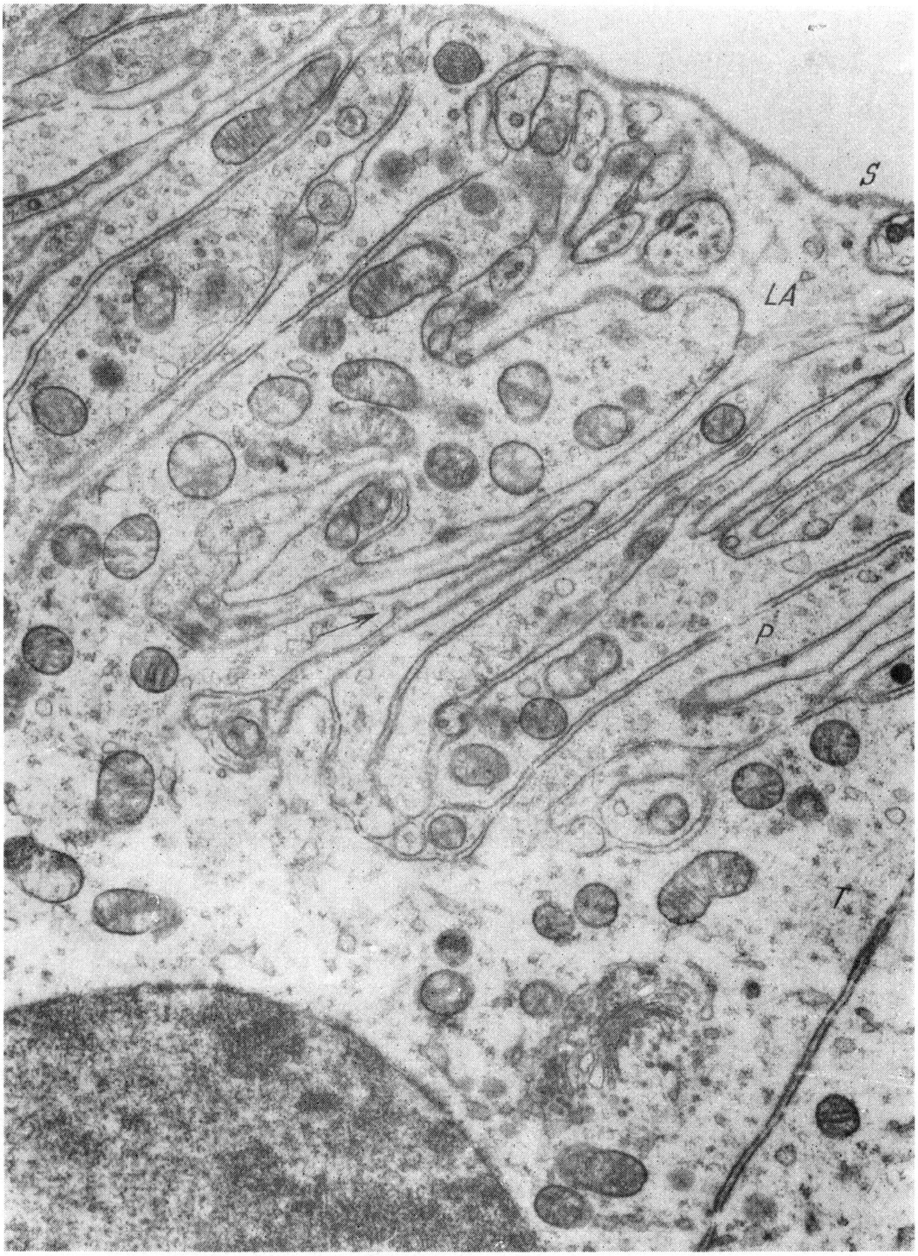

Plate 66. Fig. 8. Surface view of prothoracic gland *(Leucophaea*, sixth instar female, 20 days after moult). The connective tissue matrix forming the thin sheath *S* extends into subsurface lacunae *LA*, distentions of intercellular clefts with which they communicate. In this compartment the matrix becomes separated into sheets or threads. *P* process of parenchymal cell, (*arrow)* micropinocytotic pit, *T* tonofilament. X 16,000 (From Scharrer, 1964b)

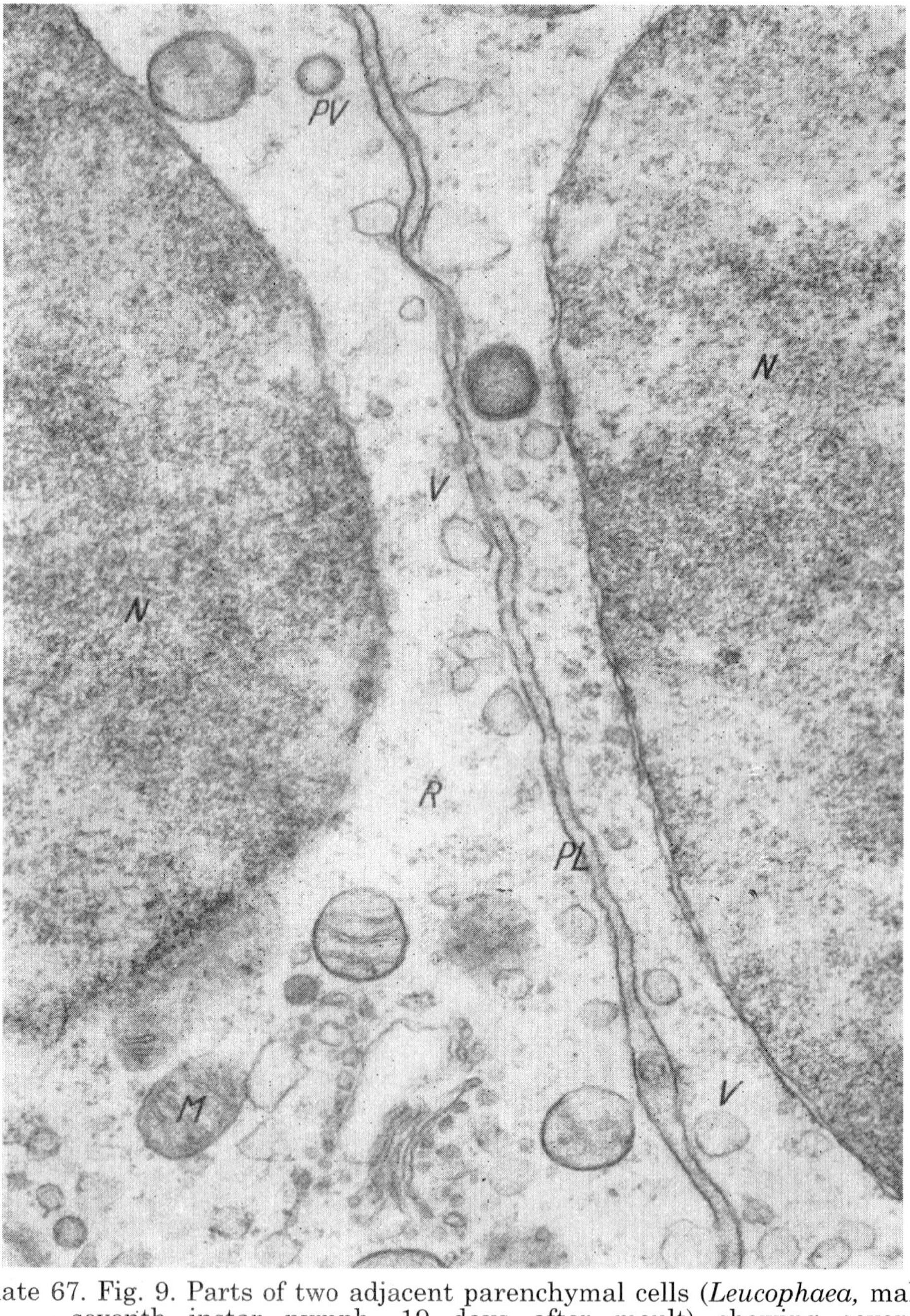

Plate 67. Fig. 9. Parts of two adjacent parenchymal cells (*Leucophaea,* male seventh instar nymph, 19 days after moult) showing several cytoplasmic vesicles *V* in close proximity to, and probably arising from, the plasmalemma *PL*. *PV* coated micropinocytotic vesicle, *M* mitochondrion, *N* nucleus, *R* free ribosomes. X 26,000 (From Scharrer, 1964b)

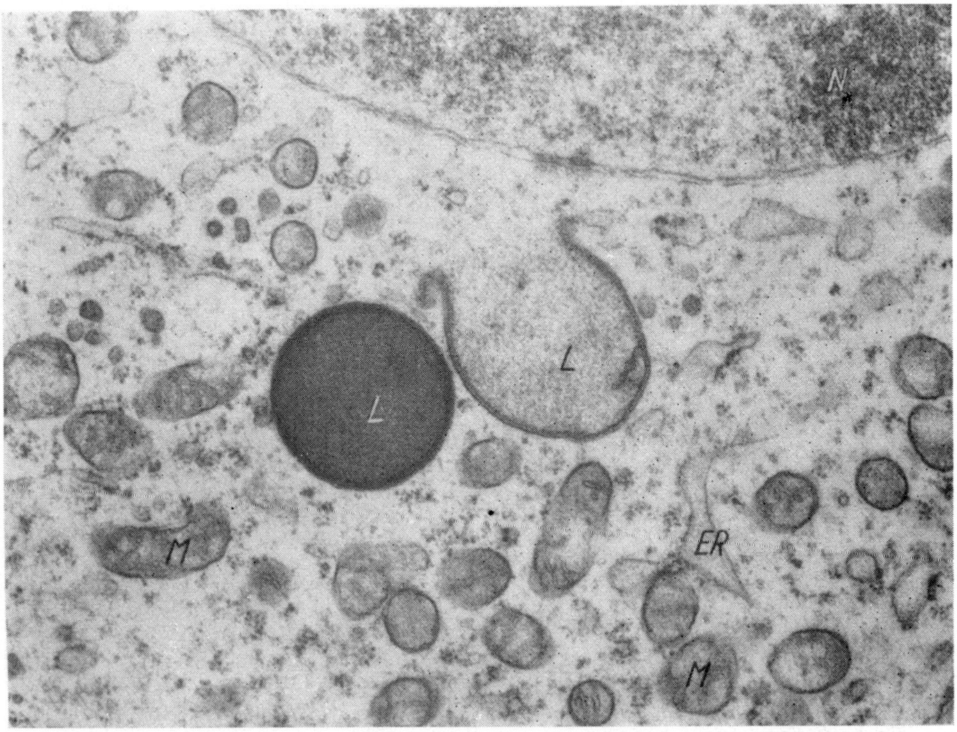

Plate 68. Fig. 10. Detail of parenchymal cell (*Blaberus,* male last instar nymph, 16 days after moult). Note two `lysomoes' *L*, one with electron dense content and concentric membranes, the other with less dense interior and discontinuous membrane. *ER* cistern studded with relatively few ribosomes, *M* mitochondrion, *N* nucleus. X 24,000 (From Scharrer, 1964b)

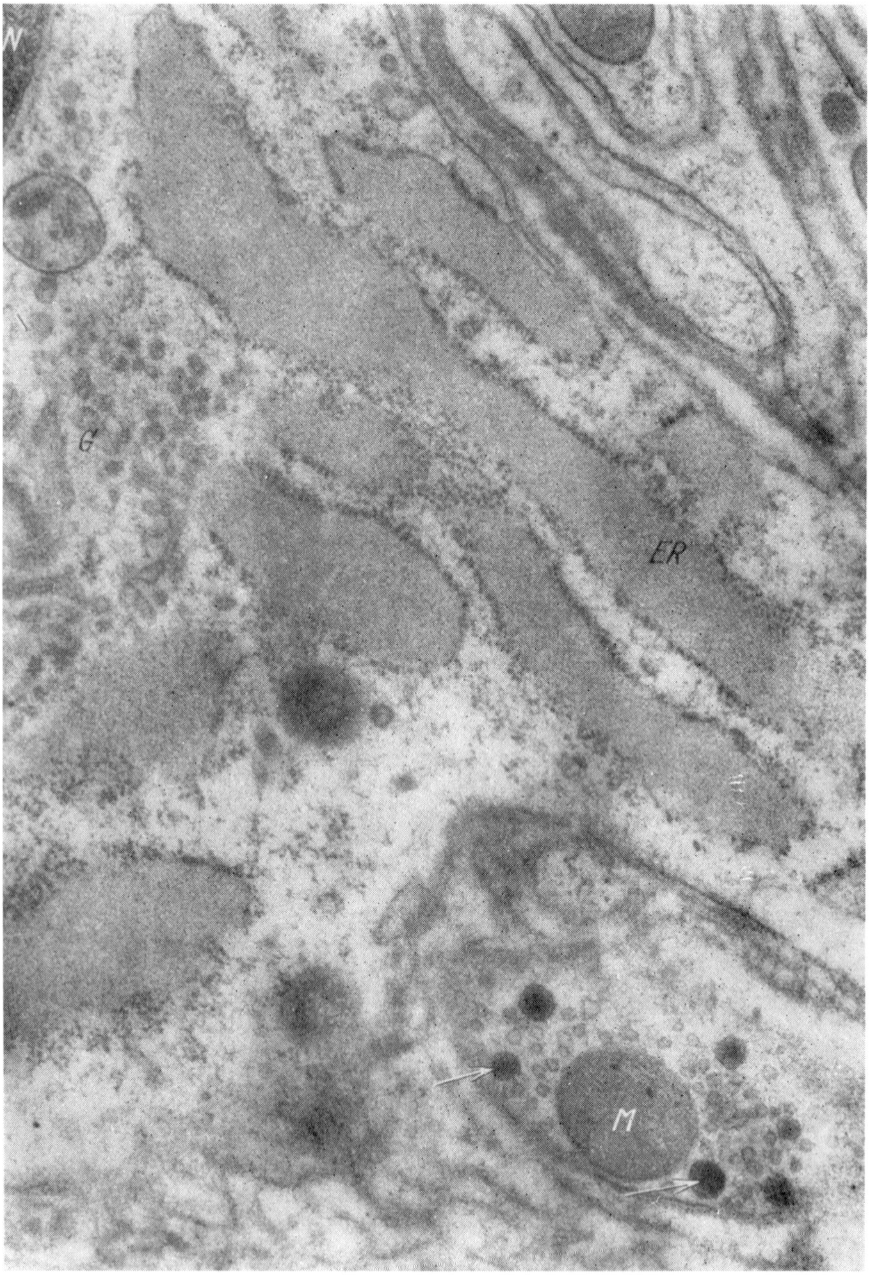

Plate 69. Fig. 11 Part of cell (*Leucophaea,* male seventh instar nymph, 4 days after moult) showing prominent ergastoplasm with abundant intracisternal content of moderate electron density *ER*, and nerve terminal with `synaptic vesicles', electron dense neurosecretory granules (*arrows*) and mitochondrion *M*. *G* Golgi vesicles, *N* nucleus. X 36,000 (From Scharrer, 1964b)

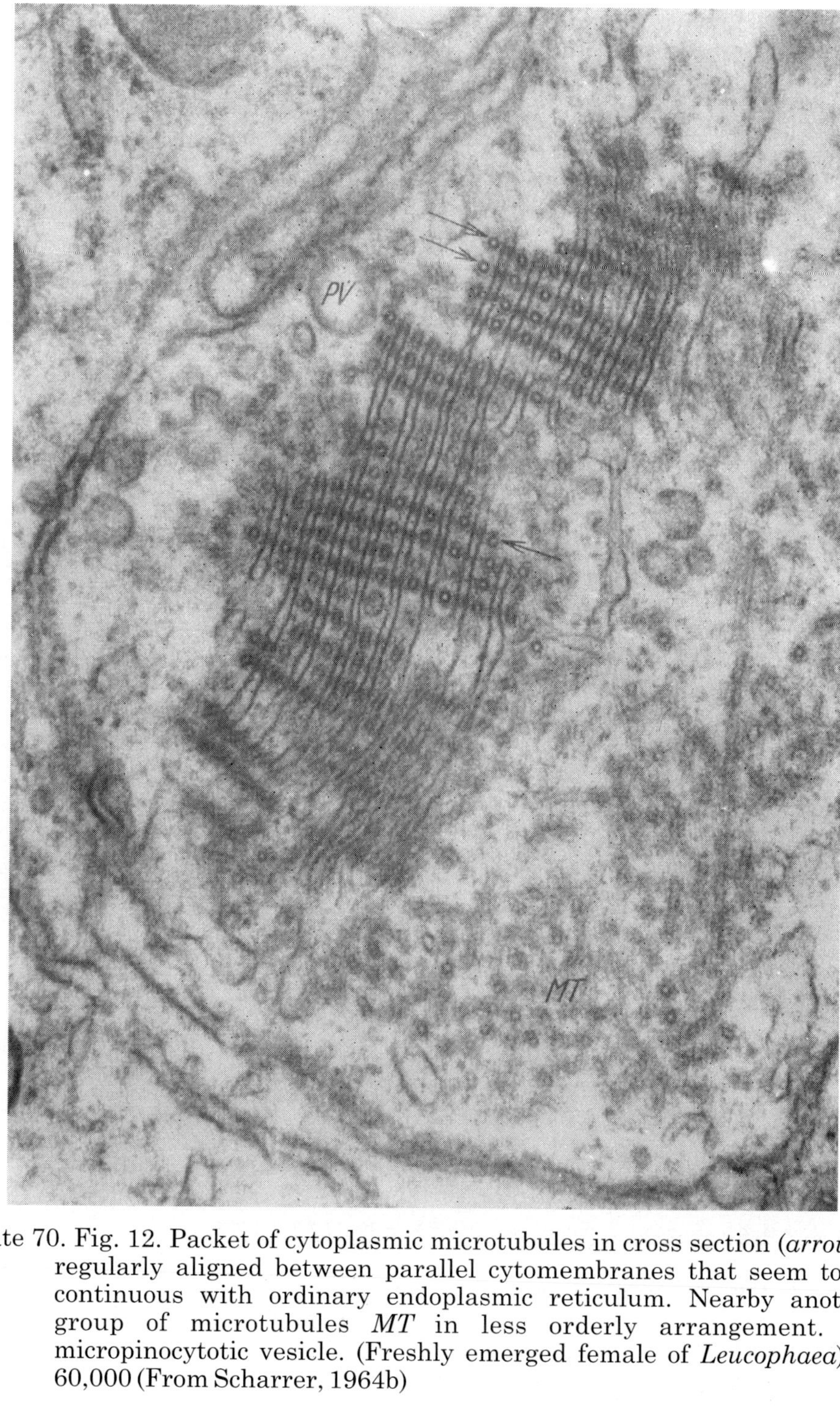

Plate 70. Fig. 12. Packet of cytoplasmic microtubules in cross section (*arrows*), regularly aligned between parallel cytomembranes that seem to be continuous with ordinary endoplasmic reticulum. Nearby another group of microtubules *MT* in less orderly arrangement. *PV* micropinocytotic vesicle. (Freshly emerged female of *Leucophaea*). X 60,000 (From Scharrer, 1964b)

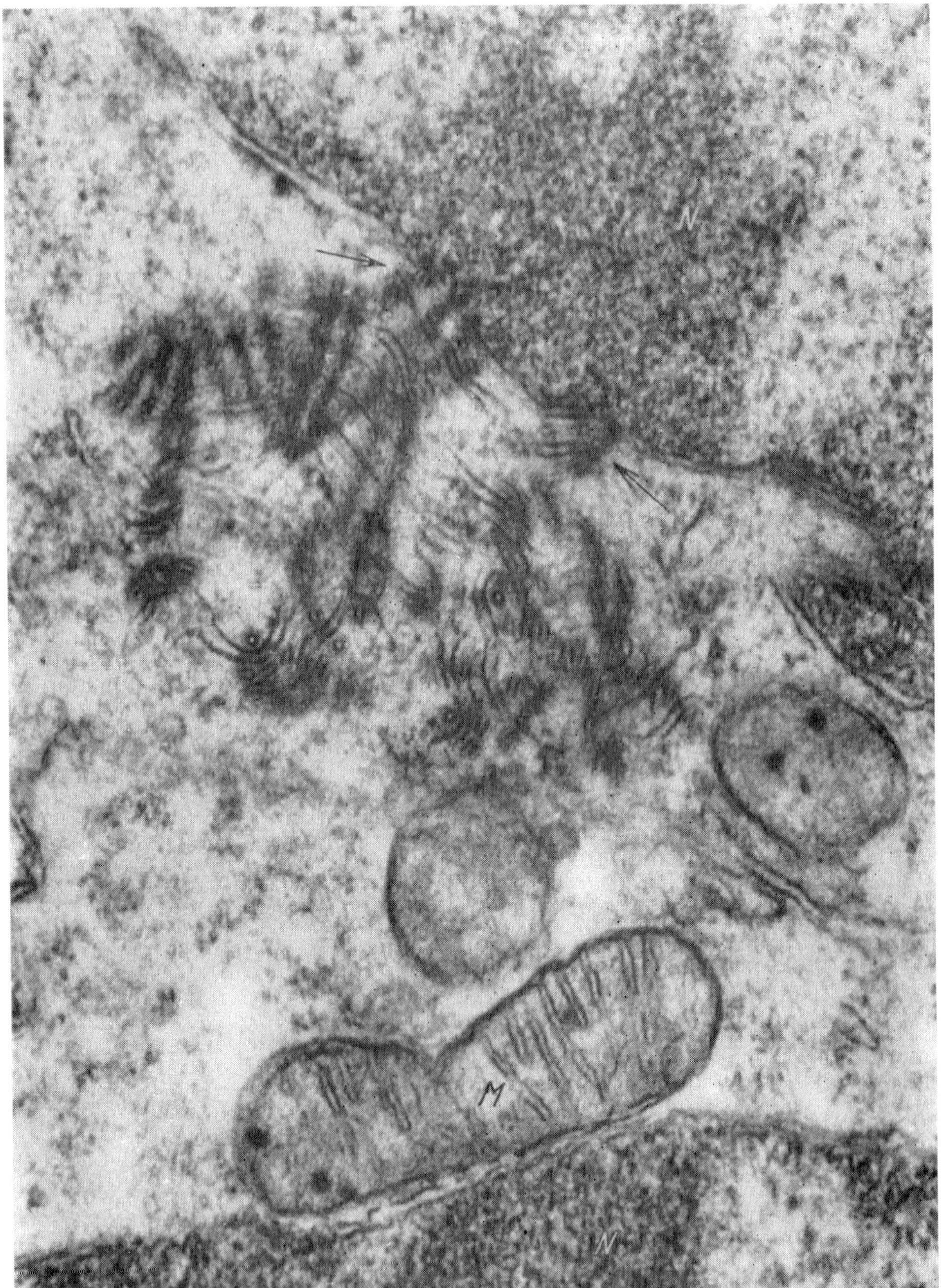

Plate 71. Fig. 13. Another area of microtubules associated with cytoplasmic lamellae from same case as in Fig. 12. Note intimate relationship to nuclear envelope (*arrows*). *M* mitochondrion containing dark granules. *N* nucleus. X 60,000 (From Scharrer, 1964b)

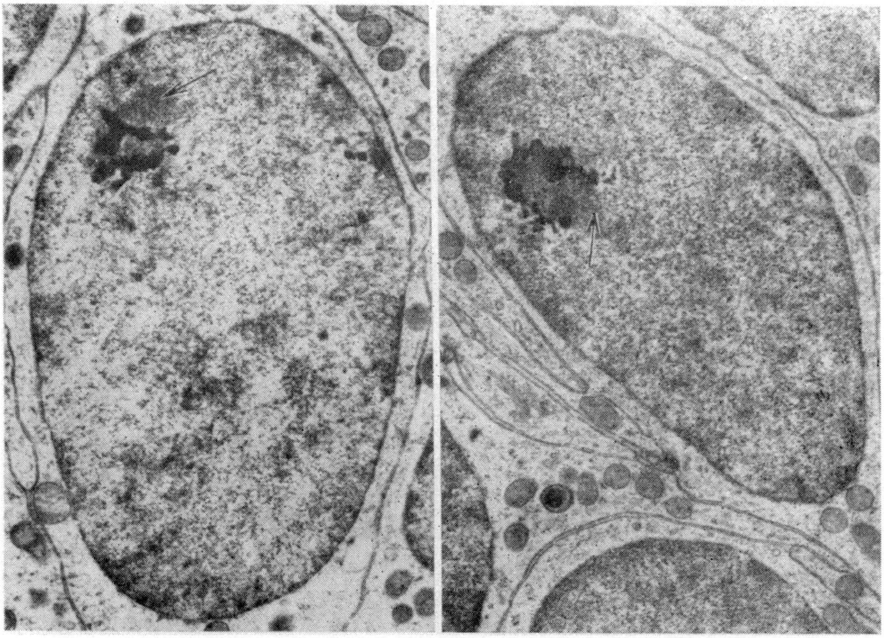

Plate 72. Fig. 14. Two nuclei exhibiting dispersed type of nucleoli, characteristic of early part of intermoult period. Also note spatial relationship with chromatin (*arrows*). At left : cell from *Leucophaea* female, sixth instar nymph, 12 days after moult. X 9,600. At right : cell from *Leucophaea*, last instar female, 8 days after moult. X 8,600 (From Scharrer, 1964b)

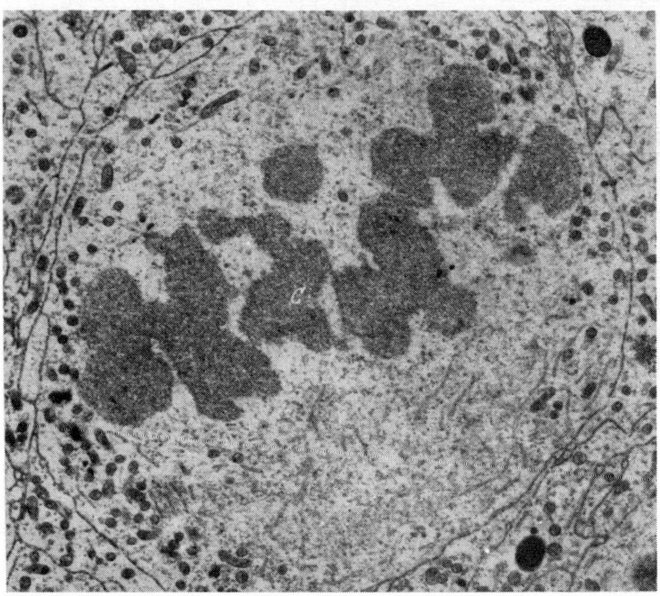

Plate 73. Fig. 15. Mitotic figure in prothoracic gland cell of *Blaberus,* male nymph, 16 days after moult. The cytoplasm contains mitochondria and cytomembranes some of wich may be parts of spindle fibres. *C* chromosomes. X 9,000 (From Scharrer, 1964b)

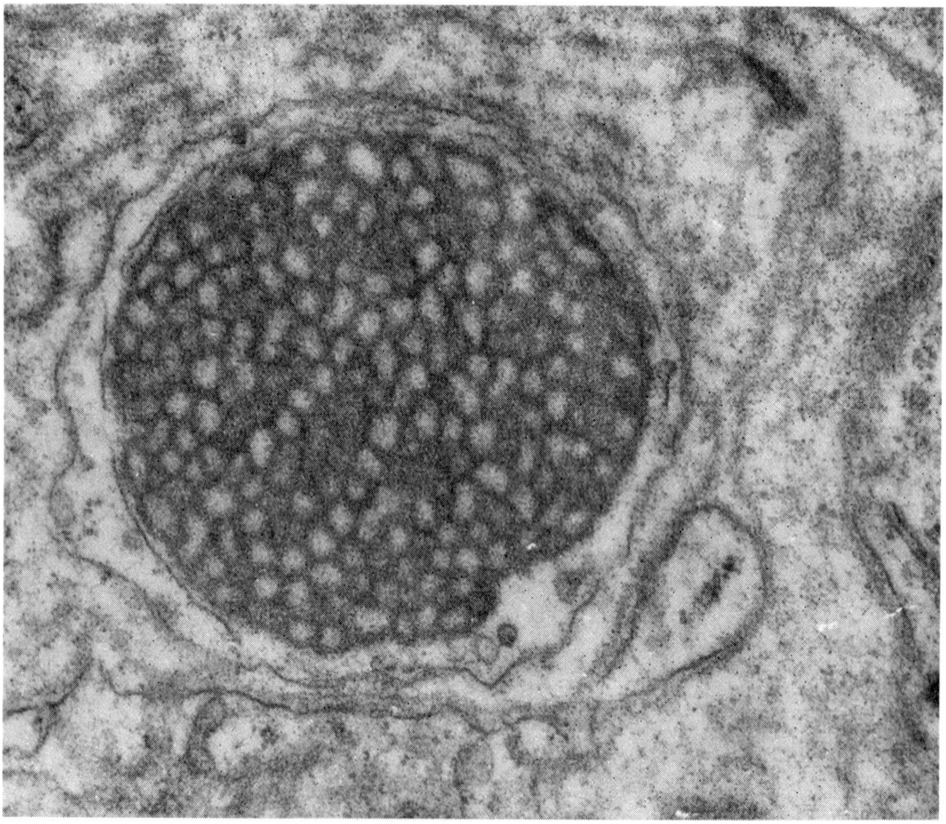

Plate 74. Fig. 16. Inclusion body of unknown nature filling a lacunar space within parenchyma of prothoracic gland of *Leucophaea* X 38,400 (From Scharrer, 1964b)

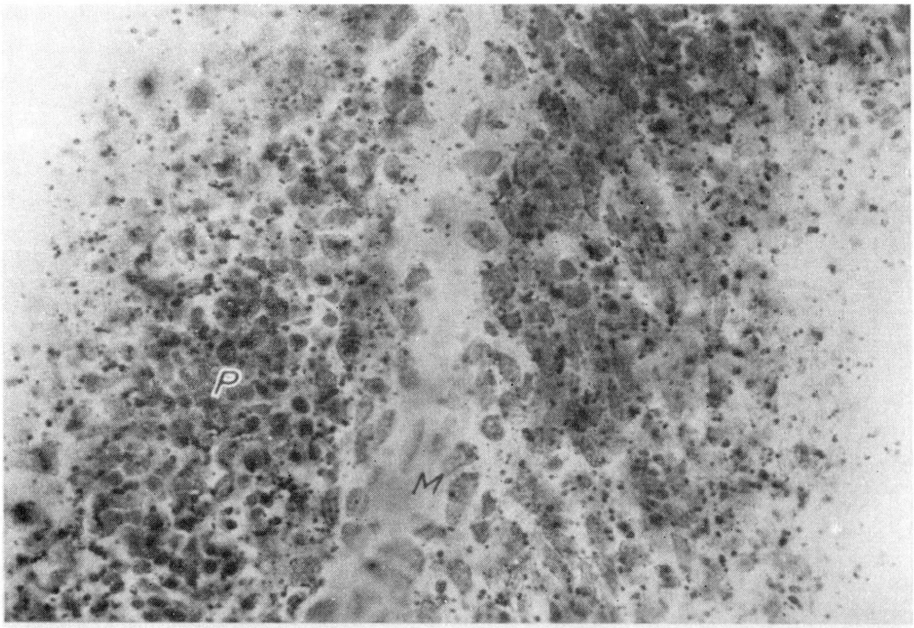

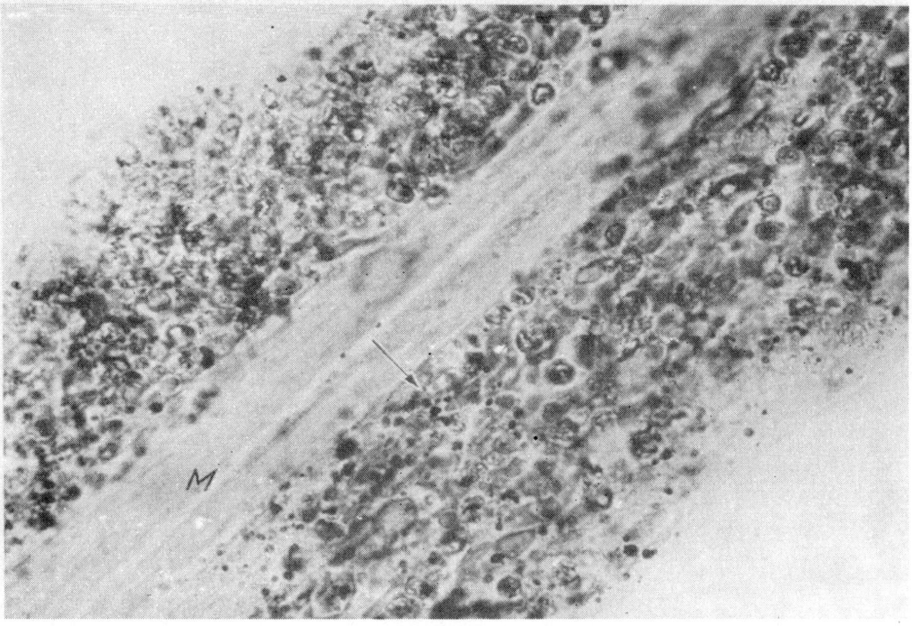

Plate 75. Fig. 1a and b. Prothoracic glands of adult *Leucophaea*, Whole mounts, Gomori's acid phosphatase reaction a Male, day of emergence, gland still largely intact. Note broad layers of parenchymal cells *P* on either side of central muscle *M*. Numerous dark granules are reaction sites, presumed to be lysosomes. X 500. b Female, 2½ days old. Amount of parenchymal tissue in relation to diameter of muscle has markedly diminished. Degenerating cells show reaction product, probably in autophagic vacuoles (arrow). X 630 (From Scharrer, 1966)

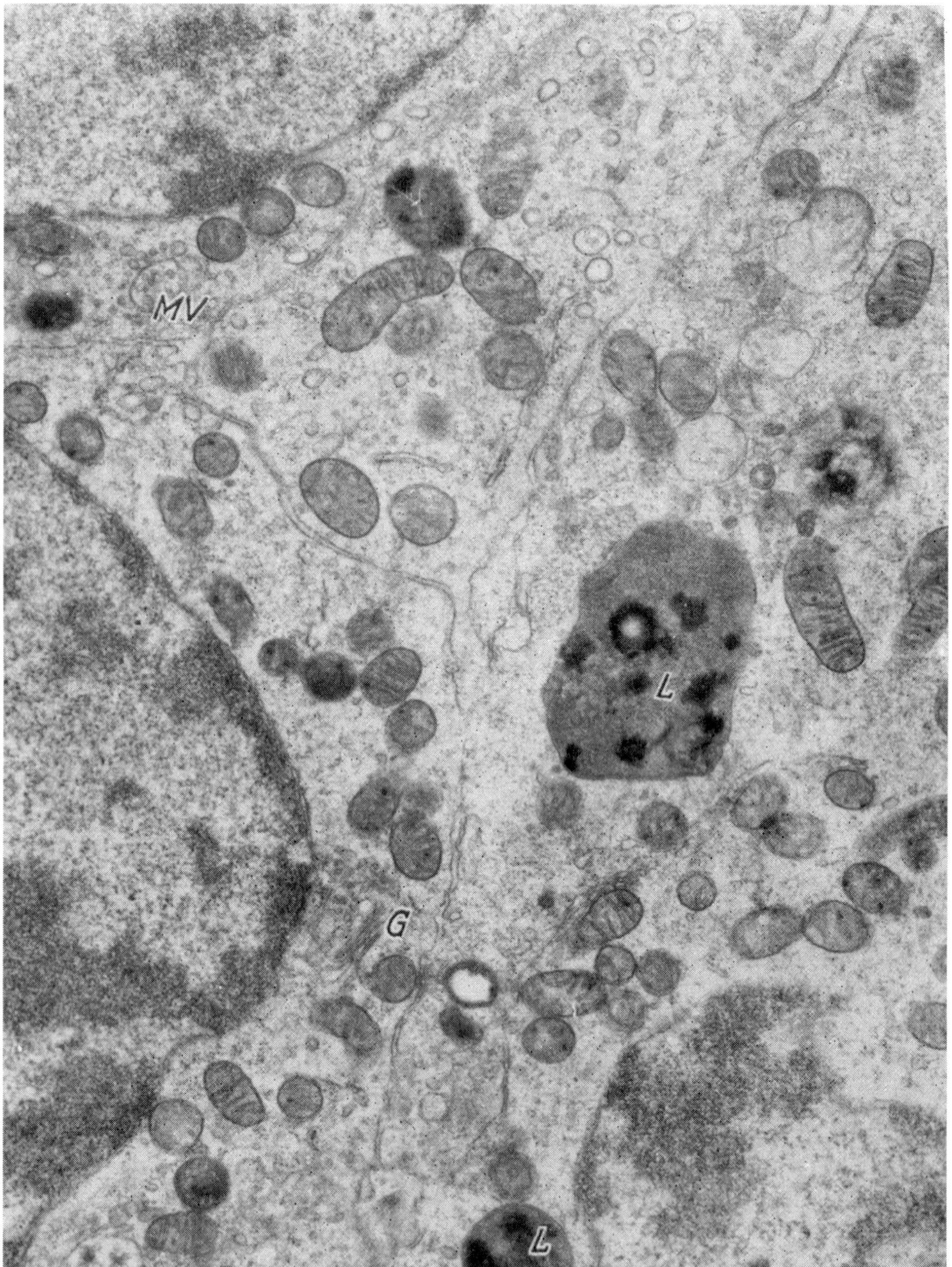

Plate 76. Fig. 2. *Leucophaea* adult female, day of emergence. Electron micrograph of parts of prothoracic gland cells that are still practically intact. Note lysosomes *L*, multivesicular body *MV*, Golgi apparatus *G*. 1% OsO_4. X 16,500 (From Scharrer, 1966)

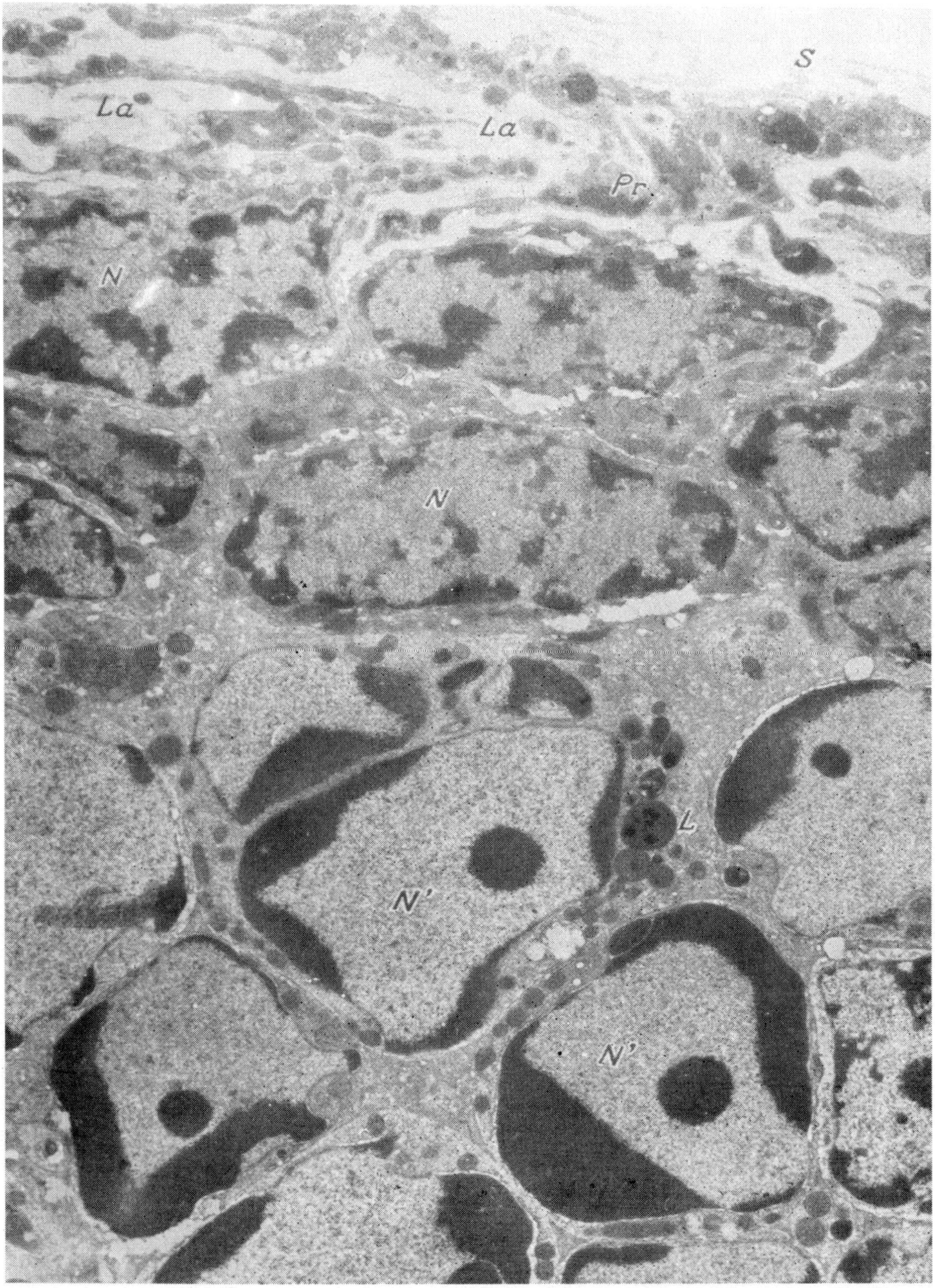

Plate 77. Fig. 3. *Leucophaea* female, one-day-old adult. Part of prothoracic gland near surface (top) shows cells practically still normal *N* with their processes *Pr*, next to a group of degeneratng cells with pycnotic nuclei *N' L* lysosomes, *La* lacunae in stroma, *S* thin connective tissue sheath of gland. Glutaraldehyde. X 6,800 (From Scharrer, 1966)

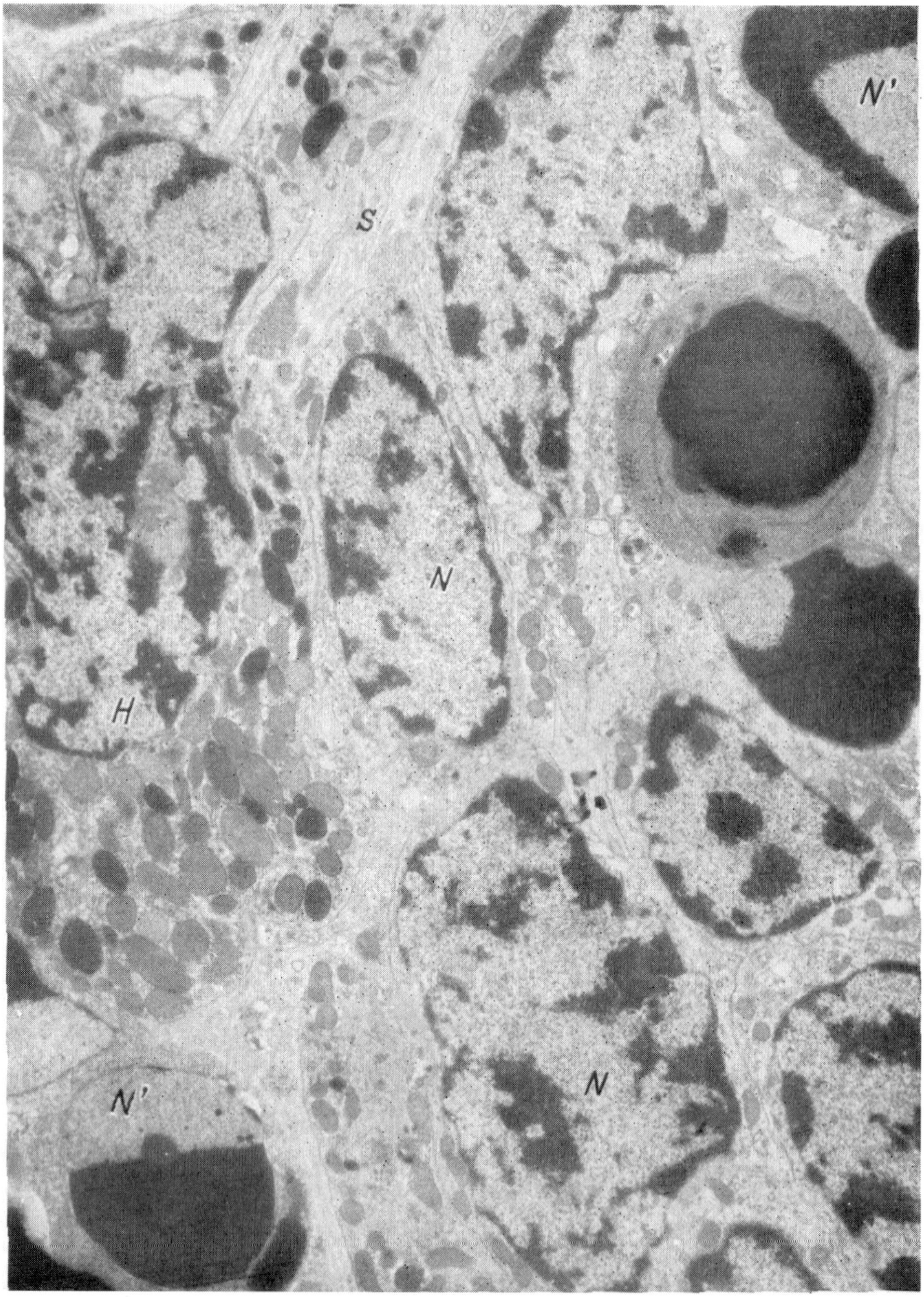

Plate 78. Fig. 4. *Leucophaea,* female, same age and technique as in Fig. 3. Note haemocyte *H* with characteristic cytoplasmic inclusions among normal *N* and regressing parenchymal cells *N'*. *S* connective tissue septum. X 6,800 (From Scharrer, 1966)

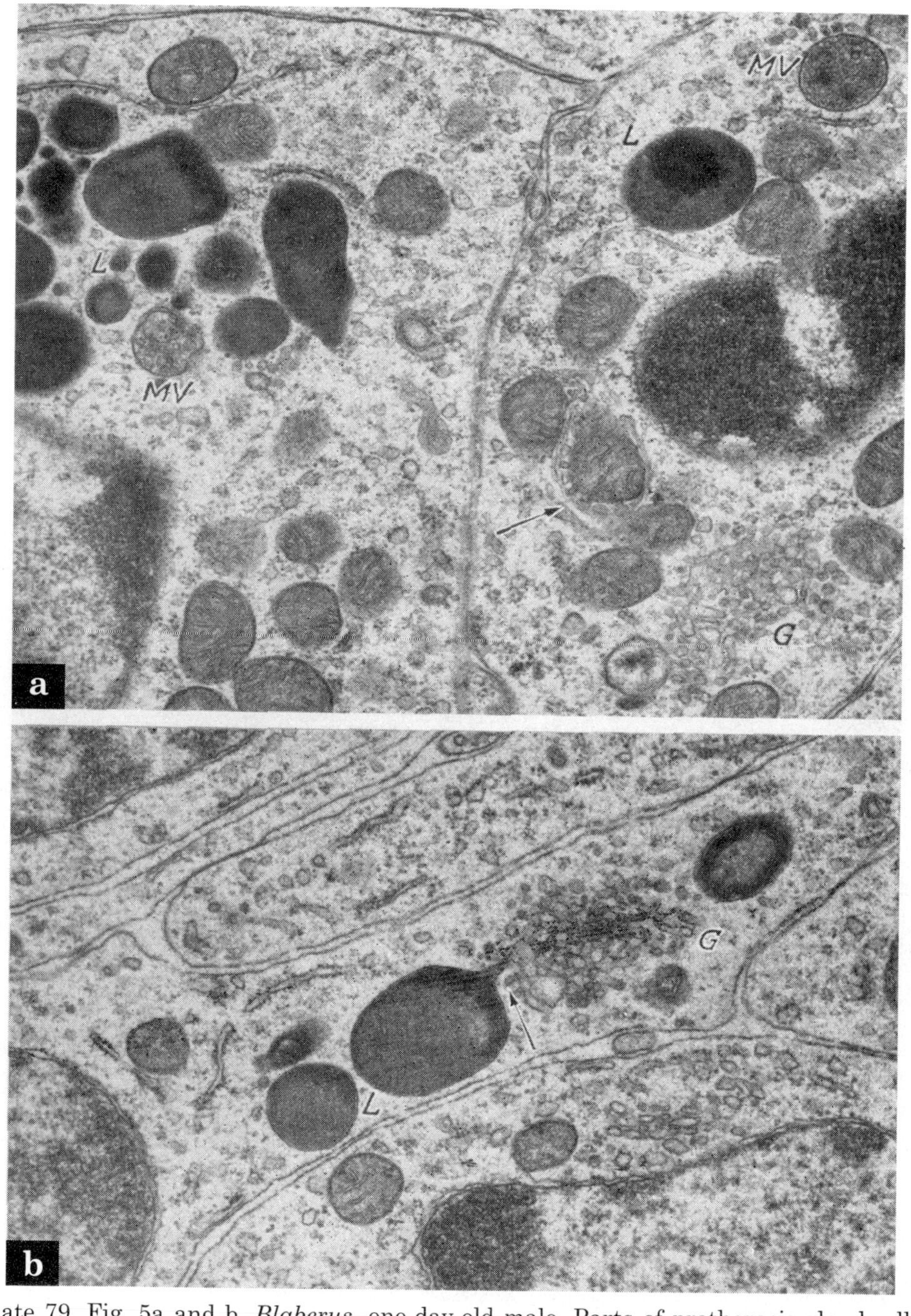

Plate 79. Fig. 5a and b. *Blaberus*, one-day-old male, Parts of prothoracic gland cells showing Golgi elements *G*, multivesicular bodies *MV*, and lysosomes *L*. Arrow in 5a shows corona of small vesicles and membrane surrounding mitochondrion. Arrow in 5b shows relationships between flask-shaped `lysosome' and Golgi system. 2% OsO_4. X 23,200 (From Scharrer, 1966)

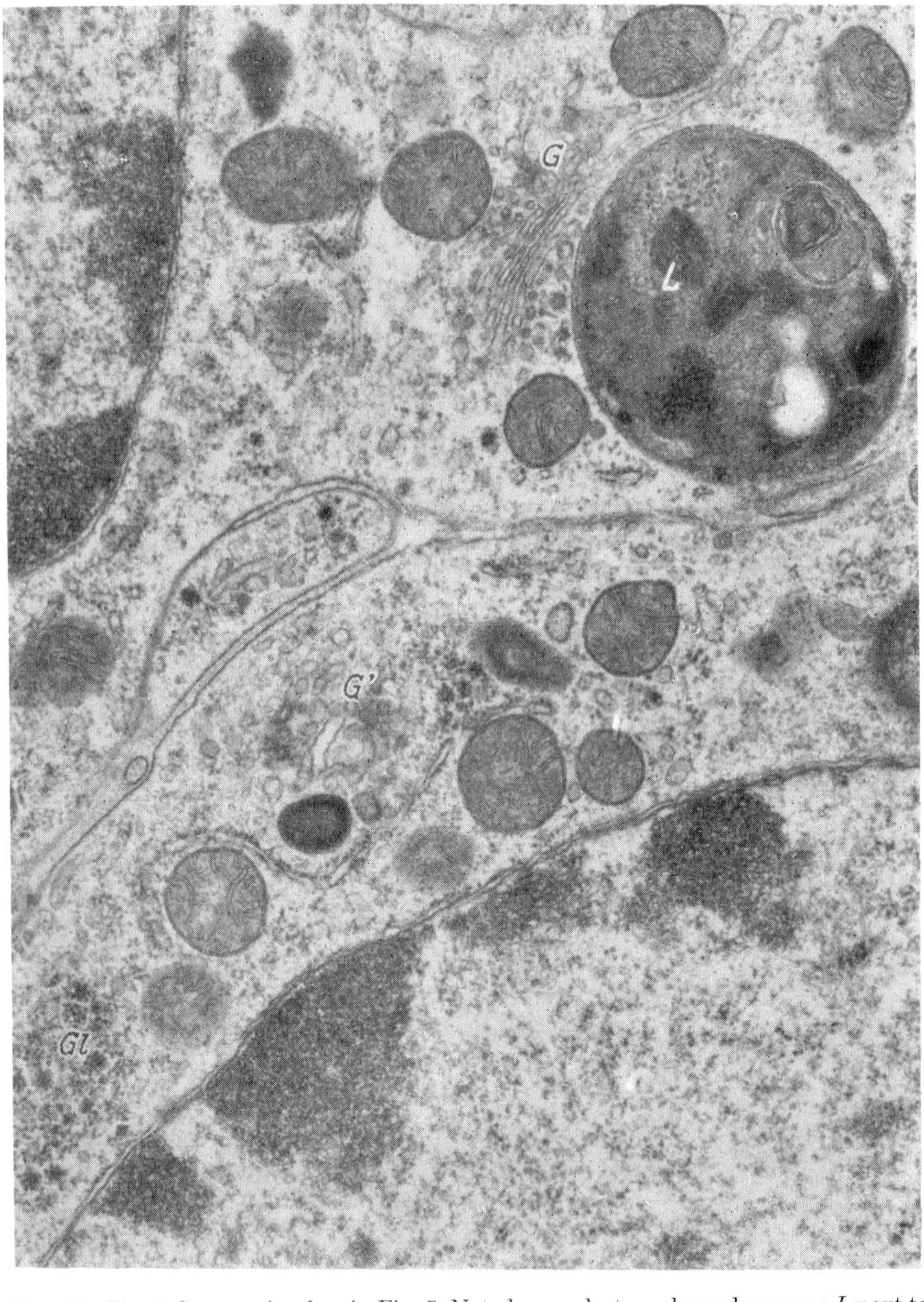

Plate 80. Fig. 6. Same animal as in Fig. 5. Note large, electron dense lysosome *L* next to Golgi element *G* some of whose vesicles have electron dense contens. *G'* modified Golgi elements, perhaps in process of forming membranes for autophagic vacuoles. *Gl* glycogen. X 24,600 (From Scharrer, 1966)

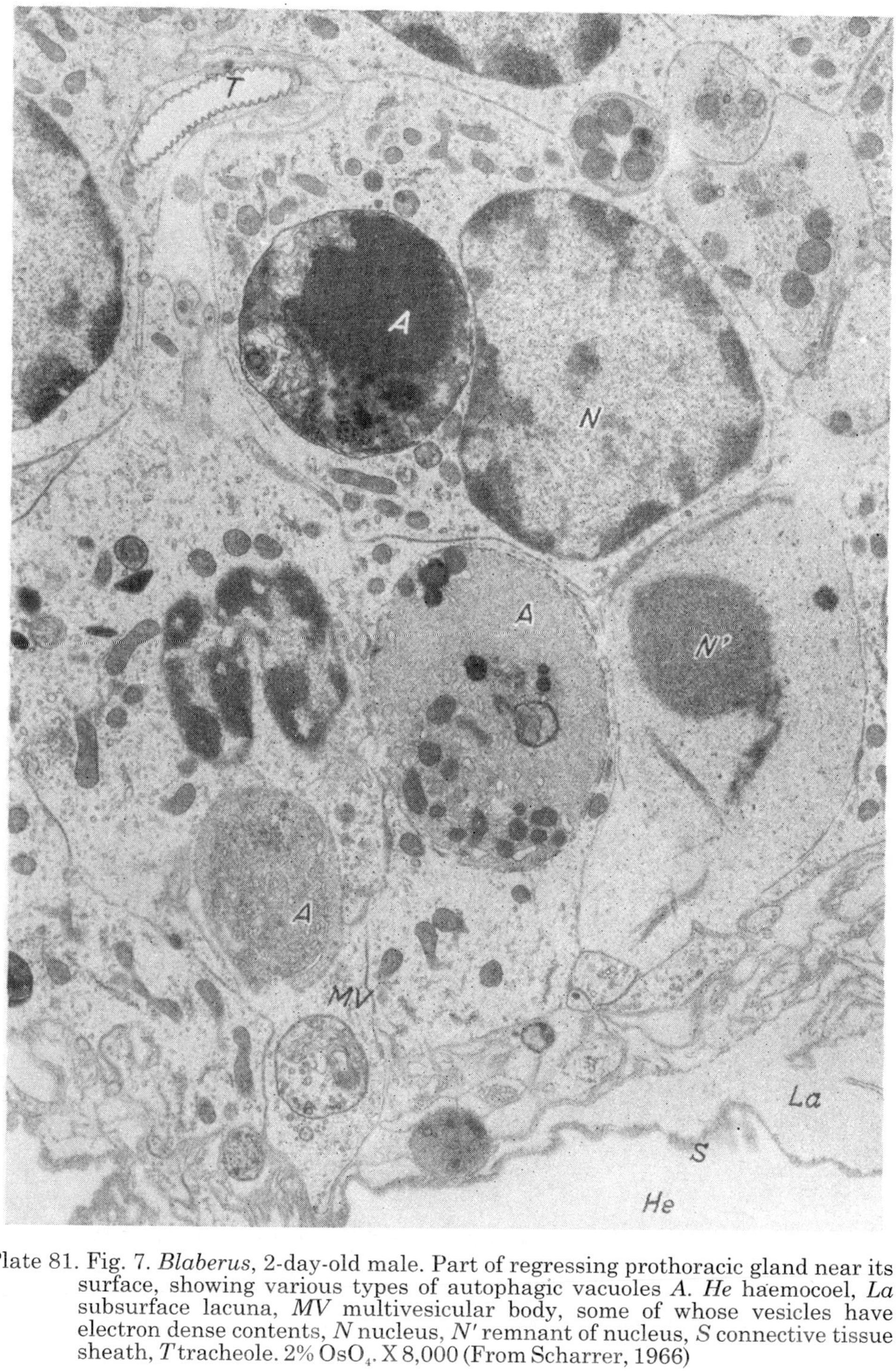

Plate 81. Fig. 7. *Blaberus*, 2-day-old male. Part of regressing prothoracic gland near its surface, showing various types of autophagic vacuoles *A*. *He* haemocoel, *La* subsurface lacuna, *MV* multivesicular body, some of whose vesicles have electron dense contents, *N* nucleus, *N'* remnant of nucleus, *S* connective tissue sheath, *T* tracheole. 2% OsO_4. X 8,000 (From Scharrer, 1966)

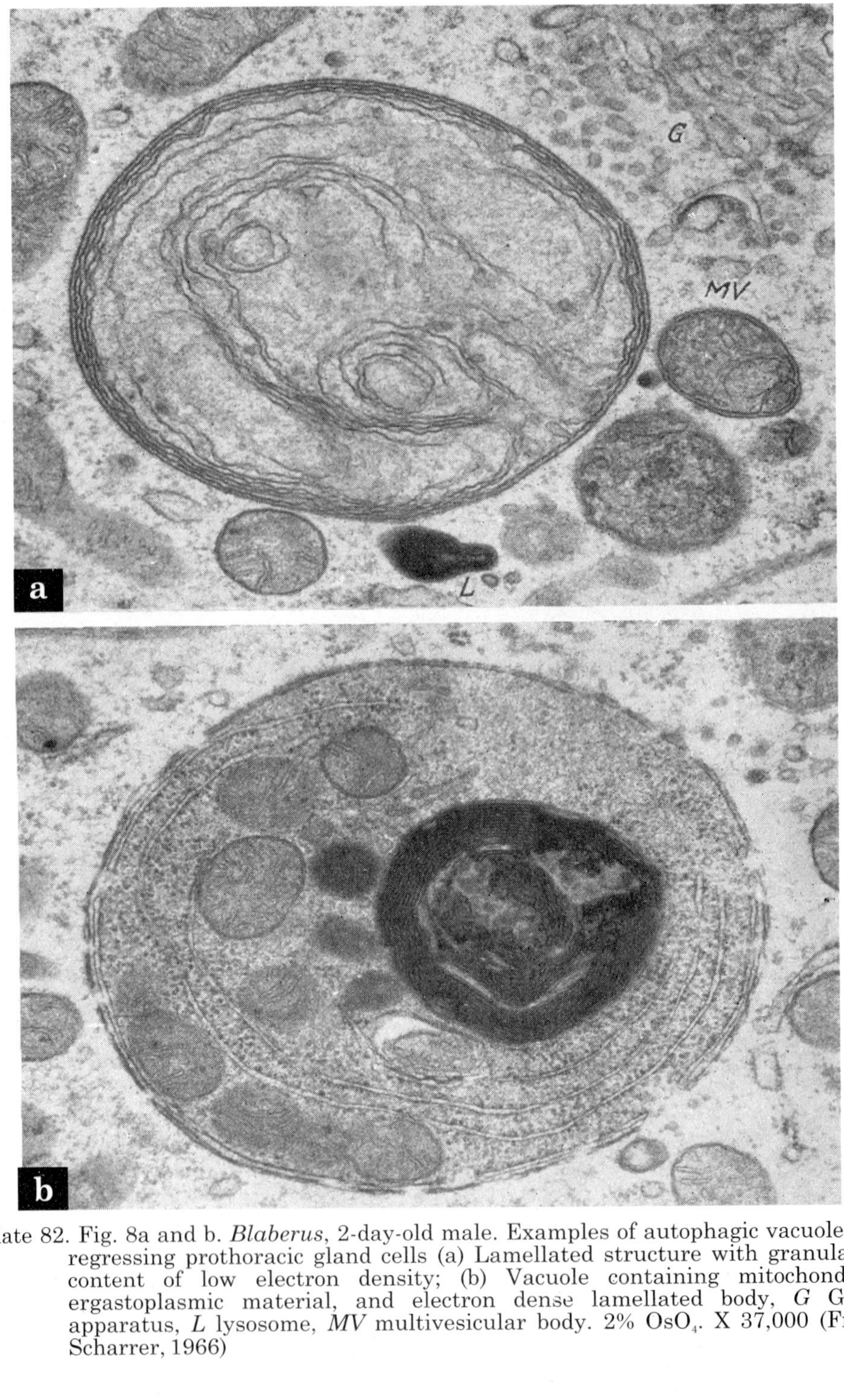

Plate 82. Fig. 8a and b. *Blaberus*, 2-day-old male. Examples of autophagic vacuoles in regressing prothoracic gland cells (a) Lamellated structure with granulated content of low electron density; (b) Vacuole containing mitochondria, ergastoplasmic material, and electron dense lamellated body, *G* Golgi apparatus, *L* lysosome, *MV* multivesicular body. 2% OsO_4. X 37,000 (From Scharrer, 1966)

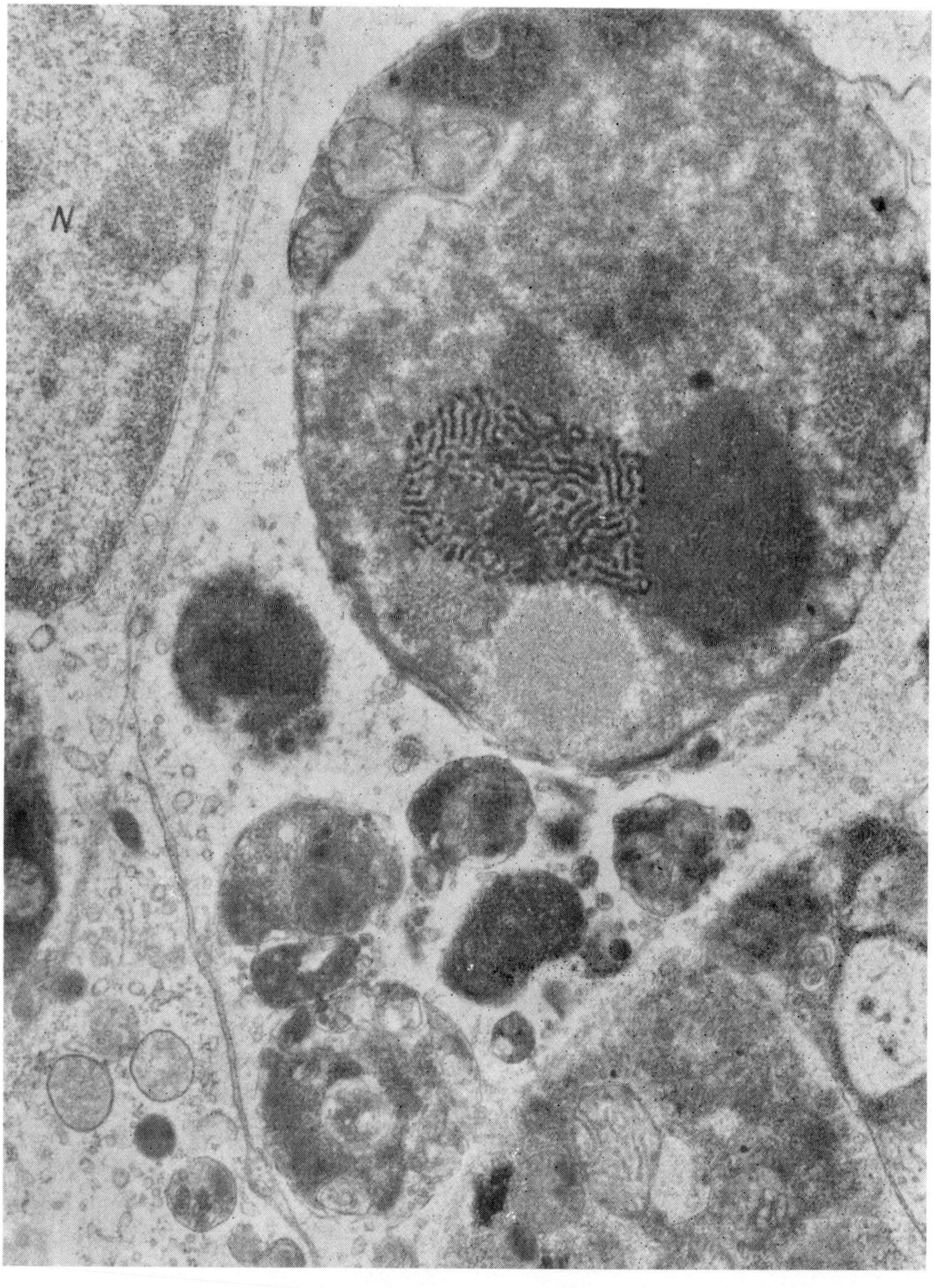

Plate 83. Fig. 9. *Leucophaea,* 3-day-old female. Variously sized structures resembling autophagic vacuoles in disrupted cellular area. Note orderly pattern of dark residual elements in the upper vacuole. The nucleus *N* of the cell to the left is still intact. 2% OsO_4. X 17,000 (From Scharrer, 1966)

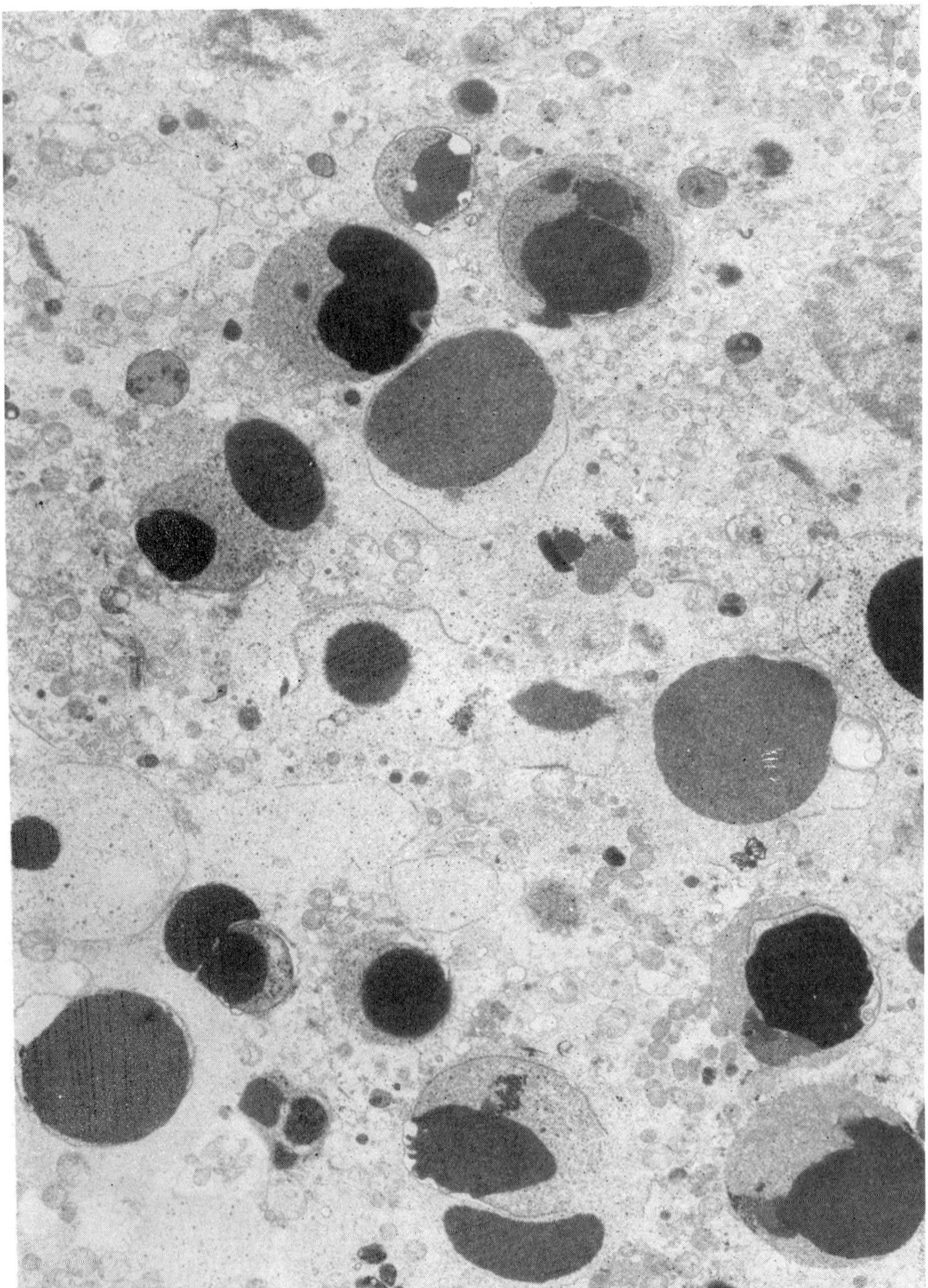

Plate 84. Fig. 10. *Leucophaea,* 5-day-old female. Survey picture of prothoracic gland in advanced stage of regression. Electron dense globular bodies are nuclear remnants. Cellular integrity is now completely lost. 1% OsO_4. X 5,600 (From Scharrer, 1966)

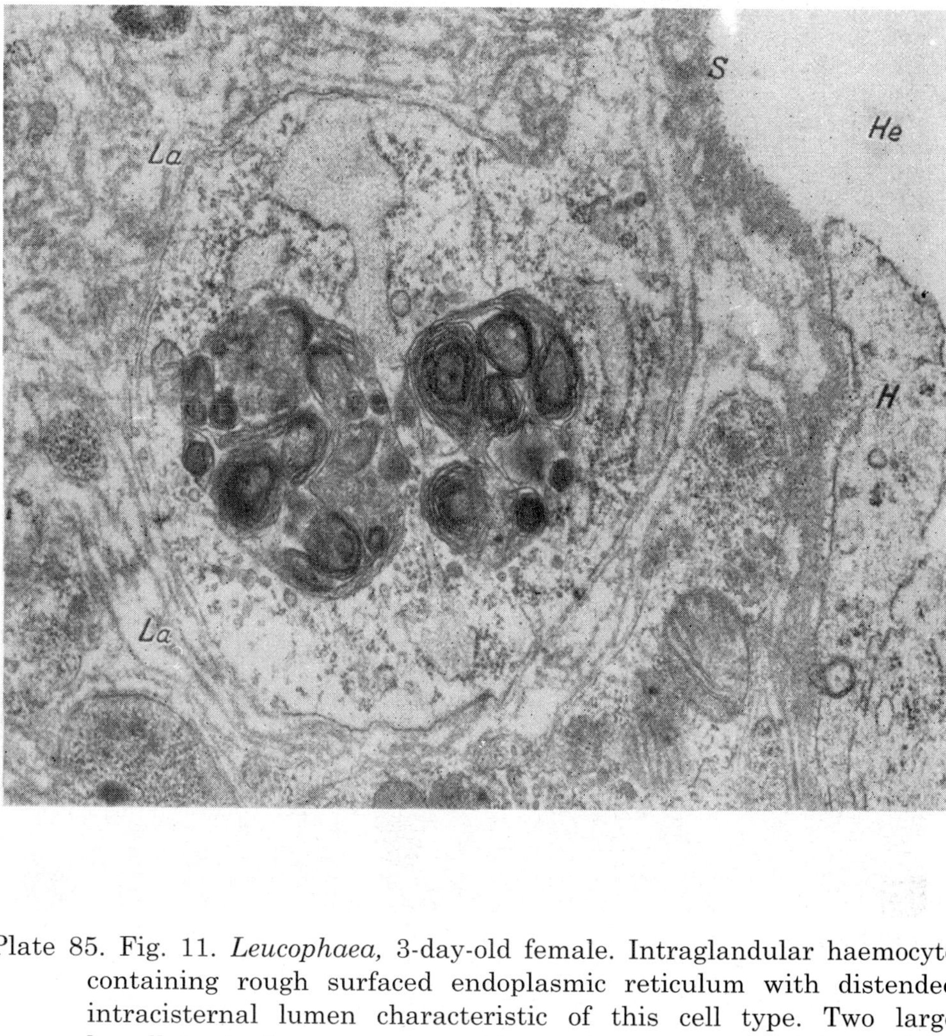

Plate 85. Fig. 11. *Leucophaea,* 3-day-old female. Intraglandular haemocyte containing rough surfaced endoplasmic reticulum with distended intracisternal lumen characteristic of this cell type. Two large lamellated bodies seem to be residues of degeneratng prothoracic gland cells that have been engulfed by the blood cell. *H* part of extraglandular haemocyte, *He* haemocoel, *La* lacunar spaces containing layers of connective tissue matrix, *S.* connective tissue sheath. 1% OsO_4. X 23,400 (From Scharrer, 1966)

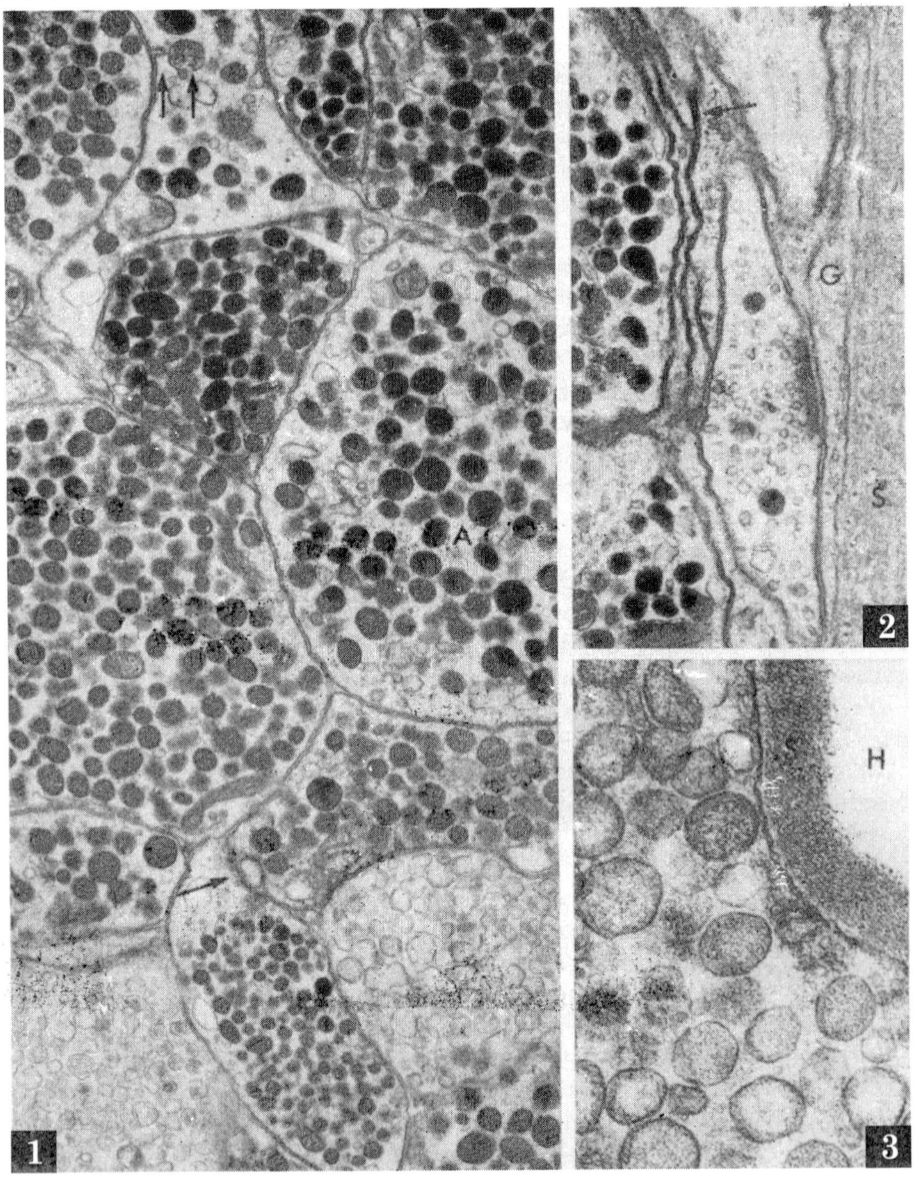

Plate 86. Fig. 1. Low power survey of corpus cardiacum of adult male of *Periplaneta* showing profiles of contiguous neurosecretory neurons varying as to size, shape and electron density of contents of secretory vesicles. Note ``synaptoid'' contact area (arrow) between axon with small granules (B fibre?) and axon with large granules (A fibre), variously shaped electron-lucent vesicles interspersed with A-type granules (*A*), multivesicular body and invagination of plasma membrane (double arrow). X 17,600 Fig. 2. Synaptoid structure in perisympathetic neurohaemal organ of adult female of *Blaberus*. Note electron-dense deposit below plasma membrane facing thin glial process (G), numerous small electron-lucent and few dense-core vesicles. S: extracellular stroma. Arrow : septate desmosome. X 26,000. Fig. 3. Part of neurosecretory axon, containing large vesicles with moderately dense granular or lucent content, facing sheath (S) of corpus cardiacum of male adult *Periplaneta*. Small cluster of synaptoid vesicles and dense material subjacent to axolemma. H : haemocoel. X 41,600 (From Scharrer, 1968)

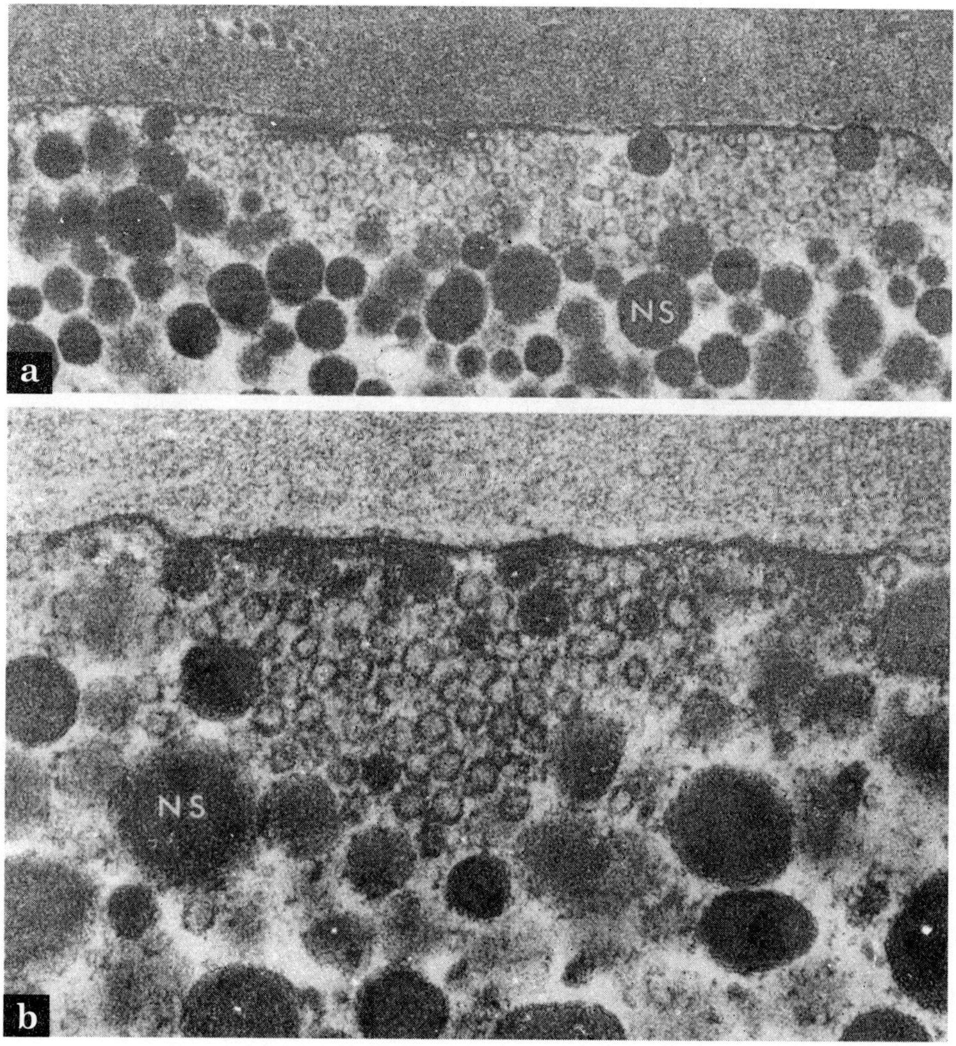

Plate 87. Fig. 4a and b. Sites of release of neurosecretory substance facing extracellular stroma in corpus cardiacum of adult female of *Leucophaea*, fixed on day of emergence. Note A-type neurosecretory granules (NS), and numerous small ``clear" and few dense-core vesicle plus diffuse opaque material pressing against axolemma. a X 48,000; b X 74,000 (From Scharrer, 1968)

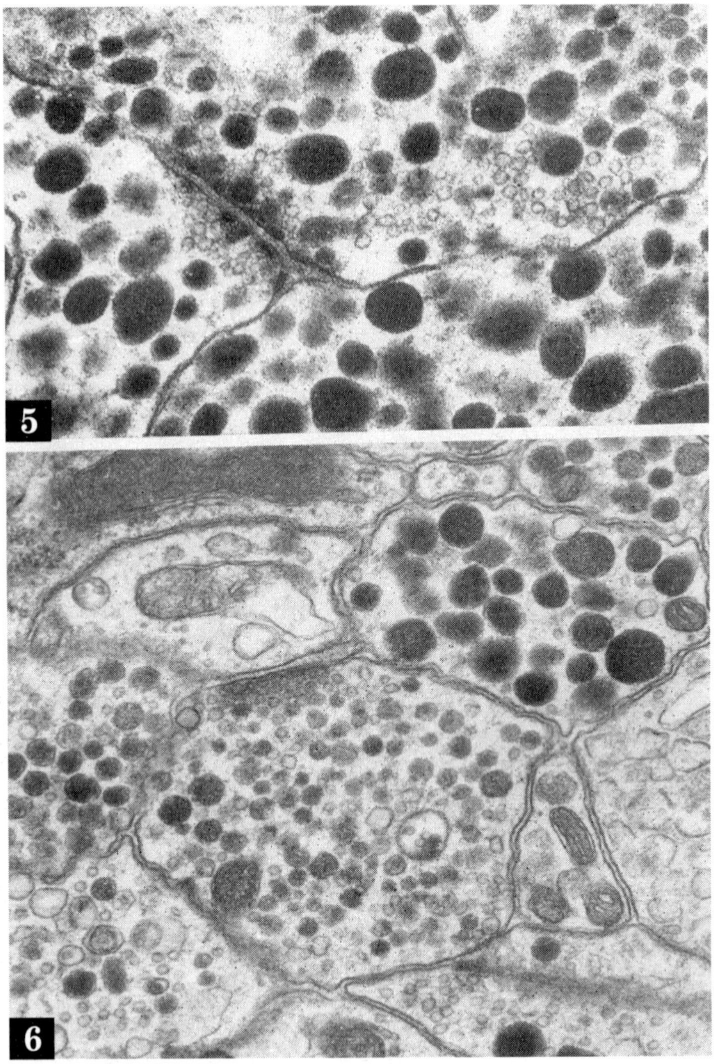

Plate 88. Fig. 5. Paired sites of release of neurosecretory material facing each other across narrow extracellular gap that separates two contiguous A-type fibres and contains a granular material of moderate density. Note synaptoid vesicles and small, more or less rounded masses of opaque material close to plasma membranes on either side. Corpus cardiacum of newly emerged female of *Leucophaea*. X 40,000 Fig. 6. Neurosecretory fibres of corpus cardiacum of adult male of *Periplaneta*. Profile in centre shows small (B-type ?) dense-core vesicles, a multivesicular body, and a prominent cluster of presynaptic vesicles. Fibre profile at lower left shows various presumed stages in the transformation of neurosecretory vesicles. X 33,600 (From Scharrer, 1968)

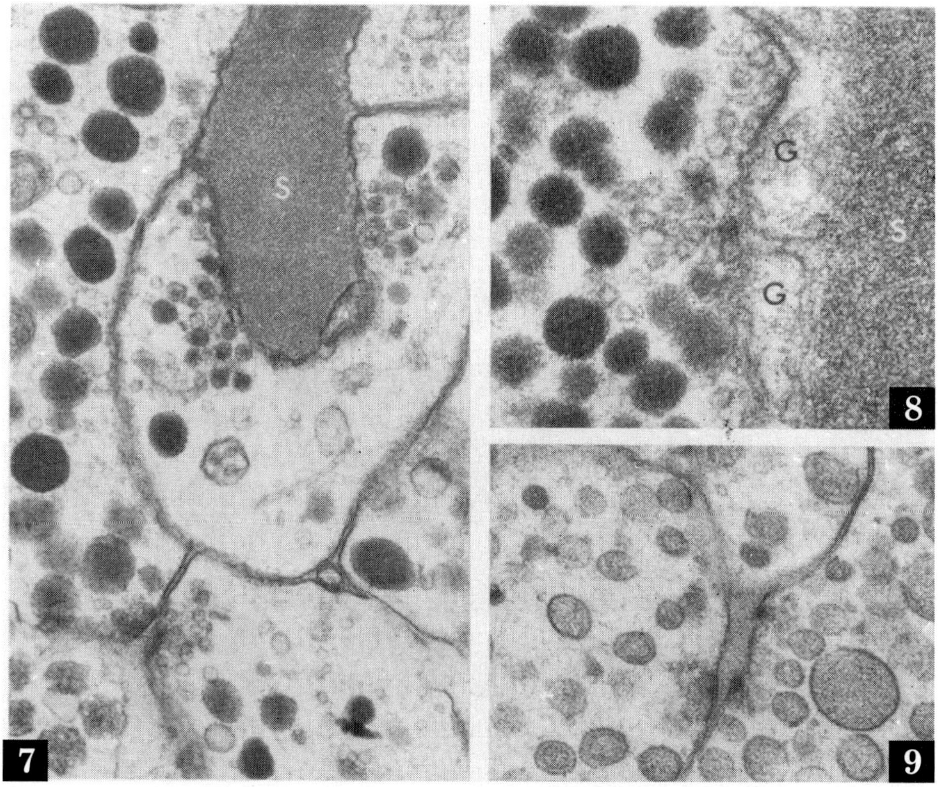

Plate 89. Fig. 7. Connective tissue partition (S) in corpus cardiacum of male adult *Periplaneta*. Small variously filled vesicles accumulate near the axon-stroma boundary. At lower left a pair of small sites of release. X 33.600 Fig. 8. Site of release vis-à-vis narrow extracellular channel separating two glial processes (G) and thus giving access to stroma (S) *Leucophaea*, freshly moulted last instar male nymph. X 58,000 Fig. 9. Paired sites of release of neurosecretory material from two adjacent fibres that contain moderately dense granulated neurosecretory vesicles. Note material of the same appearance filling widened interneuronal gap. Corpus cardiacum of newly emerged female of *Leucophaea*. X 36,000 (From Scharrer, 1968)

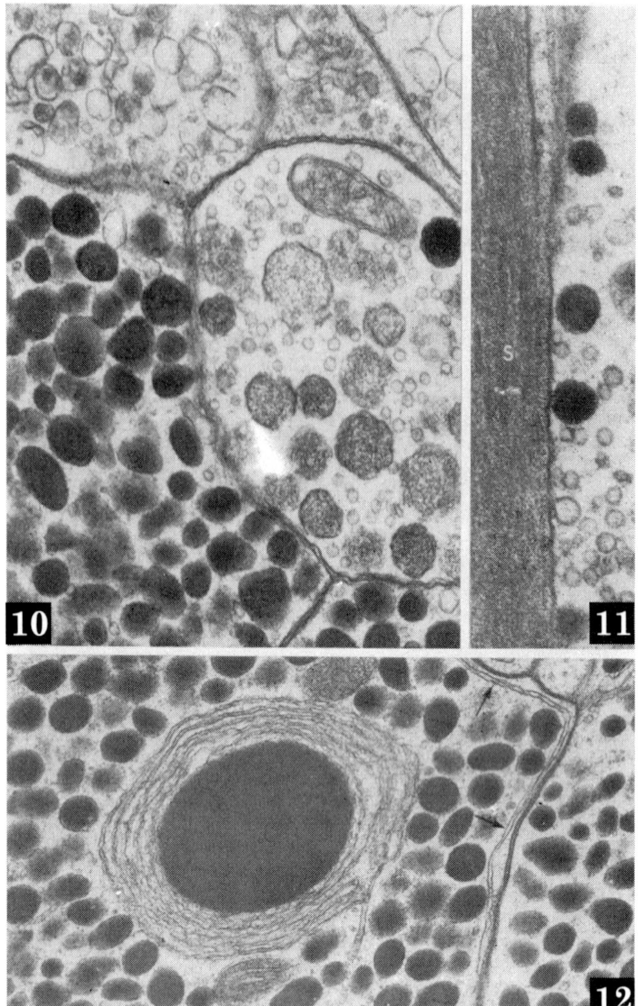

Plate 90. Fig. 10. Axon profiles in corpus cardiacum of adult female of *Periplaneta*. Neurosecretory granules in fibre at right appear to be in process of vesiculation prior to release of their contents. Additional stages of transformation at upper left. X 41,600 Fig. 11. Boundary between neurosecretory axon and stromal sheath (S) in perisympathetic organ of *Blaberus*. Note small clear vesicles bounded by trilaminar membrane and larger dense-core vesicles close to the palsma membrane. X 60,000 Fig. 12. Part of granule-filled A-type fibre in corpus cardiacum of male adult of *Periplaneta,* showing large ovoid mass of material of same appearance as that within neurosecretory vesicles. It is enwrapped by several layers of smooth endoplasmic reticulum that also appears in the form of a subsurface cistern (arrows). X 33,600 (From Scharrer, 1968)

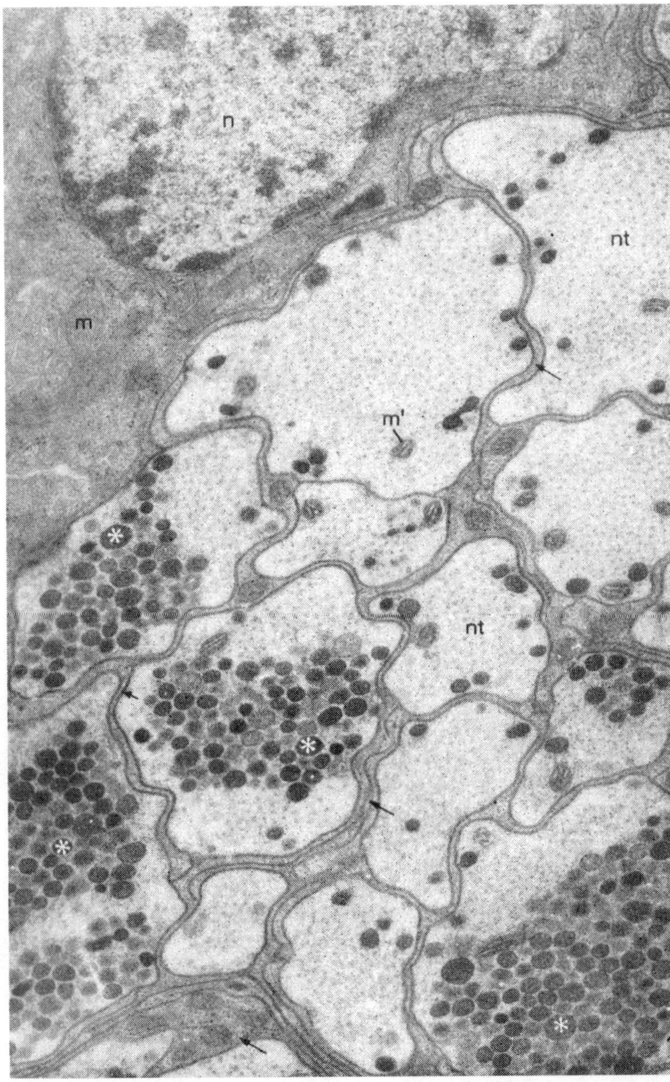

Plate 91. Fig. 1. A transverse section of a portion of one of the nerve trunks (nervi corporis cardiaci) of *Carausius* entering the corpus cardiacum from the brain. The upper region of the field is occupied by the nucleus (n) of a glial cell and surrounding cytoplasm containing elongated mitochondria (m) and other cell inclusions. Narrow extensions of the glial cell envelope (arrow) extend through the nerve fibre, insinuated between the axon surfaces. The axons contain transverse profiles of neurotubules (nt), mitochondria (m'), and, in several instances, large numbers of droplets, exhibiting various degrees of electron opacity (white asterisks). X 28000. (From Smith and Smith, 1966)

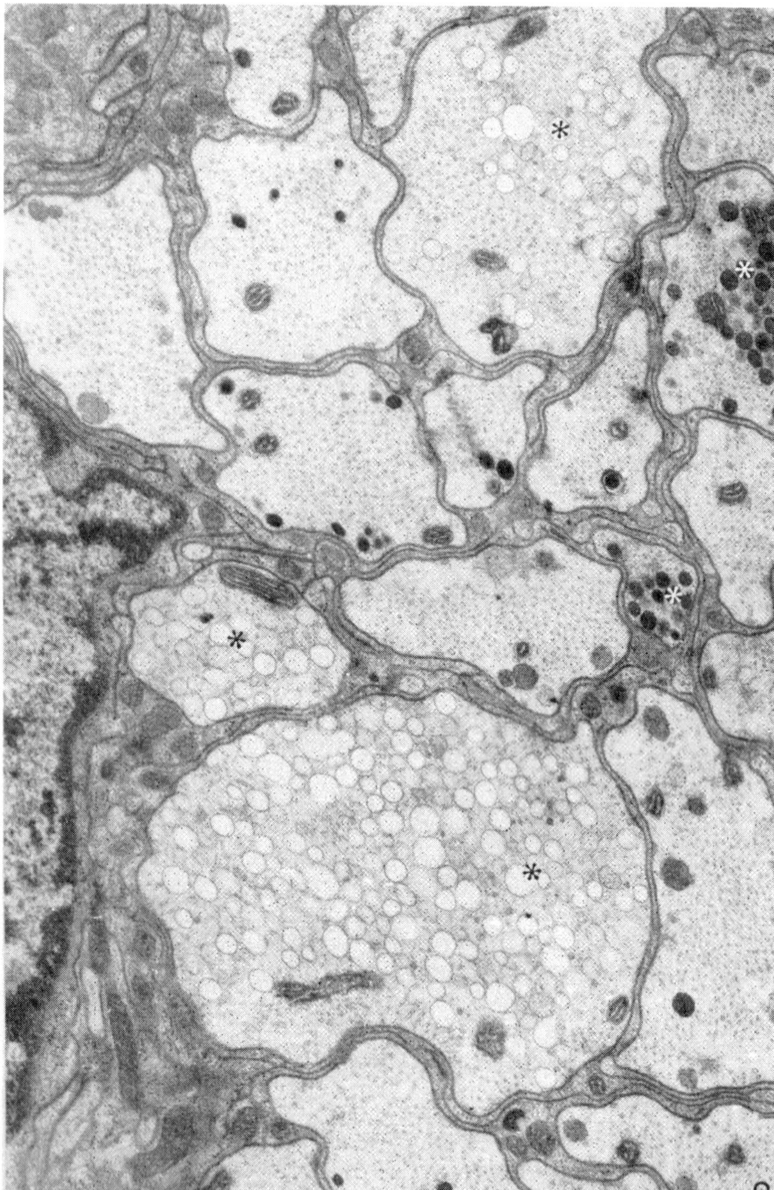

Plate 92. Fig. 2. A field similar to that reproduced in Fig. 1, but in which three axons contain numerous membrane-limited vesicles with light contents (black asterisks), while adjoining profiles contain the dense axoplasmic droplets (white asterisks) also seen in Fig. I. X 23000. (From Smith and Smith, 1966)

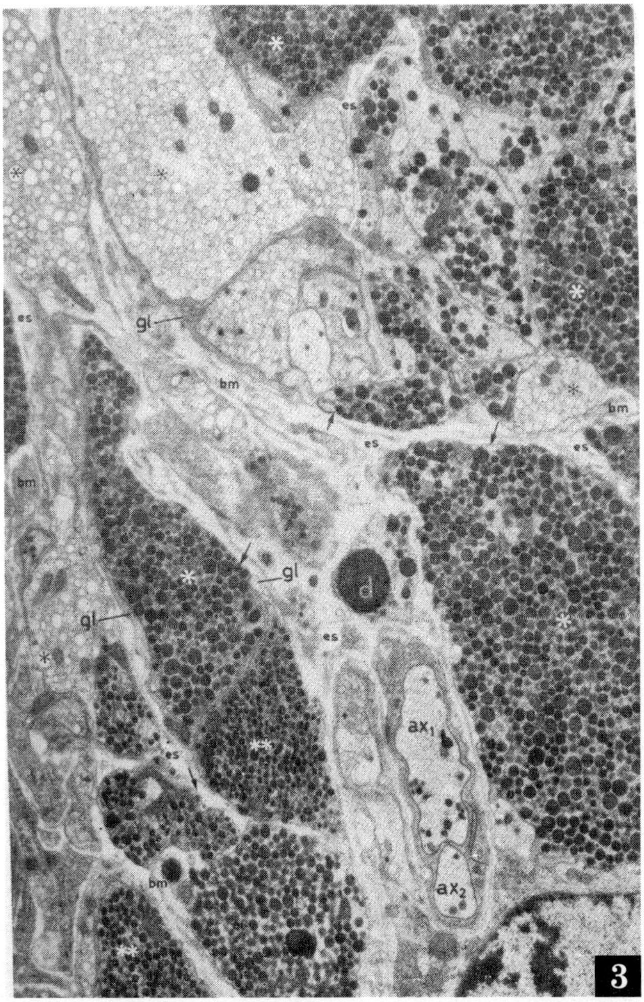

Plate 93. Fig. 3. A survey field illustrating some of the features of organization of the corpus cardiacum of *Carausius*. The bulk of the field is occupied by profiles of nerve cell processes containing vesicular structures. The latter may here be generally categorized into vesicles with more or less electron-transparent contents (black asterisks) and dense electron-opaque droplets. In each instance, within a single nerve process, the diameter of the axoplasmic inclusions varies considerably, though amongst the second category the dense droplets in certain regions (as in the nerve processes indicated with double white asterisks) appear to be predominantly smaller than elsewhere (single white asterisks). Many of these nerve cell processes are invested with a narrow glial sheath (gl) though elsewhere this sheath is absent (as at arrows) where the surfaces of the nerve cell processes directly adjoin extracellular spaces (es), which contain basement membrane-like material (bm) (compare Figs. 5,6,8 and 9). A portion of the glial cell nucleus (n) and surrounding cytoplasm is included at lower right. The large dense body in the centre of the field (d) is perhaps derived by fusion of smaller droplets. Two small axons, containing neurotubules and a few dense droplets, and surrounded by a glial sheath, are included (ax_1, ax_2). X 14,000. (From Smith and Smith, 1966)

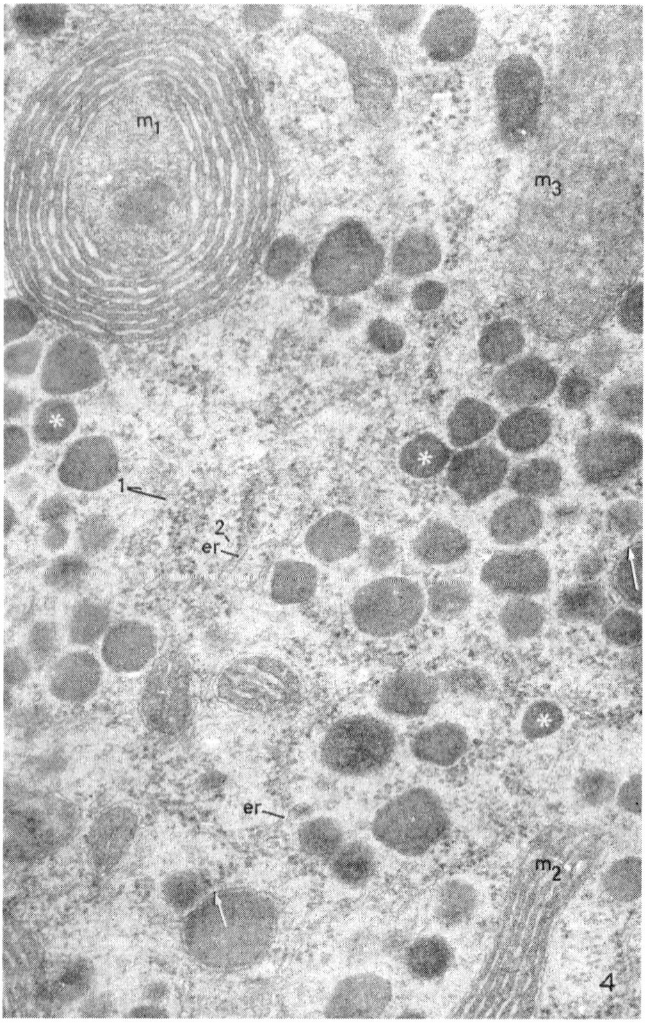

Plate 94. Fig. 4. A field illustrating some of the cytological features of the cell body (perikaryon) of an intrinsic secretory neuron within the corpus cardiacum of *Carausius*. The mitochondria are prominent and variable in their organization, frequently displaying a concentric (m1) or longitudinally oriented (m2) disposition of the cristae; the profile at m3 probably represents a mitochondrion sectioned obliquely with respect to the cristae. The ground cytoplasm is rich in free ribosomes and ribosome-bearing cisternae of the endoplasmic reticulum (er), and in Golgi bodies or dictyosomes (not illustrated in this field). Longitudinal (I) and transverse (2) profiles of neurotubules also occur. Numerous more-or-less electron-opaque secretion droplets are present (asterisks), initially elaborated in association with the Golgi bodies (Scharrer, 1963). During formation (as indicated in two instances here by arrows) the limiting agranular membrane surrounding these droplets is easily distinguished, but in most profiles within the perikaryon and in the extensions of the secretory neurons (compare Figs. 5, 6) the secretion products apparently fill the vesicle, obscuring the limiting membrane. X 62000. (From Smith and Smith 1966)

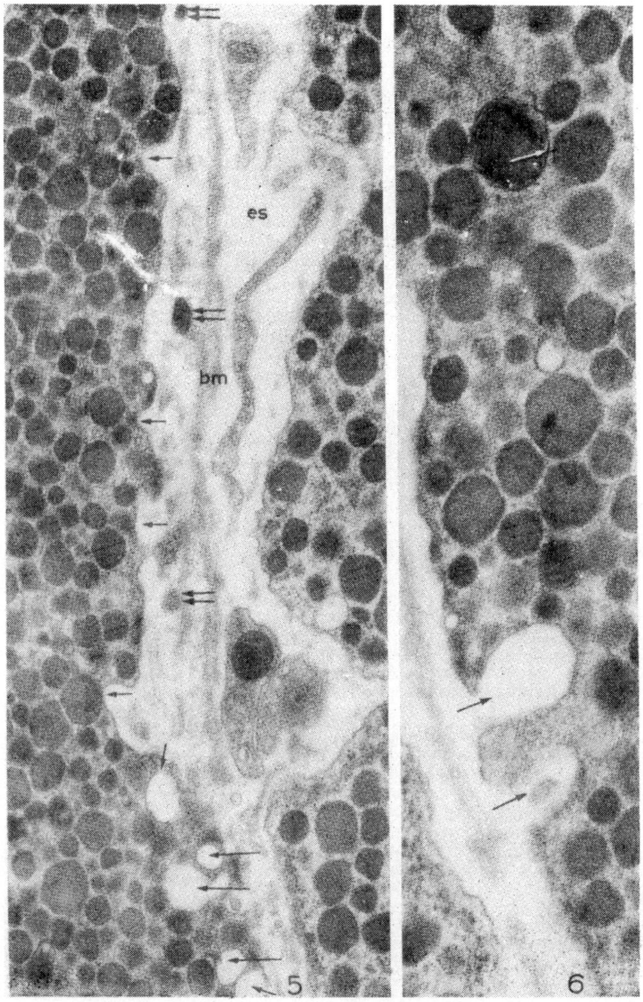

Plate 95. Fig. 5. A field within the corpus cardiacum of *Carausius,* including portions of two axons containing dense granules, separated by an extracellular space (es), containing strands of finely fibrillar basement membrane-like material (bm). Attention is drawn to the surface of the left-hand axon, which is believed to illustrate stages in the release of secretory droplets. At several points (short arrows) the axon surface is indented to form cavities, continued within the extracellular system, which are interpreted as sites of granule release, while the electron-transparent 'vesicles' (long arrows) may represent similar cavities, confluent with the surface at a point out of the plane of section. Some evidence for the presence of cohesive secretion droplets within the extracellular space is provided by the dense bodies indicated with double arrows. X 38000. Fig. 6. A field similar to that shown in Fig. 5, at higher magnification. Two omega-shaped depressions of the cell surface are present (black arrows), possibly representing discharged-droplet membranes. A distinct limiting membrane surrounding the secretion droplets, such as is commonly seen in the perikaryon, is shown at the place indicated by a white arrow. Note the diversity of the material lying between the droplets; small particles (probably ribosomes) are abundant, as in the perikaryon. X 72000. (From Smith and Smith, 1966)

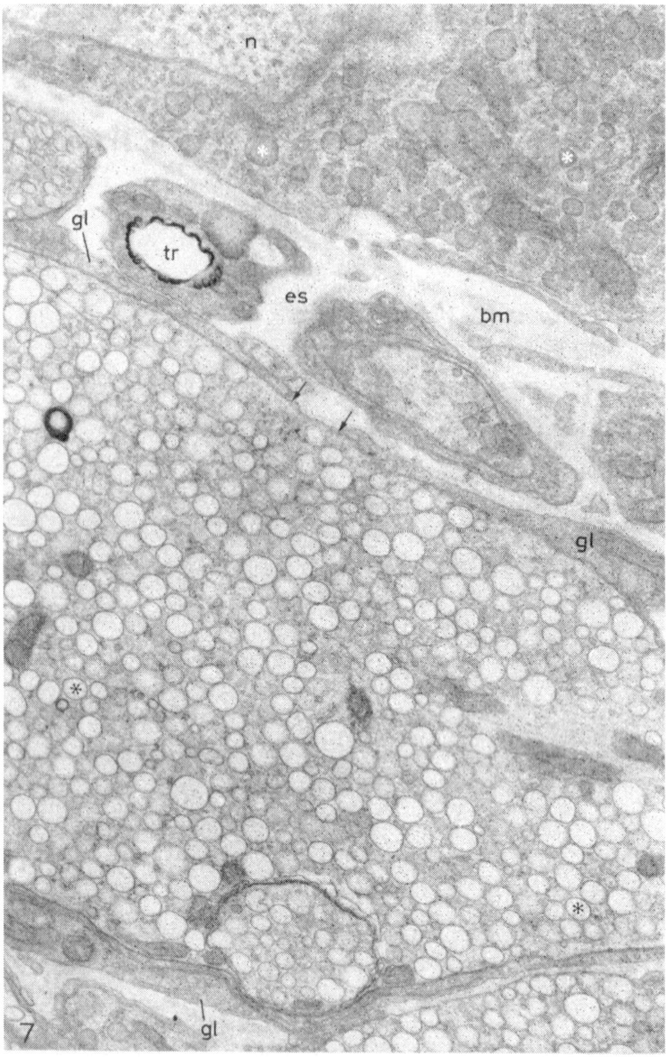

Plate 96. Fig. 7. The upper portion of this field includes the edge of a cell body of an intrinsic secretory neuron, in the corpus cardiacum of *Carausius*. Note the nucleus (n) and the electron-dense secretion droplets within the surrounding cytoplasm (white asterisks). The bulk of the field is occupied by a nerve cell process containing large numbers of membrane-limited vesicles with faint flocculent contents (black asterisks). This process is for the most part invested with a narrow glial sheath (gl), which is however interrupted over a short distance (between arrows), where the cell membrane directly adjoins the neighbouring extracellular space (es) intervening between the perikaryon and nerve cell process. This space contains a tracheole (tr) and strands of basement membrane material (bm) (compare Fig. 3). X 26000. (From Smith and Smith, 1966)

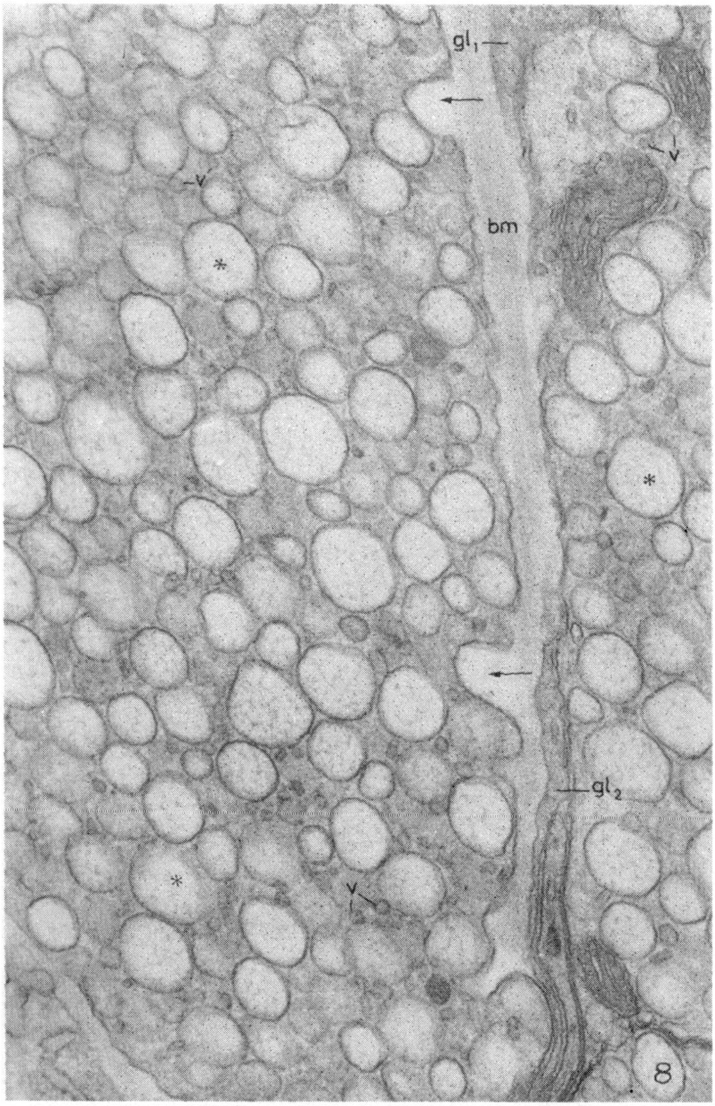

Plate 97. Fig. 8. A micrograph further illustrating the postulated mechanism of release of the contents of vesicles from nerve cell processes into the extracellular environment, in the corpus cardiacum of *Carausius* (compare Figs. 5, 6). Portions of two cell processes are included. That on the right is partly invested with glial prolongations (gl_1, and gl_2); these are absent from the parallel surface of the adjacent nerve cell process. Note the two omega-shaped depressions of the latter (arrows), similar in diameter to the adjoining membrane-limited vesicles (asterisks); these lack the flocculent content of the intact vesicles. It is suggested that these represent sites of release of vesicle contents, achieved by momentary confluence between vesicle membranes and the cell membrane of the axon. In this instance, the extracellular gap between the two nerve processes is largely filled with finely fibrillar material (bm) resembling that present in basement membranes. X 67000. (From Smith and Smith, 1966).

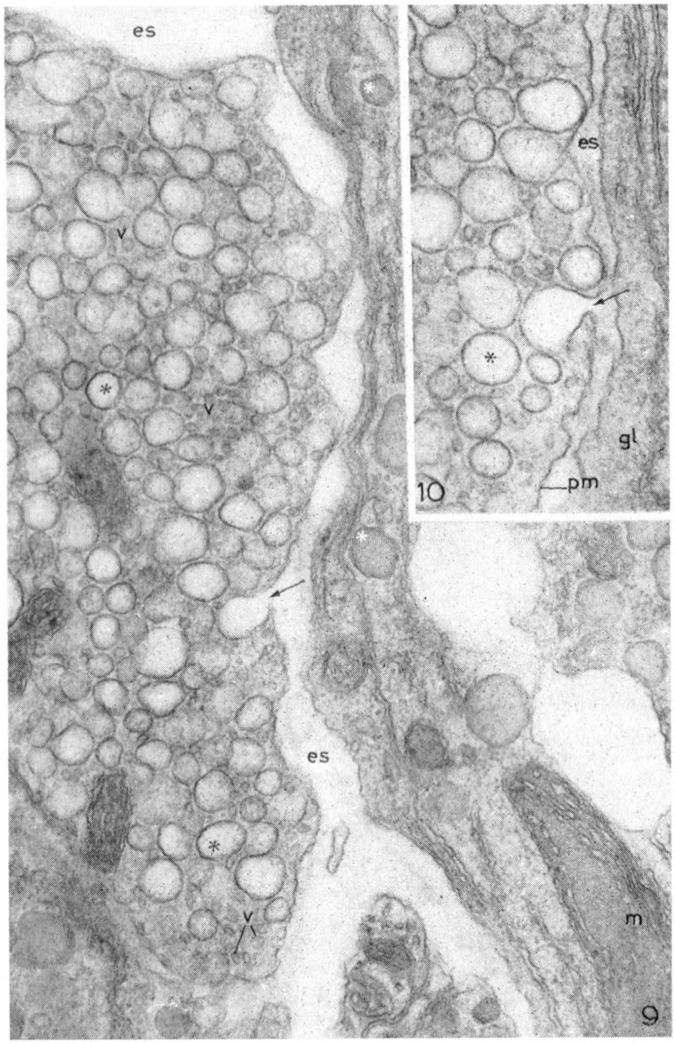

Plate 98. Fig. 9. Including a nerve cell process similar to that shown in Fig. 7, but divested of a glial sheath (compare Fig. 3). Note the vesicles with flocculent contents (black asterisks), one of which appears to have fused with the membrane of the nerve cell process at the point indicated with an arrow, establishing continuity between the interior of the vesicle and the adjoining extracellular space (es). In addition to these structures considerable numbers of smaller vesicles (300-500 Å in diameter) are present within the nerve process (v). The edge of a secretory neuron perikaryon is included on the right: this contains secretion droplets (white asterisks) and an elongated mitochondrion (m) with longitudinally oriented cristae and an extensive matrix (compare Fig 4.). X 43000. Fig. 10. A field similar to that shown in Fig. 9, illustrating another exmple of the confluence between a vesicle and a plasma membrane (pm) limiting a nerve cell process. On the right, a glial cell prolongation (gl) is present, separated from the latter by an extracellular space. (Other labelling as in Fig. 9). X 64000 (From Smith and Smith, 1966).

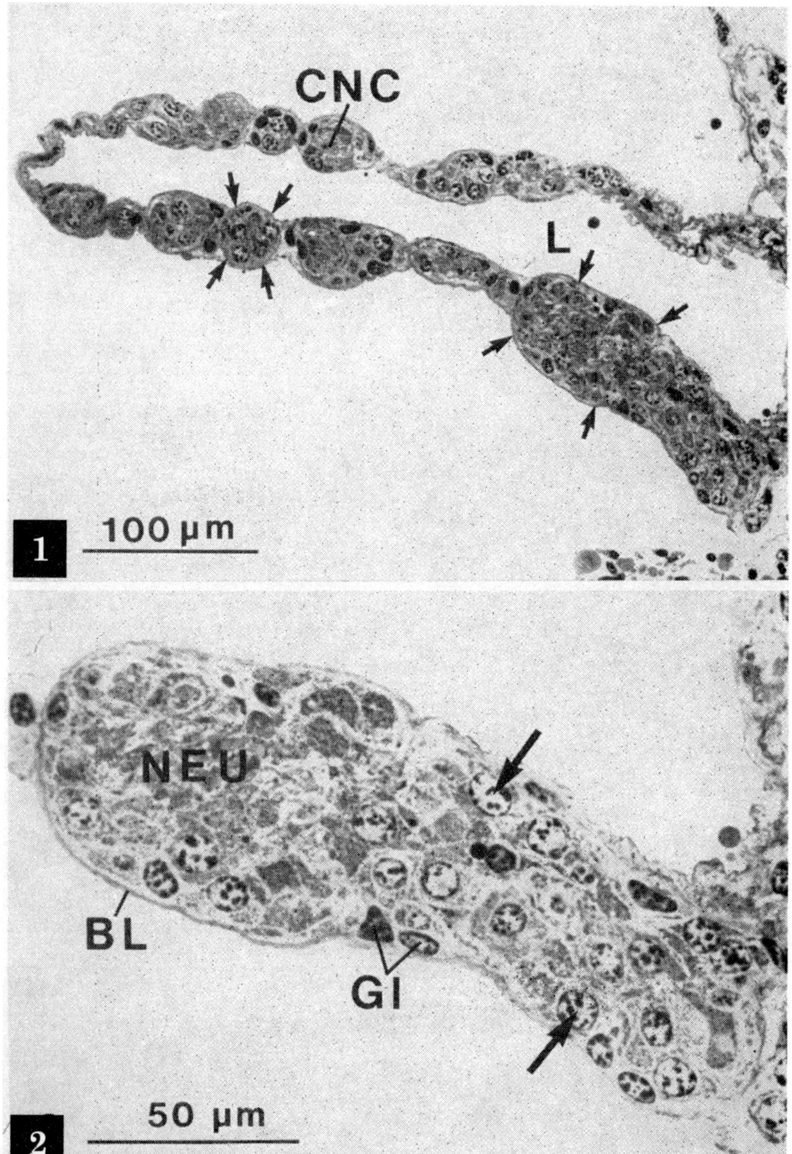

Plate 99. Fig. 1. The glandular lobe of corpus cardiacum of *Schistocerca*. Cross section 60 µm from the posterior end of the organ. The neurosecretory cells (CNC) are forming lobules *(arrows)* which make up the lateral parts of the aorta wall. (L lumen of the aorta). X 310. Fig. 2. The central part of a lobule of *Schistocerca* is devoid of nuclei. Cell projections form a neuropile (NEU). The CNC nuclei (*arrows*) in the periphery are larger than the glial nuclei (Gl), which are irregularly shaped and have more condensed chromatin. A basal lamina (BL) covers the organ. X 770. (From Krogh and Normann, 1977)

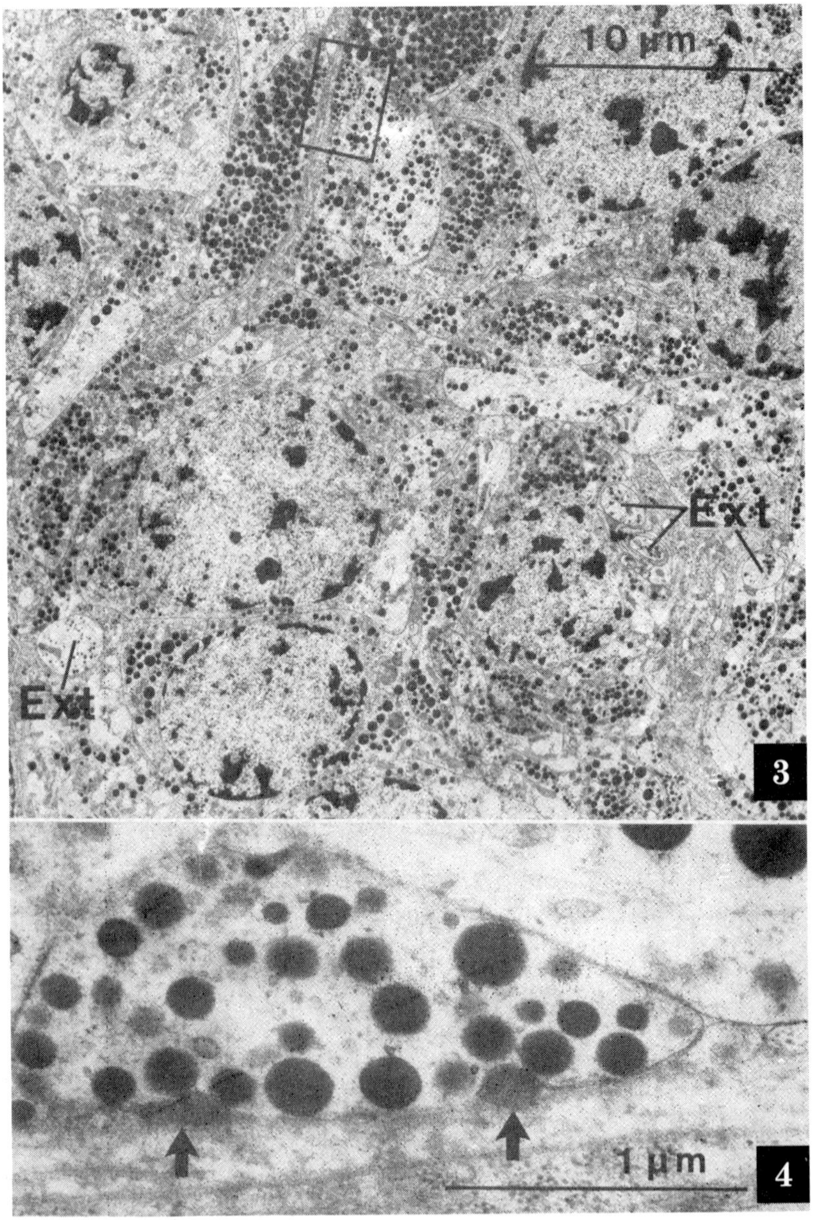

Plate 100. Fig. 3. Low power electron micrograph of corpus cardiacum with CNC perikarya and axonal projections containing neurosecretory granules; some axons are of extrinsic origin (Ext) with small neurosecretory granules. The axon terminal (in rectangle, in the upper part of the figure) is shown at higher magnification in Fig. 4. X 4300. Fig. 4. Detail of Fig. 3. Extrusion by exocytosis of two CNC granules (*arrows*). The locust was anaesthetized by immersion in water. X 50,000. (*Schistocerca*) (From Krogh and Normann, 1977)

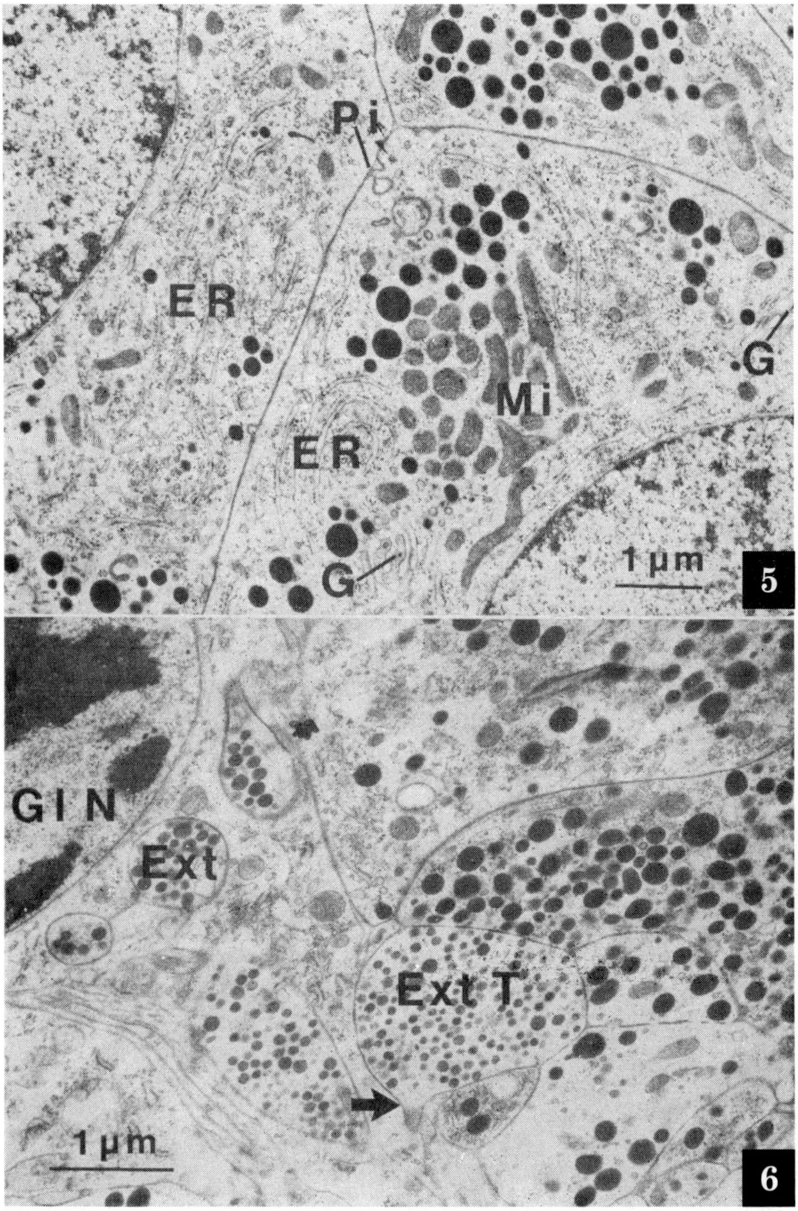

Plate 101. Fig. 5. CNC perikarya from an acetylcholine treated specimen of *Schistocerca*. Pinocytotic invaginations (Pi); granular endoplasmic reticulum (ER); Golgi complexes (G). Mitochondria (Mi) as well as the secretory granules occur in clusters. X 15,000. Fig. 6. Terminal swelling of an extrinsic axon (Ext T) with small granules, in direct contact with CNC. The axonal neurotubules apparently end abruptly (*arrow*). Other extrinsic axons (Ext) are embedded in glioplasm. Gl N gliocyte nucleus. X 21,000. (From Krogh and Normann, 1977).

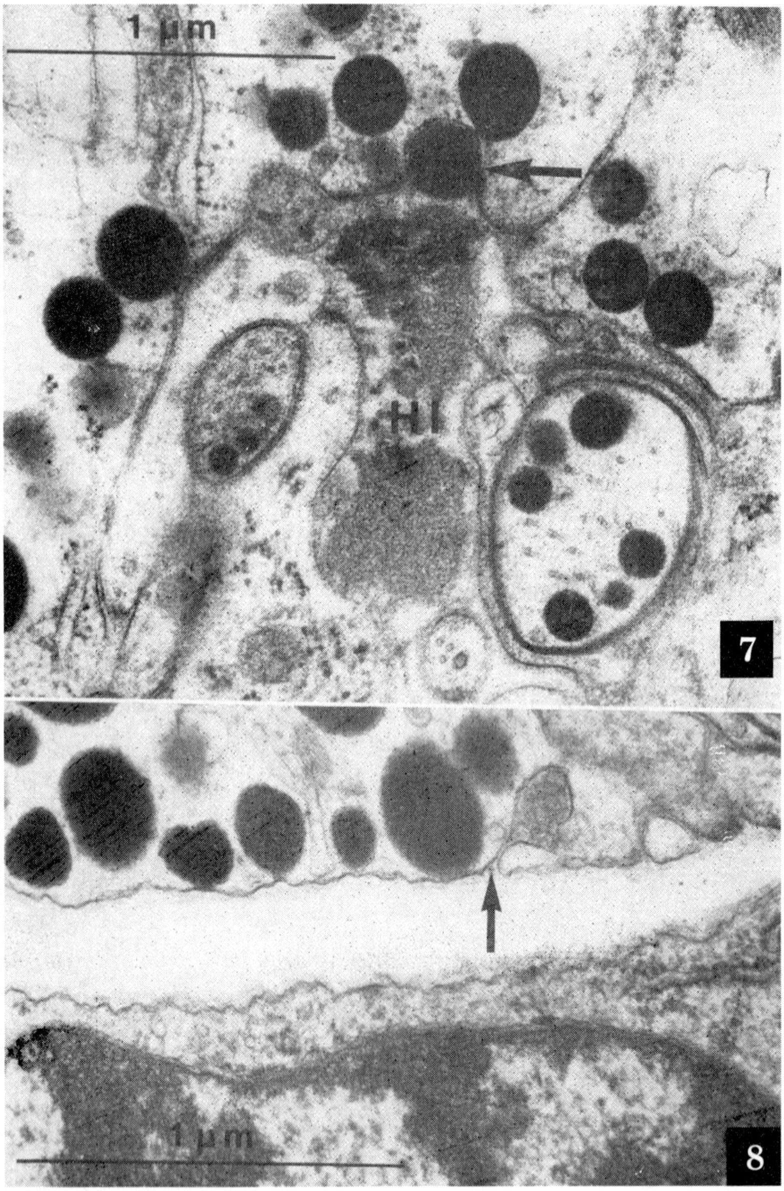

Plate 102. Fig. 7. Omega profile, indicative of exocytosis of a CNC granule (*arrow*). The density of the extruded granule begins to decrease. The nature of the greyish material in the haemocoel indentation (HI) is unknown. X 57,000. Fig. 8. Later stage of exocytosis (*arrow*). The content of secretory material, which has remained in the collapsing vesicle, shows signs of a sub-structure, as a dense component seems to have disappeared. X 67,000. (*Schistocerca*) (From Krogh and Normann, 1977).

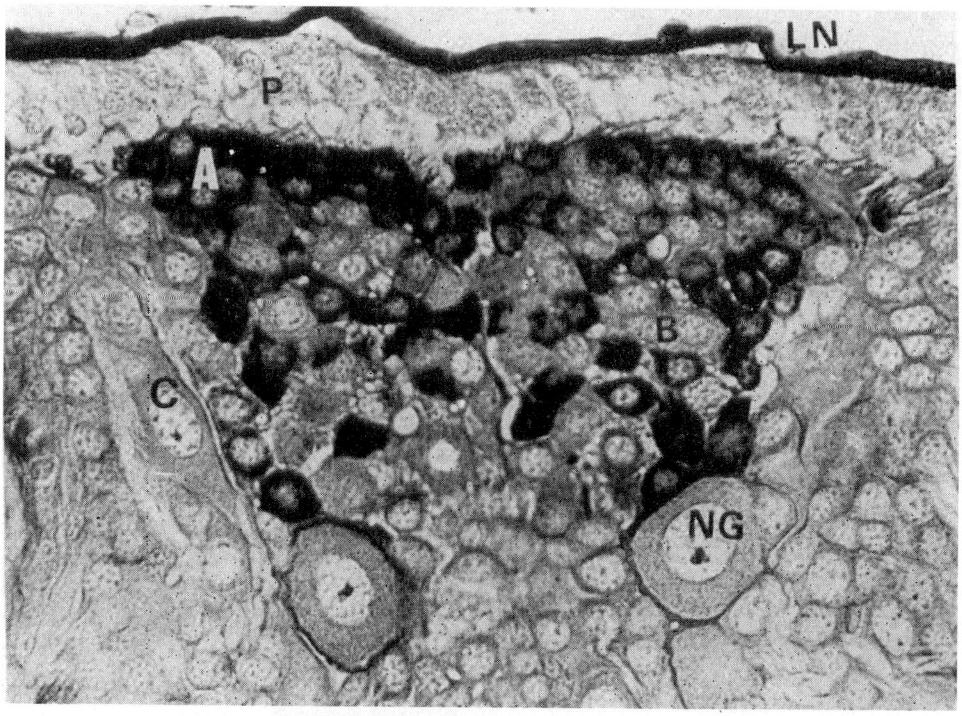

Plate 103. Fig. 1. Coupe transversale de la PI de male mature montrant la repartition des CNS A (A), des CNS B (B), des CNS C (C) et des neurones geants (NG), LN lamelle neurale. P perilemme. Paraffine. FP. Gr X 350 (*Locusta*) (From Girardie and Girardie, 1967)

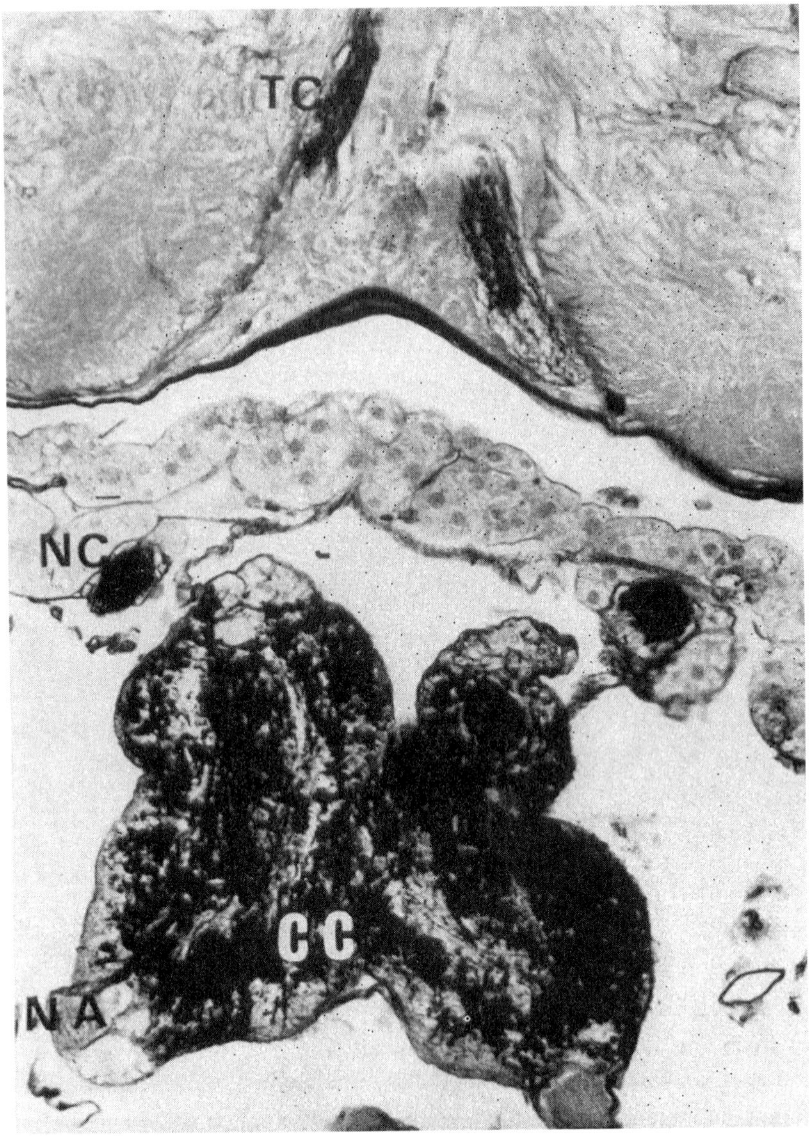

Plate 104. Fig. 2. Coupe transversale des corps paracardiaques et de la partie ventrale du cerveau de male mature *Locusta*. Accumulation de la neurosecretion dans le tractus cardiaque (TC), dans les nerfs paracardiaques internes (NC) et dans les corps cardiaques (CC). Absence de neurosecretion dans les nerfs allato-cardiaques (NA). Paraffine. FP. Gr. X 130 (From Girardie and Girardie, 1967).

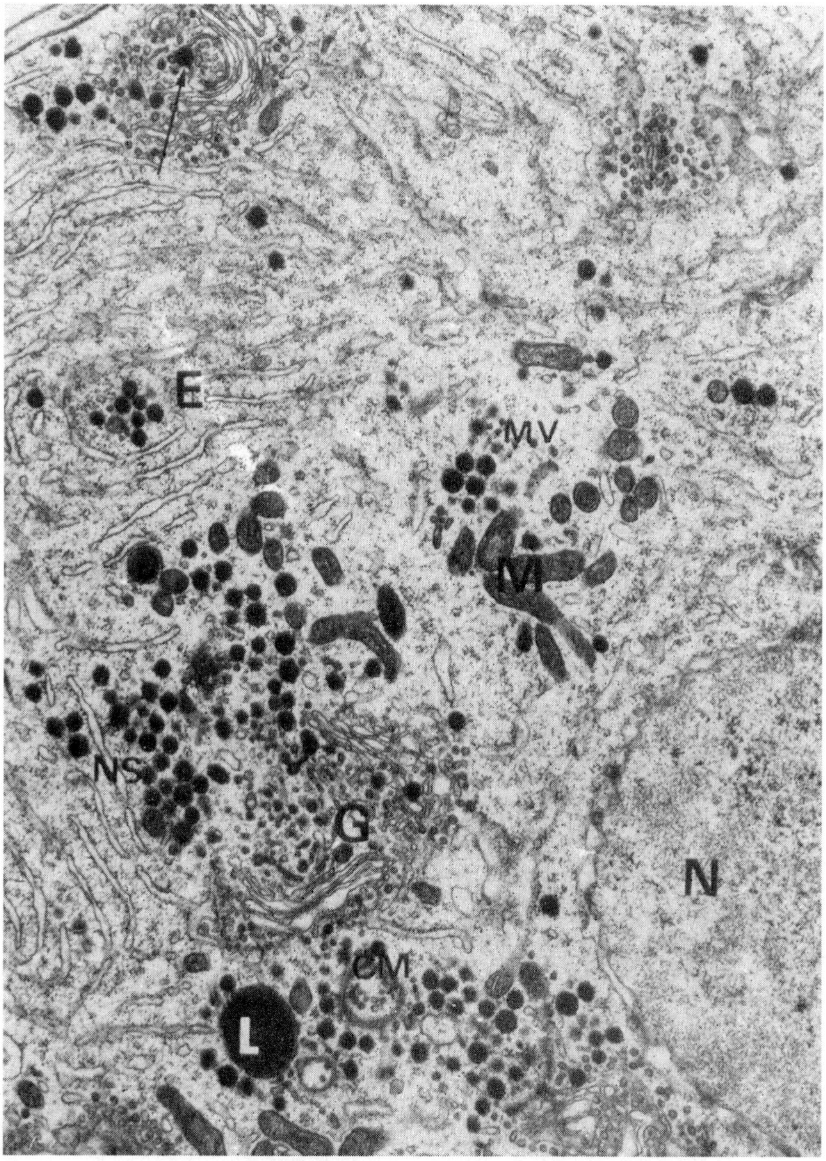

Plate 105. Fig. 3. Coupe ultra-fine de PI de femelle mature *Locusta*. CNS A riche en grains de neurosecretion (NS) s'individualisant dans les microvesicules golgiennes (MV) ou dans l'extremite des saccules (fleche). CM corps multivesiculaire. E. ergastoplasme, G. Golgi, L. lysosomes, M. mitochondrie, N. noyau. Gr. X 15000. (From Girardie and Girardie, 1967)

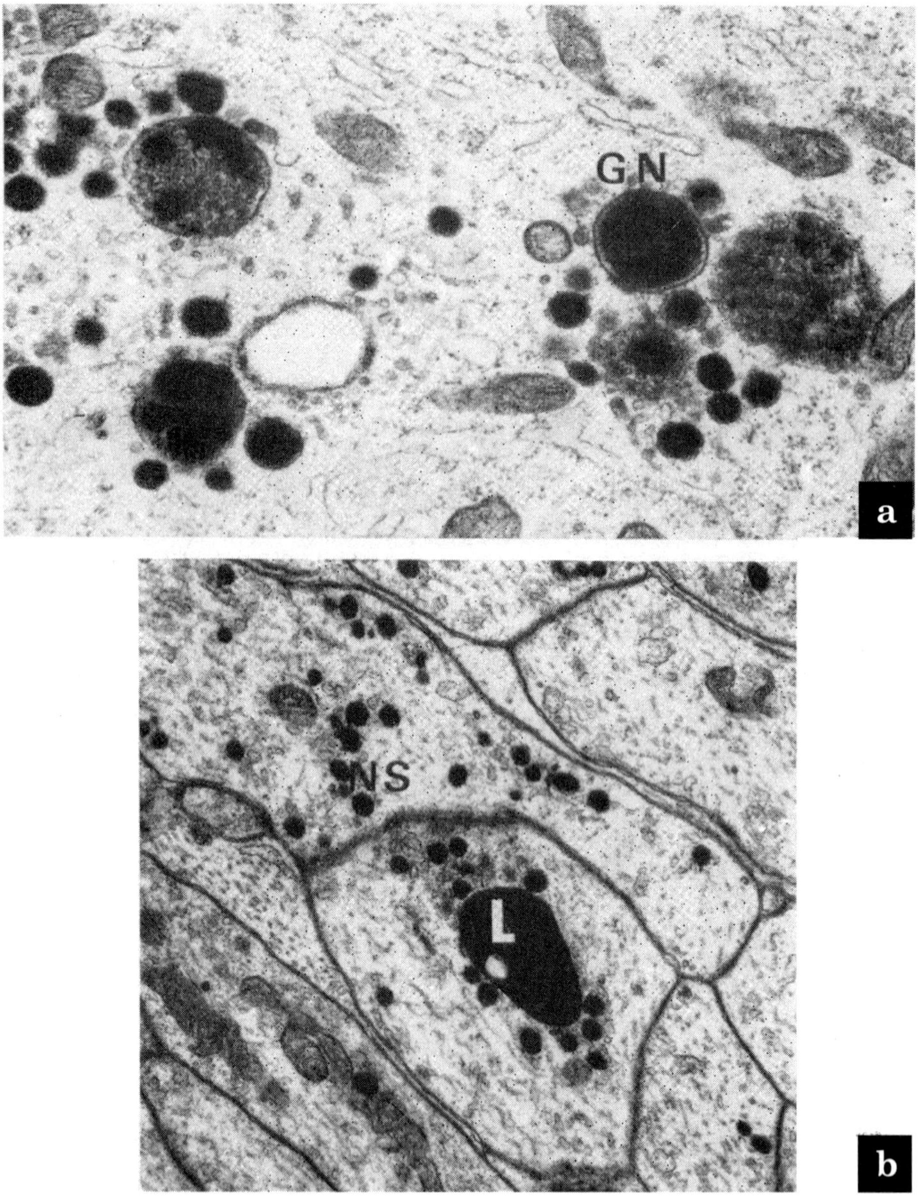

Plate 106. Fig 4a. Coupe ultra-fine de PI de femelle mature *Locusta* Differents stades de la formation des gros grains de neurosecretion (GN) dans les CNS A presentant des analogies avec les lysosomes. Gr. X 40,000. Fig. 4b. Coupe ultra-fine du tractus cardiaque. Cheminement des grains de neurosecretions (NS) et des lysosomes (L.). Gr. X 15000 (*Locusta*) (From Girardie and Girardie, 1967)

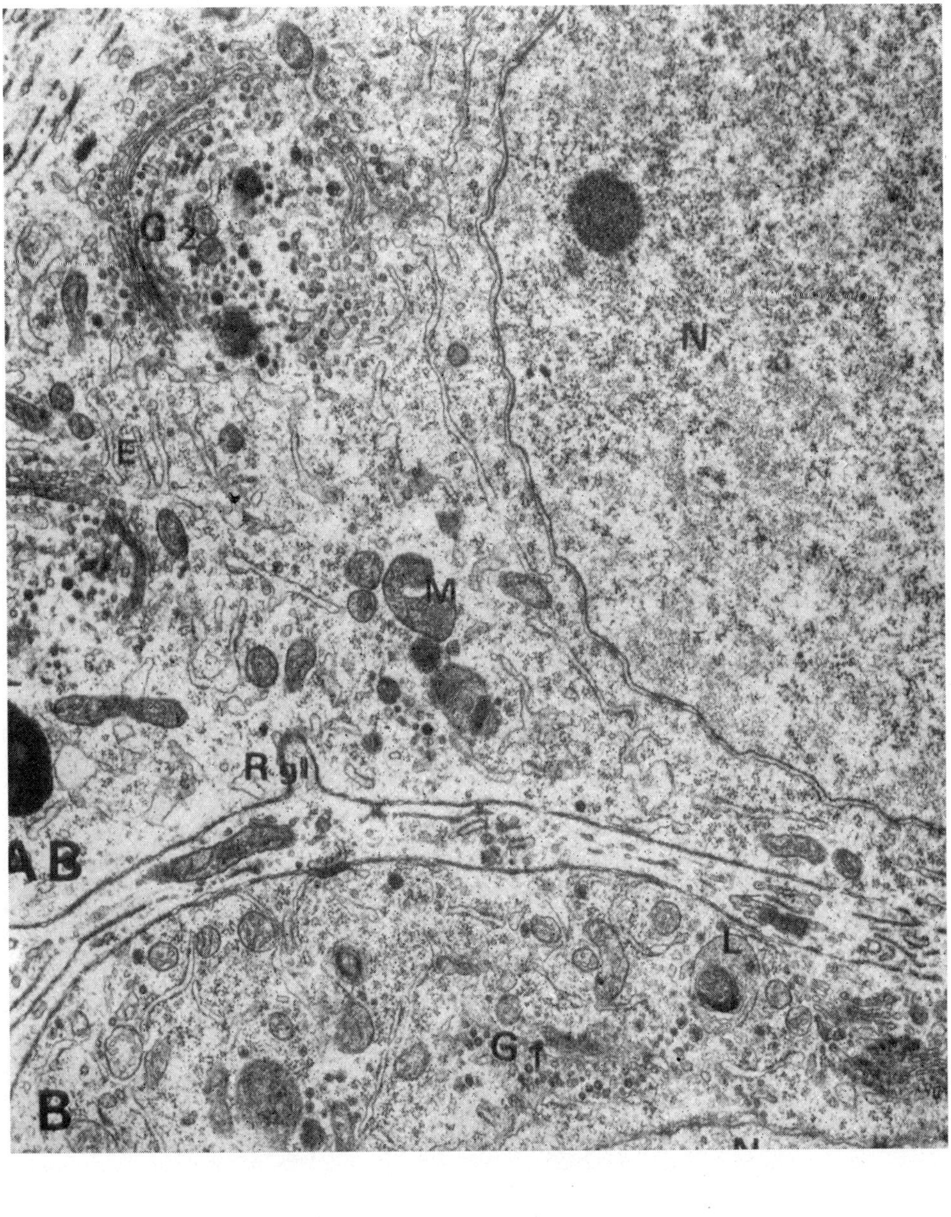

Plate 107. Fig. 5. Coupe ultra-fine de PI de femelle mature *Locusta*. CNS B (B) a Golgi quiescent (gl) et a nombreux lysosomes parfois membranaires (L.). CNS intermediaire (AB) entre le type A et le type B avec reprise d'activite secretoire du complexe golgien (G2). R gl repli glial. Pour les autres legendes voir Fig. 3. Gr. X 15000 (From Girardie and Girardie, 1967).

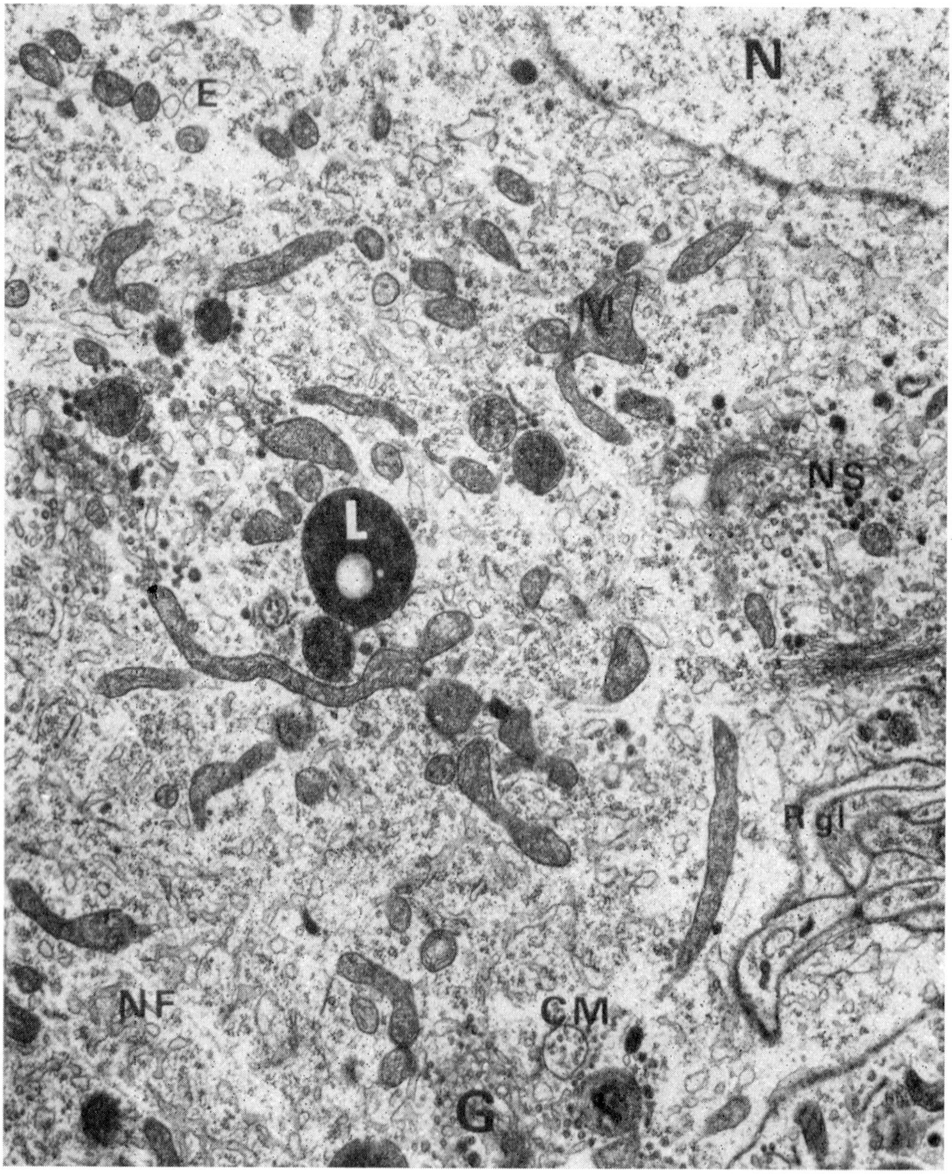

Plate 108. Fig. 6. Coupe ultra-fine de PI de femelle mature *Locusta*. CNS C avec replis gliaux (R gl) et segregation de la neurosecretion dans le compexe golgien (G). NF neurofibrille. Pour les autres legendes voir Fig. 3. Gr. X 13000. (From Girardie and Girardie, 1967).

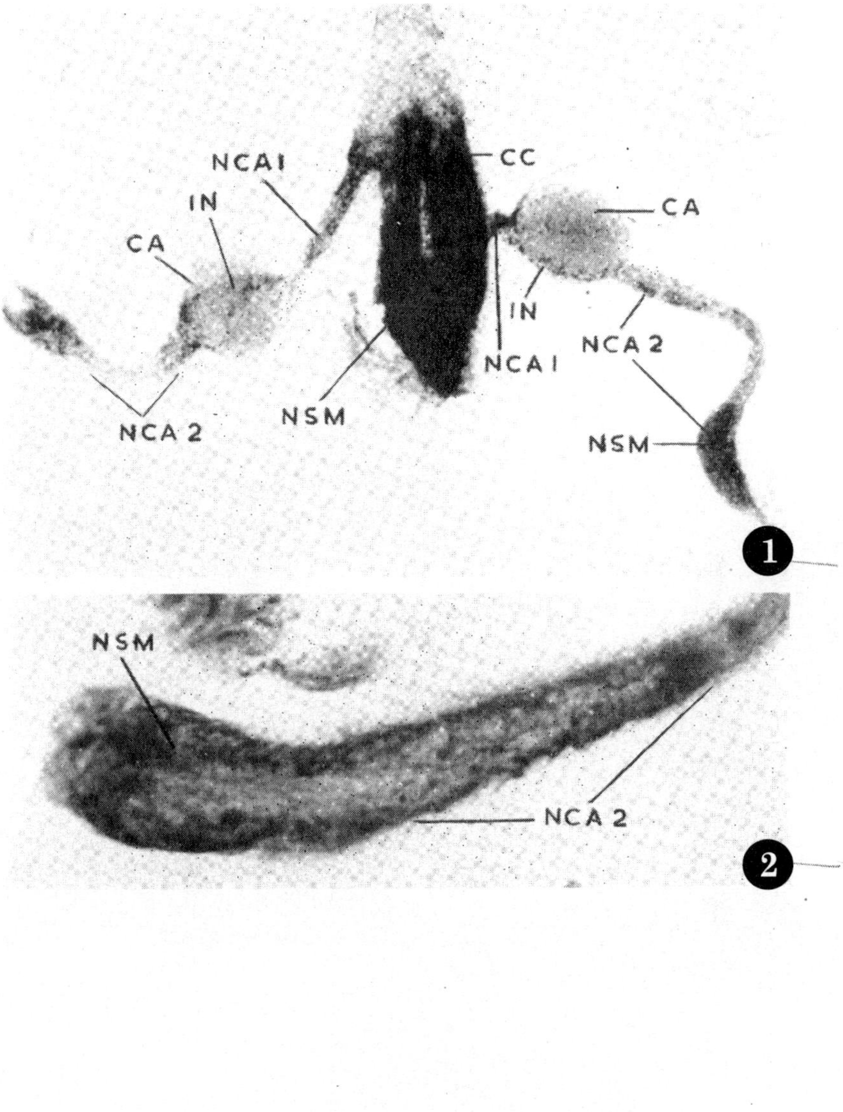

Plate 109. Fig. 1. Whole mount of the corpus cardiacum-corpus allatum complex of *Gryllodes sigillatus* showing the occurrence of two patterns of innervations at a time. Also note the identical material stored in the corpora cardiaca and the NCA2. X 240. Fig. 2. Longitudinal section of NCA 2 of *Gryllodes sigillatus* showing the stored NSM in the dilated part. X 2000. (From Awasthi, 1968).

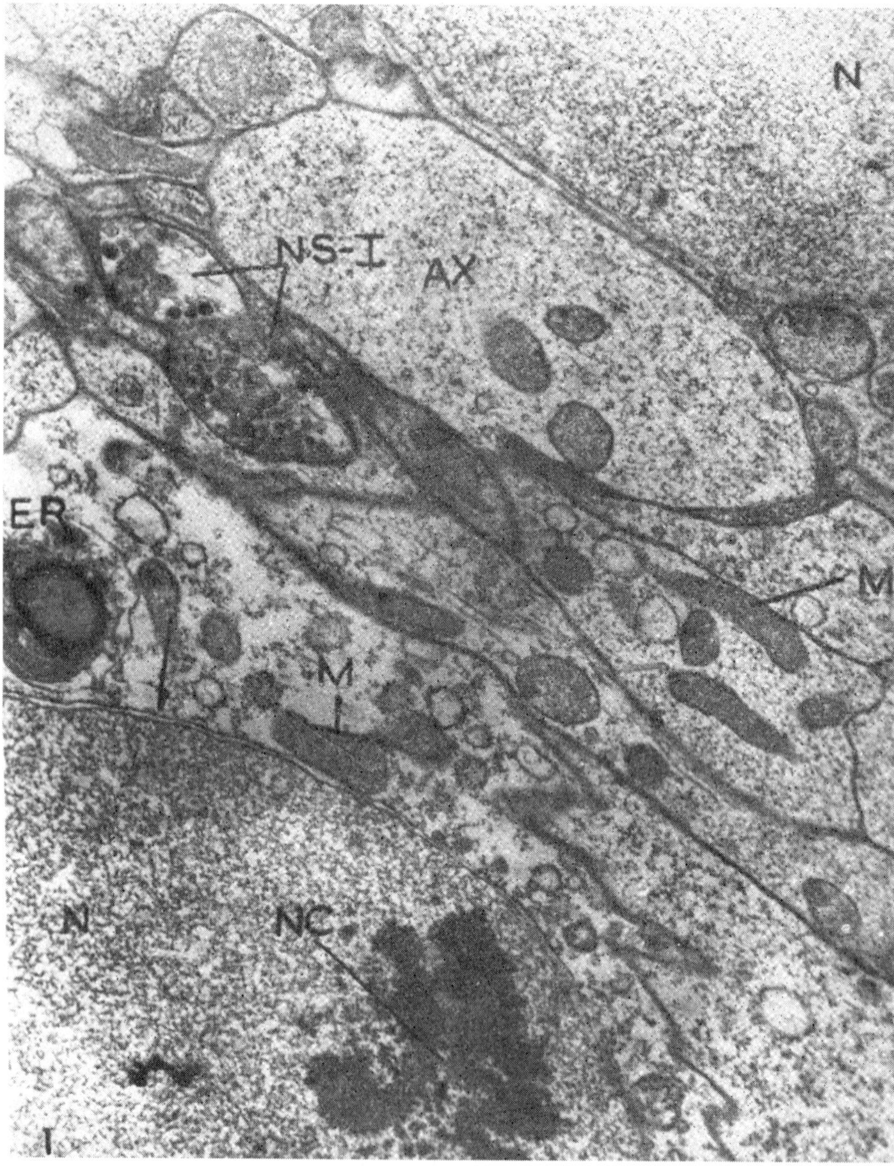

Plate 110. Fig. 1. Electron micrograph of a part of corpus cardiacum of *Gryllodes sigillatus*. The intrinsic neurosecretory cells have a double layered nuclear membrane (arrow). Electron dense nucleolus is clearly marked out. The axons containing neurosecretory granules (NS type I) are also visible. Both the cells and the axons contain a large number of mitochondria. X 18000. (Ax axon, ER endoplasmic reticulum, M mitochondria, N- nucleus, NC Nucleolus. (From Awasthi, 1977).

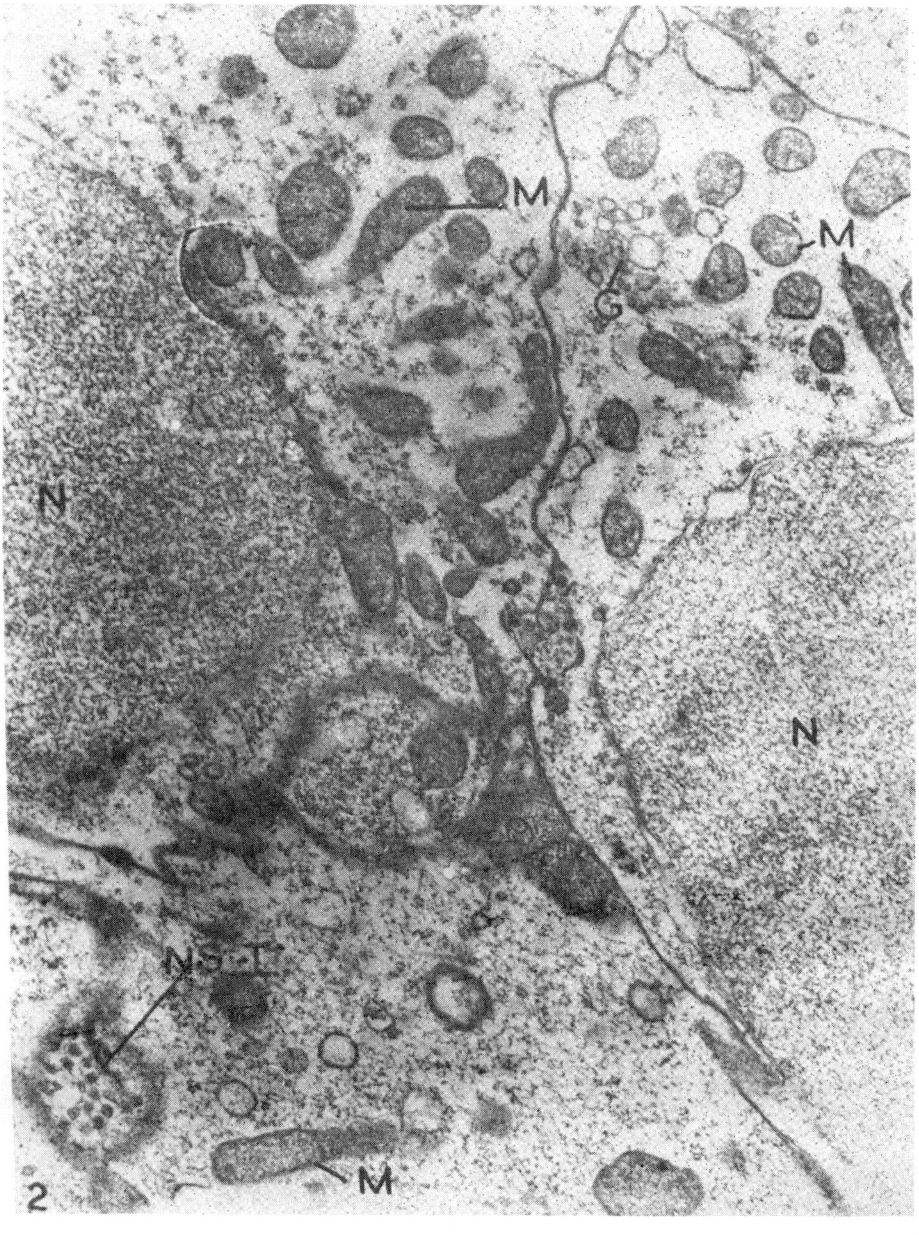

Plate 111. Fig. 2. Electron micrograph of a part of corpus cardiacum of *Gryllodes sigillatus* showing intrinsic neurosecretory cells. A few neurosecretory granules type I (NS-I) in the cells are clearly seen. Note numerous mitochondria in the perikaryon. (G. Golgi zone) X 18000. (From Awasthi, 1977)

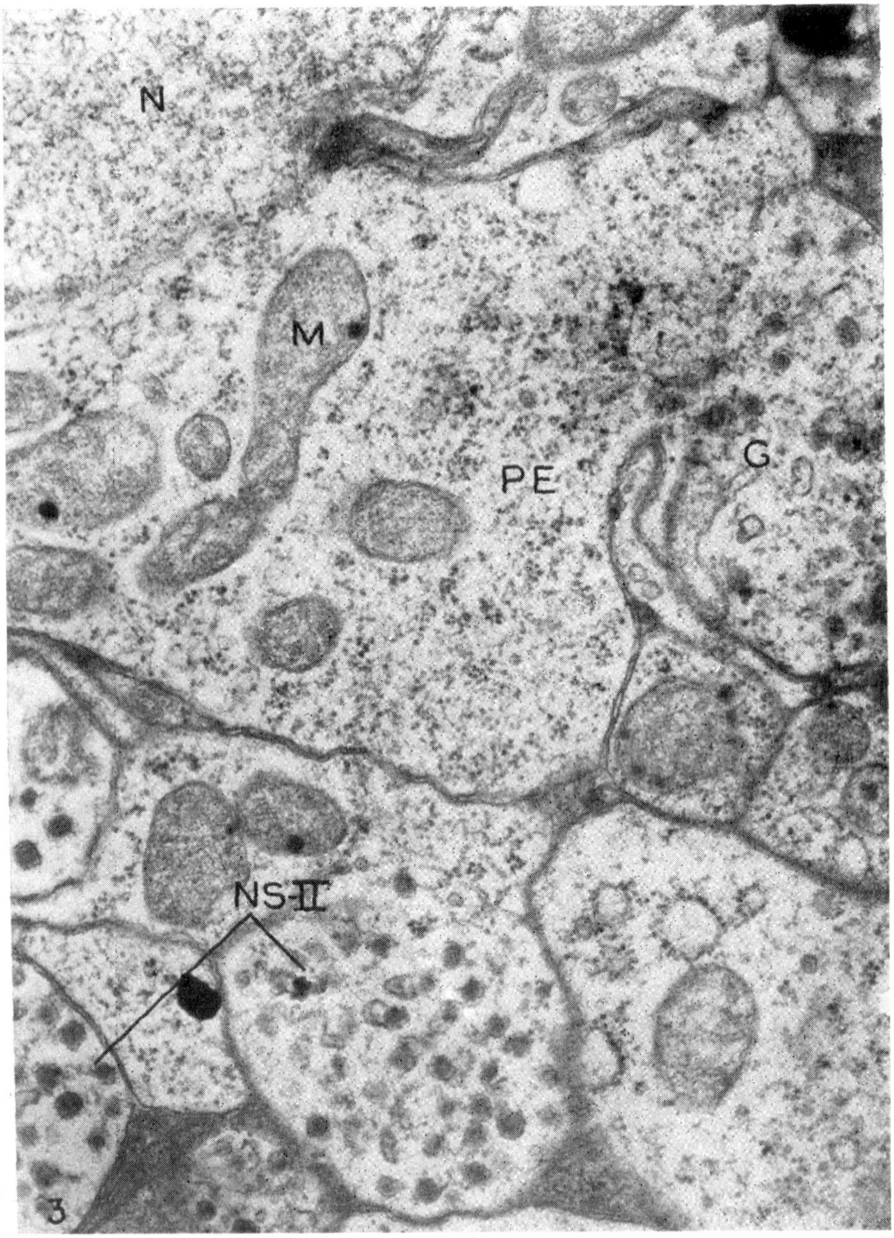

Plate 112. Fig. 3. Electron micrograph of a part of corpus cardiacum of *Gryllodes sigillatus* showing an intrinsic cell with numerous mitochondria and plentiful free ribosomes in the perikaryon. The Golgi zone with neurosecretory granules having dense core vesicles are clearly marked out. Axons with neurosecretory granules type II (NS-II) with dense core vesicles are also visible. X 30000. (From Awasthi, 1977).

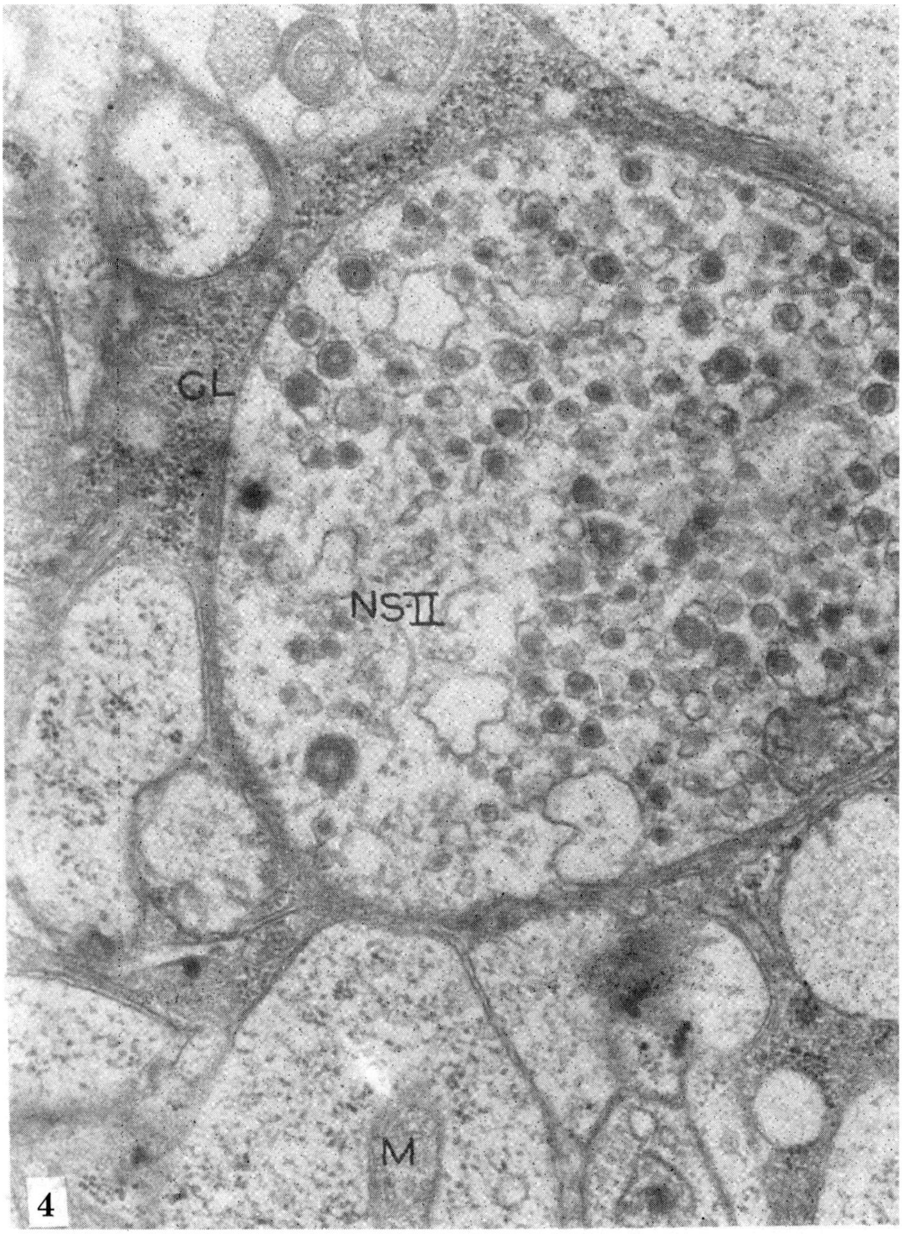

Plate 113. Fig. 4. Electron micrograph of a part of corpus cardiacum of *Gryllodes sigillatus* showing axon containing neurosecretory granules type II (NS-II) with electron lucent space between the boundary membrane and the matrix. The axon is poorly covered by glial sheath. In addition ordinary axons having mitochondria are also visible. X 30000. (From Awasthi, 1977).

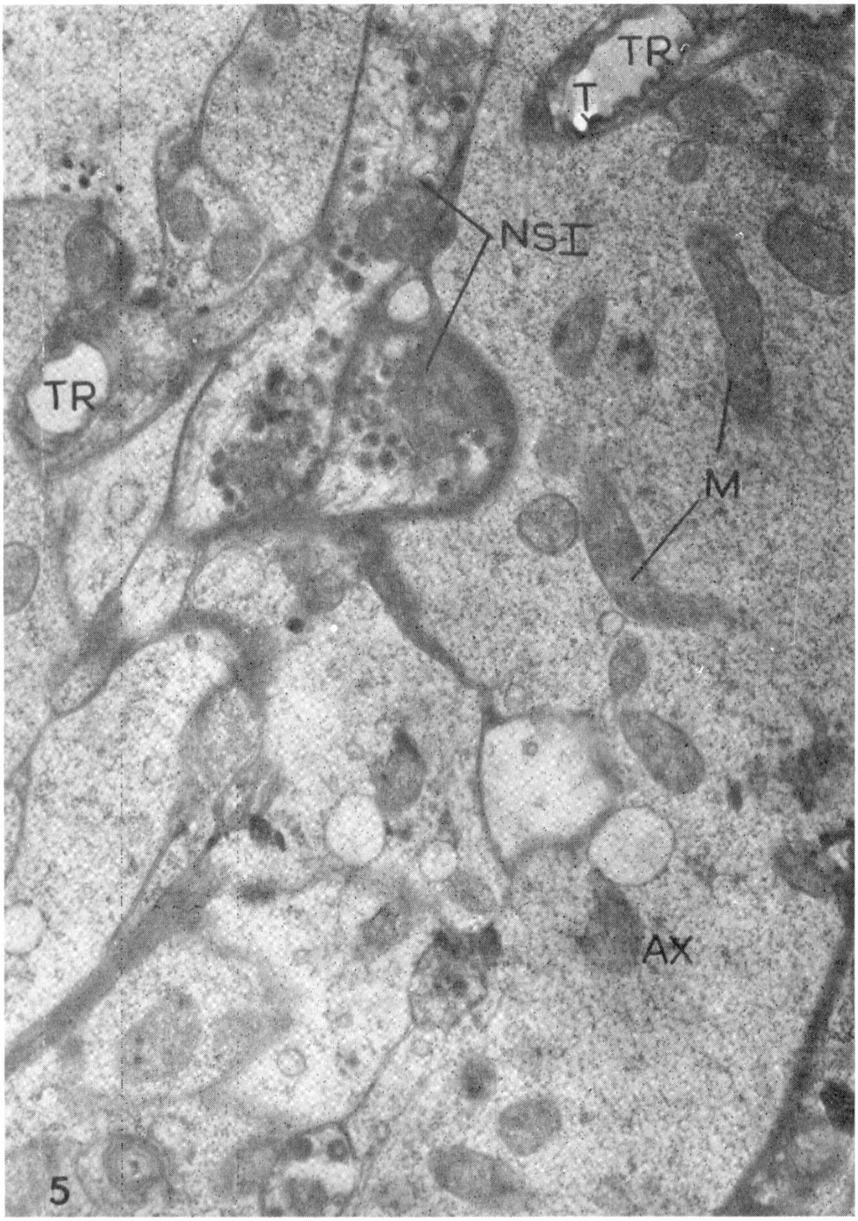

Plate 114. Fig. 5. Electron micrograph of a part of corpus cardiacum of *Gryllodes sigillatus* showing the axons (Ax) containing neurosecretory granules type I (NS-I). 2 tracheoles (with small and smooth taenidia), are also clearly seen on the upper side. Note the axon with a large number of mitochondria (M). TR- Tracheole, T- taenidia X 18000. (From Awathi, 1977).

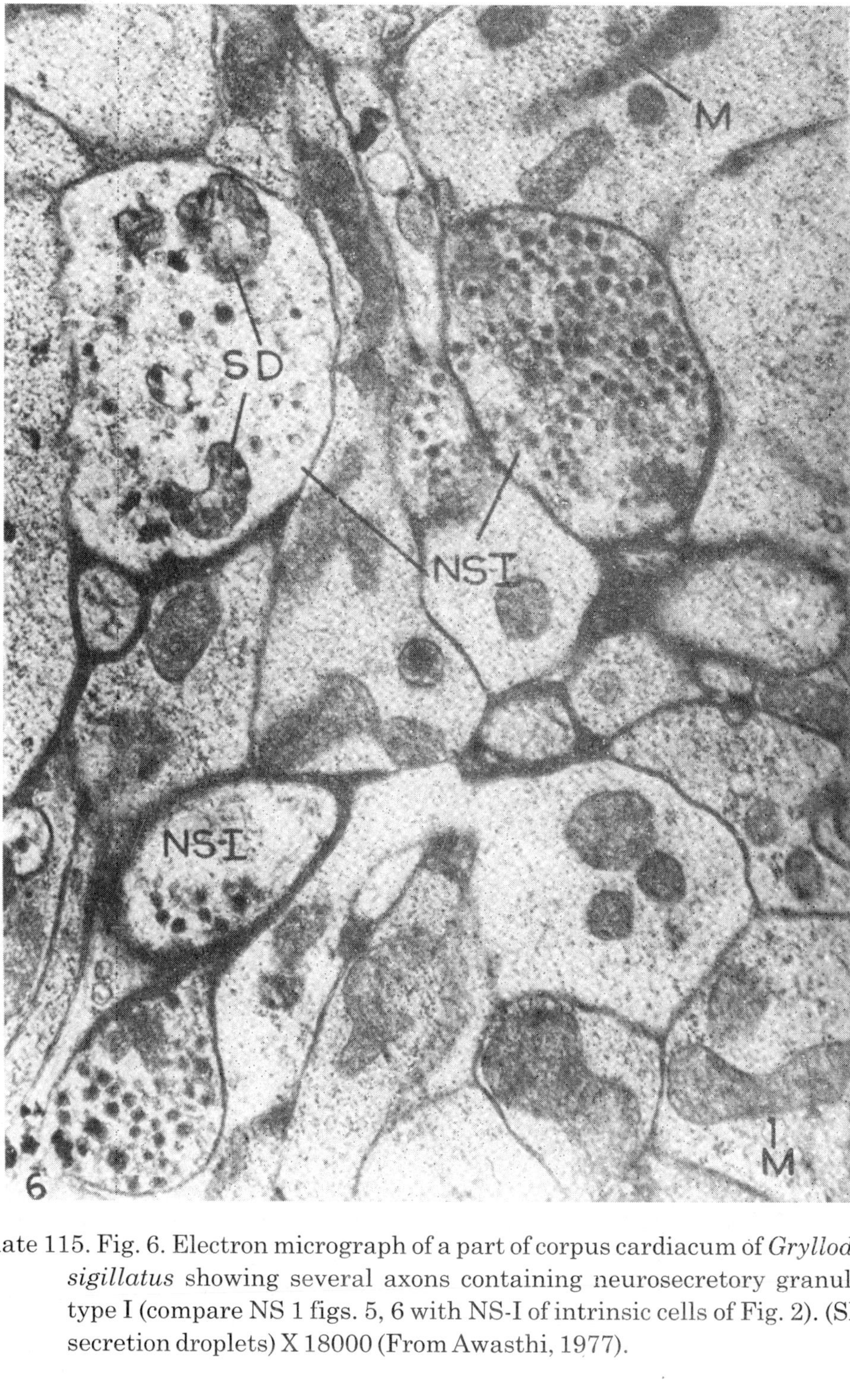

Plate 115. Fig. 6. Electron micrograph of a part of corpus cardiacum of *Gryllodes sigillatus* showing several axons containing neurosecretory granules type I (compare NS 1 figs. 5, 6 with NS-I of intrinsic cells of Fig. 2). (SD-secretion droplets) X 18000 (From Awasthi, 1977).

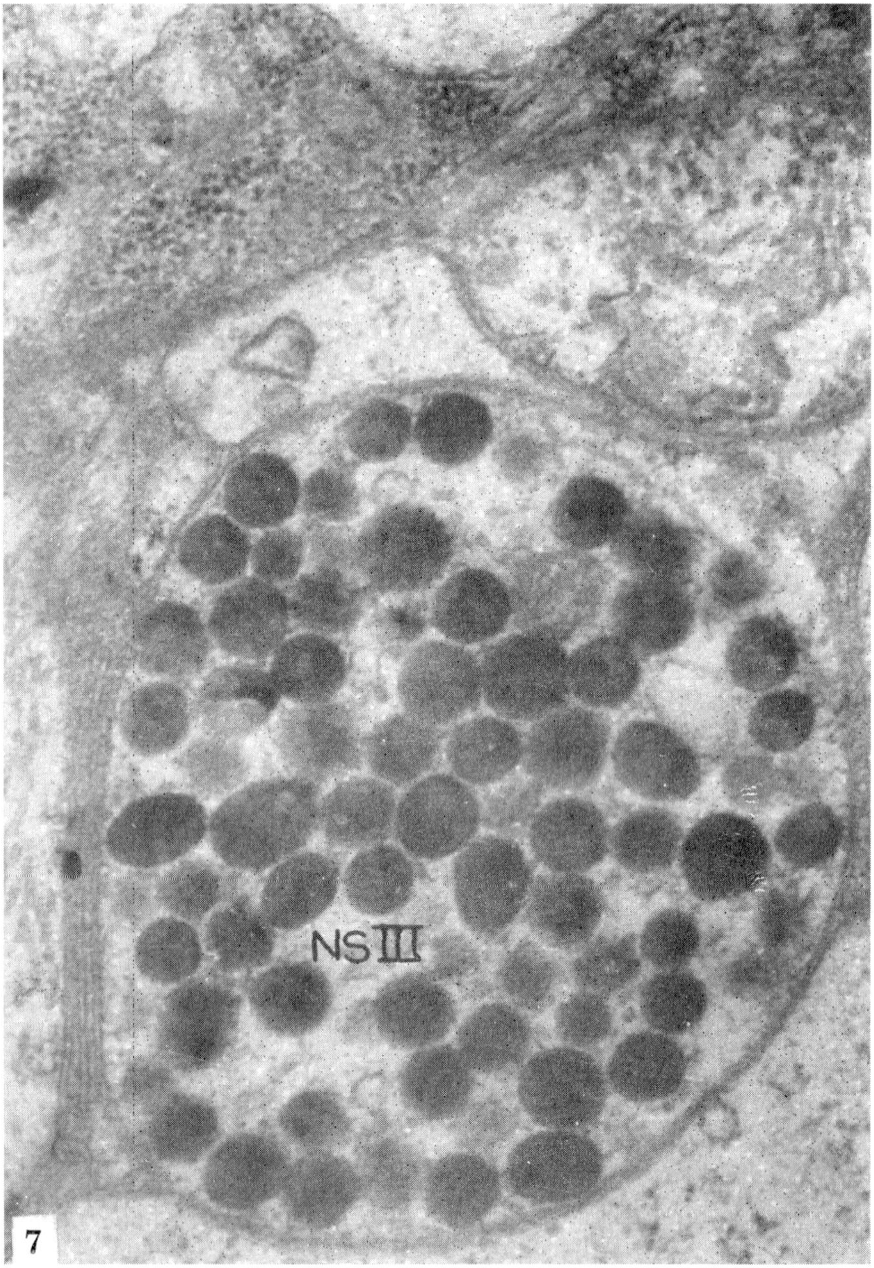

Plate 116. Fig. 7. Electron micrograph of a part of corpus cardiacum of *Gryllodes sigillatus* showing the axon containing electron dense, oval or spherical neurosecretory granules type III (NS-III) the boundary membranes of which cannot be marked out. X 48000. (From Awasthi, 1977).

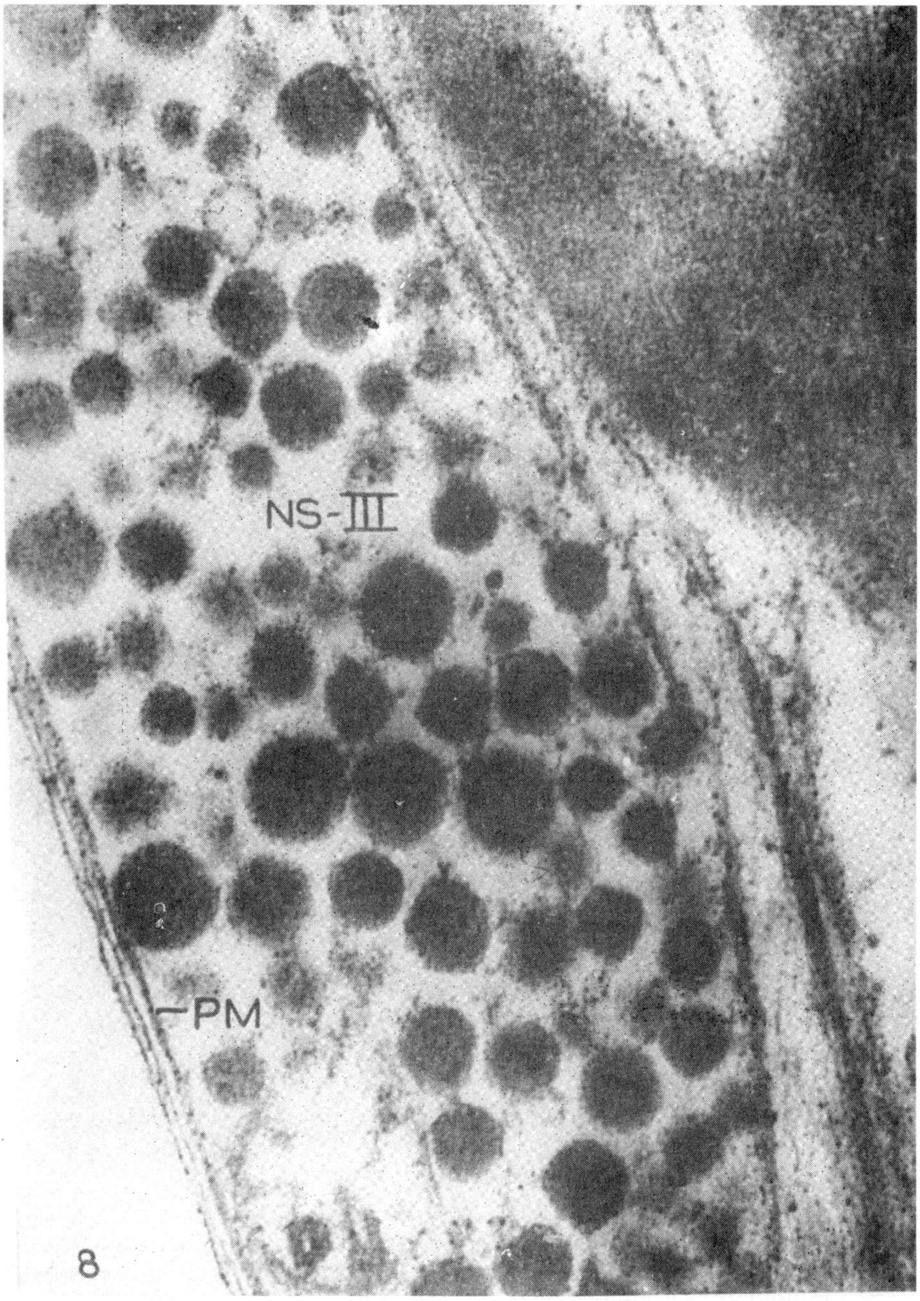

Plate 117. Fig. 8. Electron micrograph of a part of corpus cardiacum of *Gryllodes sigillatus* showing electron dense neurosecretory granules type III (NS-III) with indistinct boundary membranes. The axon is covered by extremely this glial sheath. (PM plasma membrane) X 6000. (From Awasthi, 1977).

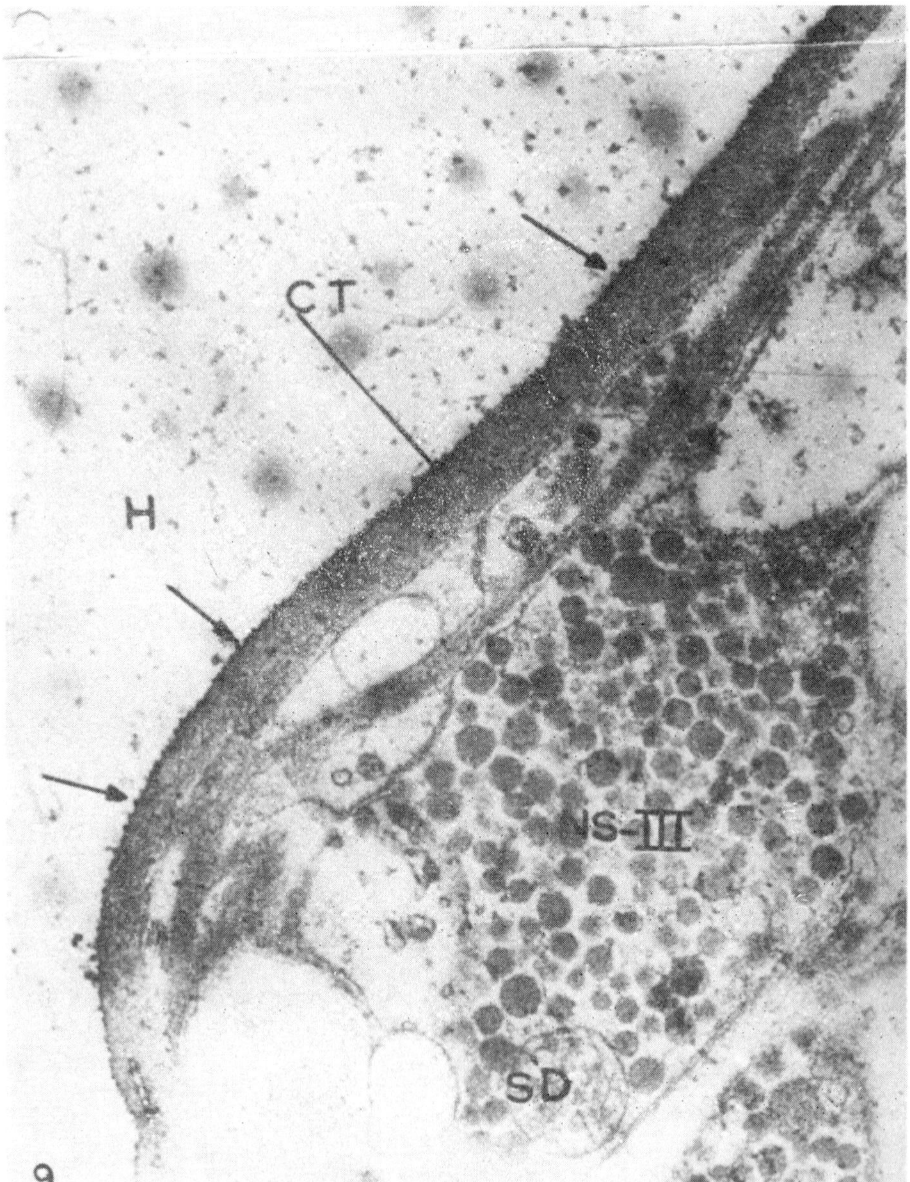

Plate 118. Fig. 9. Electron micrograph of a part of corpus cardiacum of *Gryllodes sigillatus* showing a fairly thick connective tissue sheath enveloping the gland. The fibrous character of the sheath is clearly marked out. The outer surface is provided with a large number of electron dense bodies (arrows). Also note the axons containing spherical, electron dense granules type III. A large secretion droplet is seen on the lower side formed probably by fusion of the small granules. (CT connetive tissue, H - haemocoel, SD secretion droplet) X 30000. (From Awasthi, 1977).

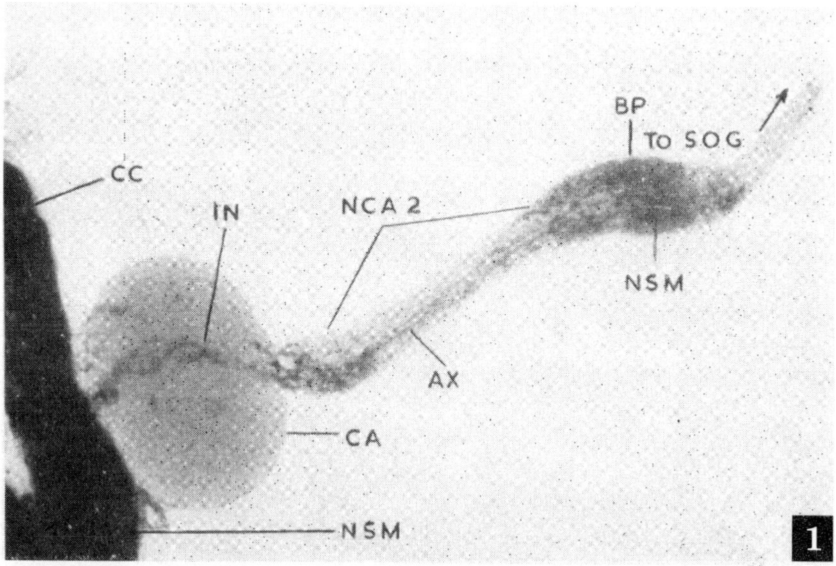

Plate 119. Fig. 1. Part of the whole mount of the corpus cardiacum - corpus allatum complex of *Gryllodes sigillatus* showing the mid-dorsal innervation of corpus allatum (IN). Note the abundant stored NSM in the dilated part (BP) of NCA 2, which is identical with the NSM in the corpus cardiacum. Note the absence of the NSM in the NCA 2 after the bulbous portion and which proceed to join the suboesophageal ganglion. X 400. (From Awasthi, 1968)

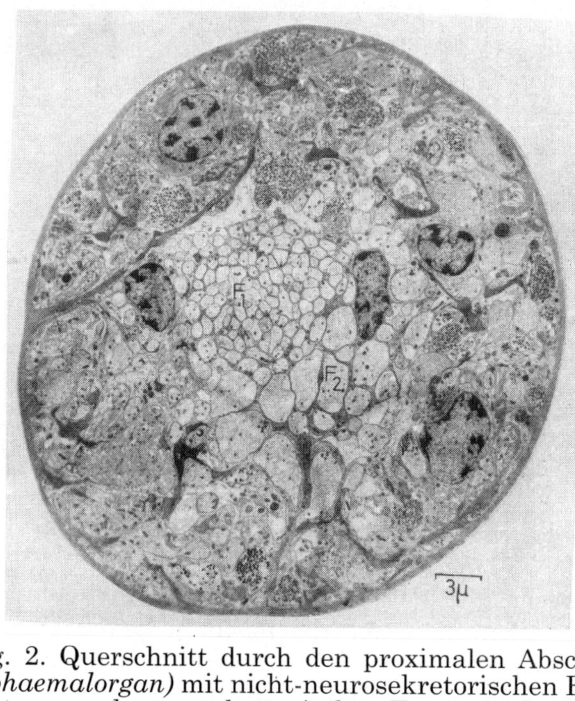

Plate 120. Fig. 2. Querschnitt durch den proximalen Abschnitt des Nca II (*Neurohaemalorgan)* mit nicht-neurosekretorischen Fasern (F_1 and F_2) im Zentrum und neruosekretorischen Fasern in der Peripherie. Vergr. 3400fach (*Acheta*) (From Weber and Gaude, 1971).

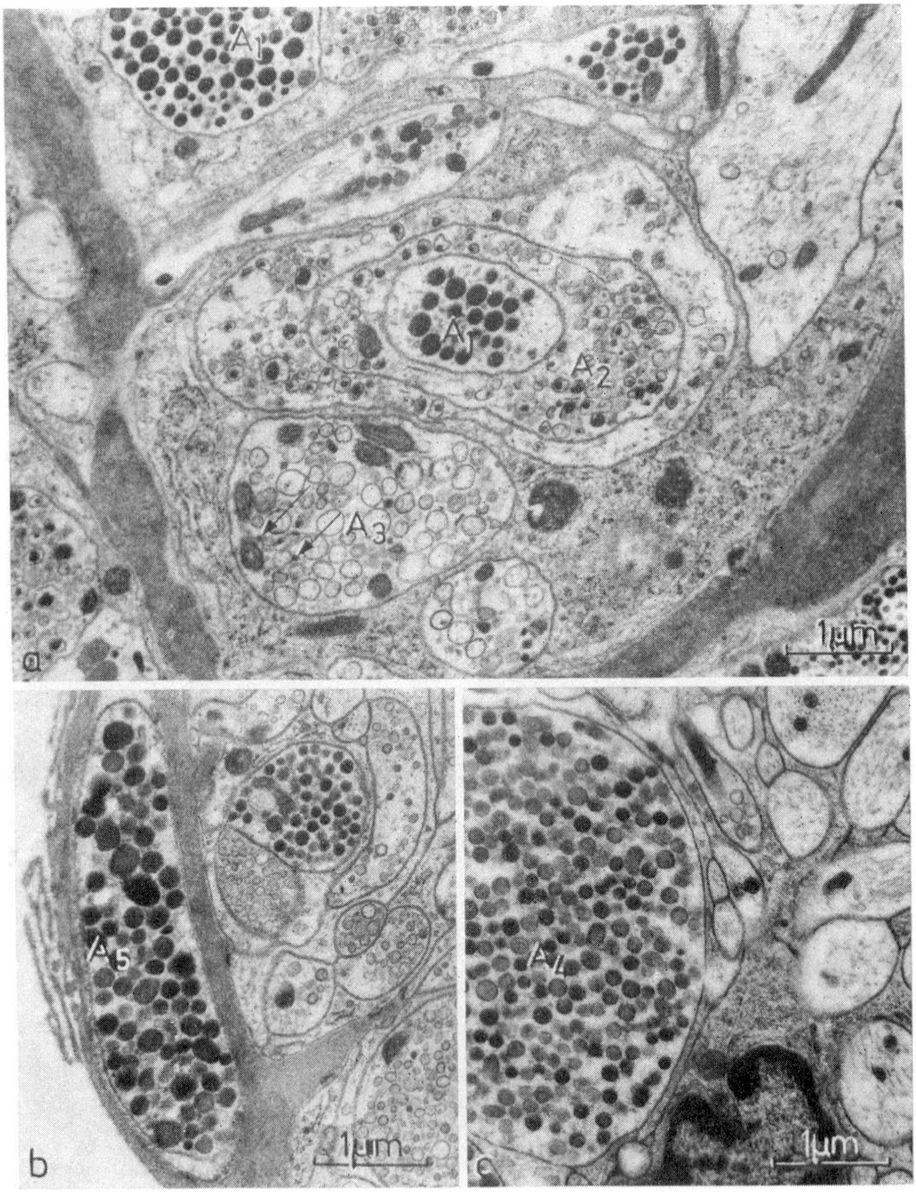

Plate 121. Fig. 3a c. Ausschnitte aus dem Nca II. a Neurosekretorische Fasern mit elektronendichten Vesikeln (A_1), mit, dense core``-Vesikeln (A_2) und hellen Blaschen (A_3). Abschnurung elektronendichten Materials (↑). b Axonprofil mit groBen elektronendichten Granula (A_5). c Faser mit Vesikeln unterschiedlicher Elektronendichte (A_4). Vergr. 14000 fach (*Acheta*) (From Weber and Gaude, 1971).

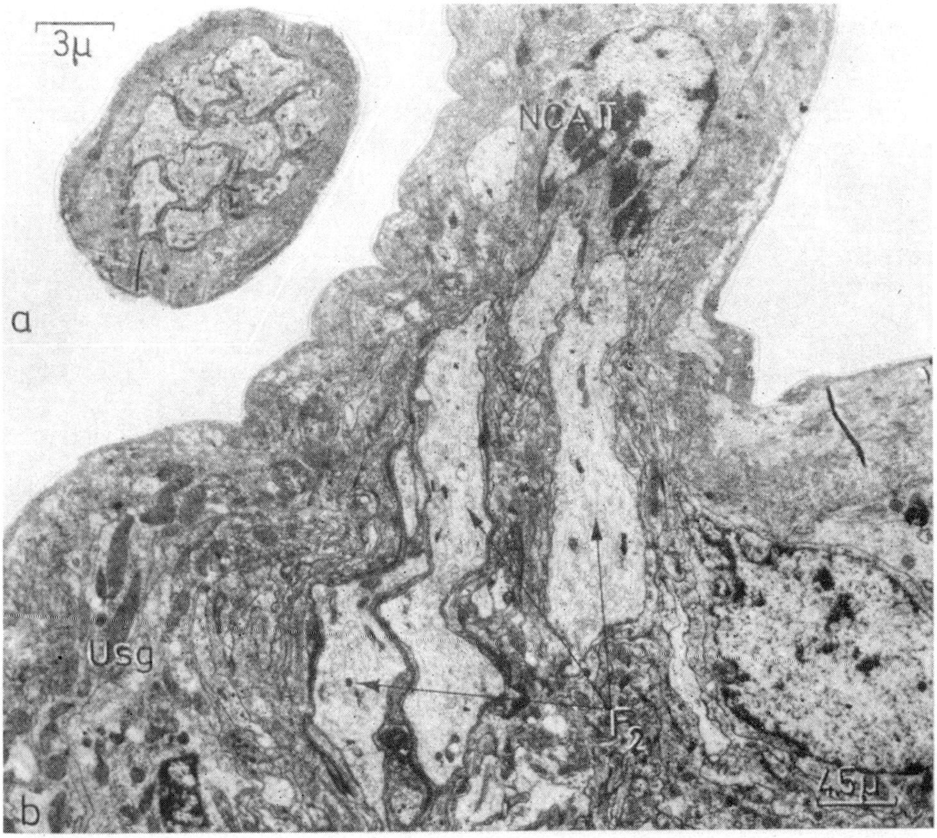

Plate 122. Fig. 4a u. b. Distale Region des Nca II. a Querschnitt. 9 Axone von Perineurium und Glia-hullen umgeben. b Langsschnitt durch Nca II und Unterschlundganglion (Usg) mit Fasern der F_2 Gruppe. Vergr. 5100 fach (*Acheta*) (From Weber and Gaude, 1971)

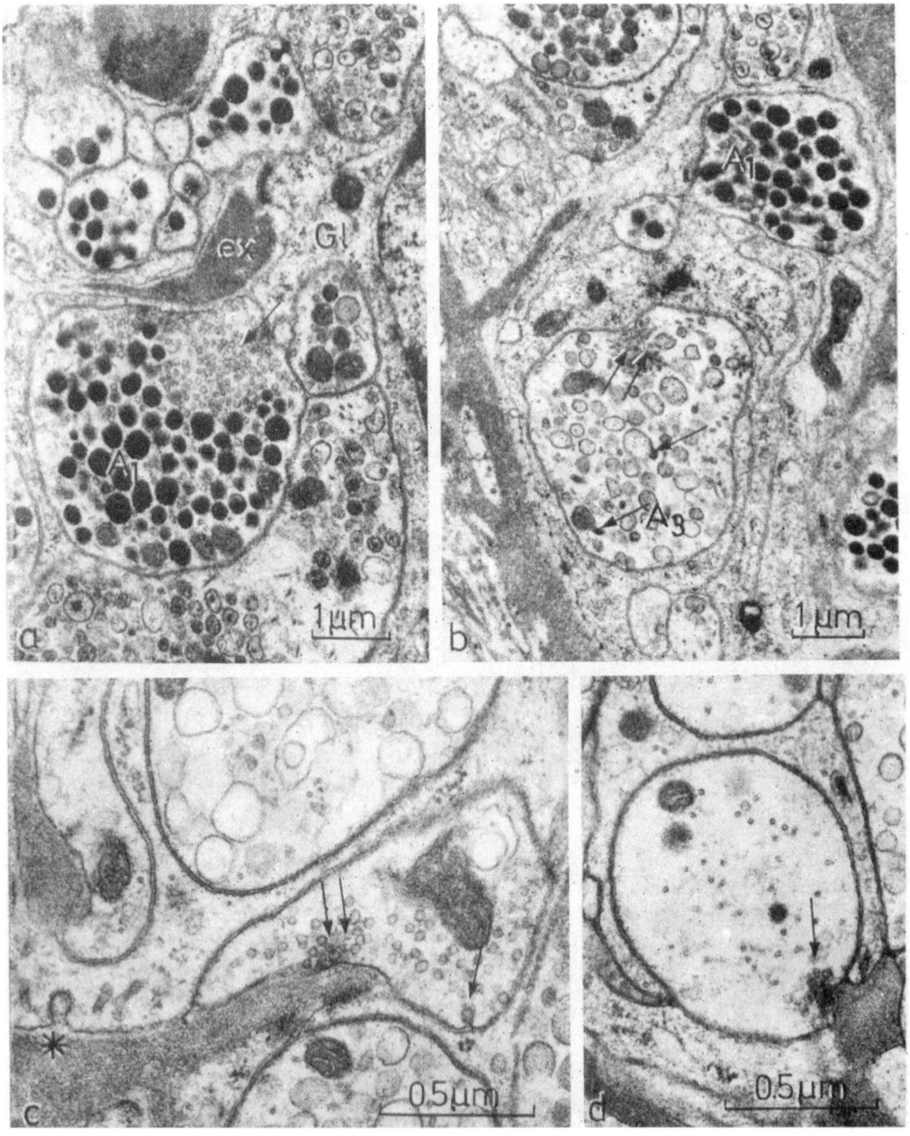

Plate 123. Fig. 5 a-d. Bildung von., Synaptoiden``. a Axon mit Elementargranula (A1) und einer Akkumulation kleiner Vesikel (↑). Kontakt mit Gliafortsatz (Gl) und extrazellularem Stroma (*ex*). Vergr. 14000 fach. b Axon mit hellen vesikeln (A3) und einer Ansammlung kleiner Blaschen (↑↑) an der Membran in Kontakt mit Gliafortsatz. Abschnurung elektronendichten Materials von hellen Vesikeln (↑) Vergr. 14000 fach. c Ansammlung synaptoider Blaschen (↑↑) in Kontakt mit extrazellularem Stroma. (↑) Exocytosevesikel. (*) Stachelsaumblaschen. d An-sammlung synaptoider Vesikel (↑). Vergr. 40000 fach (*Acheta*) (From Weber and Gaude, 1971)

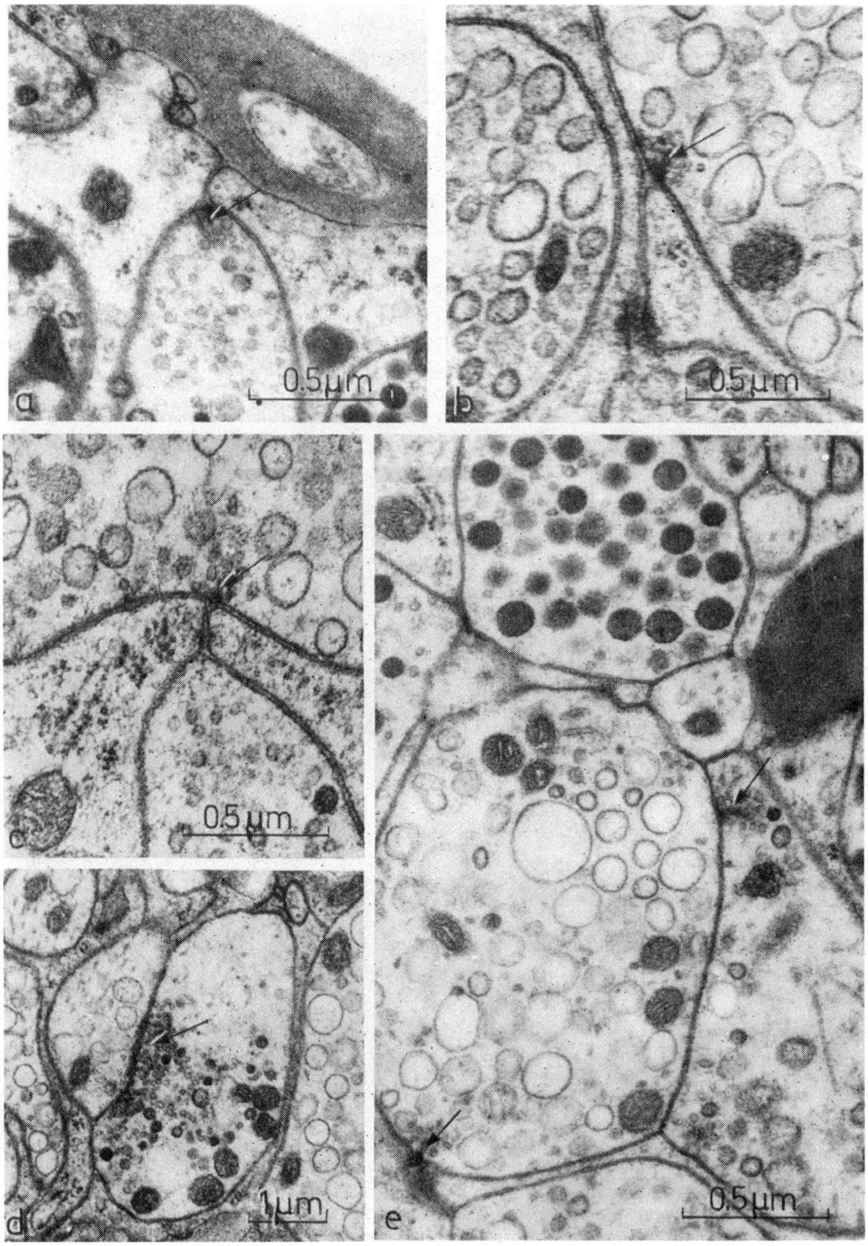

Plate 124. Fig. 6 a-e. Membrankontakte. a c Stabchnformige Kontaktstellen (↑) mit Interzellular spalten angrenzender Gliafortsatze. Vergr. 40000 fach. e Dasselbe (↑) im axo-axonalen Kontakt. Vergr. 40000 fach. d Synaptoid (↑) im axo-axonalen Kontakt. Vergr. 40000 fach (*Acheta*). (From Weber and Gaude, 1971)

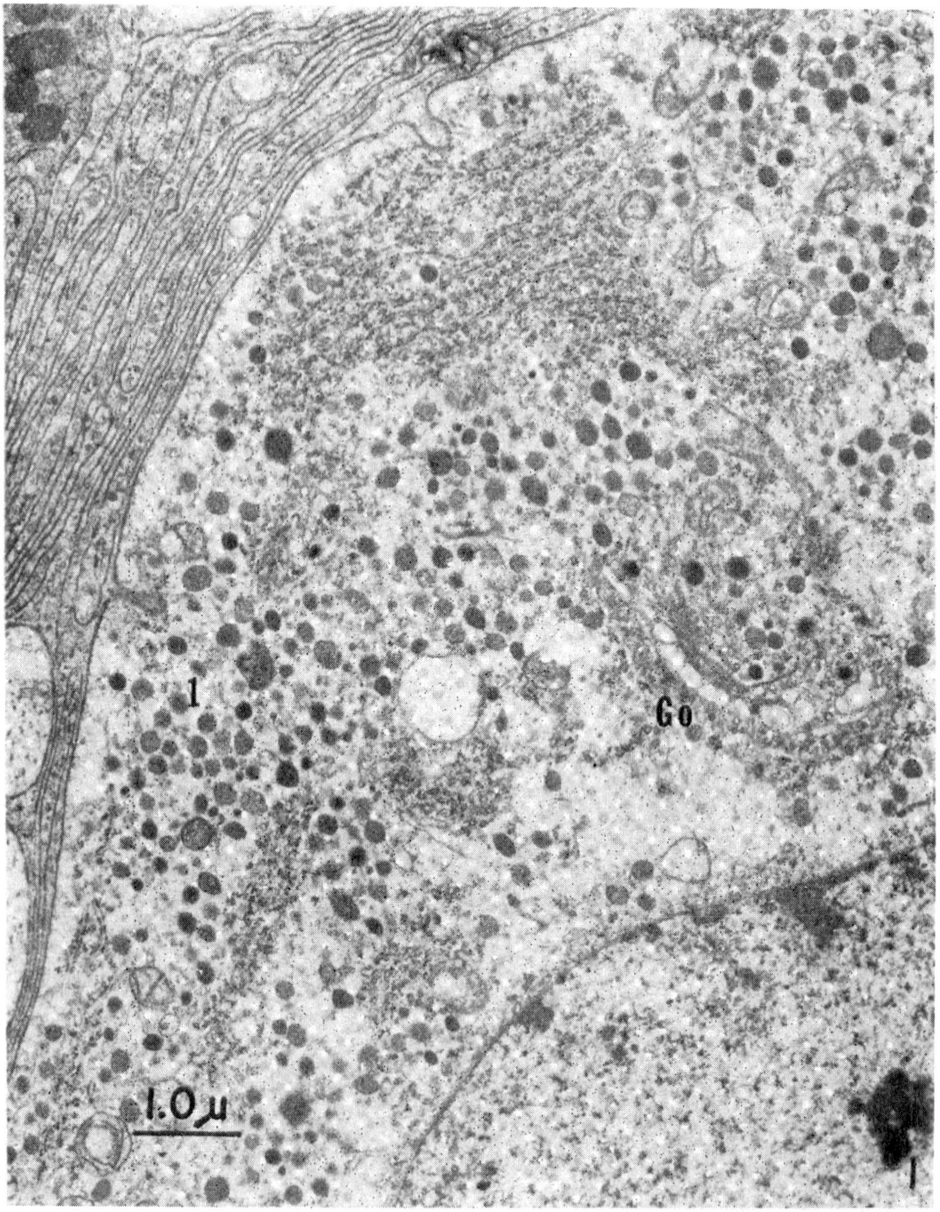

Plate 125. Fig. 1 Perikaryon of a neurosecretory cell in the perineurium of the sixth abdominal ganglion of *Acheta domestica*. Type I neurosecretory droplets and Golgi apparatus (Go) are seen. (From Stratton and Booth, 1975)

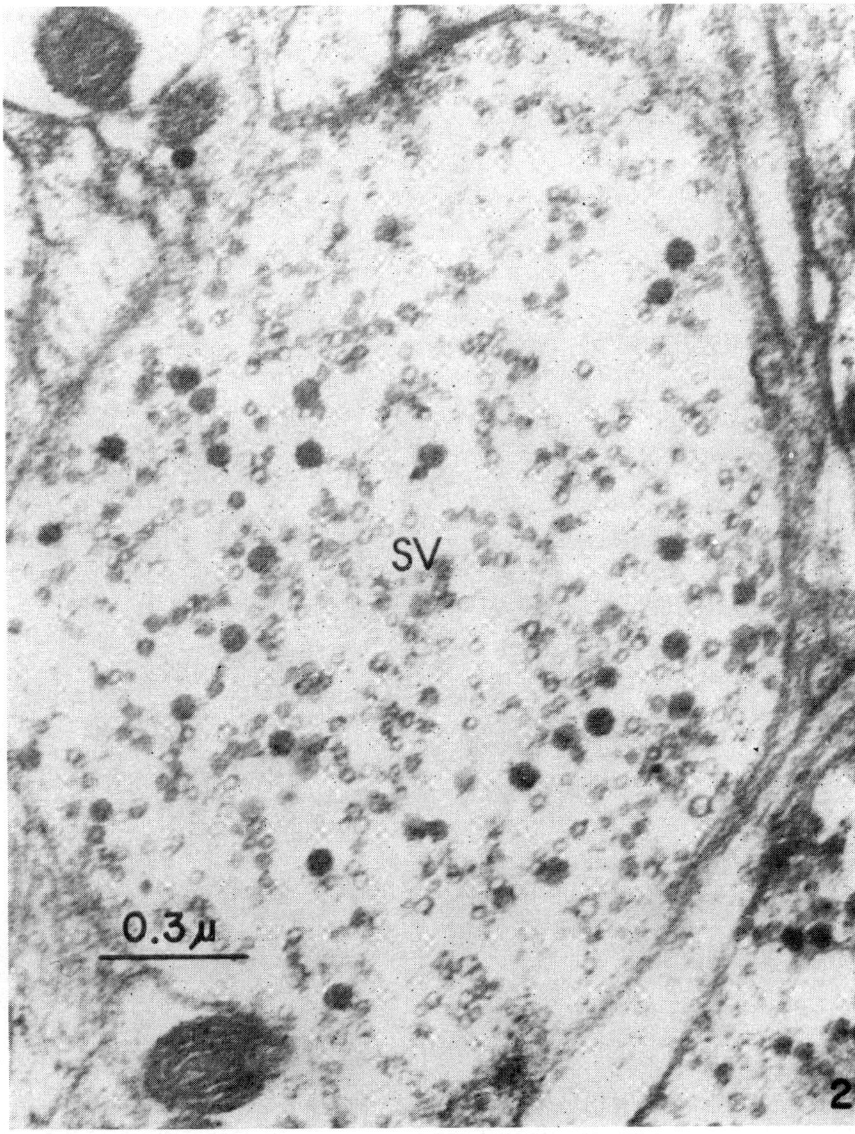

Plate 126. Fig. 2. Terminating axon in neuropile showing synaptic vesicles and Type I neurosecretory droplets. (*Acheta*) Agranular synaptic vesicles (SV) are all `spheroid'. (Stratton and Booth, 1975).

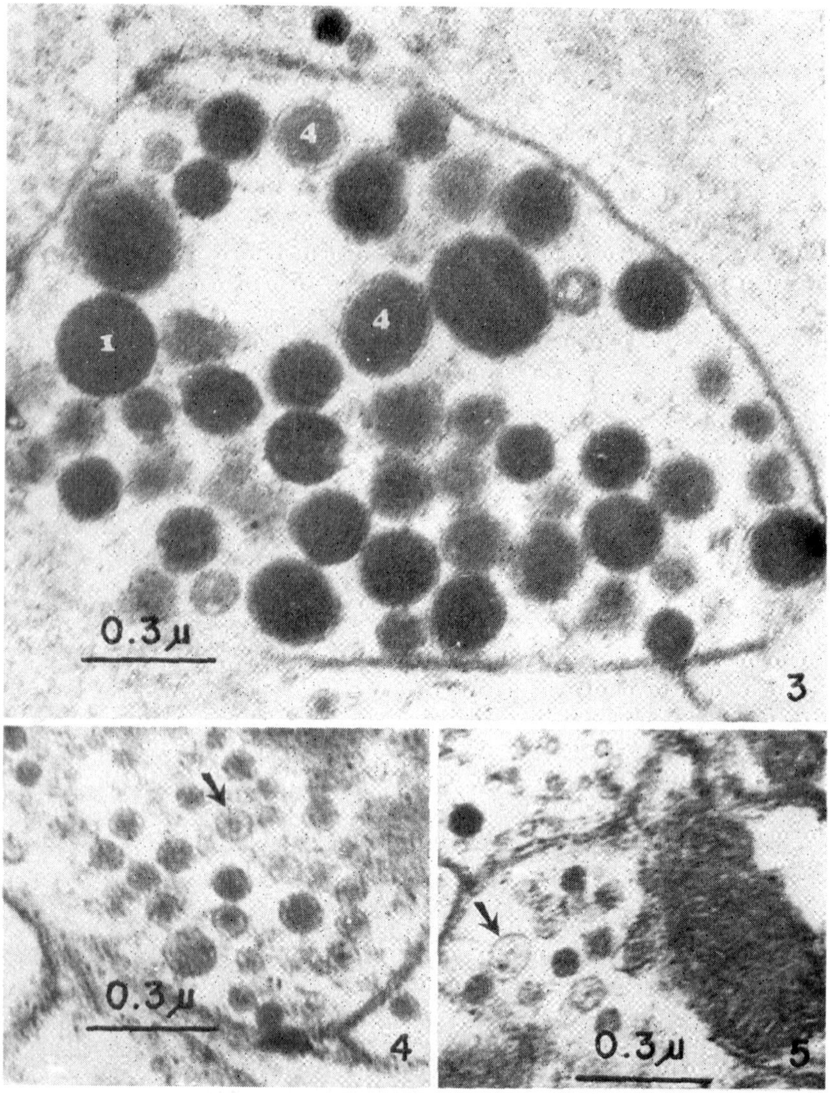

Plate 127. Fig. 3. Axon in neuropile showing what is thought to be two types of neurosecretory droplets. Type I (1) and Type IV (4) are labelled. Type IV droplets are granular in the centre surrounded by a small electron opaque area. Note that both appear in various sizes. (*Acheta*) Fig. 4. Axon in neuropile showing Type II neurosecretory droplets. Several Type II droplets are present. Type II have a small granular centre surrounded by much agranular material. All are about the same size. (*Acheta*) Fig. 5. Axon in neuropile showing Type III neurosecretory droplets. Type III droplets are significantly larger than the synaptic vesicles seen in the adjacent axon, yet appear mostly agranular. (*Acheta*) (From Stratton and Booth, 1975).

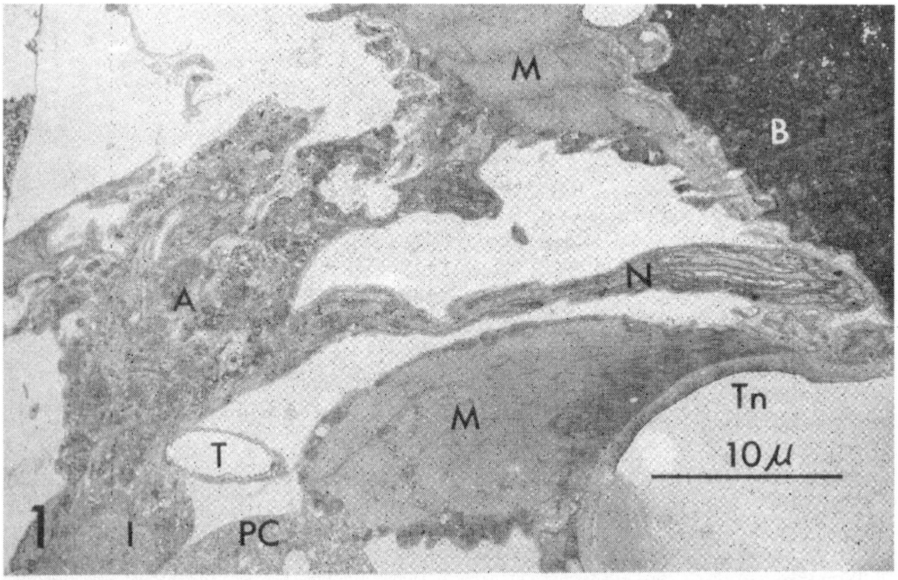

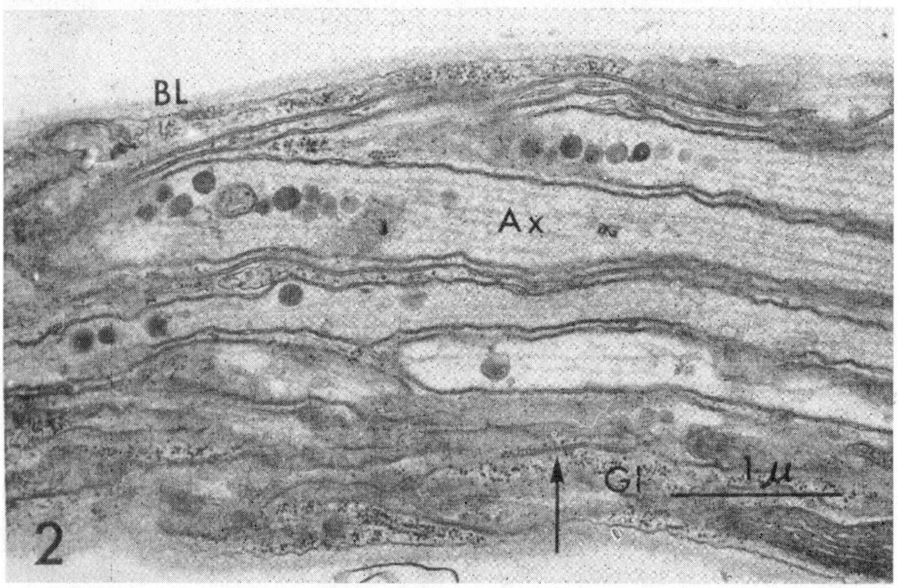

Plate 128. Fig. 1. Low-power electron micrograph of a sagittal section through the head of an adult *M. persicae*, showing a portion of the nervus corporis cardiaci II (N) passing ventrally into the axonal area of the corpus cardiacum (A). An intrinsic cell (I) is at the lower left. B, brain; M, muscle; T, trachea, Tn, tentorium; PC, pericardial cell. Fixed in glutaraldehyde. Fig. 2. Longitudinal section of the nervus corporis cardiaci II, showing portions of five axons (Ax). The axons contain microtubules, and three show large granules about 150 mμ in diameter. Glial processes (Gl) wrap the axons. The glial processes contain microtubules and ribosomes and are joined to each other in places by septate desmosomes (arrow). BL, basal lamina. Glutaraldehyde. (From Browers and Johnson, 1966).

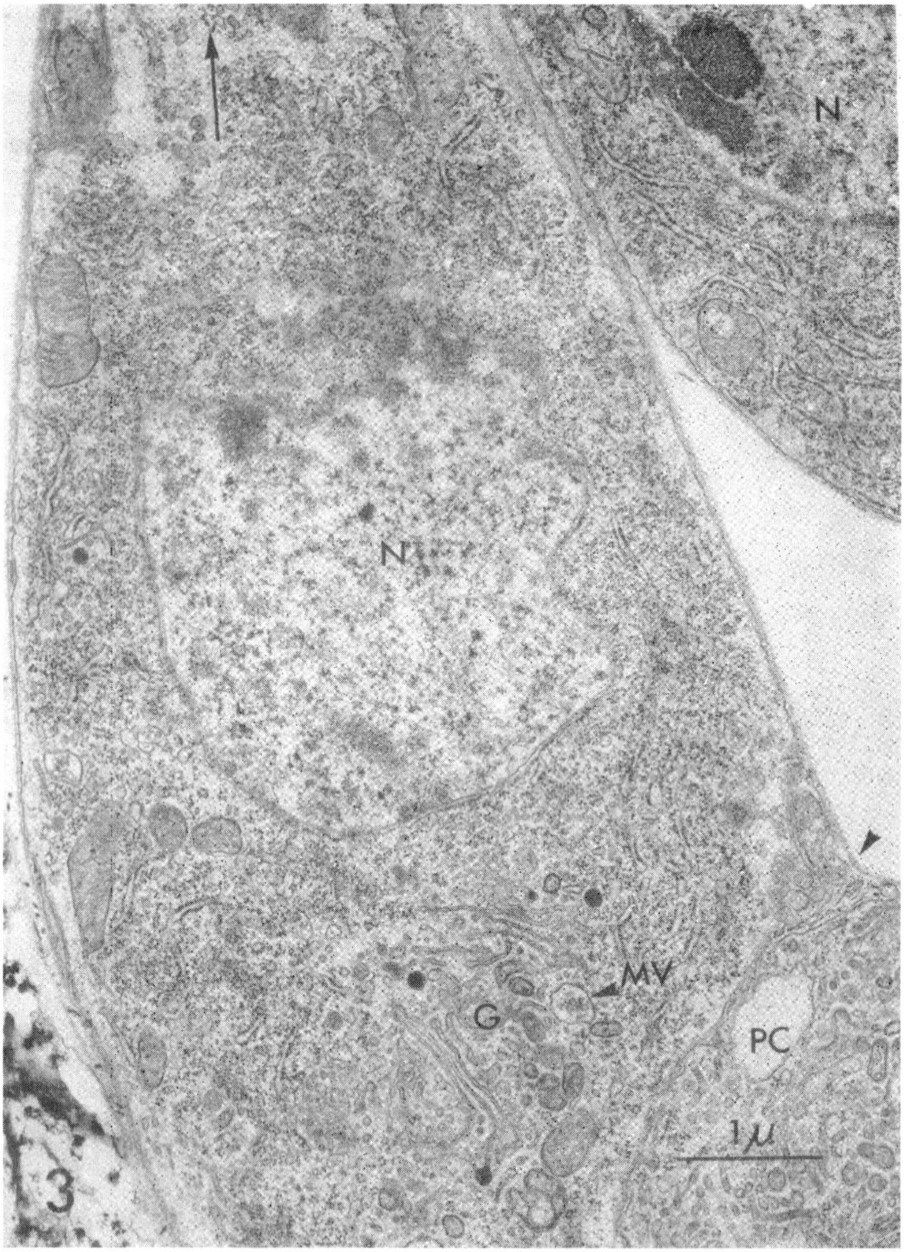

Plate 129. Fig. 3. Intrinsic cells of the corpus cardiacum of *Myzus*. Note the absence of enveloping glial cells and the continuation of the basal lamina around a contiguous pericardial cell (PC). Dense secretory granules are present in the Golgi apparatus (G) and in the cytoplasm. The intrinsic cell joins the axonal region in the direction of the arrow. N, nucleus; MV, multivesicular body. Fixed in Osmium tetroxide. (From Bowers and Johnson, 1966).

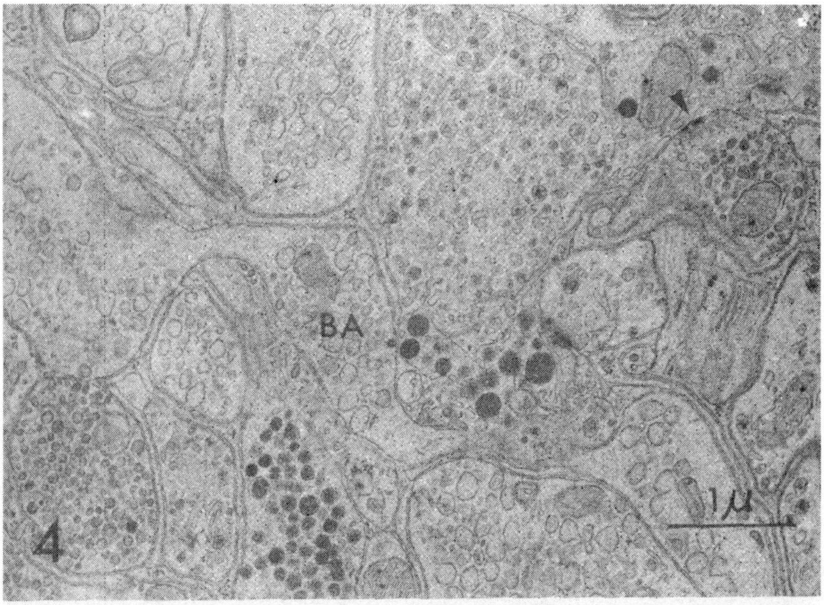

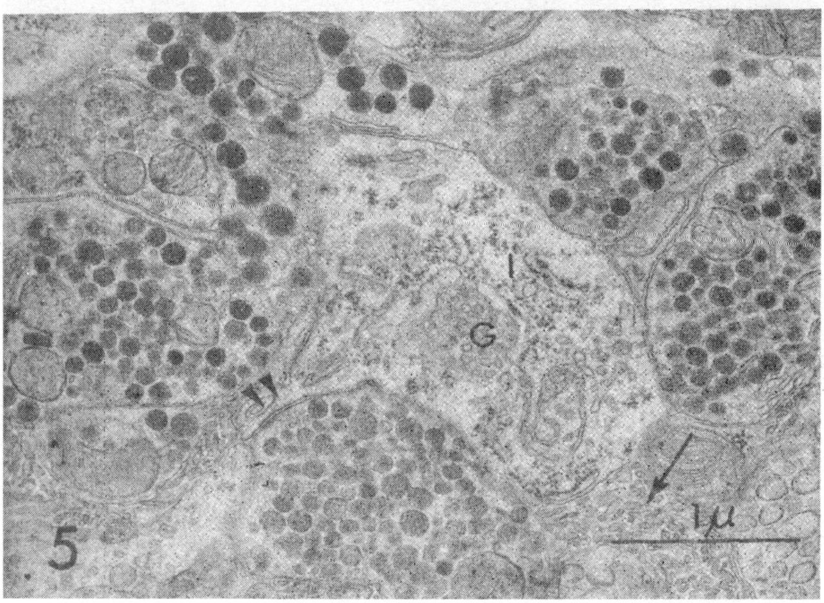

Plate 130. Fig. 4. A portion of the axonal region of the corpus cardiacum of *Myzus*. At the upper right is an axon with characteristics of a synapse. Apparently ending on an intercellular space (arrow). An axon with granule-filled dilatations crosses the field diagonally (BA). Osmium tetroxide. Fig. 5. Intrinsic cell process (I) in the axonal region of the corpus cardiacum. One of the granule-filled axons surrounding the process appear to synapse with it (double arrows). Intercellular spaces are filled with tubular extensions of the plasma membranes (arrow). G, Golgi apparatus. Osmium tetroxide. (From Bowers and Johnson,

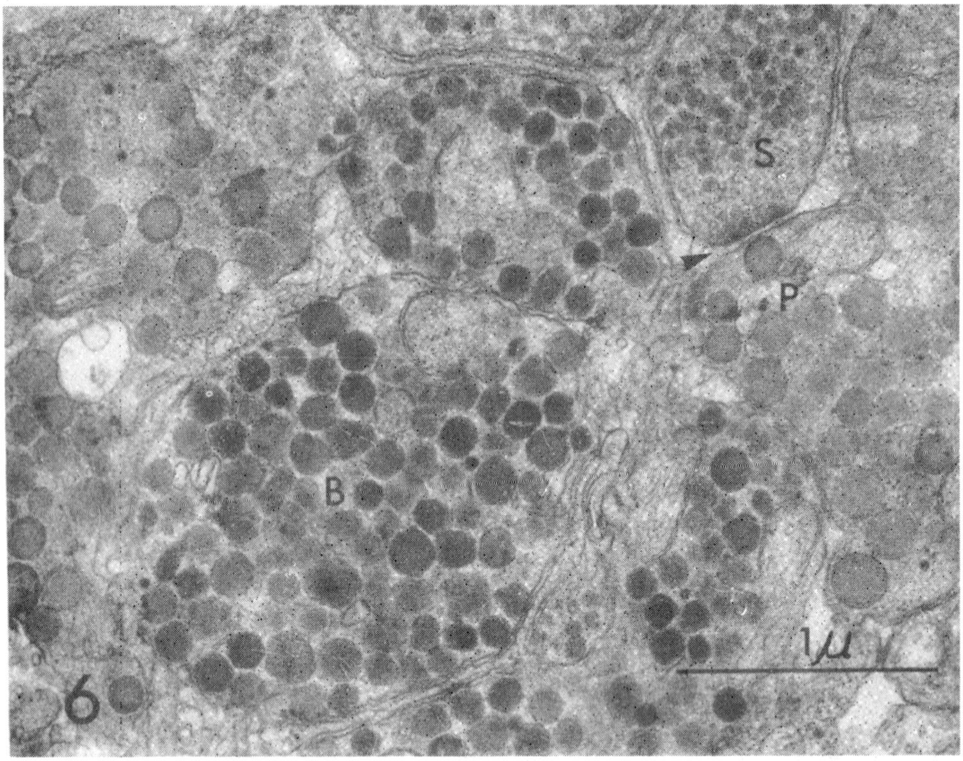

Plate 131. Fig. 6. Axonal region of *Myzus*. At least three types of granules can be identified here : small granuels in a presynaptic ending (S); large granules which range from black to grey (B); and large ``pale'' granules in the post-synaptic axon (P). Glutaraldehyde. (From Bowers and Johnson, 1966).

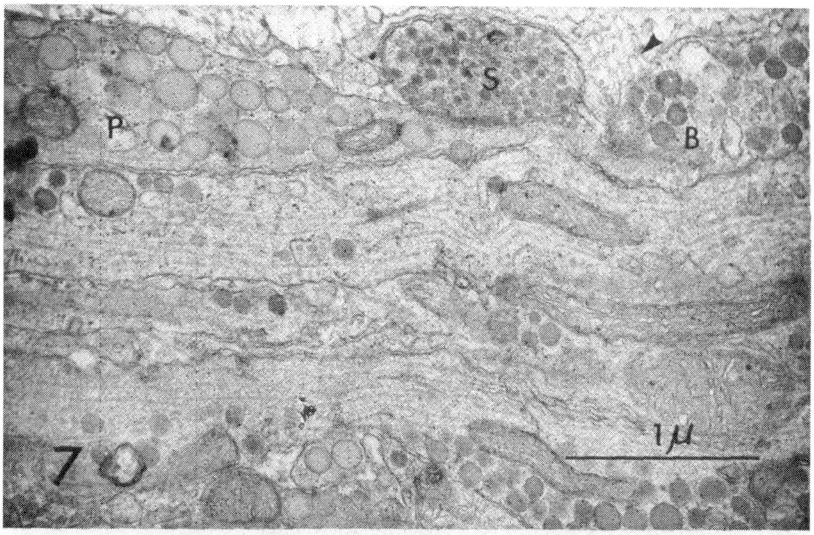

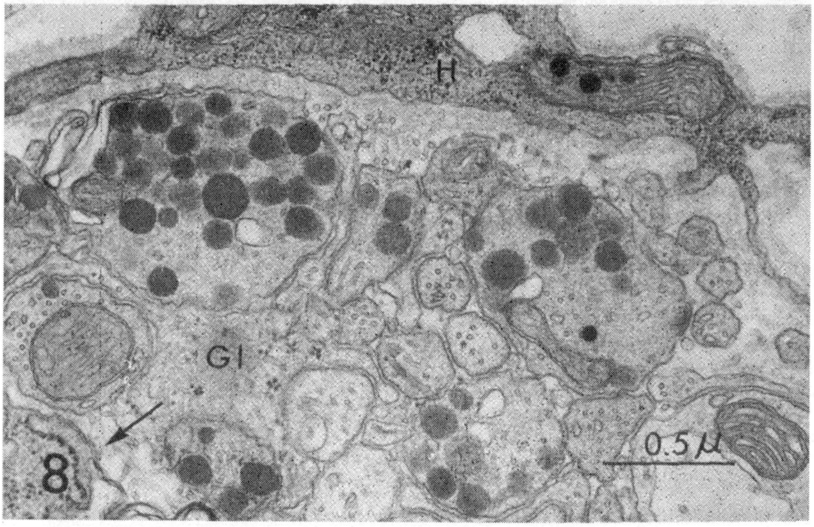

Plate 132. Fig. 7. A longitudinal section of the lateral nerve from the corpus cardiacum of *Myzus*. The axons contain the same three types of secretory granules observed in the corpus cardiacum. Microtubules and mitochondria are also visible in these axons. Note the synapse of an axon containing small granules (S) with an axon containing large ``pale" granules (P). A tubular mesh fills the intercellular space, as in the corpus cardiacum. B, axon with ``black" granules. Glutaraldehyde. Fig. 8. A cross section of the medial nerve from the corpus cardiacum. The nerve is closely attached to both heart muscle (H) and pericardial cells (arrow). Some of the axons contain ``black" secretory granules. A stellate glial process (GI) partially wraps several axons. Note that most axons are in contact with at least the tip of a glial process. Cross sections of microtubules are evident in both axons and glial cells. Glutaraldehyde. (From Bowers and Johnson, 1966)

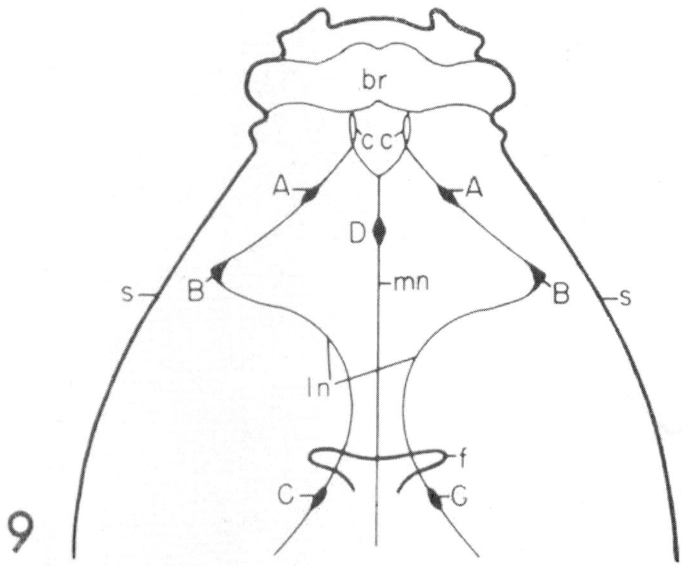

Plate 133. Fig. 9. Diagram of the anterior half of an apterous adult of *M. persicae* showing the course of the lateral and medial nerves which leave the corpora cardiaca and the positions of various cells (A-D) along them. Cells B and C have been examined with the electron microscope and have been found to be glandular cells. br. Brain; cc, corpora cardiaca; mn, medial nerve; In, lateral nerve; f, furca; s, position of the prothoracic spiracle. (From Bowers and Johnson, 1966)

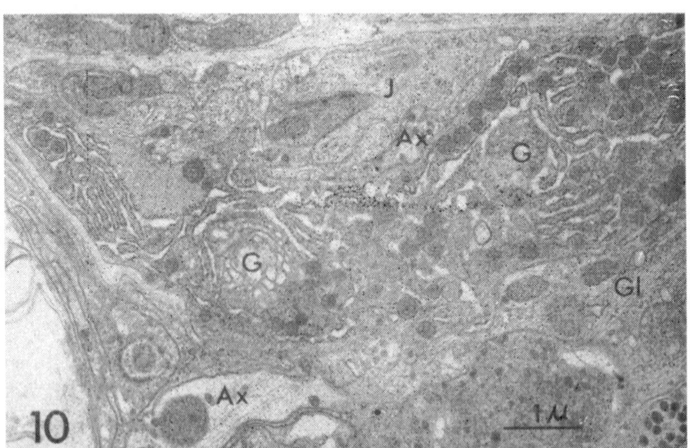

Plate 134. Fig. 10. Glandular cell at the level of the prothoracic spiracle (B, in Fig. 9). The cell lies between two muscles (not shown in this micrograph) and is surrounded by many granule-containing axons (Ax). The cytoplasm contains numerous secretory granules (290 mμ). Two Golgi regions (G) and many dilated cisternae of the ergastoplasm are also evident. Gl glial cell; J, neuromuscular junction. Glutaraldehyde. (From Bowers and Johnson 1966).

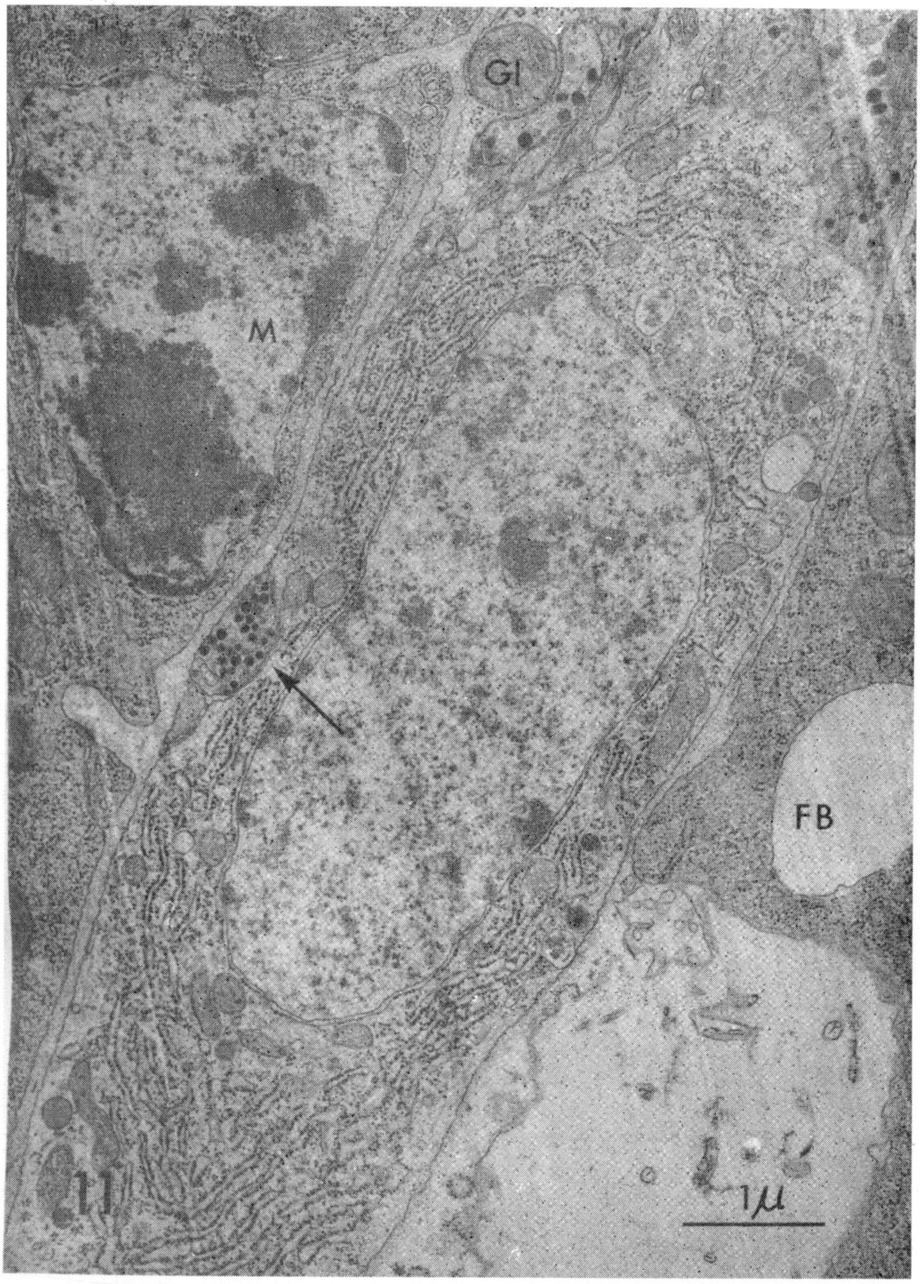

Plate 135. Fig. 11. Glandular cell just posterior to the furca of *Myzus* (C, in Fig. 9). The cell body has an ultrastructure similar to that of intrinsic cells of the corpus cardiacum. No glial cytoplasm surrounds the cell and granule-containing axons are found in intimate association with the cell body (arrow). M, muscle; FB fat body; Gl, glial process. Glutaraldehyde. (From Bowers and Johnson, 1966).

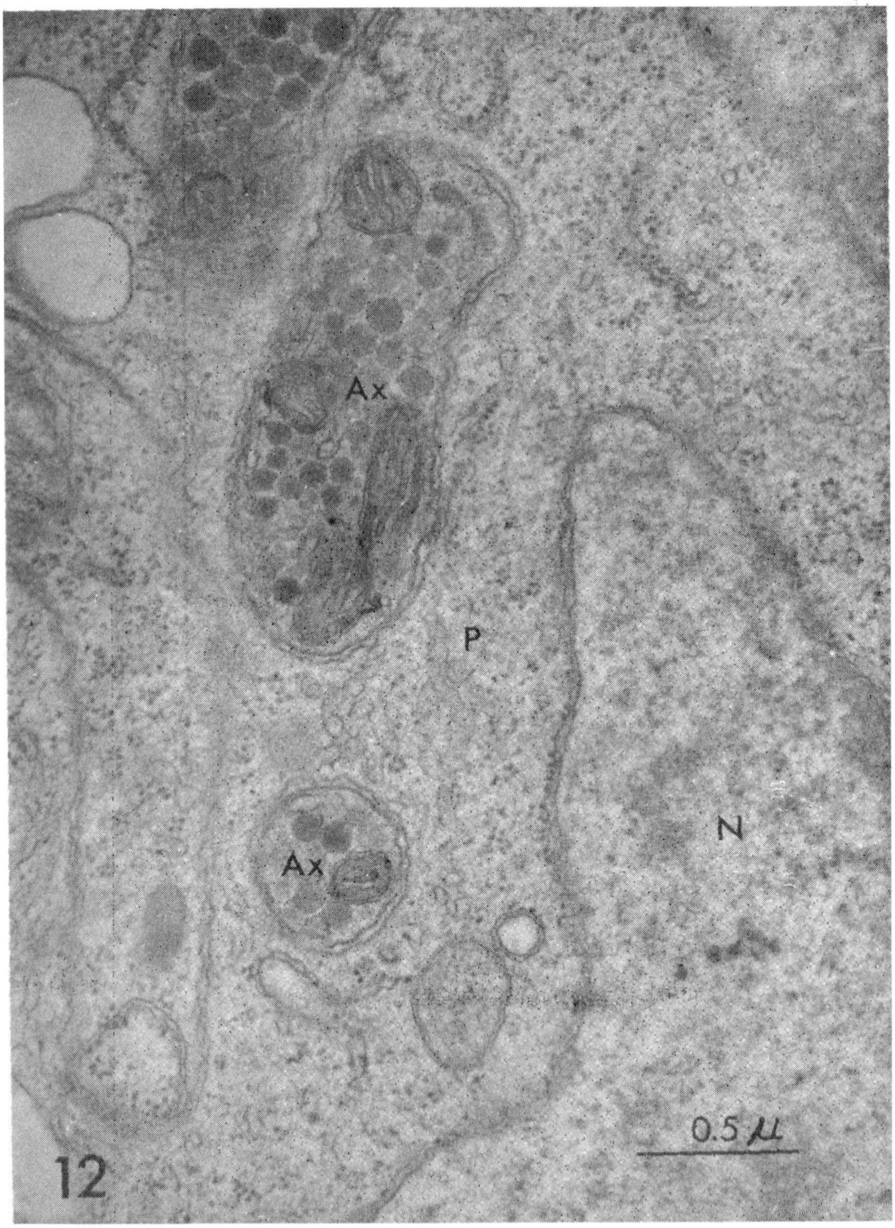

Plate 136. Fig. 12. Portion of the corpora allata of *Myzus* showing two axons (Ax) which appear to have pushed deeply into a parenchymal cell (P). These processes are surrounded by two layers of plasma membrane, the inner one belonging to the axon and the outer one being the infolded plasma membrane of the parenchymal cell. Note that the processes lack glial sheaths, contain mitochondria and ``synaptic vesicles" as well as granules. N, parenchymal cell nucleus. Glutaraldehyde. (From Bowers and Johnson, 1966)

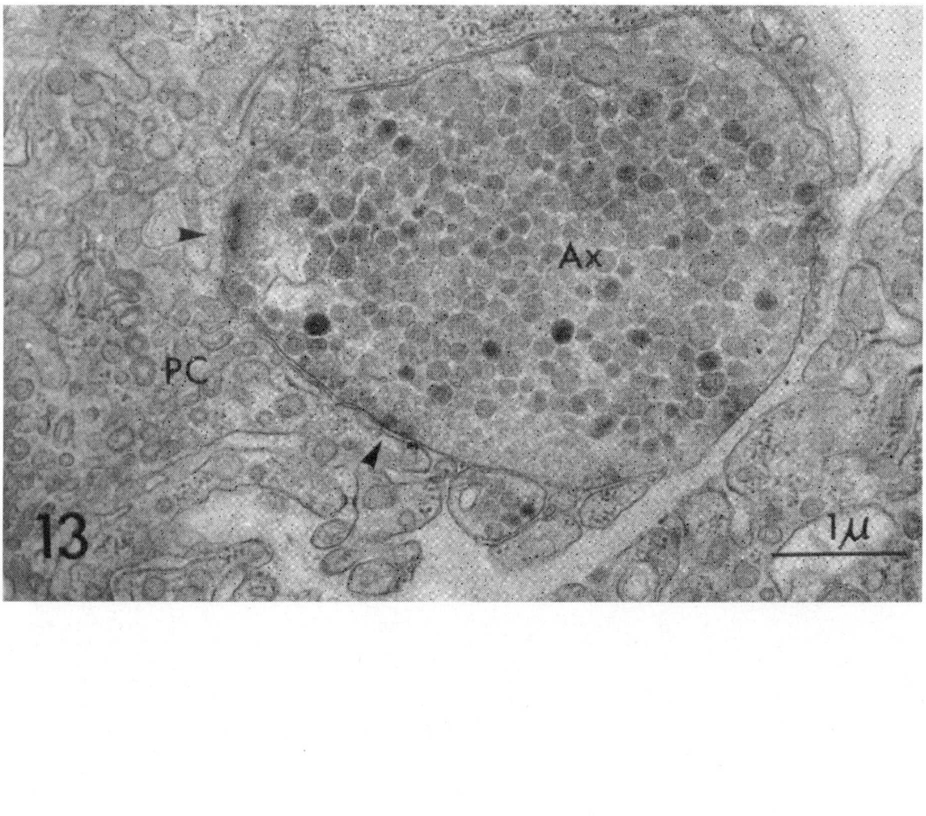

Plate 137. Fig. 13. An axon (Ax) from the medial nerve is closely apposed to a pericardial cell (PC) of *Myzus*. The axon is differentiated at two places (arrows) and at one of these small vesicles are also present. Glutaraldehyde. (From Bowers and Johnson, 1966).

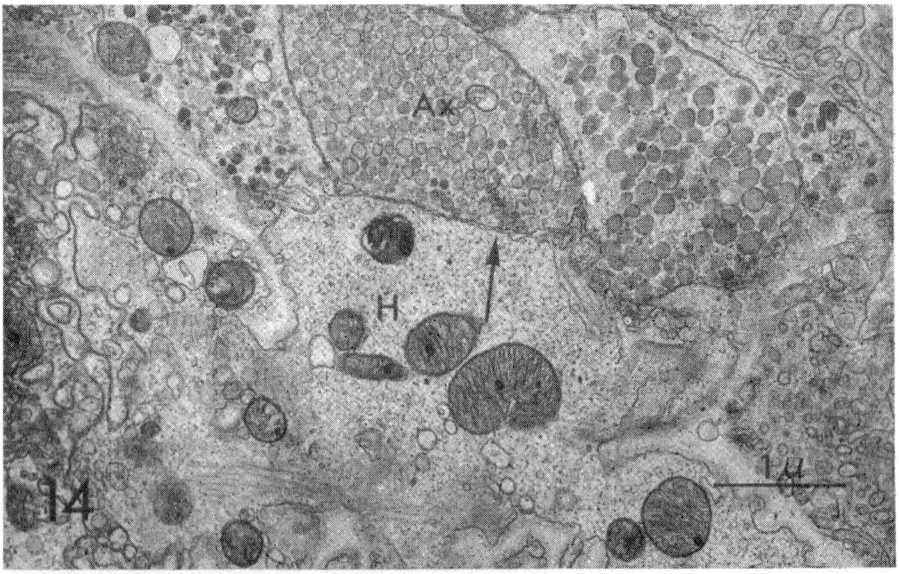

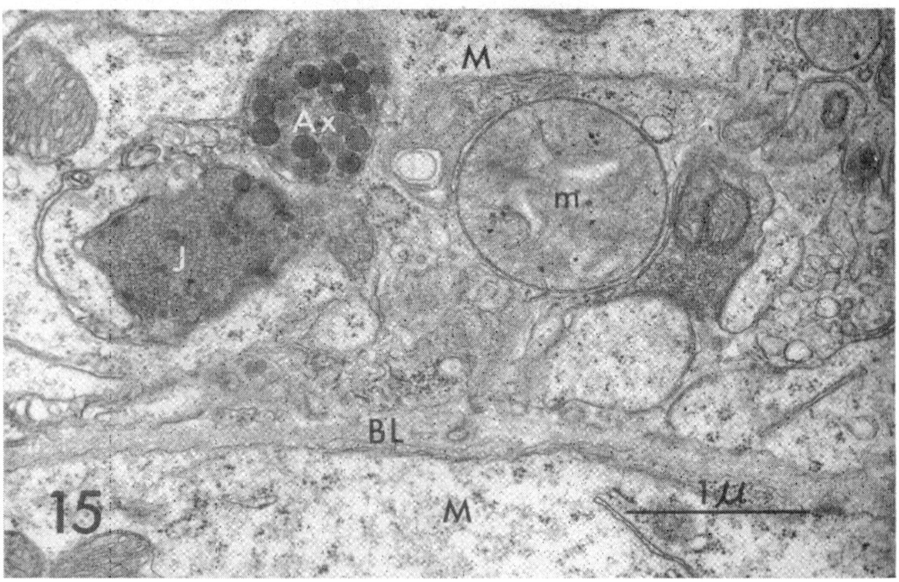

Plate 138. Fig. 14. A section of the medial nerve of *Myzus* showing the contact of an axon (Ax) with heart muscle (H). The small clear vesicles (arrow) suggest that this region may be close to a site of granule release. Fig. 15. Section of somatic muscle (M) in the prothorax of *Myzus*. An axon (Ax) containing granules roughly 150 mμ in diameter is in contact with the muscle cytoplasm. The nearby neuromuscular junction (J) is packed with clear vesicles and contains one 110 mμgranule. A glial cell process is recognizable by its characteristic mitochondrion (m) and microtubules. BL, basal lamina. Glutaraldehyde. (From Bowers and Johnson, 1966)

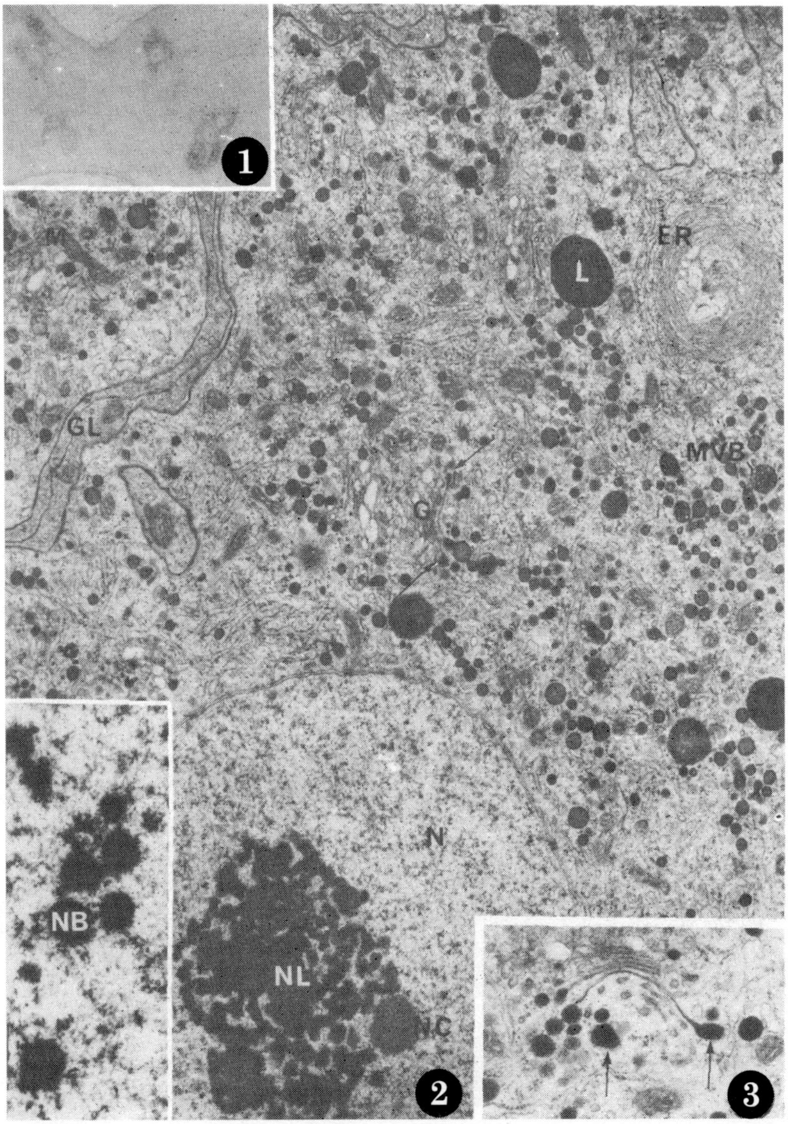

Plate 139. Fig. 1. Photomicrograph of section of pars intercerebralis of the brain of *Oncopeltus* showing seven neurosecretory A cells; 1 μm thick section stained with paraldehyde fuchsin after removing the embedding resin. X 600. Fig. 2. Electron micrograph of A cell from 4-day-old *Oncopeltus* showing general cytomorphology. Note the electron-dense material associated with Golgi cisternae (arrows) and the concentric layers of endoplasmic reticulum (ER). X 16,000. The inset on the left shows extranucleolar bodies. Note the ribosome like particles in close proximity of these bodies (arrow). X 23,000. Fig. 3. Higher magnification of a Golgi zone showing newly-formed neurosecretory granules and dense material in the Golgi cisternae (arrows). X 30,000. (From Unnithan and Nair, 1977).

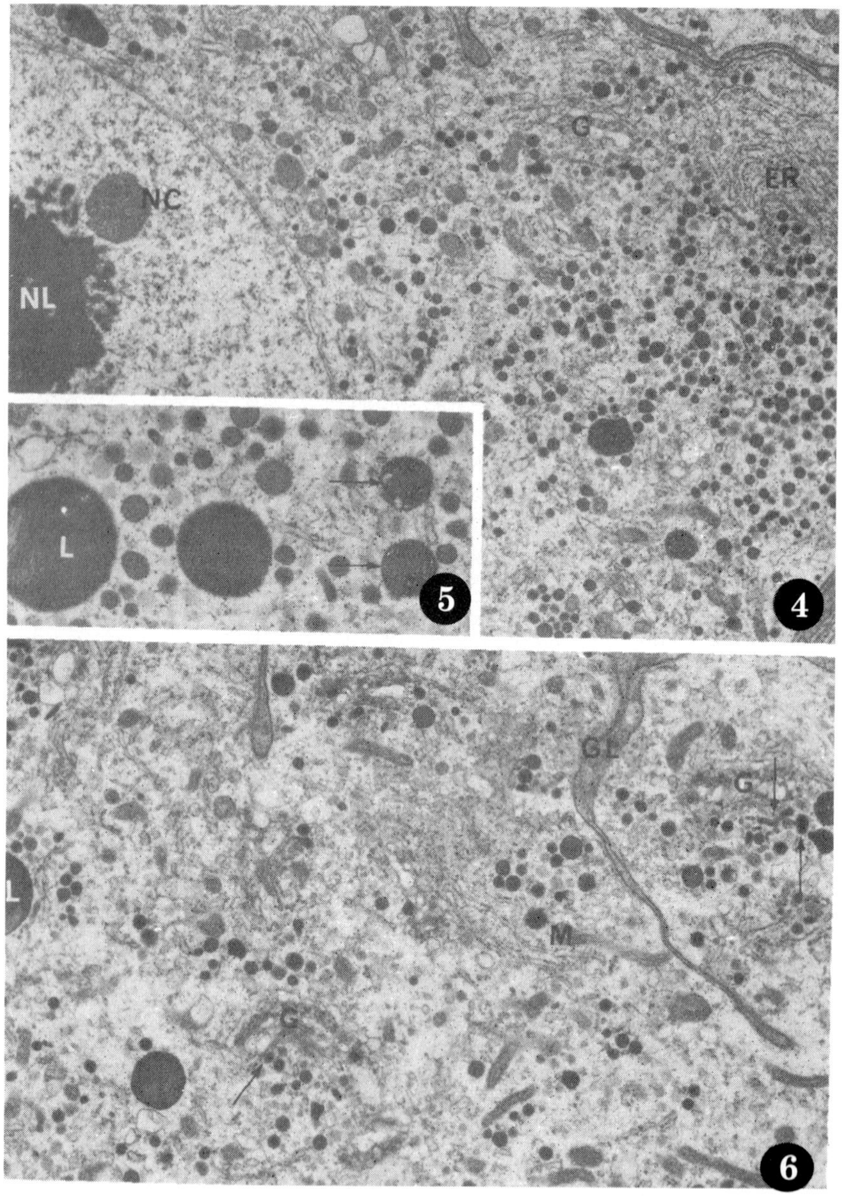

Plate 140. Fig. 4. Electron micrograph of A cell from O-day-old *Oncopeltus* Golgi bodies in these cells are rare and inconspicuous. X 13, 600. Fig. 5. Higher magnification of a different region of the same cell in figure 4 showing lysosomes (L) and degenerating neurosecretory granules (arrows). X 25,000. Fig. 6. Portion of A cell from 2-day-old *Oncopeltus*. Note the abundance of active Golgi bodies (G) with dense material associated with the cisternae and newly-formed NSG in close proximity (arrows). Also note the sparse distribution of NSG. X 13,600. (From Unnithan and Nair, 1977).

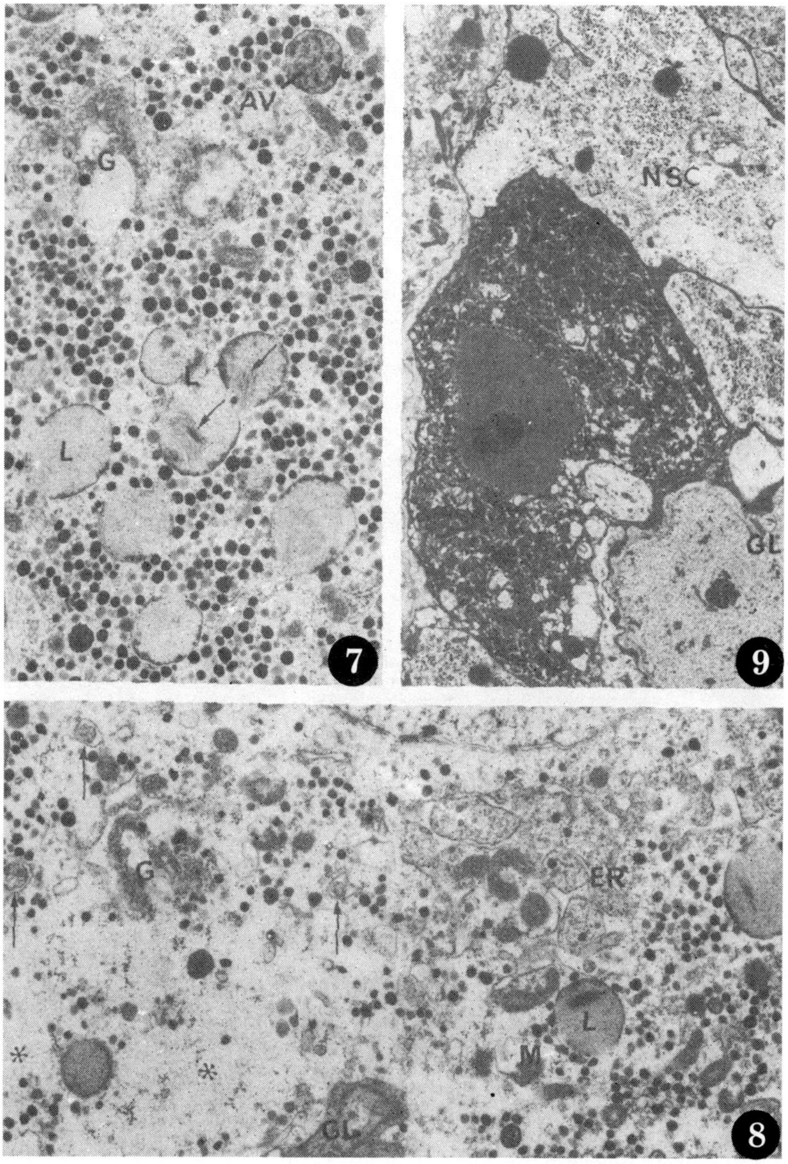

Plate 141. Figs. 7-8. A cells from 0-day-old *Oncopeltus* four hours after irradiation. Fig. 7. Portion of the A cells of *Oncopeltus* showing lysosomes (L) with ruptured limiting membrane and membrane arrays inside (arrows). Also note the swollen Golgi cisternae (G) and autophagic vesicle (AV). X 15,400. Fig. 8. Portion of the A cell of *Oncopeltus* where the cytoplasm appears to have hydrolized partially (*). Note the dilated ER and ruptured lysosomes (L). Some mitochondria (M) show partial dissolution of the matrix. Arrows indicate regions of the cytoplasm isolated by membranes of the ER. X 11,500 Fig. 9. A necrotic A cell, surrounded by other type of neurosecretory cell (NSC) and glial cell (GL), from 2-day-old irradiated insect X 4,000. (From Unnithan and Nair, 1977).

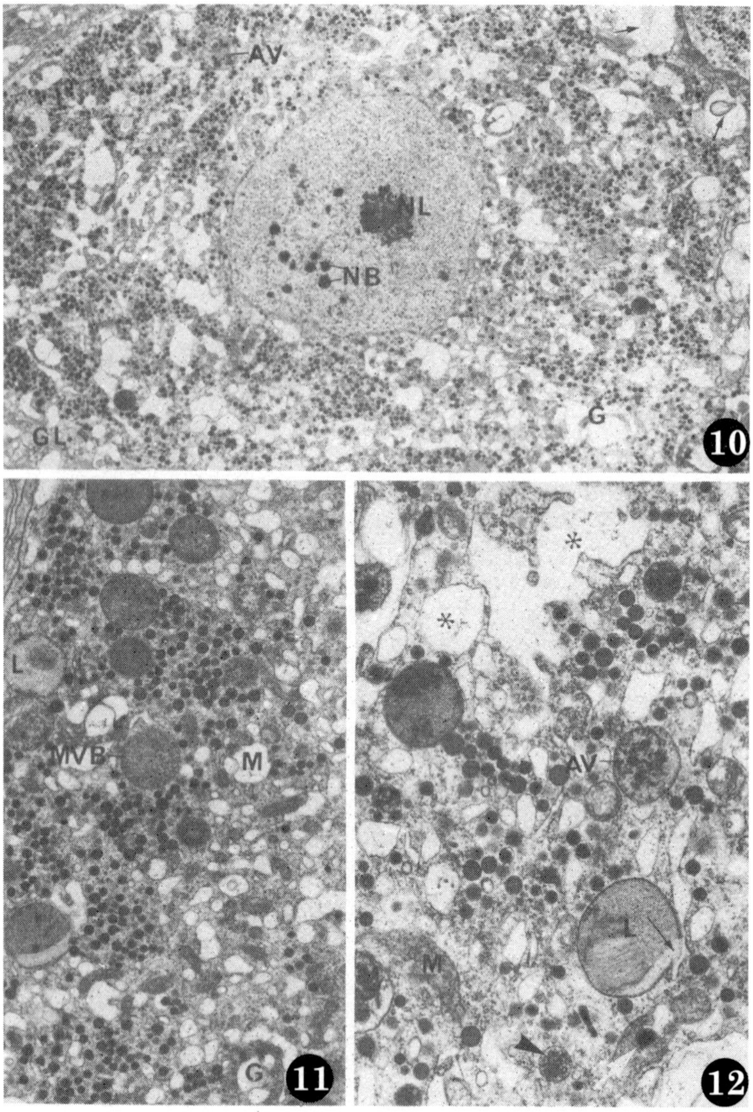

Plate 142. Figs. 10-12 A cells from 6-day-old irradiated *Oncopeltus*. Fig. 10. Low magnification electron micrograph of an A cell showing channels and vacuolar spaces in the cytoplasm, damaged Golgi (G), and frequently distributed autophagic vesicles (AV). Arrows indicate remnants of membranes in the vacuolar space. X 6,600. Fig. 11. Portion of the cell showing ruptured lysosomes (L) multivesicular body (MVB) and vacuolated mitochondria (M). Vacuolated mitochondria and dilated cisternae of the ER appear to coalesce. X 11,500. Fig. 12. Higher magnification of portion of the A cell. Note the vacuolar spaces (*) autophagic vesicles (AV) and dilated tubules of ER. The ruptured portion of the lysosome is in close proximity with ER (black arrow). White arrow indicates a neurosecretory granule-like inclusion in the mitochondria. One multivesicular body encloses particle similar to neurosecretory granule (arrow head). X 18,750. (From Unnithan and Nair, 1977).

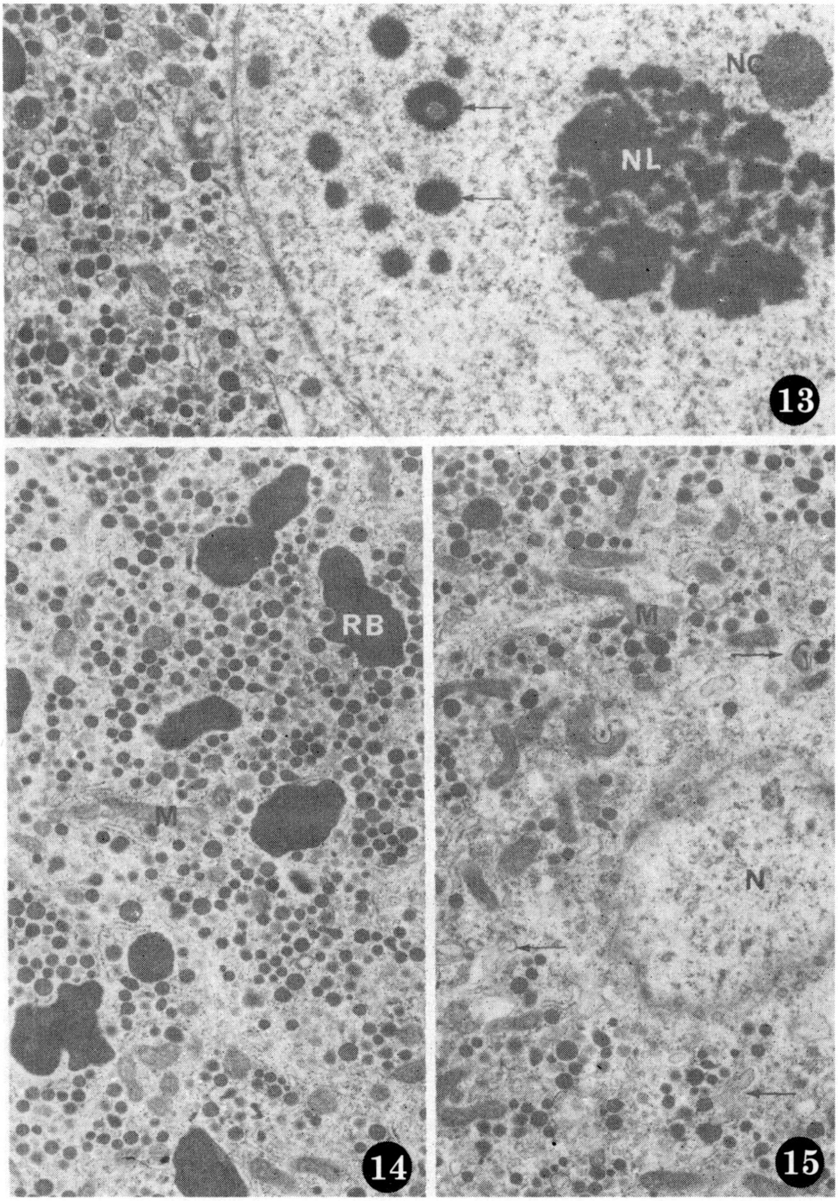

Plate 143. Figs. 13-15. A cells which have apparently recovered from radiation damage from 8-day-old irradiated *Oncopeltus*. Fig. 13. Part of cell showing conspicuous nucleolus and extranucleolar bodies (arrows). X 15,400. Fig. 14. Portion of A cell showing accumulation of NSG and a number of residual bodies (RB), x 15,400. Fig. 15. Part of a cell showing lamellate remnants of membranes (arrows) and increased distribution of mitochondria in the perinuclear area. X 15,400. (From Unnithan and Nair, 1977).

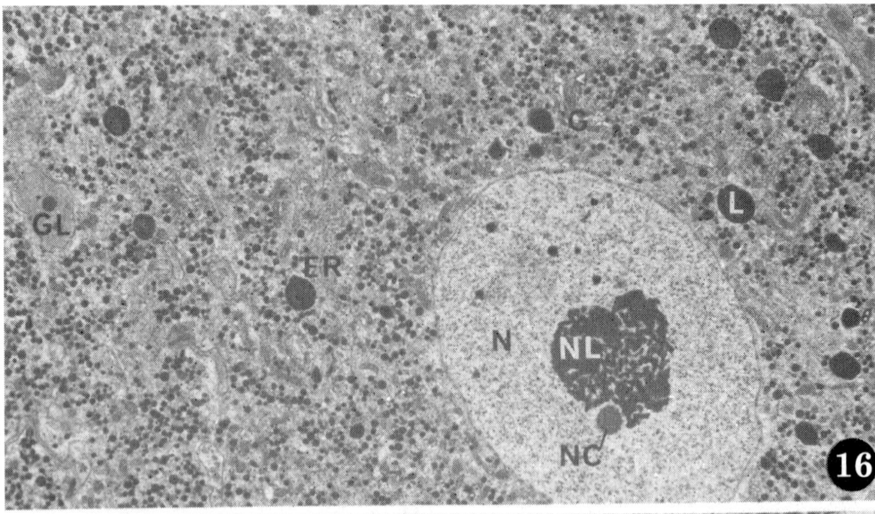

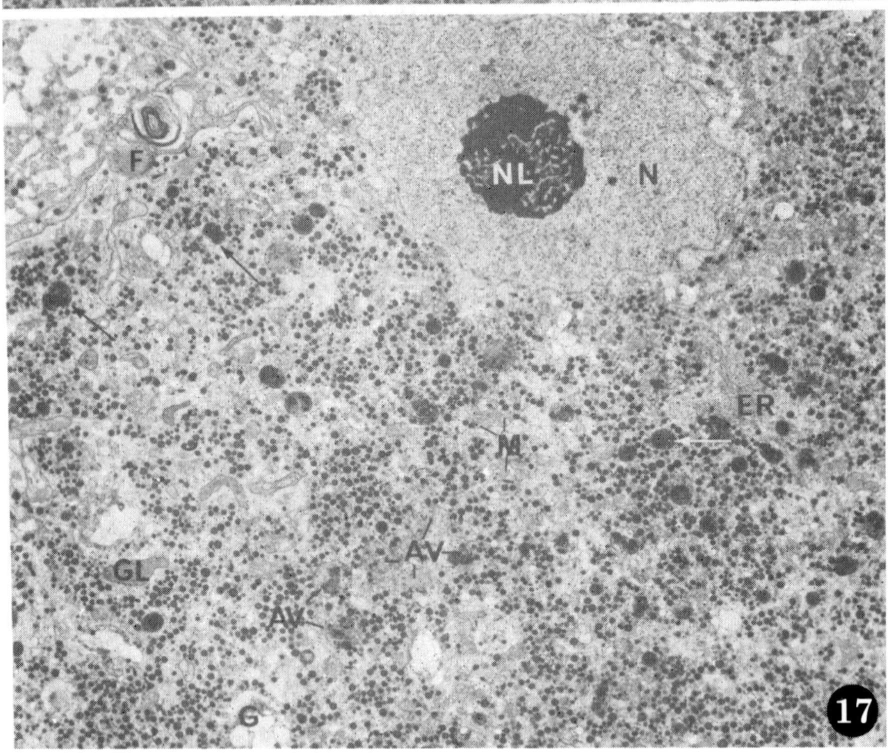

Plate 144. Fig. 16. A cell from 17-day-old *Oncopeltus*. Note the absence of autophagic vesicles in the perikaryon. X 6,600. Fig. 17. A cell from 17-day-old irradiated *Oncopeltus*. The nuclear membrane is corrugated. Note the frequently distributed autophagic vescles (AV), disintegrating NSG (arrows), and swollen Golgi body (G). Mitochondria (M) are less dense and swollen. X 6,600. (From Unnithan and Nair, 1977).

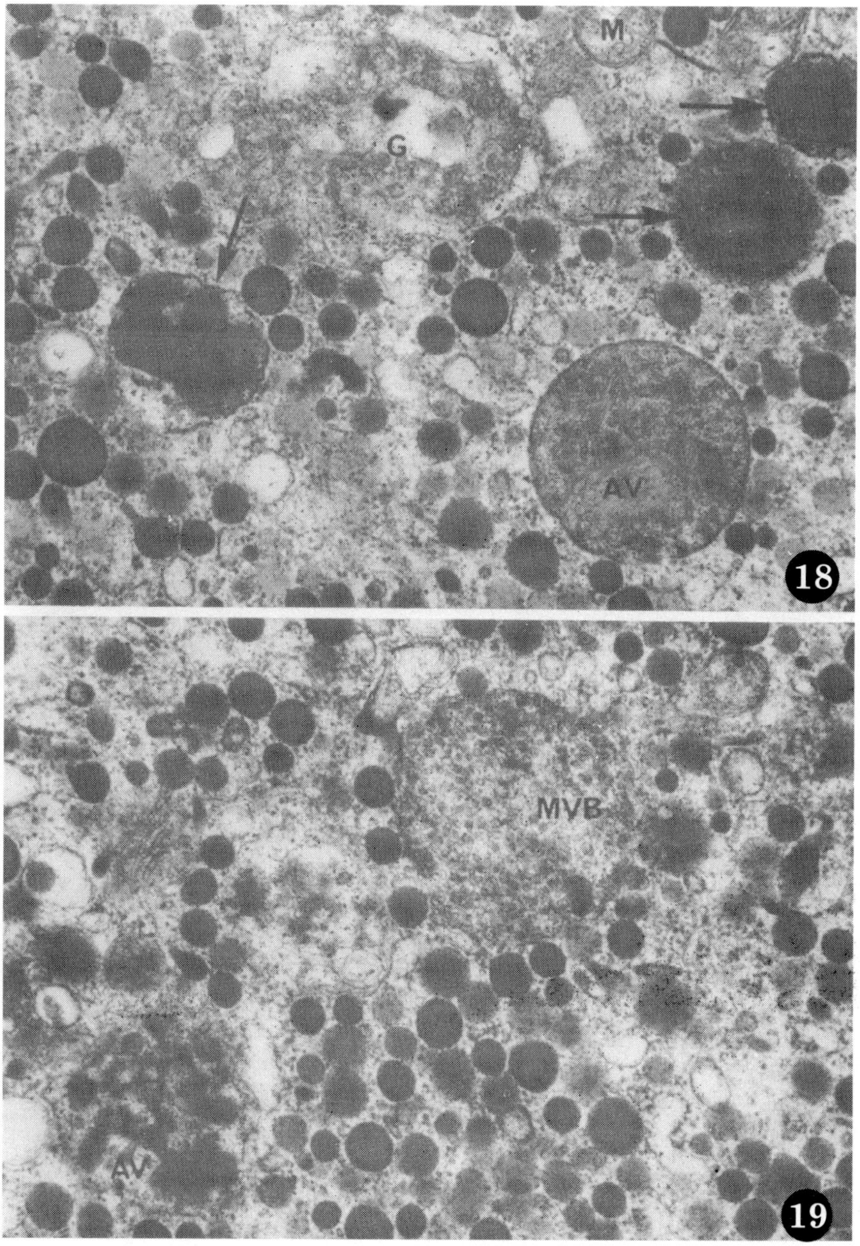

Plate 145. Figs. 18-19. Higher magnification of parts of cell shown in figure 17. Fig. 18. Note the autophagic vesicles and isolated groups of NSG which are at different stages of degradation (arrows). X 40,500. Fig. 19. The small vesicles in the multivesicular body (MVB) probably are the break-down products of NSG. The dense strucutres in the autophagic vesicle (AV) appears to be degrading neurosecretory granules. X 40,500. (From Unnithan and Nair, 1977).

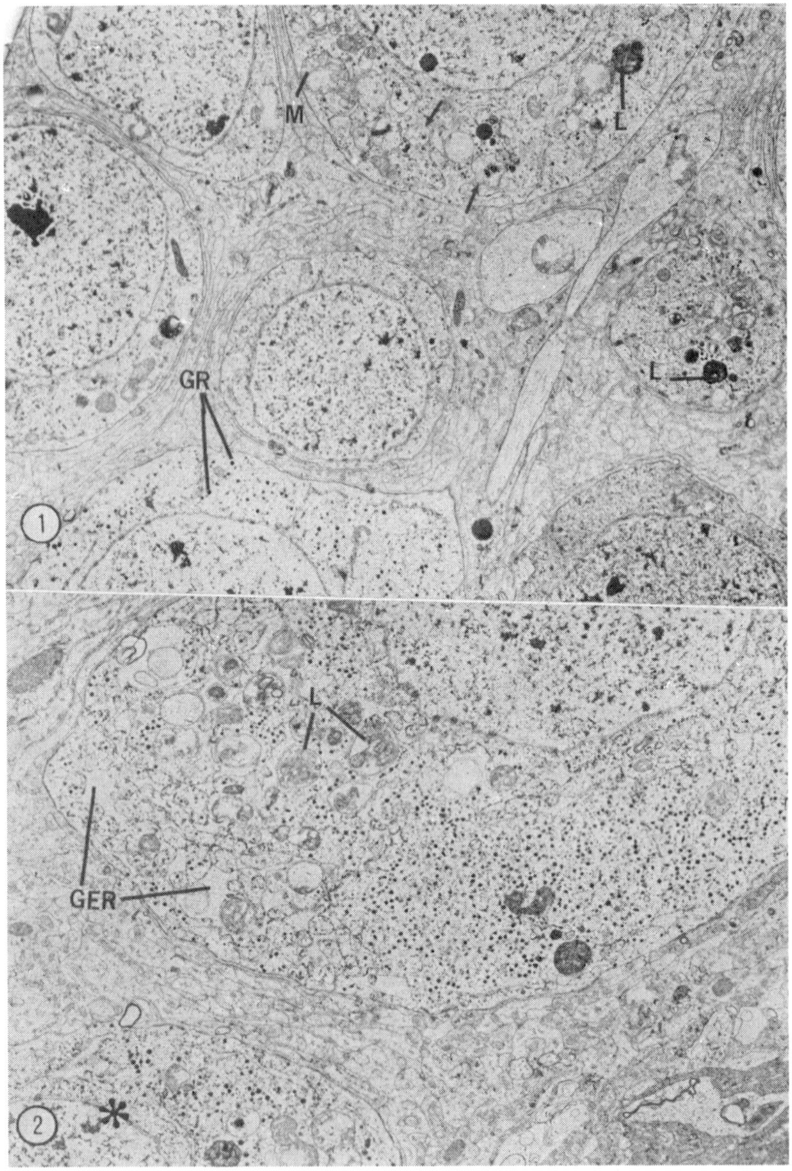

Plate 146. Fig. 1. Small Type 1 neurons of *Rhodnius*. Secretory granules (GR) can be seen in the cytoplasm of all of the cells shown. Lysosomes (L) are numerous and some appear to contain secretory granules (arrows). Swollen mitochondria (M) are common in all the neuron types. X 8500. Fig. 2. A large Type I neuron of *Rhodnius*. Numerous secretory granules and lysosomes (L) are present. Note the distended cisternae of the granular endoplasmic reticulum (GER). A portion of a smaller Type I neuron is also visible (asterisk). X 12,000. (From Morris and Steel, 1975).

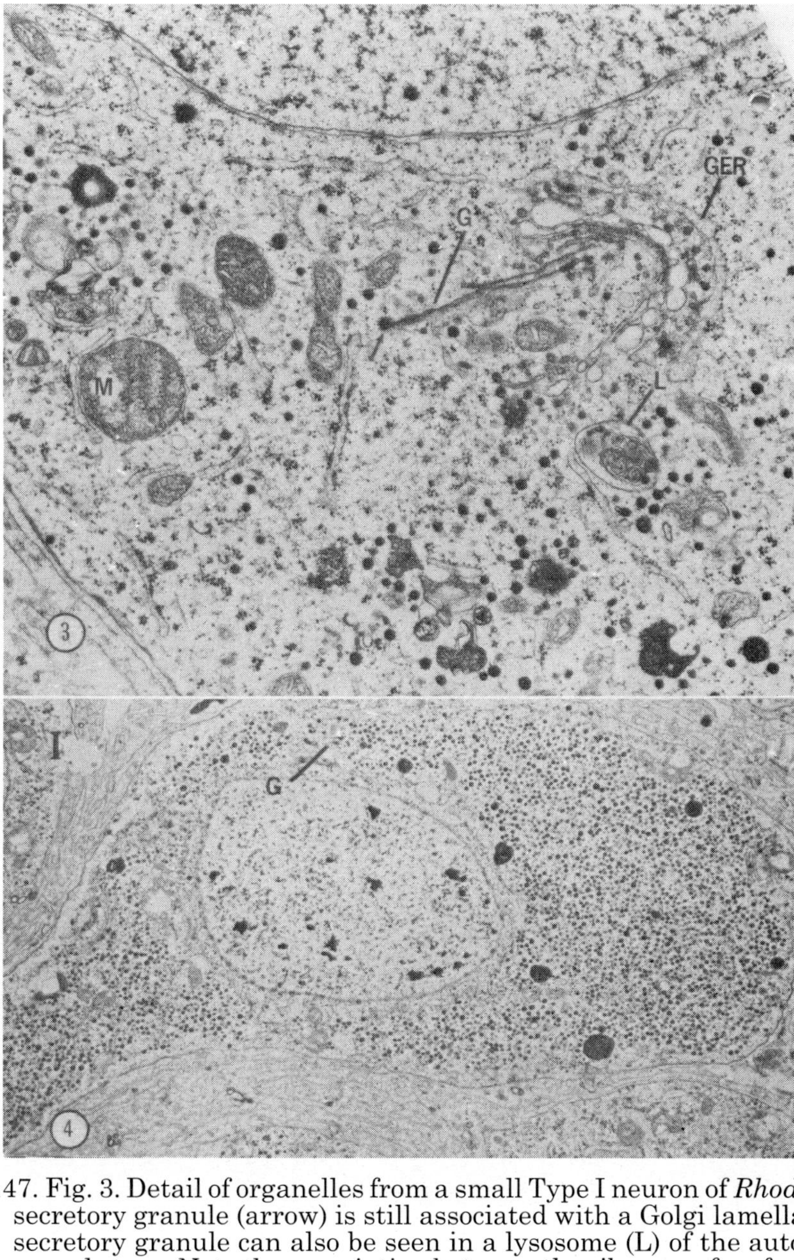

Plate 147. Fig. 3. Detail of organelles from a small Type I neuron of *Rhodnius*. A secretory granule (arrow) is still associated with a Golgi lamella (G). A secretory granule can also be seen in a lysosome (L) of the autophagic vacuole type. Note the association between the ribosome-free face of the granular endoplasmic reticulum (GER) and the Golgi complex. M, mitochondrion. X 47,000. Fig.; 4. Type II neuron of *Rhodnius*. These neurons are usually elongated with irregular nuclei. The larger size of Type II secretory granules is evident when compared to granules in the cytoplasm of a nearby Type I neuron (I). G. Golgi complex. X 9000. (From Morris and Steel, 1975).

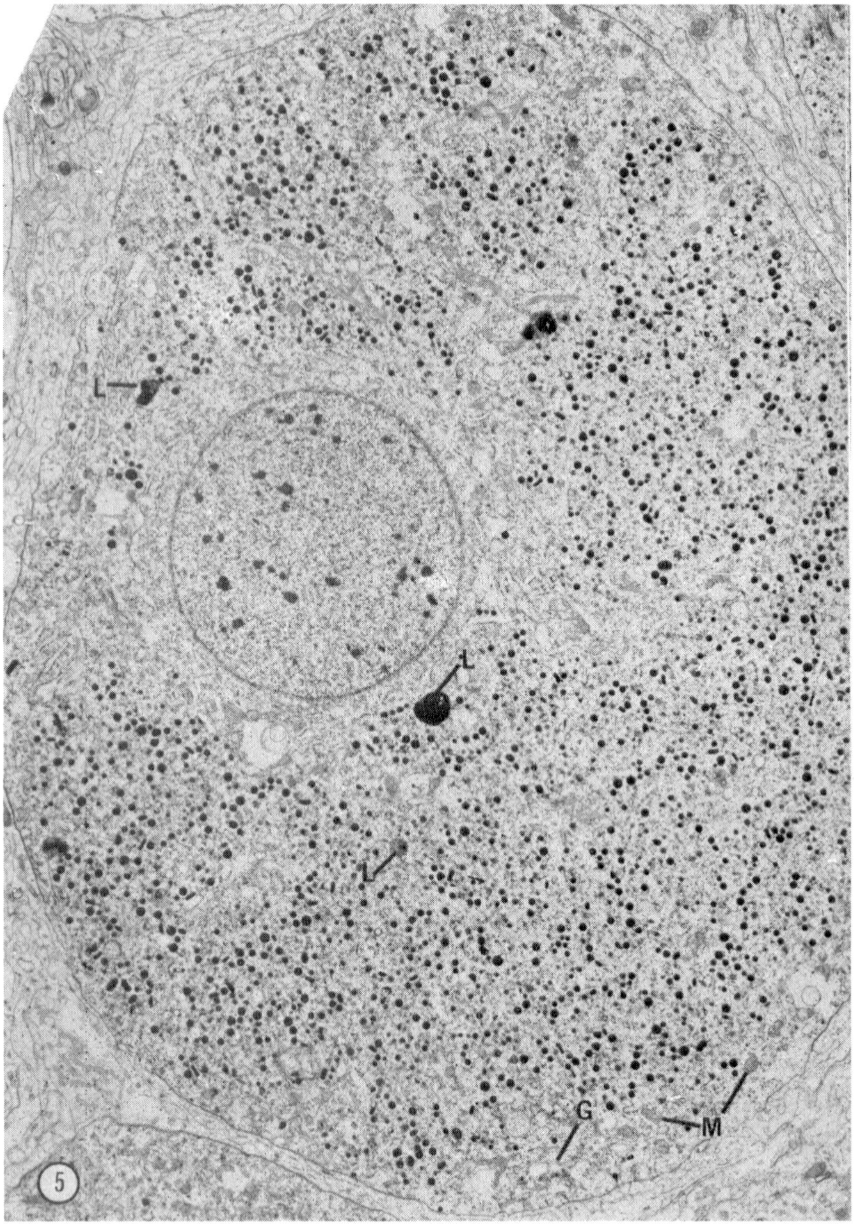

Plate 148. Fig. 5. Type III neuron of *Rhodnius*. Note the relatively dense appearance of the nucleus and cytoplasm. Circular and elongate granule profiles can be seen in the cytoplasm. Most lysosomes (L) are of the multivesicular body or dense body types. G, Golgi complex; M, mitochondria. X 10,000. (From Morris and Steel, 1975).

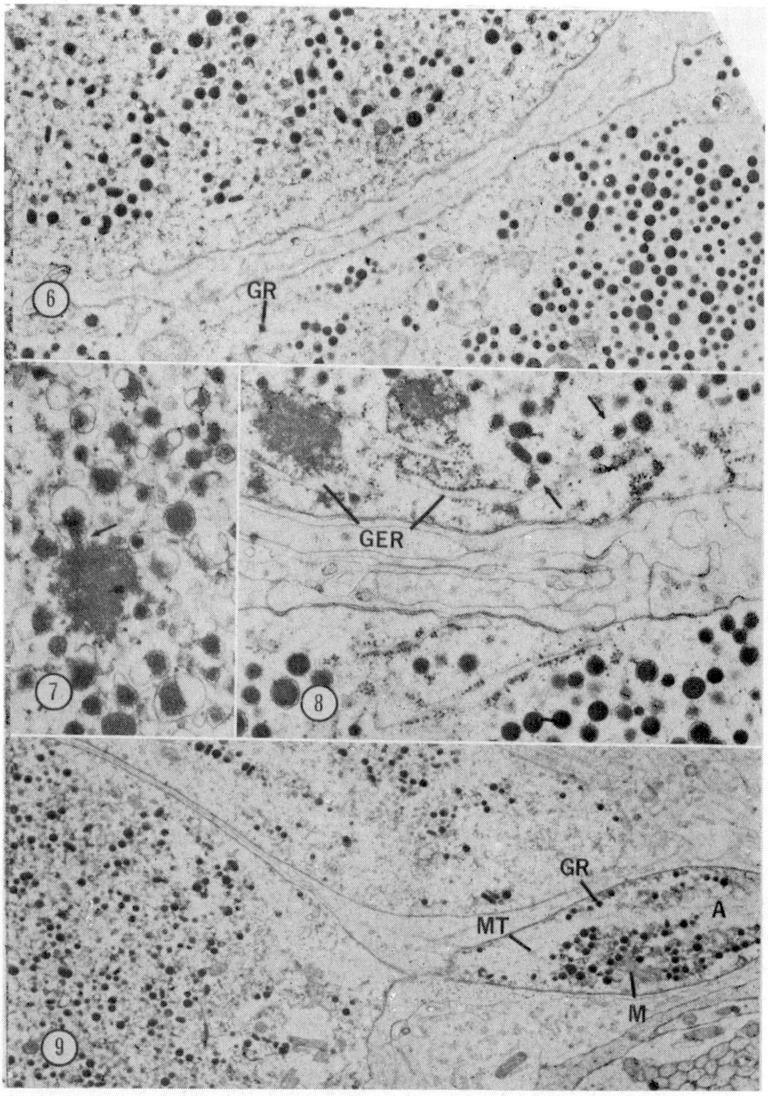

Plate 149. Fig. 6. Adjacent cytoplasms of Type III and Type IV neurons of *Rhodnius*. A secretory granule (GR) can be seen in a Golgi lamella of the Type IV neuron. Granules of the Type III neuron are more variable in shape and density and numerous irregular membrane profiles can be seen. X 18,000. Fig. 7. Detail of Type V neuron of *Rhodnius* showing possible transfer of material between membrane-delimited granule and dense cytoplasmic patch (arrow). X 50,000. Fig. 8. Adjacent cytoplasm of Types IV and V neurons of *Rhodnius*. Note the proximity of the granular endoplasmic reticulum (GER) and the dense patches in the Type V neuron. Granules of the dense-core type can also be seen in the Type V neuron (arrows). X 35,000. Fig. 9. Cytoplasm and axon (A) of a Type III, neuron of *Rhodnius* showing granules (GR) in axon. M, mitochondrion, MT, microtubules. X 14,000. (From Morris and Steel, 1975).

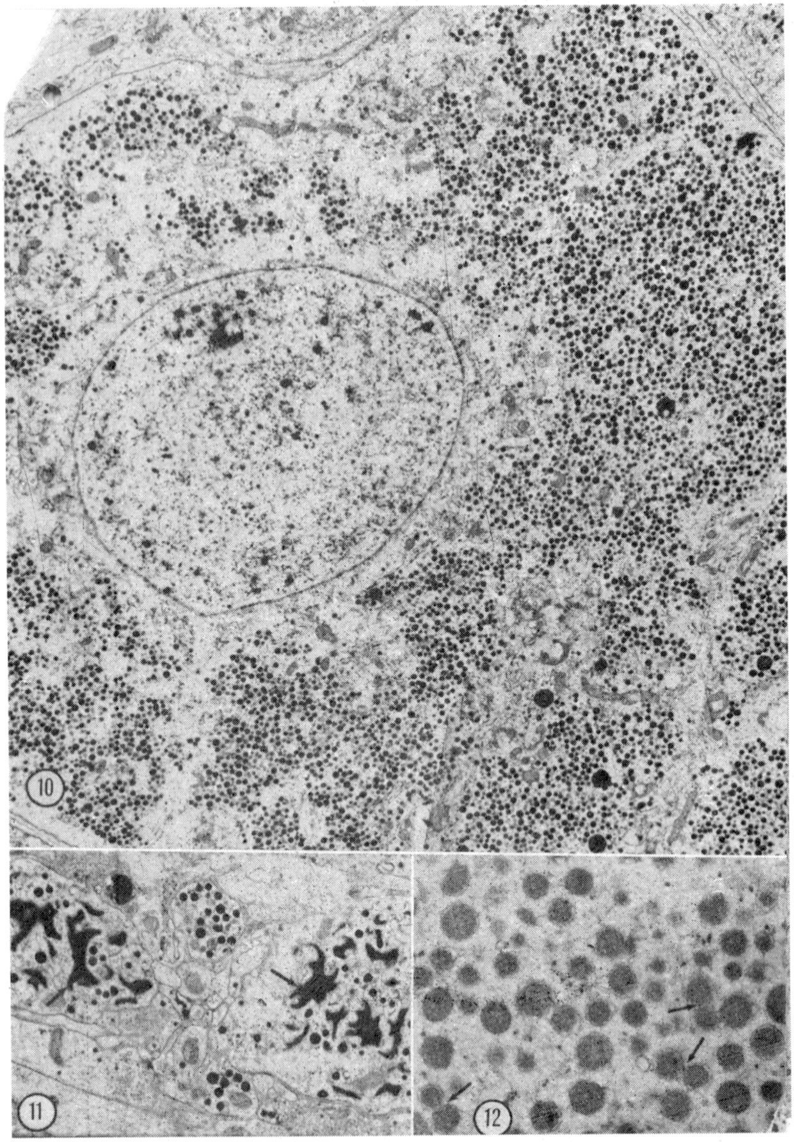

Plate 150. Fig. 10. Type IV neuron of *Rhodnius*. The secretory granules are more numerous and closely packed than in other neuron types. Organelles such as mitochondria and granular endoplasmic reticulum are concentrated in the perinuclear and peripheral regions of the cytoplasm. X 10,000. Fig. 11. Axons containing secretory granules of sizes characteristic of Type IV neurons of *Rhodnius*. The large dense masses (arrows) are membrane-delimited and could be a consequence of multiple granule fusions. X 9000. Fig. 12. Detail of secretory granules from cytoplasm of Type IV neuron. Arrows indicate configurations which may show fusion of granules. X 46,000. (From Morris and Steel, 1975)

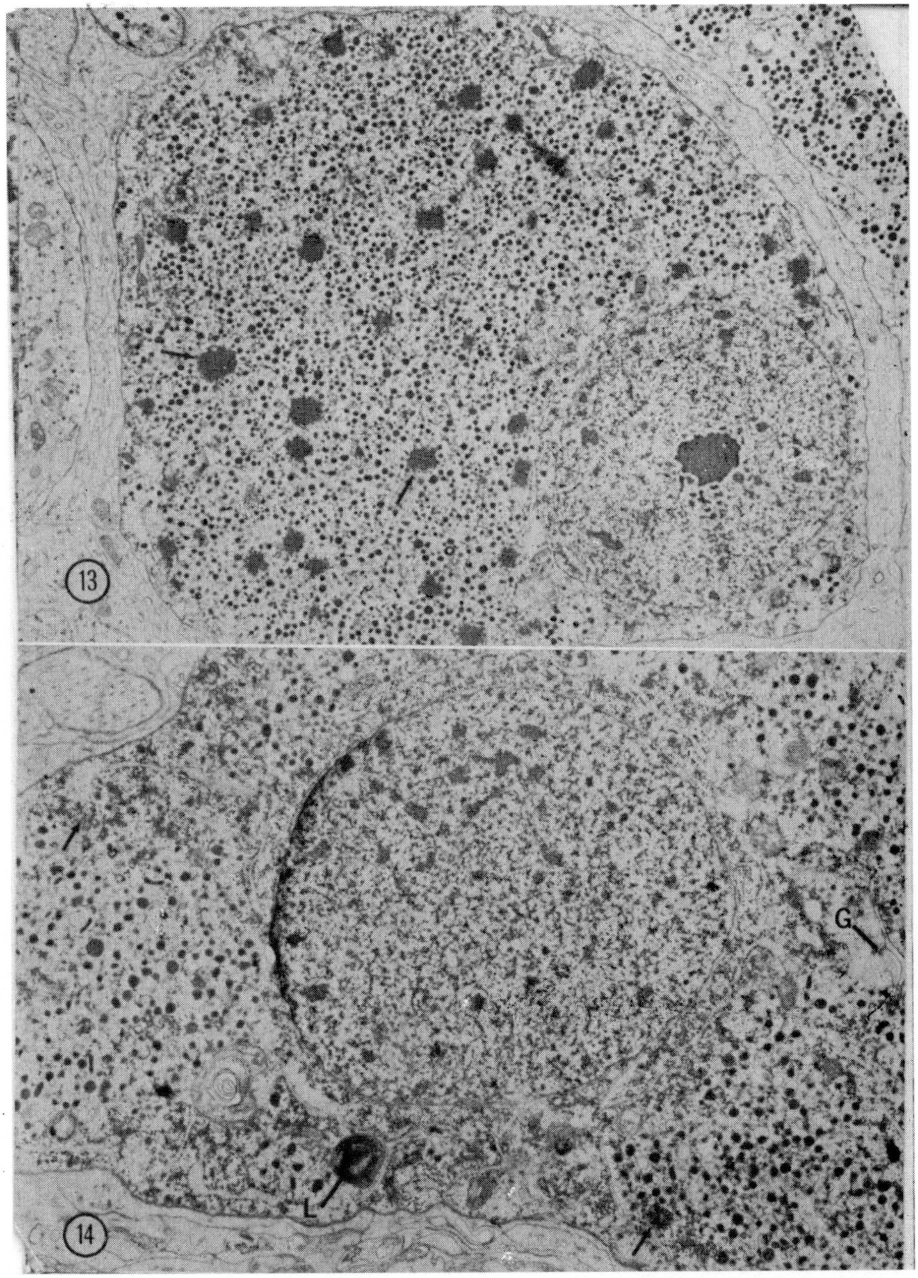

Plate 151. Fig. 13. Type V neuron of *Rhodnius* with dense cytoplasmic patches (arrows). X 7200. Fig. 14. A type V neuron without the obvious dense cytoplasmic patches. Aggregations of electron-dense material can, however, be seen near the cisternae of the granular endoplasmic reticulum (arrows). A lysosome (L) and a Golgi complex (G) are also present. X 18,000. (From Morris and Steel, 1975).

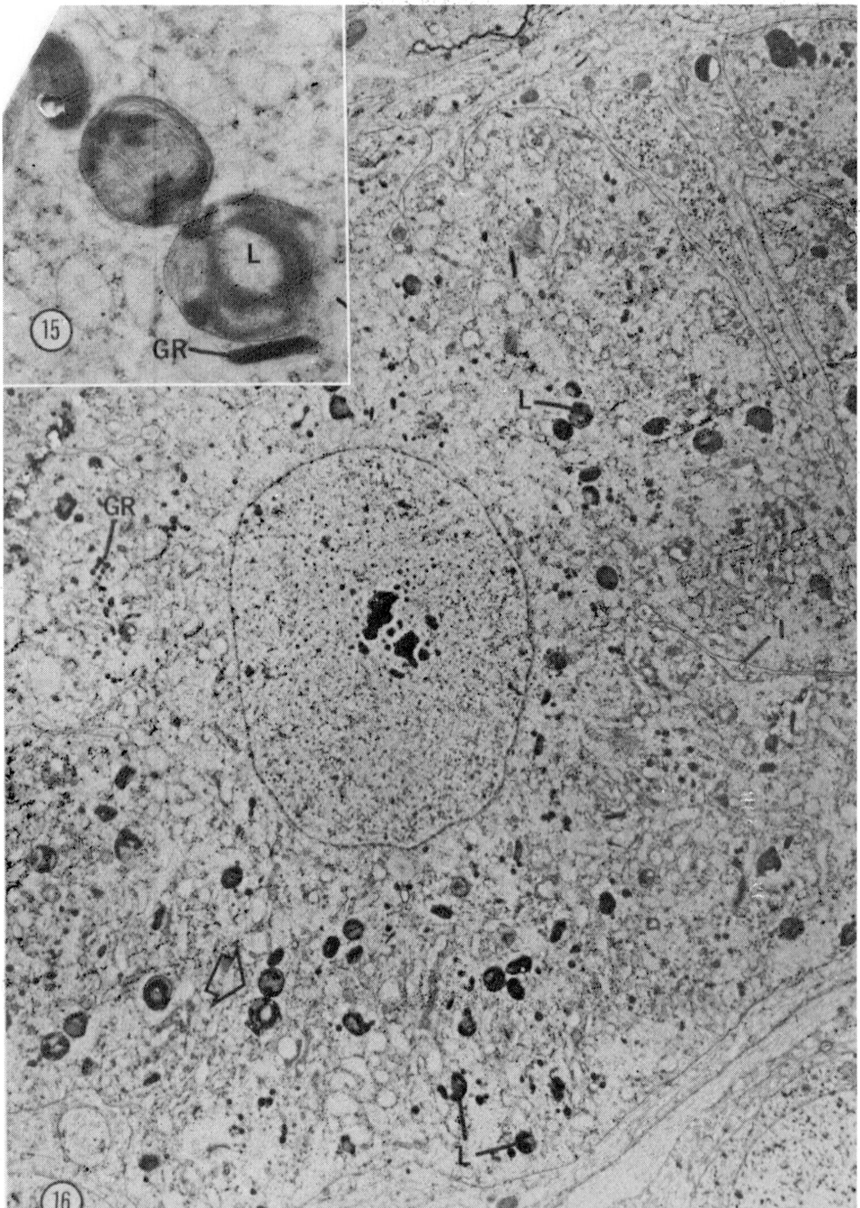

Plate 152. Fig 15. Detail of lysosomes (L) and elongate secretory granule (GR) from Type VI neuron of *Rhodnius*. X 35,000. Fig. 16. Type VI neuron of *Rhodnius*. Numerous lysosomes (L) and secretory granules (GR) of varible morphology can be seen in the cytoplasm. Note invaginations (I) of the plasma membrane. The arrow indicates the region shown in Fig. 15. X 7500. (From Morris and Steel, 1975).

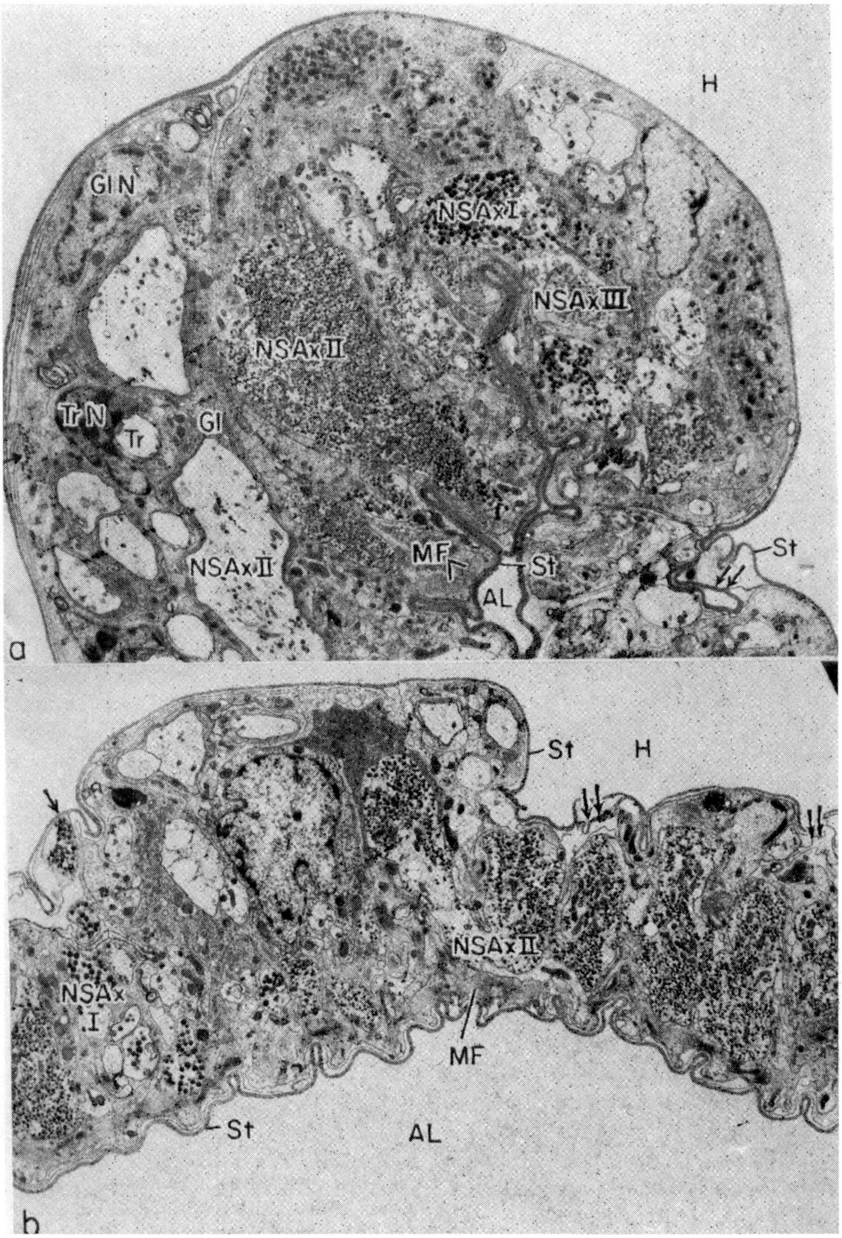

Plate 153. Fig. 1 a and b. Cross sections of parts of the aortic wall of *Oncopeltus* showing the three different types of neurosecretory axons (NSAx I, NSAx II, NSAx III). Neurosecretory axons are seen very close to the surface of the aorta (single arrows); double arrows show the invaginations of the stroma into the aortic wall. X 5500. (From Unnithan *et al.*, 1971)

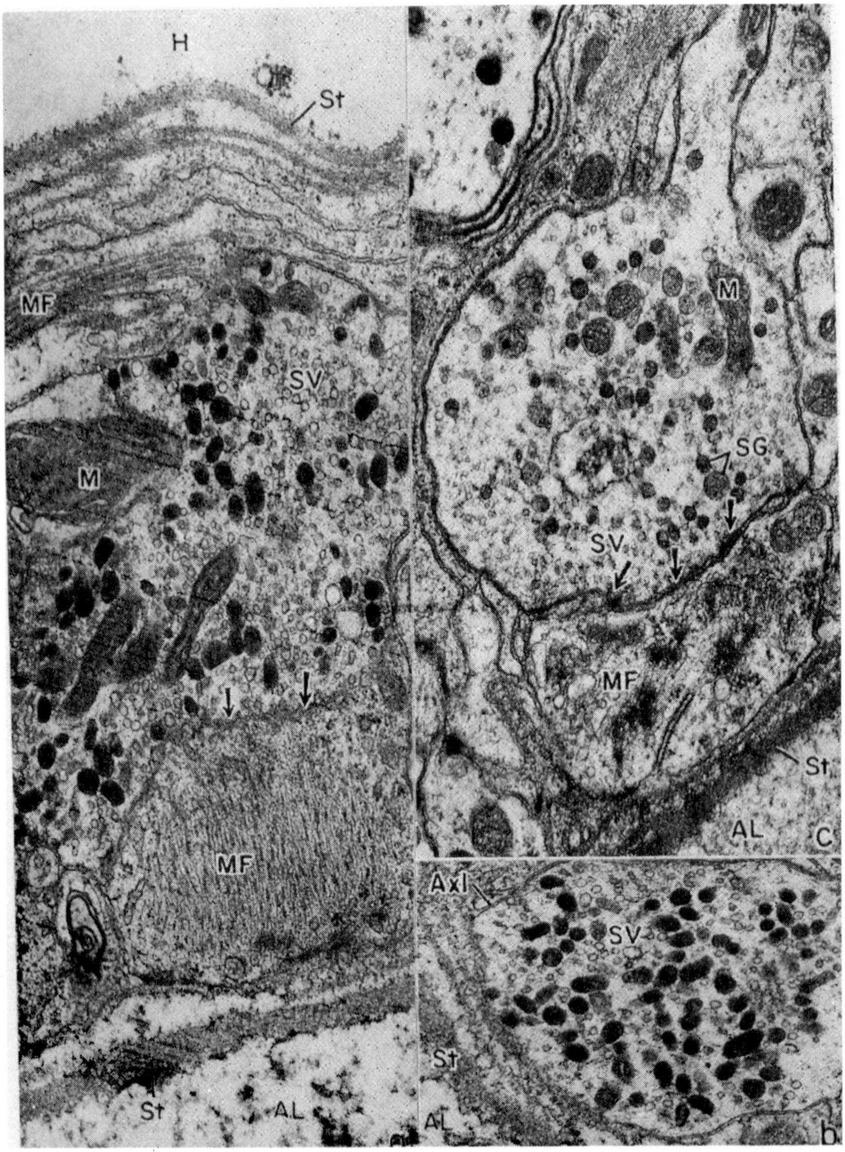

Plate 154. Fig. 2 a. Type II neurosecretory ending in the aortic wall of *Oncopeltus* in close contact with the muscle fibre (MF) (arrows). Note the small vesicles (SV). X 31,000. b. Same type of axon as in a. near the aortic lumen showing small vesicles (SV) along with electron-dense granules. The axolemma (Axl) is not distinct at the luminal side. X 31,000. c. Type III neurosecretory ending showing close contact with the aortal muscle (MF) (arrows). The small vesicles in the axoplasm may be the result of the breakdown of secretory granules (SG). X 31,000. (From Unnithan *et al.*, 1971)

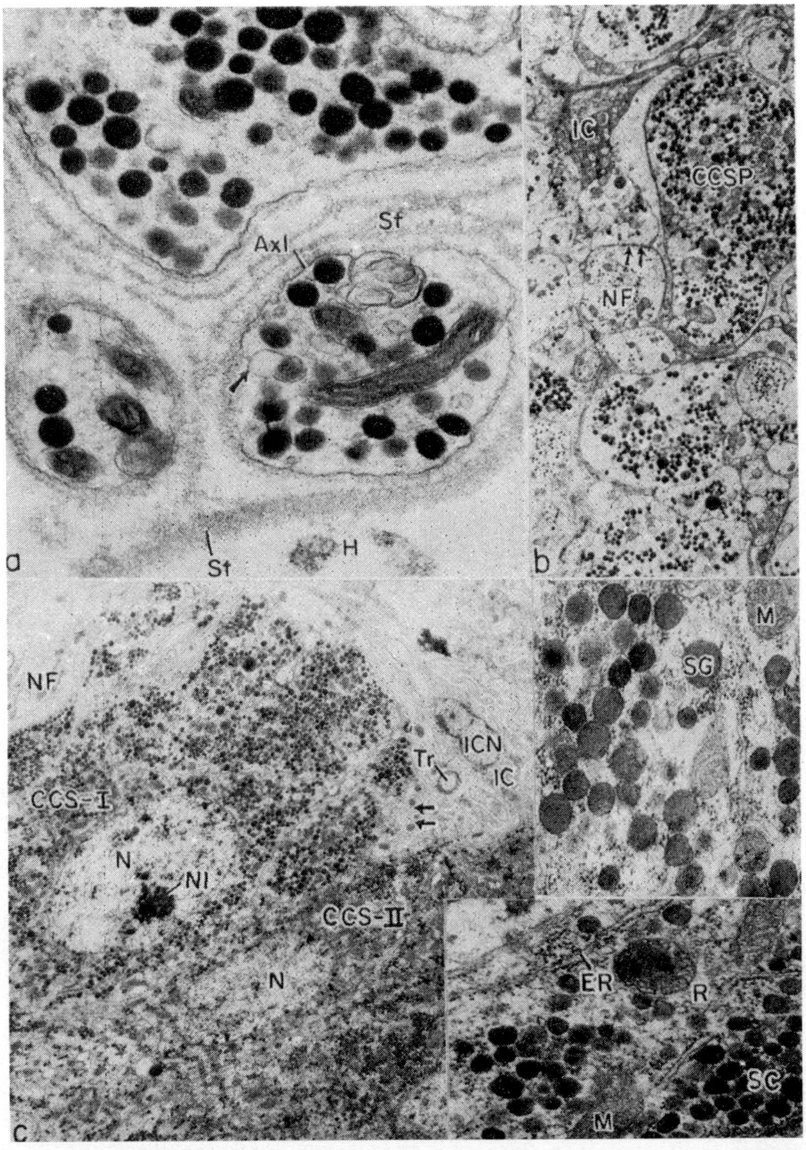

Plate 155. Fig. 3 a. Type I neurosecretory axon ending in the aortic wall of *Oncopeltus*. Note invaginated axolemma (arrow) with granule attached to it, occasional electron-lucent vesicles, and the indistinct appearance of the axolemma. X 29,000. b. Interior of the corpus cardiacum showing secretory cell processes (CCSP) and nerve fibres (NF) with small electron-dense granules. One such nerve fibre is in close contact with a CC cell process (arrows). X 6800. c. Peripheral part of the corpus cardiacum showing two types of secretory cells (CCS-I, CCS-II), interstitial cells (IC), and nerve fibres of extrinsic origin (NF). Arrows indicate a group of small nerve fibres. X 5500. The inserts at right show higher magnification of parts of CCS-I (above) and CCS-II (below). X 29,000. (From Unnithan *et al.*, 1971)

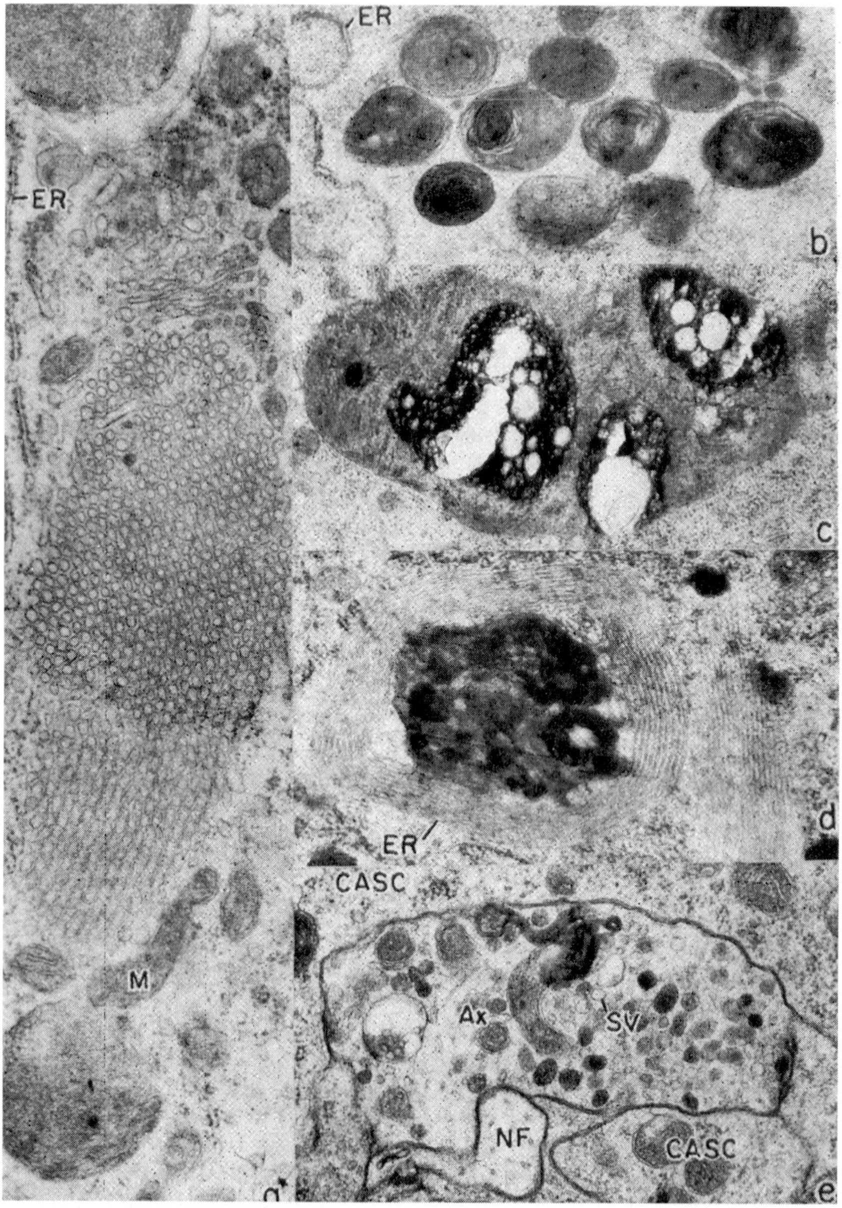

Plate 156. Fig. 4. Corpus allatum a-d of *Oncopeltus*. Some of the frequently seen cytoplasmic organelles and inclusions in the corpus allatum of cells of reproducing females. Some or all of these are presumably indentical to the PAVB-stainable inclusions. a. X 29,000; b. X 20,000; c, X 22,000; d. X 31,000. e. A granule-containing axon (Ax) seen commonly among the corpus allatum cells (CASC), probably representing neurosecretory innervation. Note granule-free axon (NF). X 31,000. (From Unnithan *et al.,* 1971)

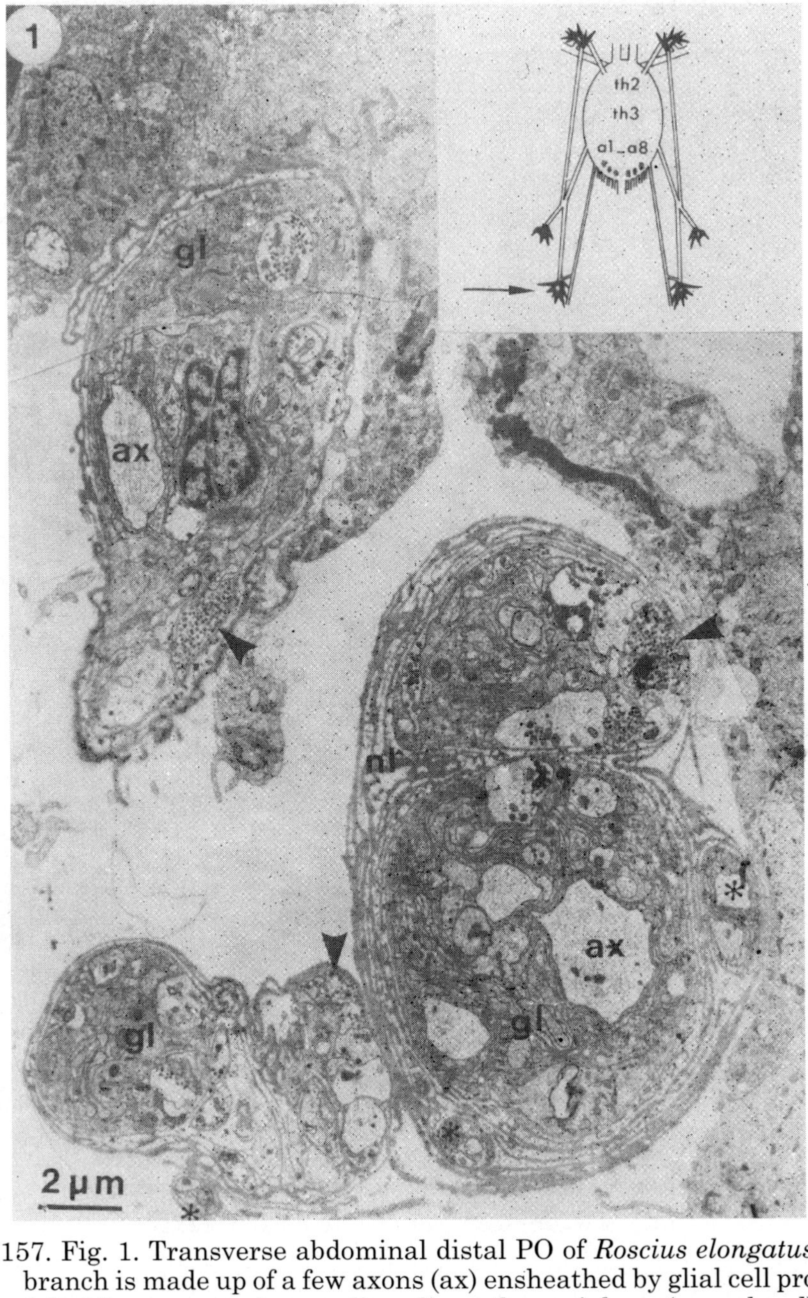

Plate 157. Fig. 1. Transverse abdominal distal PO of *Roscius elongatus*. Each branch is made up of a few axons (ax) ensheathed by glial cell processes (gl). All neurosecretory endings lie at the periphery (arrowhead) of the branches. Note the very thin branches (asterisk) with no secretory endings at the level of the section. nl, Neural lamella. X 6500. Inset : ganglionic nerve mass showing the location of an abdominal distal PO (arrow). (From Baudry - Partiaoglou, 1983)

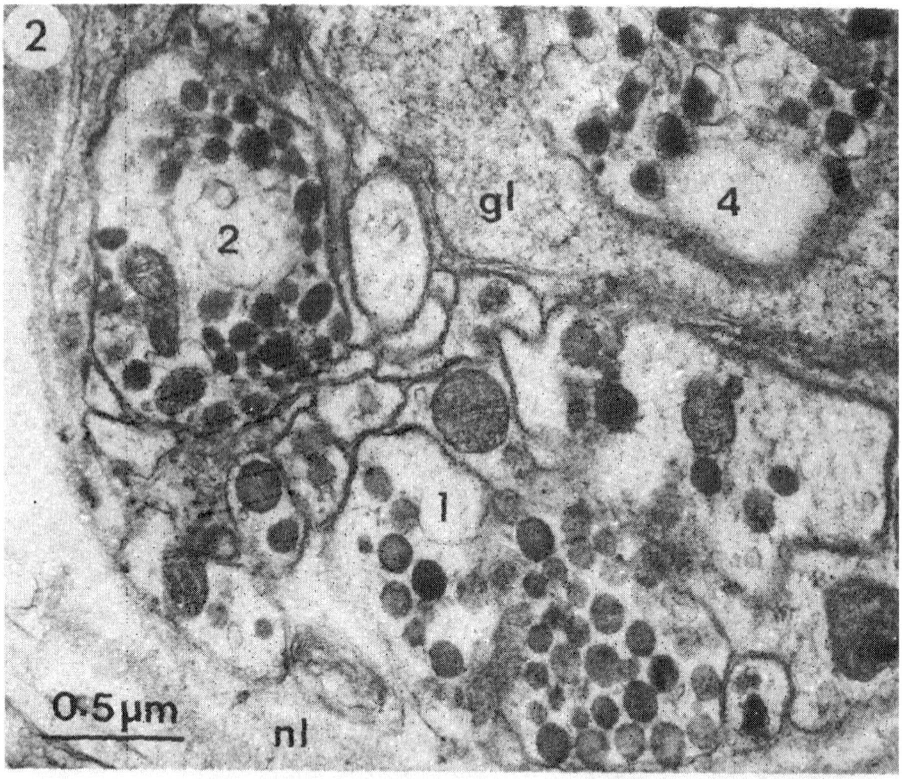

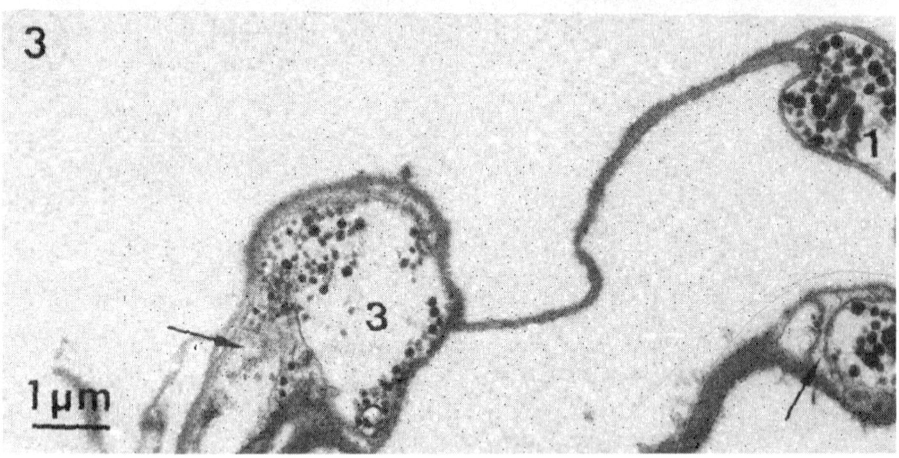

Plate 158. Fig. 2. Three types of neurosecretory ending of an abdominal distal PO (1, 2 and 4). gl, Glial cell; nl, neural lamella. X 38,000. Fig. 3. Three isolated neurosecretory endings with very thin glial processes (thin arrow). Two types of neurosecretory granule are visible (1, 3). The endings are linked by thin connective strands. X 10,000. (*Roscius*) (From Baudry - Partiaoglou, 1983)

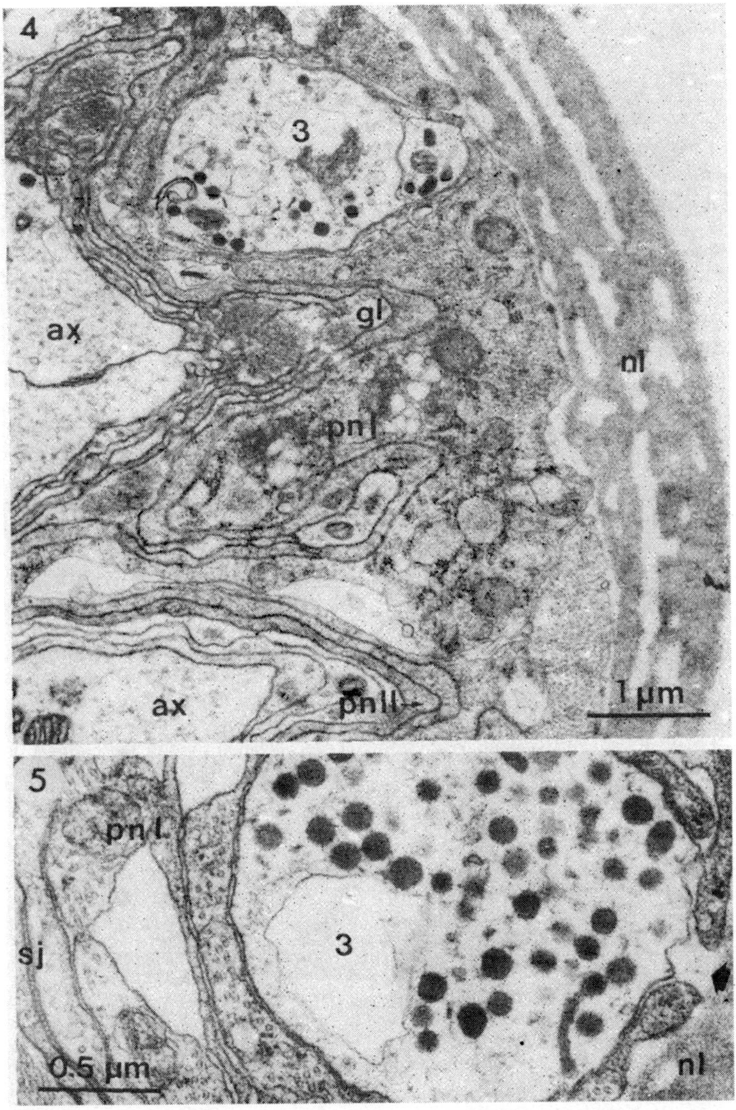

Plate 159. Fig. 4. Proximal section of the fourth abdominal nerve of *Roscius* showing a type 3 neurosecretory ending (3) located at the periphery of the perineurium. Note the abundance of microtubules and ergastoplasmic cisternae in type I perineurial cells (pn I). Glial cells proper (gl) are recognizable by their bundles of microtubules. ax, Axon; pn II, type II perineurial cell. X 21,000. Fig. 5. A type 3 neurosecretory ending (3) at the periphery of the same nerve. The type I perineurial cell (pn I) allows free access to the haemolymph through a gap (arrow), where the membrane of the ending displays indentations. Septate junctions (sj) are recognizable between type I and type II perineurial cells. (nl-Neural lamella. X 42,000. (From Baudry-Partiaoglou, 1983).

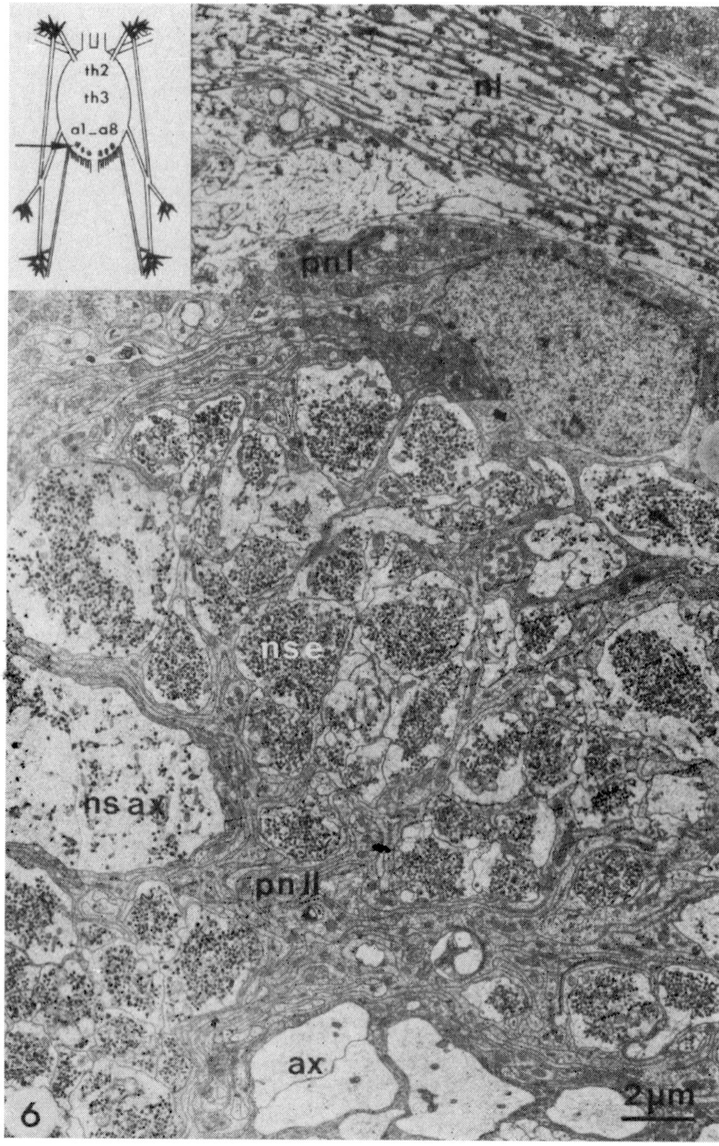

Plate 160. Fig. 6. Transeverse section in the posterior area of the ganglionic mass of *Roscius* showing part of one of the three areas of neurosecretory accumulation. All the neurosecretory endings (nse) come from the division of one or two axons (ns ax) and are ensheathed by type II perineurial cell (pn II). At the bottom of the picture, some nonneurosecretory axons (ax) form part of the bundle that enventually constitutes the second nerve. nl, Neural lamella; pn I, type I perineural cell. X 6000. Inset : ganglionic nerve mass showing the location of a subjacent PO (arrow). (*Roscius*) (From Baudry - Partiaoglou, 1983).

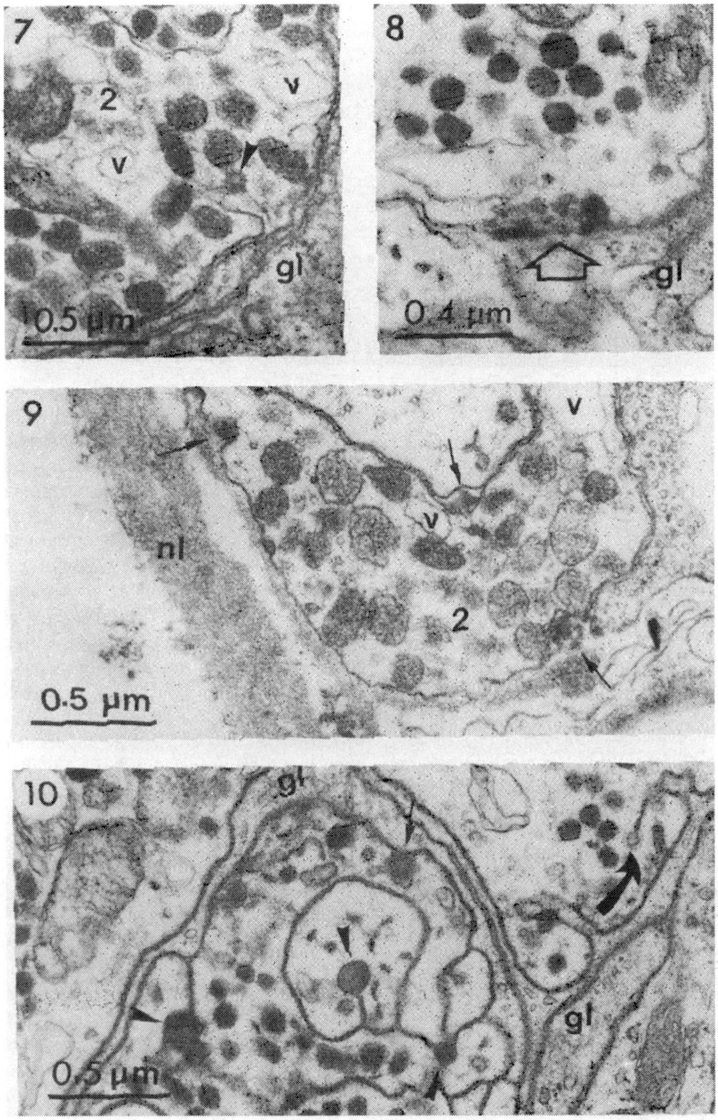

Plate 161. Fig. 7. Type 2 neurosecretory ending (2) of *Roscius* showing a coated vesicle during its formation (arrowhead) from a membrane invagination. Vacuoles (v) may be formed by the fusion of the coated vesicles. gl, Glial cell. X 42,000. Fig. 8. Synaptoid site (white arrow) in a type 3 neurosecretory ending facing a glial cell (gl). X 49,000. Fig. 9. Pictures of neurosecretory release in type 2 ending (2) of *Roscius*. Note the membrane indentations filled with dense material extruded by exocytosis (arrow). nl, Neural lamella; v. vacuole. X 42,000. Fig. 10. Neurosecretory release in subjacent PO. Note the membrane infoldings of the neurosecretory endings and the dense material at the bottom or intersection of the folds (arrowhead). Single exocytosis is visible as well as coated vesicles (thin arrow). Curved arrow indicated formation of coated vesicle from membrane infolding. gl, Glial cell. X 40,000. (From Baudry-Partiaoglou, 1983).

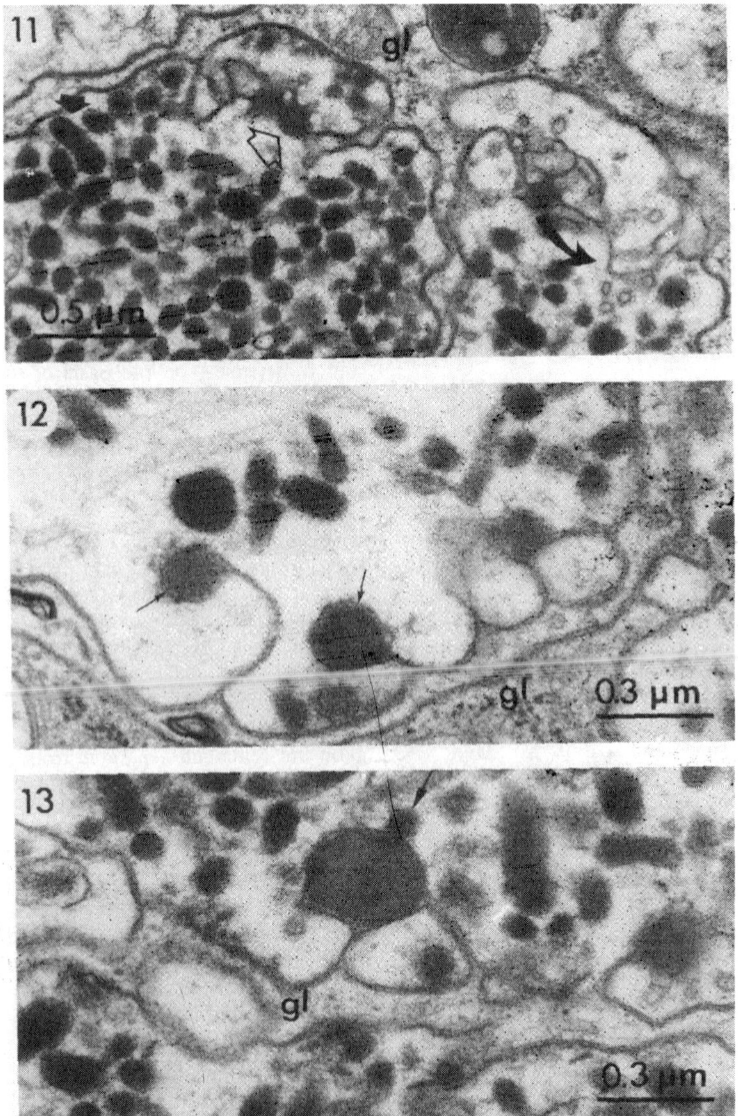

Plate 162. Figs. 11-13. Neurosecretory release in subjacent PO. Fig. 11. In the ending on the right, coated vesicles are particularly numerous (curved arrow) and are formed from membrane infoldings (white arrow). The elongated granule may constitute a fusion of two spherical ones before exocytosis (black arrow). gl. Glial cell. X 40,000. Fig. 12. Coated vesicles are only formed at released sites and membrane infoldings. The membrane of the axon ending clearly visible around a spherical formation of dense material (arrow) indicate that the material is extracellular. gl, Glial cell. X 63,000. Fig. 13. Picture of compound exocytosis : a small granule is joined to the membrane surrounding an extruded dense sphere (arrow). gl, Glial cell. X 63,000. (*Roscius*) (From Baudry-Partiaoglou, 1983).

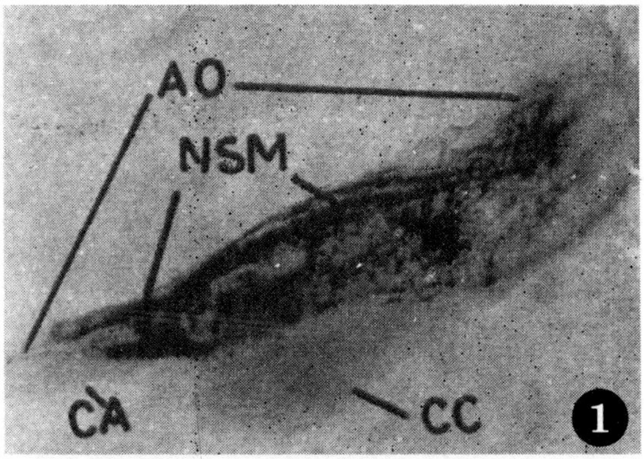

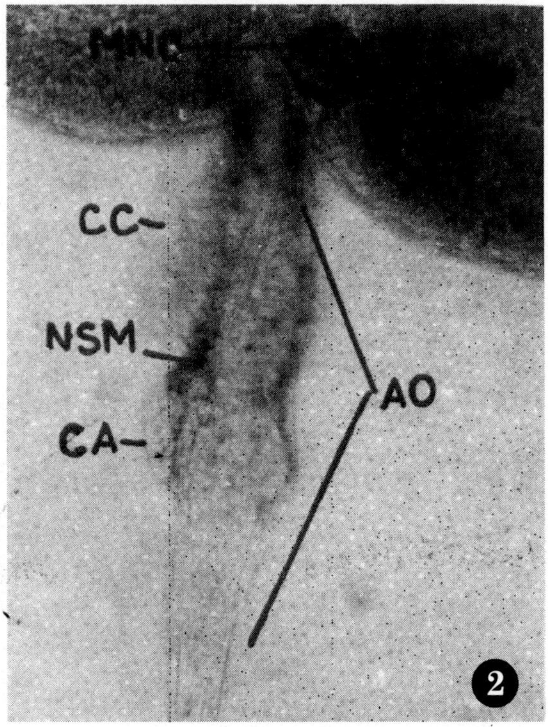

Plate 163. Fig. 1. Whole mount of retrocerebral endocrine aortal complex of the earwig *Labidura riparia* (Pallas) showing a heavy accumulation of NSM in aorta (AO). The underlying corpora cardiaca (CC) and corpus allatum (CA) are devoid of NSM (From Awasthi, 1975c). Fig. 2. Whole mount of the brain-retrocerebral complex of the earwig *Euborellia annulipes* Lucas with intact aorta (AO). The aortal walls are packed with NSM. The underlying corpora cardiaca (CC) and corpus allatum (CA) are devoid of A-cell NSM. A group of MNC is also seen (From Awasthi, 1976d).

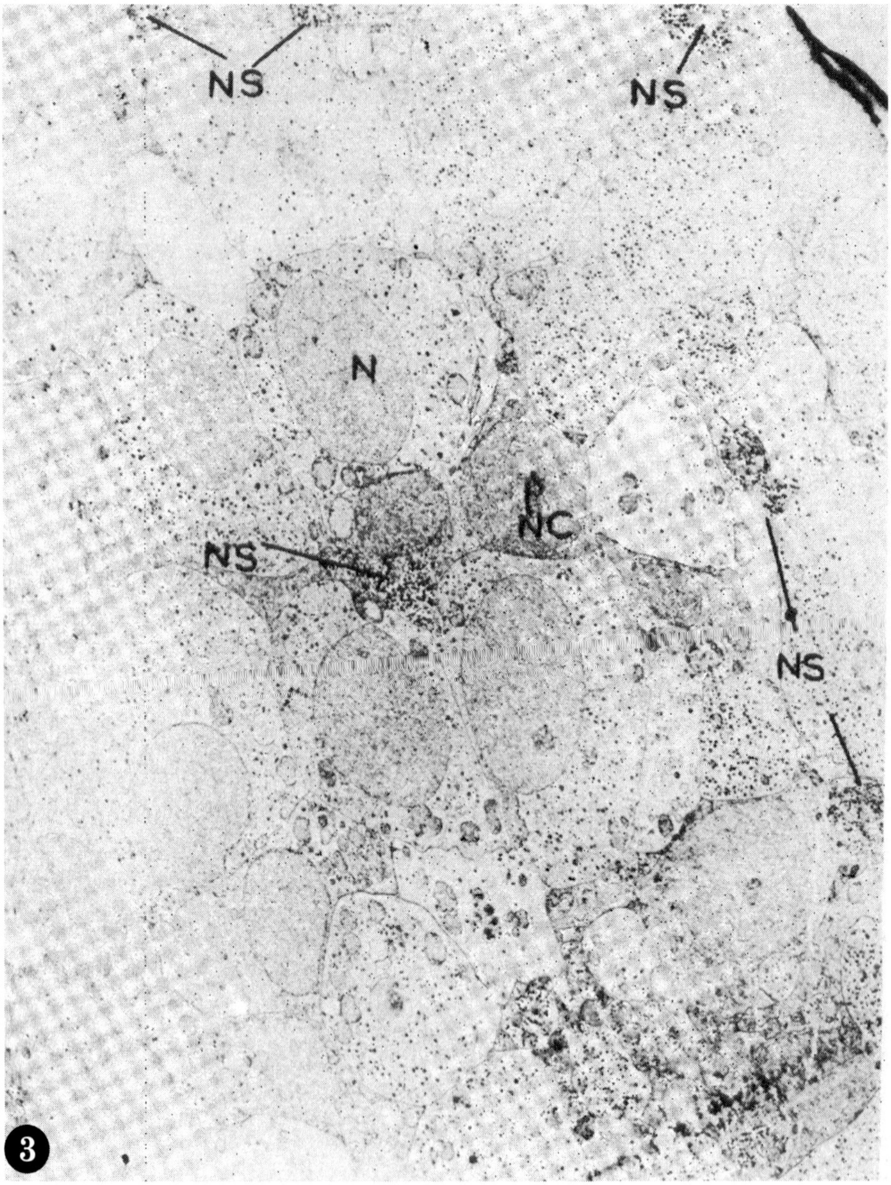

Plate 164. Fig. 3. A low power electron micrograph of *Euborellia annulipes* showing the organisation of corpus allatum cells. Axons containing neurosecretory granules are clearly marked out. X 2700 (From Awasthi, 1979b).

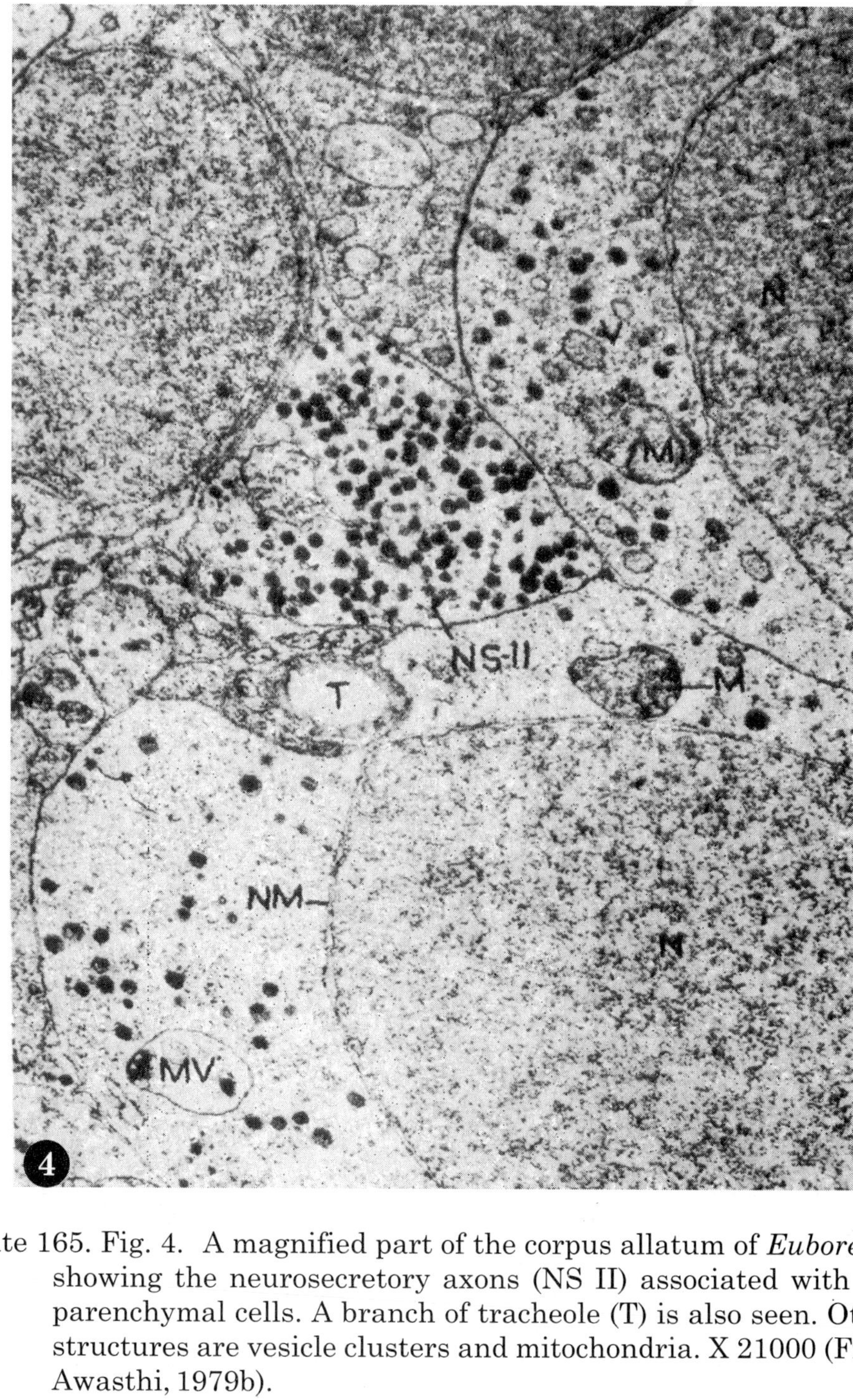

Plate 165. Fig. 4. A magnified part of the corpus allatum of *Euborellia* showing the neurosecretory axons (NS II) associated with the parenchymal cells. A branch of tracheole (T) is also seen. Other structures are vesicle clusters and mitochondria. X 21000 (From Awasthi, 1979b).

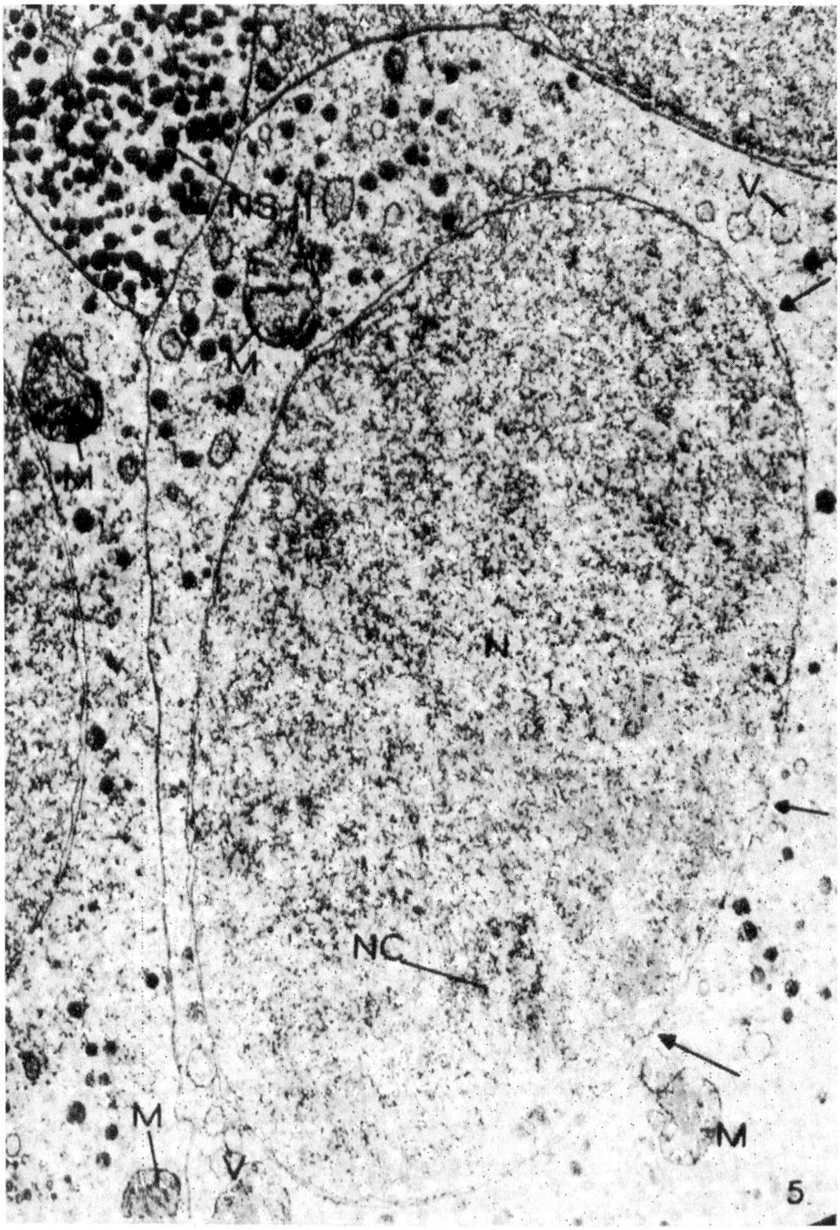

Plate 166. Fig. 5. A magnified parenchymal cell with an oval nucleus (N) and a nucleolus (NC). The double layered nuclear membrane (NM) is clearly seen. Note the neurosecretory axon is devoid of glioplasm-sheet. Mitochondria (M) and vesicular bodies (V) are clearly marked out. X 21000 (From Awasthi, 1979b).

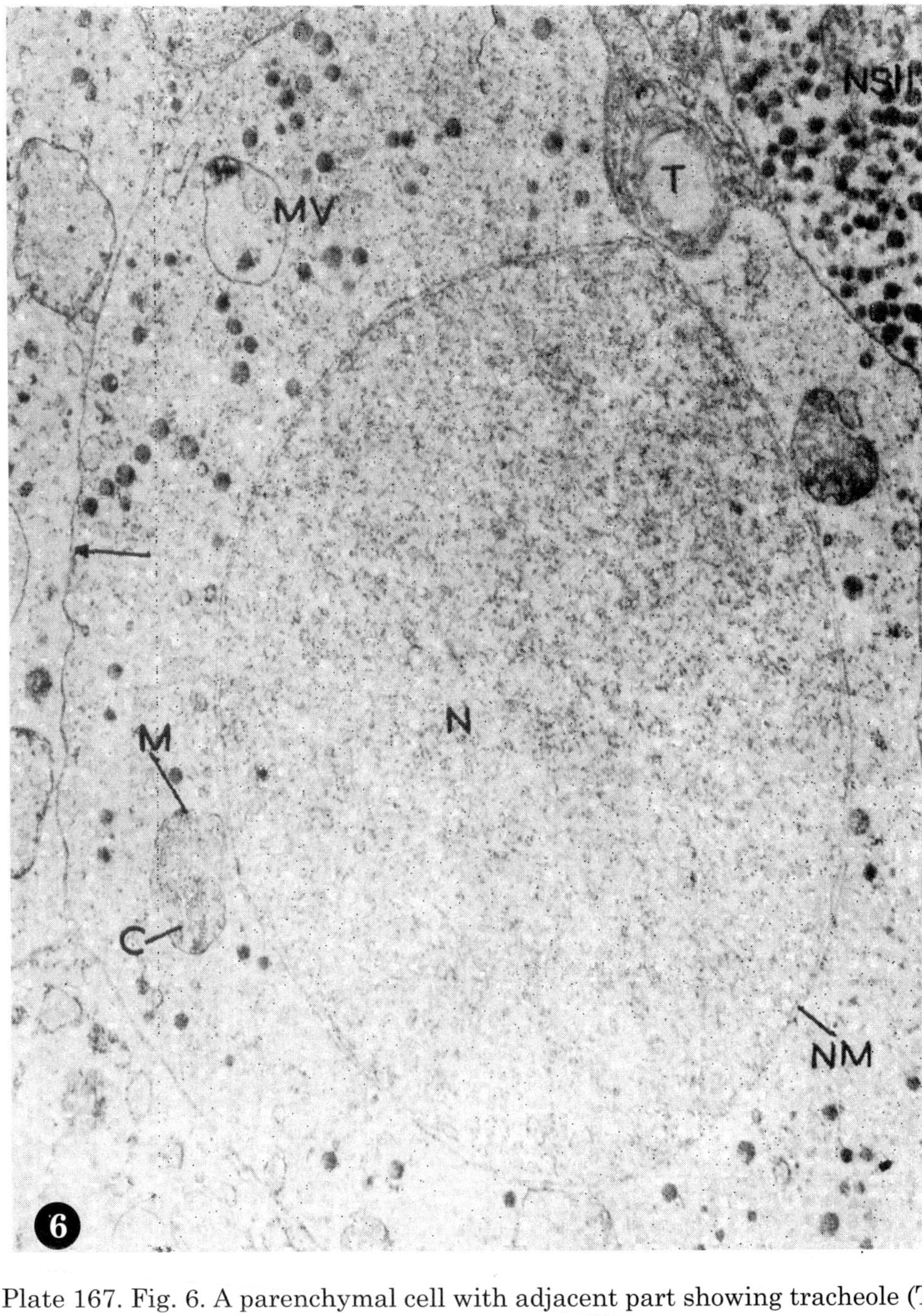

Plate 167. Fig. 6. A parenchymal cell with adjacent part showing tracheole (T) with small and smooth taenidia, multivesicular bodies (MV) and mitochondria (M). The two nuclear boundary membranes are also clearly marked out. X 21000 (From Awasthi, 1979b).

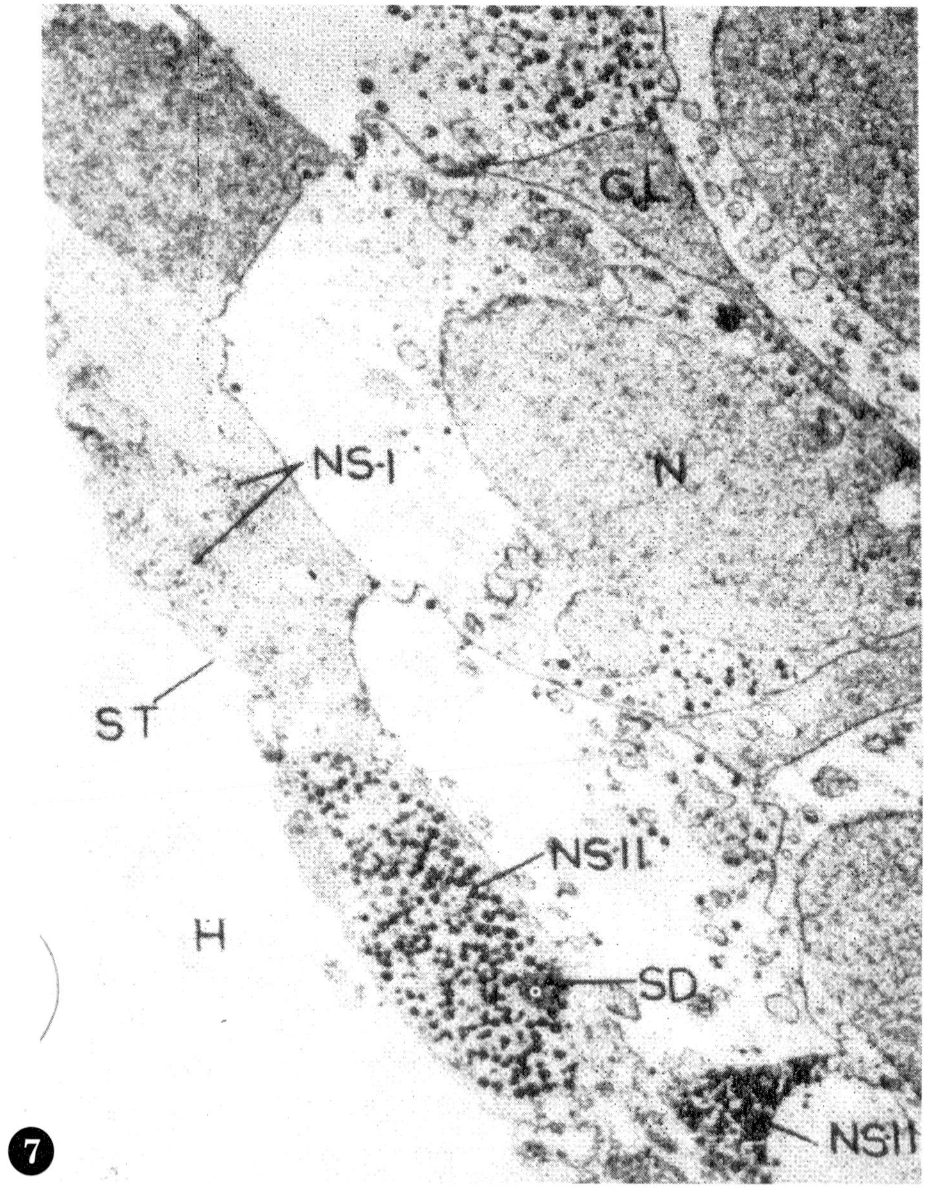

Plate 168. Fig. 7. Electron micrograph of a part of corpus allatum of *Euborellia annulipes*. Note the distribution of two types of axons containing neurosecretory granules in the peripheral part and other portion of the gland (NS-I and NS-II). The stromal layer (ST) at the axon terminal that bath the haemocoel (H) is extremely thin. A large secretion droplet (SD), probably formed by the fusion of neurosecretory granule is also seen X 10500. (From Awasthi, 1979b).

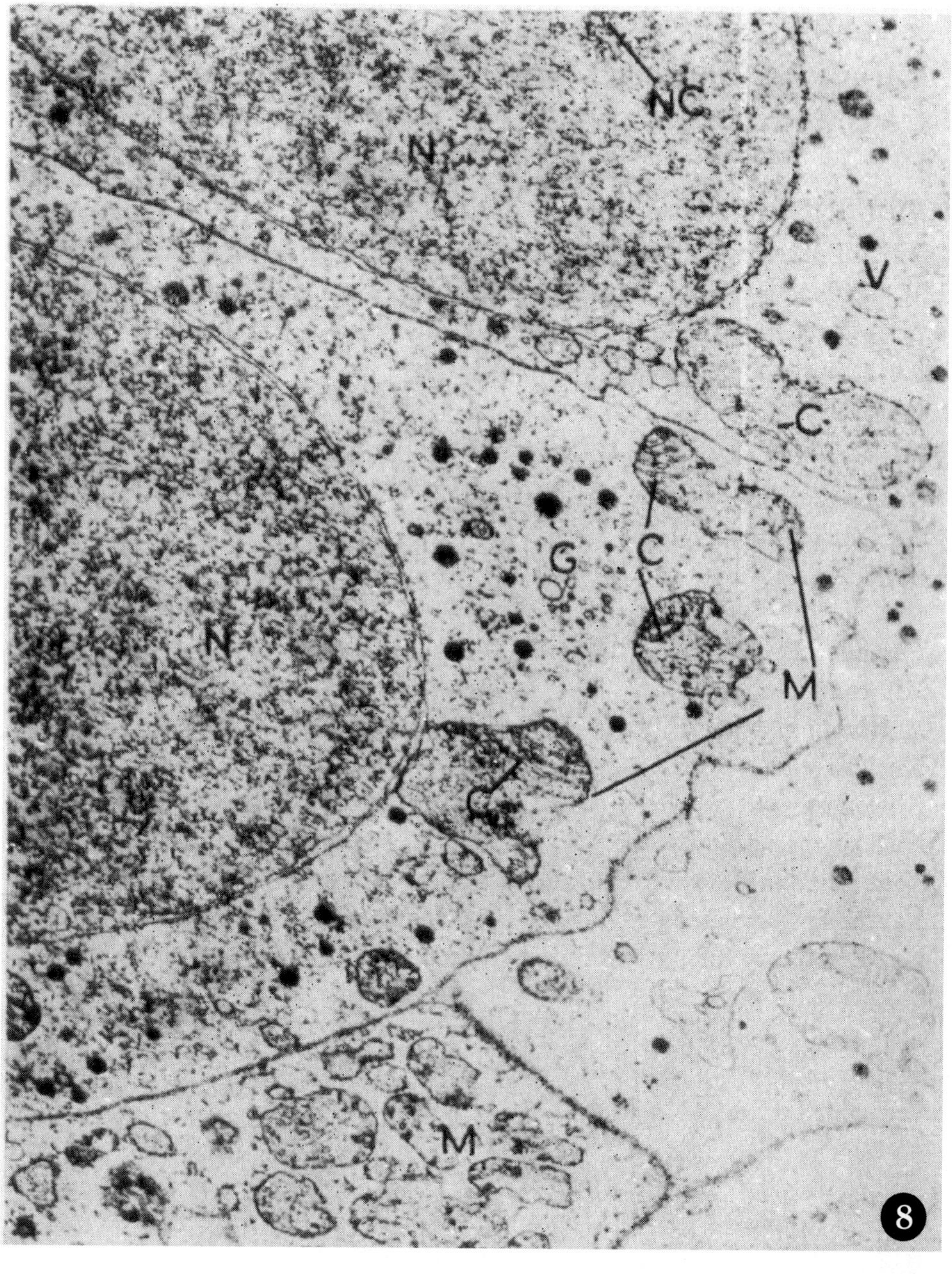

Plate 169. Fig. 8. A magnified part of corpus allatum of *Euborellia* showing a large number of mitochondria. The various types of mitochondria (M) with distinct cristae (C) are clearly seen. X 21000 (From Awasthi, 1979b).

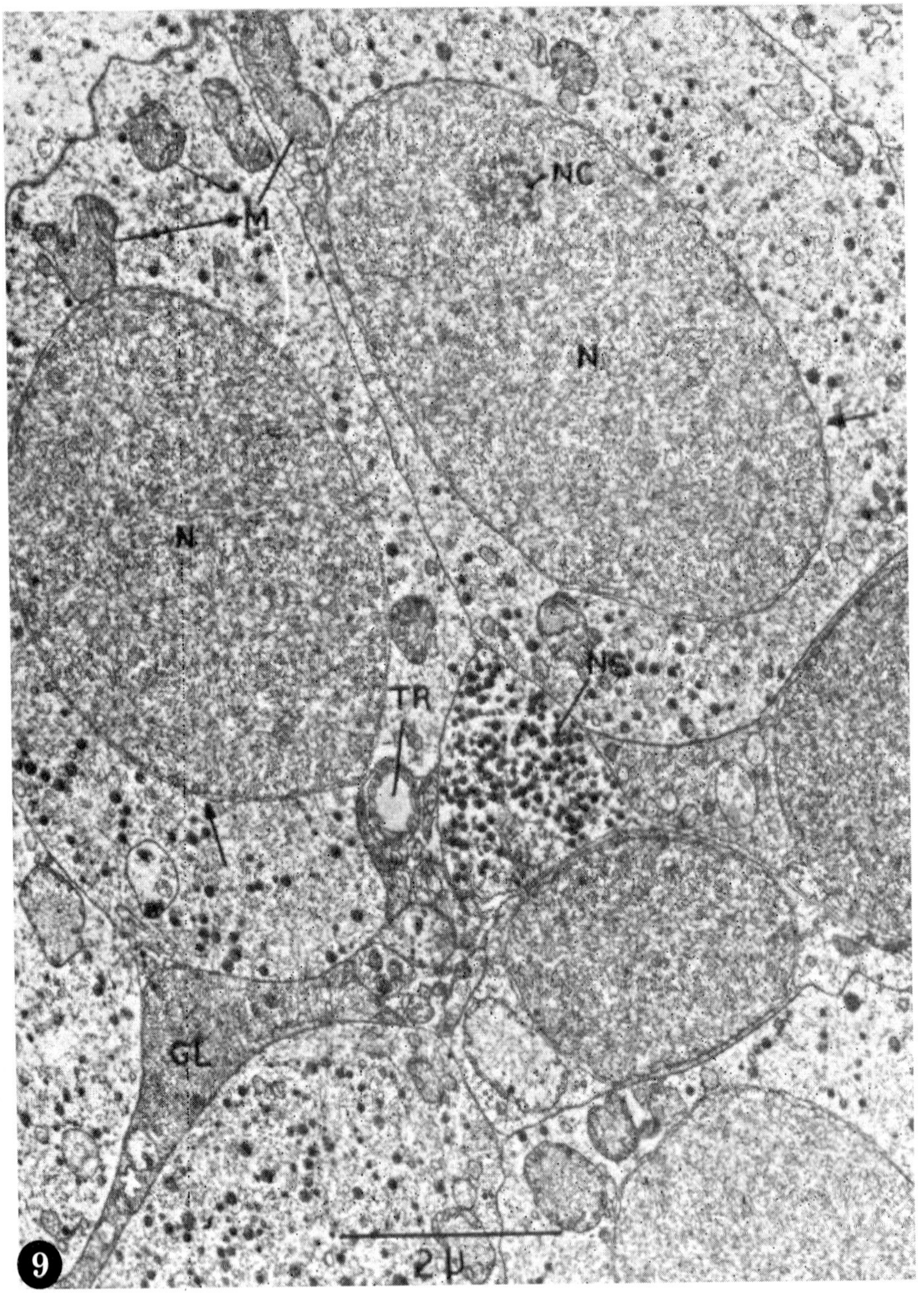

Plate 170. Fig. 9. Electron micrograph of a part of corpus allatum of *Euborellia* showing axon containing electron dense neurosecretory granules. Tracheole, mitochondria and glial cell process are also visible. (From Awasthi, 1976c).

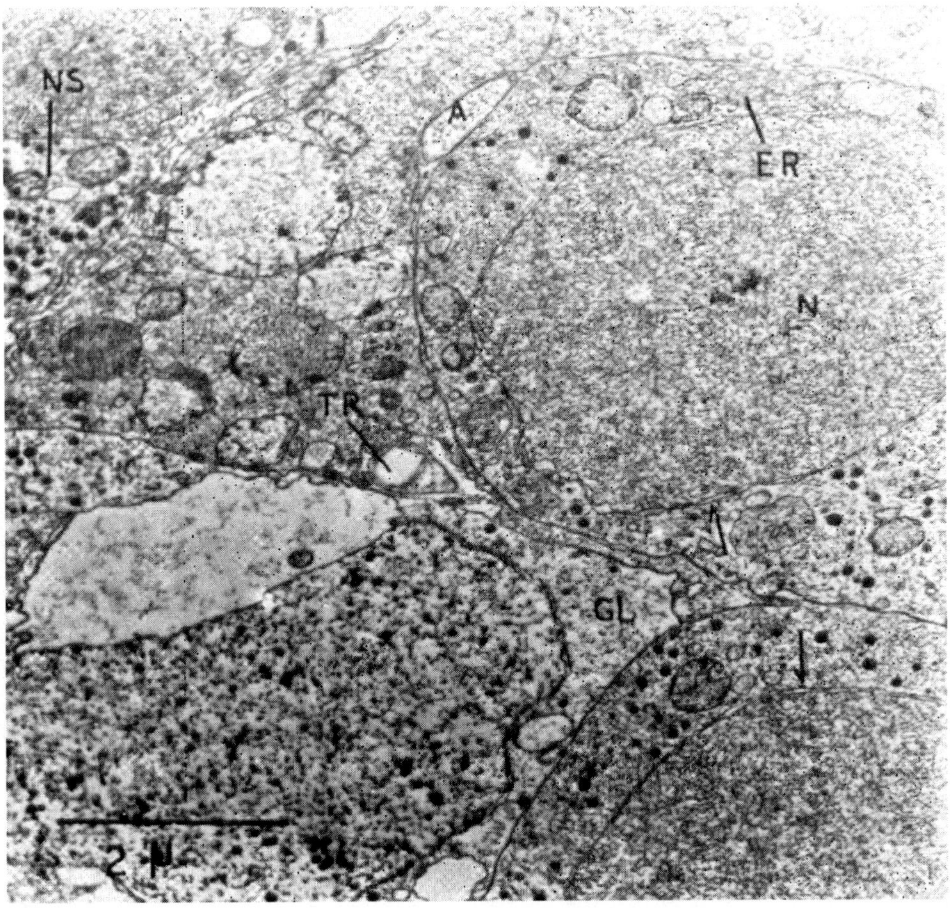

10

Plate 171. Fig. 10. Electron micrograph of a part of corpus allatum of *Euborellia* showing endoplasmic reticulum and ordinary axons. On the left side a neurosecretory axon and glial cell processes are also visible. (From Awasthi, 1976c).

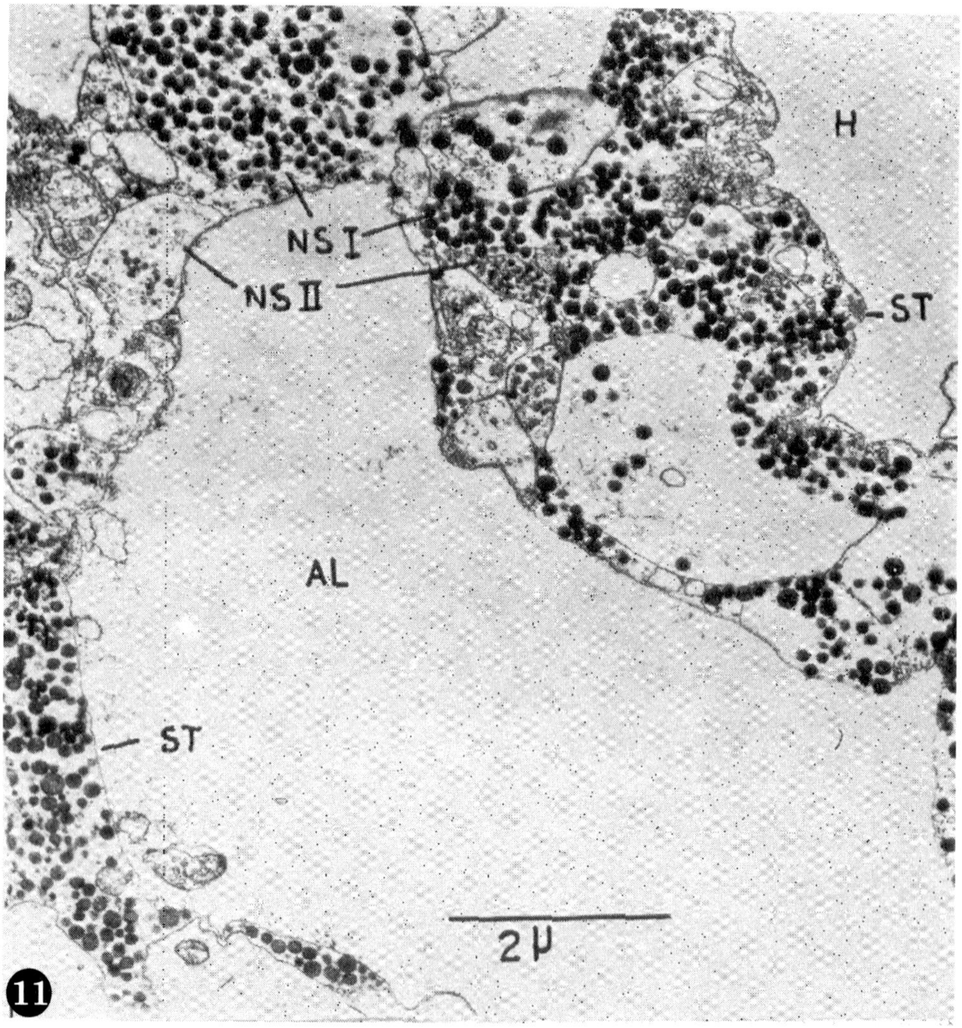

Plate 172. Fig. 11. Electron micrograph of cephalic aorta of *E. annulipes* showing a large number of axon profiles of A-fibres containing electron dense large granules (NS-I), which are confined to aorta only and differ in size as well as electron density from NS-I and NS-II granuls of the corpus allatum. The B-fibres are a few containing less electron dense (mixed type) small granules (NS-II). The glial and muscles are poorly developed. Note single layered thin outer and inner Stroma (ST). AL-aortic lumen, H-haemocoel. (From Awasthi, 1976c).

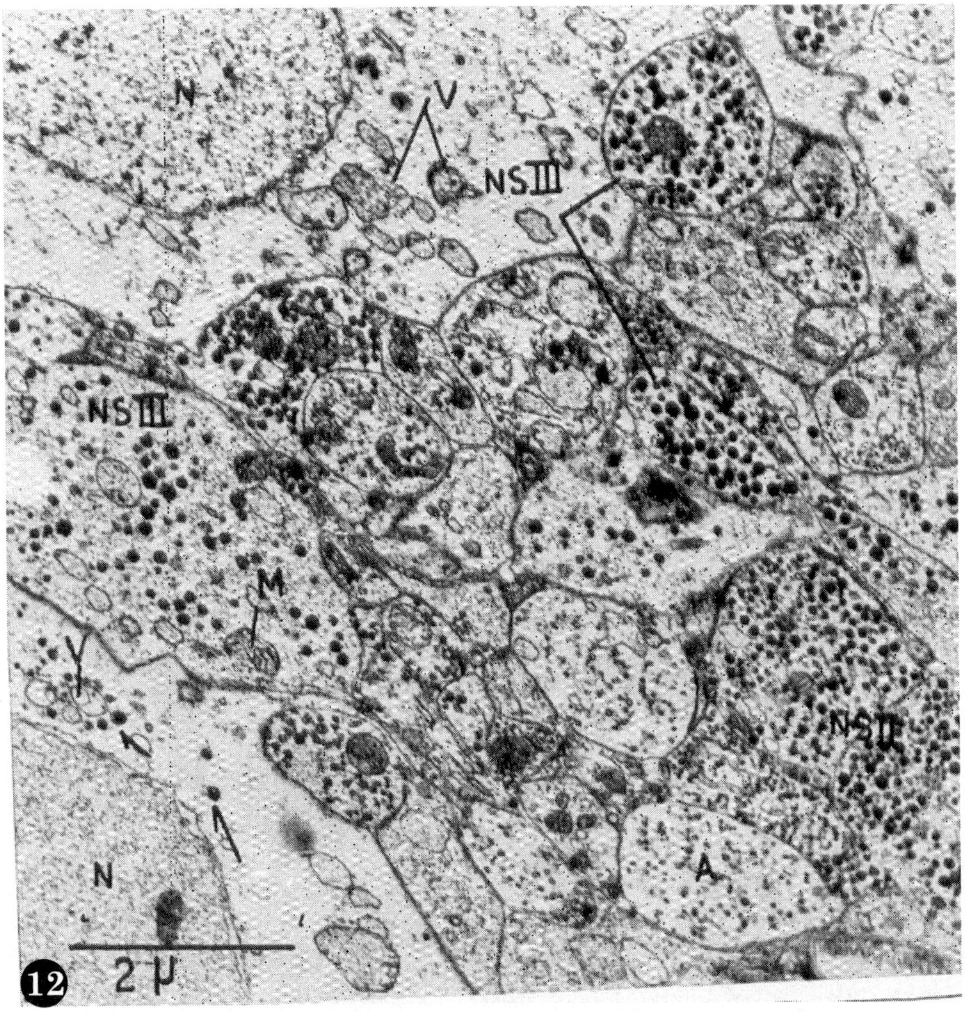

Plate 173. Fig. 12. Electron micrograph of corpus cardiacum of *E. annulipes* showing axon profiles containing two different types of neurosecretory granules (NS-II, NS-III). In both the cases mitochondria are present in large number. In addition the intrinsic cells with granules (arrows) and glial cells are also visible. Note absence of A-fibres in CC which are confined to AO only (From Awasthi, 1976c).

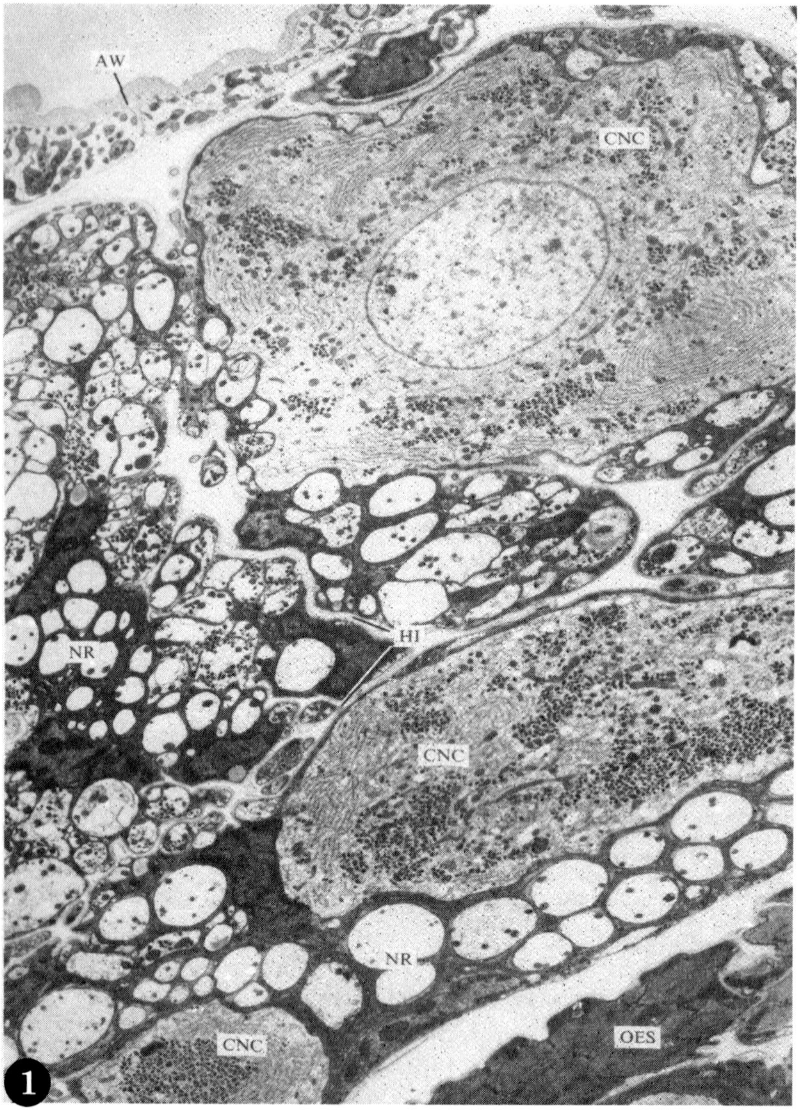

Plate 174. Fig. 1. Low magnification electron micrograph of a lateral portion of the corpus cardiacum of *Calliphora erythrocephala*, which is situated between the aorta and the oesophagus. The bodies of the corpus cardiacum neurosecretory cells, the C.N.C., are located at the ventral and lateral periphery of the ganglion-like gland. Most of the granule-containing axon profiles in the neuropile belong to the C.N.C. The axons that have few particles and are embedded in darker glioplasm belong to branches of the recurrent nerve. Note the deeply ramifying haemocoel indentations. AW, aorta wall; OES, oesophageal wall; CNC, bodies; HI, haemocoel indentations; NR, recurrent nerve. X 5850. (From Normann, 1974).

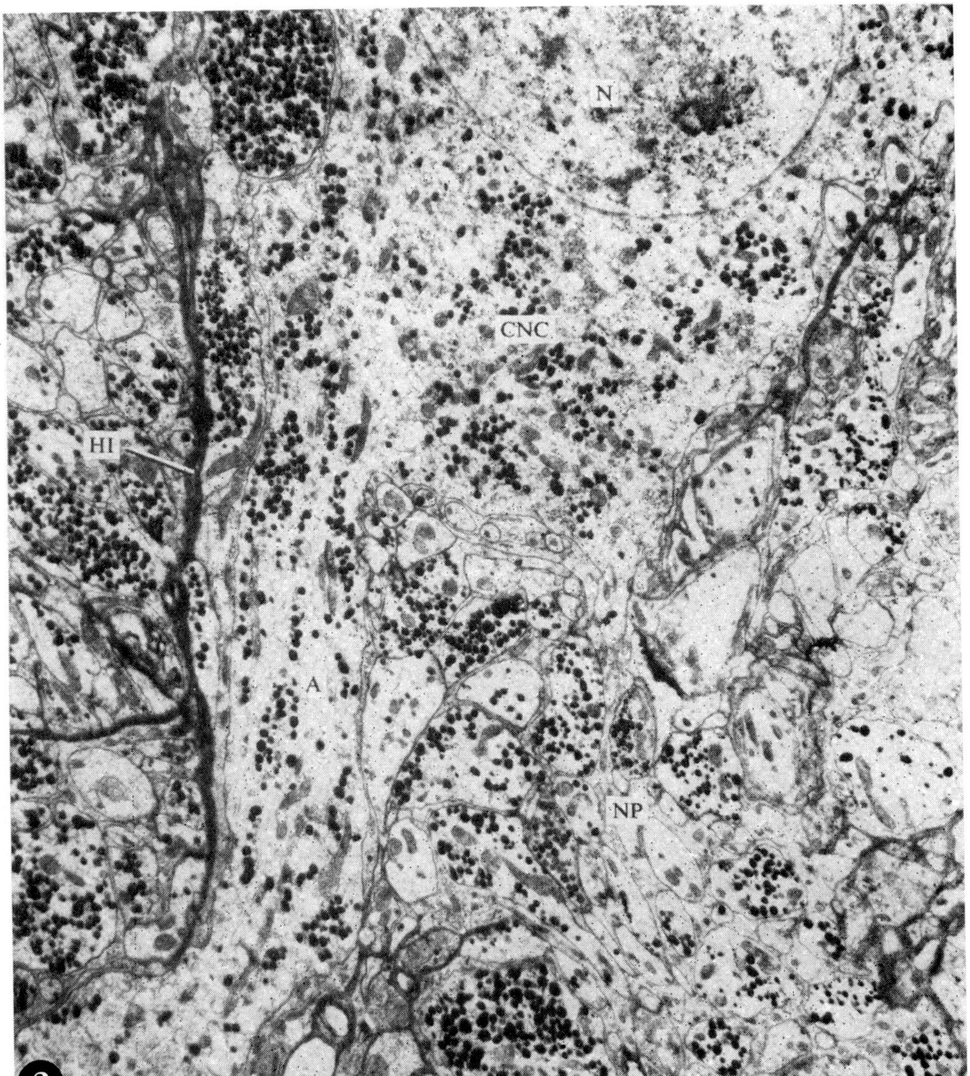

Plate 175. Fig. 2. C.N.C. with axonal projection extending into the neuropile of *Calliphora* N, nucleus of C.N.C.; A, axon, longitudinally sectioned; NP, neuropile, HI, haemocoel indentation. X 7650. (From Normann, 1974).

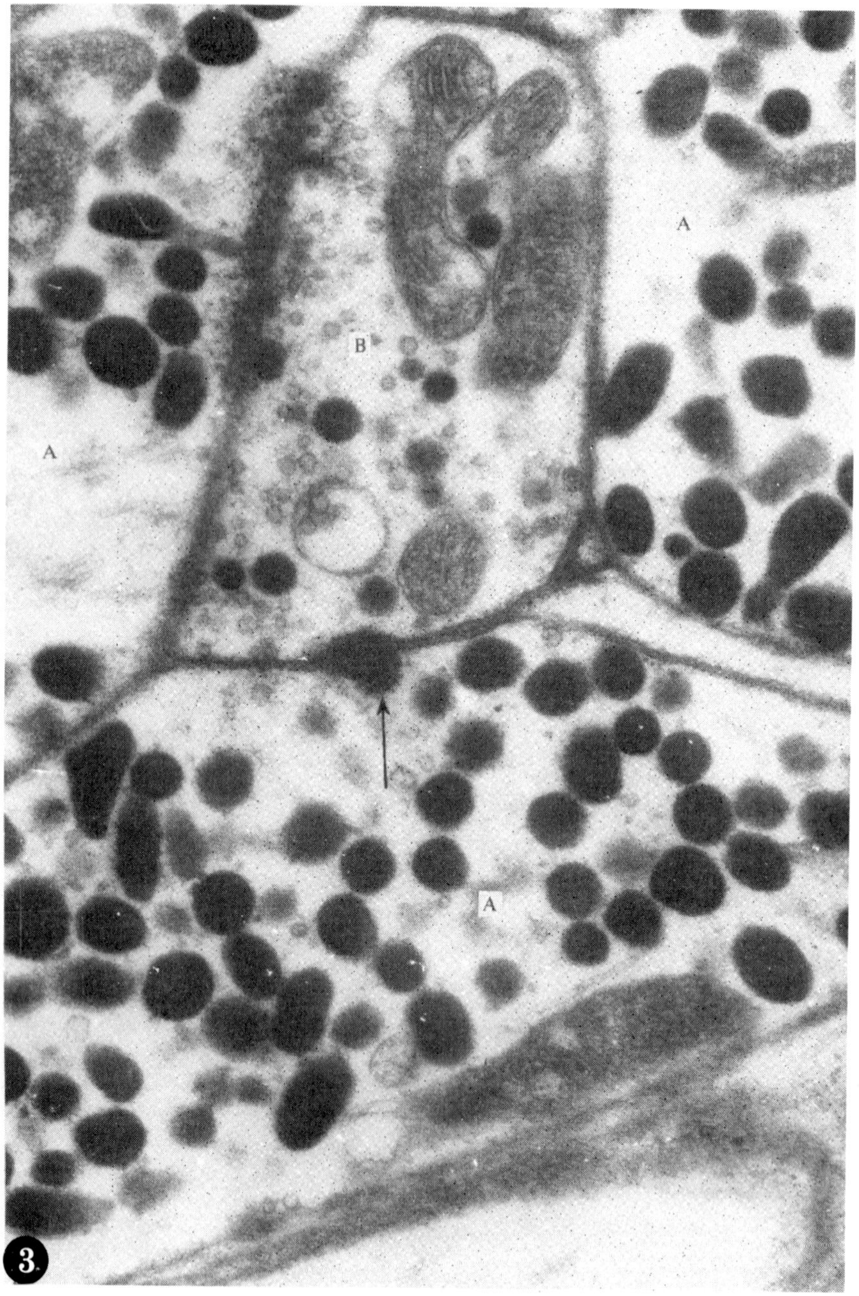

Plate 176. Fig. 3. Extrinsic microgranular axon of `B-type' of *Calliphora* with synapse on a C.N.C. axon (`A-type'). Omega figure indicating exocytosis at arrow. In this specimen, Ba^{2+} had been substituted for Ca^{2+} prior to stimulation. X 60000. (From Normann, 1974).

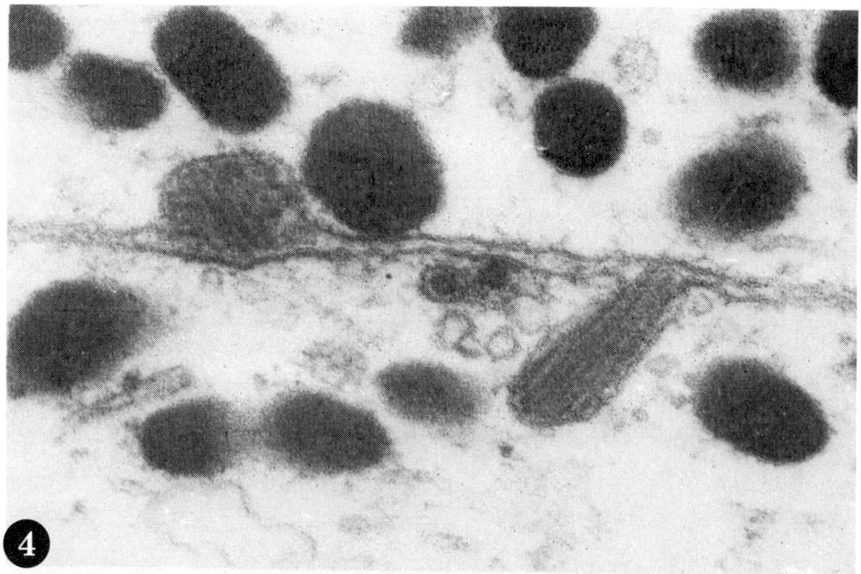

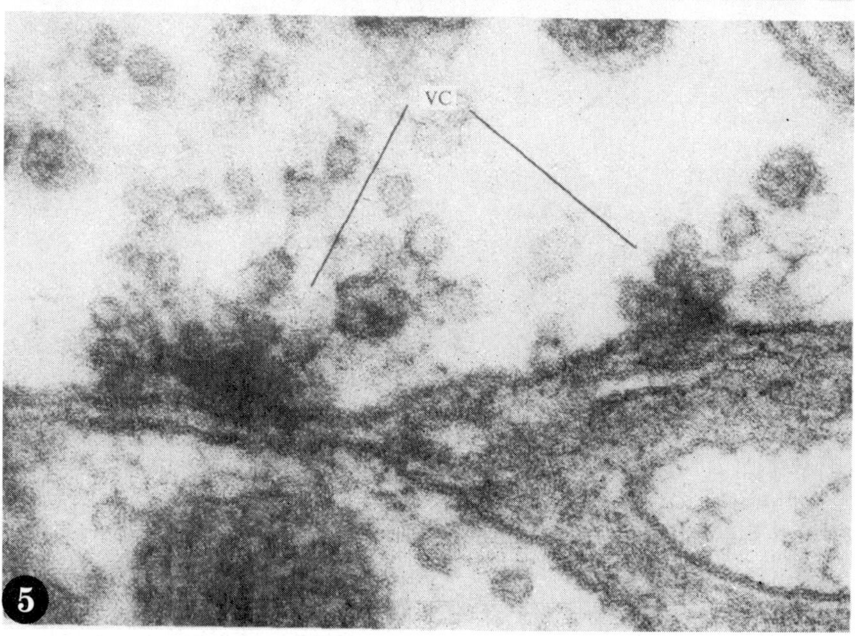

Plate 177. Fig. 4. Omega figures-exocytosis of neurosecretory granules from two adjacent axon terminals of *Calliphora*. Note the paracrystalline array of rod-like subunits, visible at the very beginning of exocytosis. This cannot be seen in intracellular (intact) granules X 98300. Fig. 5. Vesicle clusters (VC), representing the process of membrane retrieval from the axolemma at release sites. Most of such microvesicles are 30-40 nm (external diameter). X 154000. (From Normann, 1974).

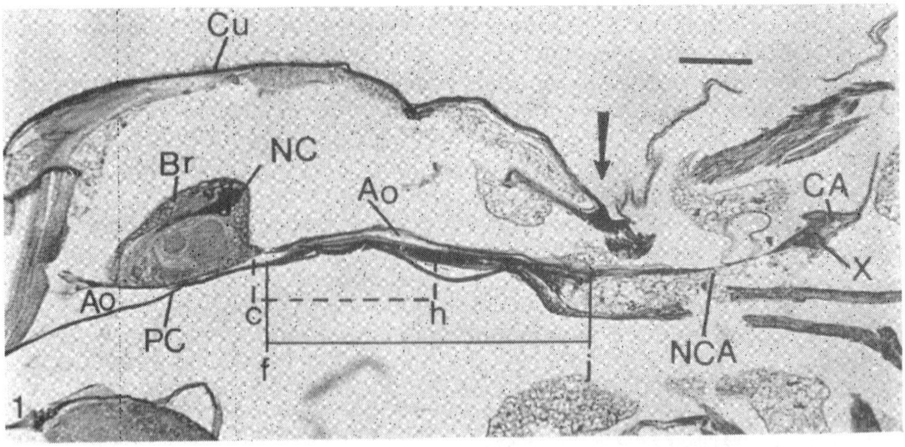

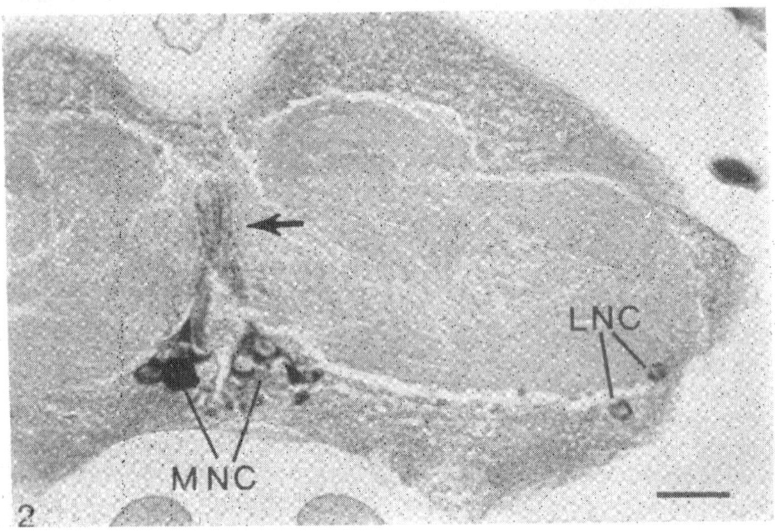

Plate 178. Fig. 1. Midsagittal section through the cerebral neursecretory system of mosquito, *Aedes*. Corpus cardiacum denoted by a solid line f-j. Dotted line c-h includes segment of aorta consisting of a double layer of muscle (internal and external aorta muscle). Ao, aorta; PC, pharyngeal cuticle; Br, brain; NC, medial neurosecretory cells; Cu, cuticle of head capsule; NCA, nervus corporis allatus; CA, corpus allatum; X, X-cells of the corpus allatum complex. Arrow, junction of head and neck. Bouin's fixative, PF stain. Scale line, 50 μ. Fig. 2. Frontal section showing the decussation (arrow) of the paired tracts of the medial neurosecretory cells (MNC) in the brain of *Aedes*. Two of the five cells of one group of lateral neurosecretory cells (LNC) are visible. Bouin's fixative, PF stain. Scale line, 25 μ. (From Meola and Lea, 1972).

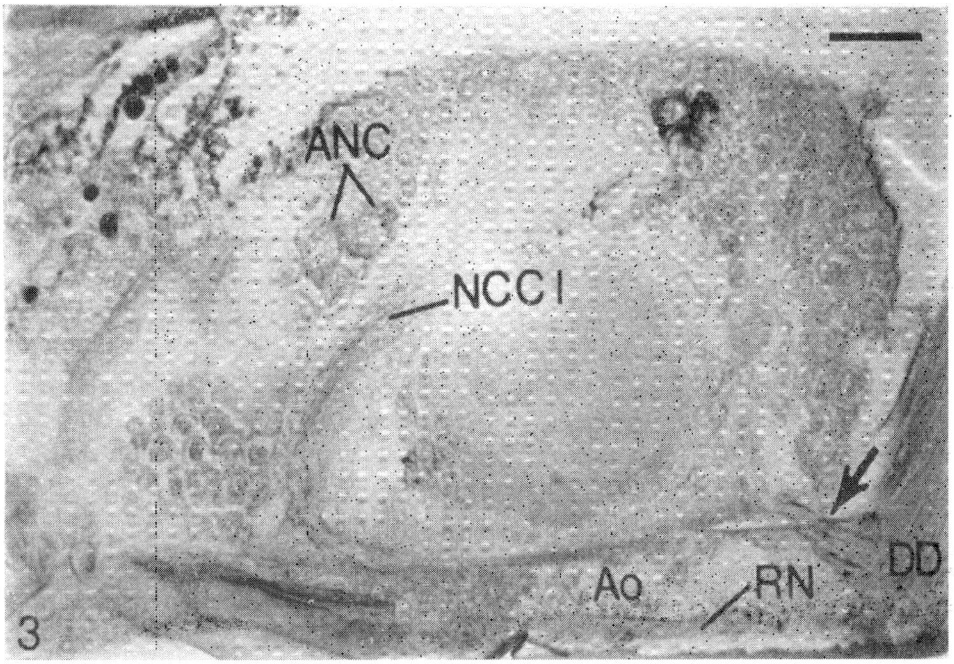

Plate 179. Fig. 3. Median section through brain of newly emerged mosquito *Aedes*. One of the paired groups of anterior neurosecretory cells (ANC) is seen dorsal to the nervi corporis cardiaci (NCC I) tract. The exit of the NCC axons from the brain into the aorta muscle is shown (arrow). DD, dorsal dilator muscle; Ao, aorta; RN, recurrect nerve. Bouin's fixative, PF stain. Scale line, 25 μ. (From Meola and Lea, 1972).

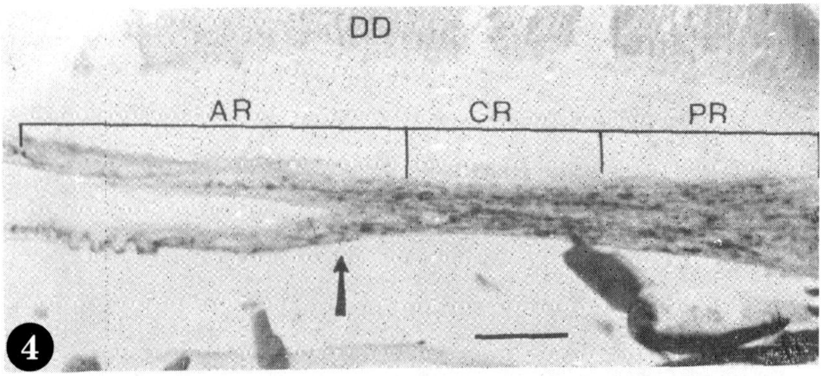

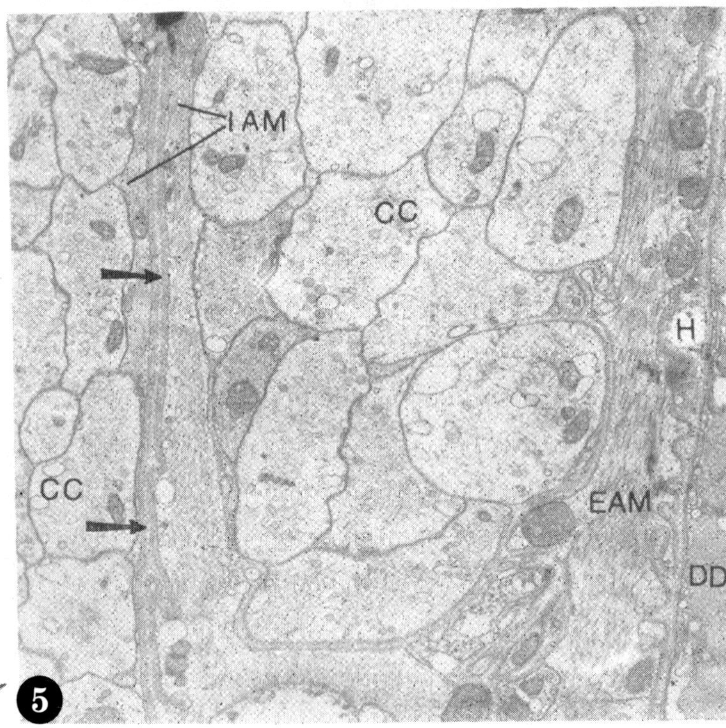

Plate 180. Fig. 4. Frontal section of the three regions of the corpus cardiacum of *Aedes*. The anterior region (AR) includes the paired and a part of the fused corpus cardiacum. The central region (CR) contains the ganglion cells; the posterior region (PR) contains the lacunar neurohaemal site. Fusion of the paired anterior region occurs about midway along the length of the corpus cardiacum (arrow). DD, dorsal dilator muscle. Bouin's fixative, PF stain. Scale 20 μ. Fig. 5. Transverse section through the area of fusion of the paired anterior corpora cardiaca (seen in Fig. 4). The internal aorta muscle (IAM) covering the medial surface of the corpora cardiaca (CC) draws together and terminates (arrows indicate muscle junction and the remnant of the lumen). A continuous layer of internal aorta muscle fibres formed by this fusion remains as the new ventral wall of the lumen. EAM, external aorta muscle; DD, dorsal dilator muscle; H, haemocoel. X 11,000. (From Meola and Lea, 1972).

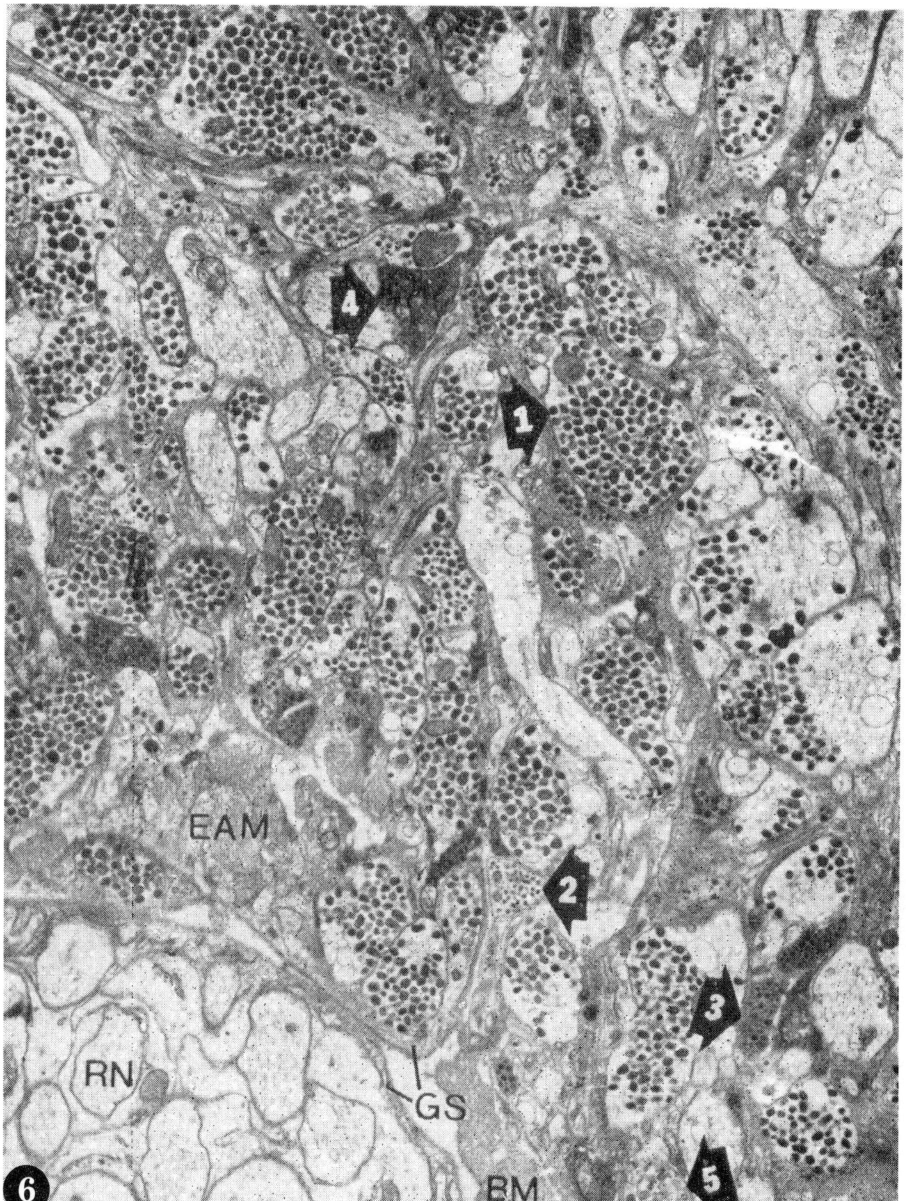

Plate 181. Fig. 6. Transverse section through the neuropile of the fused anterior region of the corpus cardiacum of *Aedes*. The compact structure of this region can be seen. The recurrent nerve (RN) has entered the hypocerebral ganglion which is separated from the corpus cardiacum by their respective glial sheaths and remnants of the external aorta muscle (EAM). The numbered arrows are the five types of neurosecretory axons described in the text. BM, basement membrane. X 7000. (From Meola and Lea, 1972).

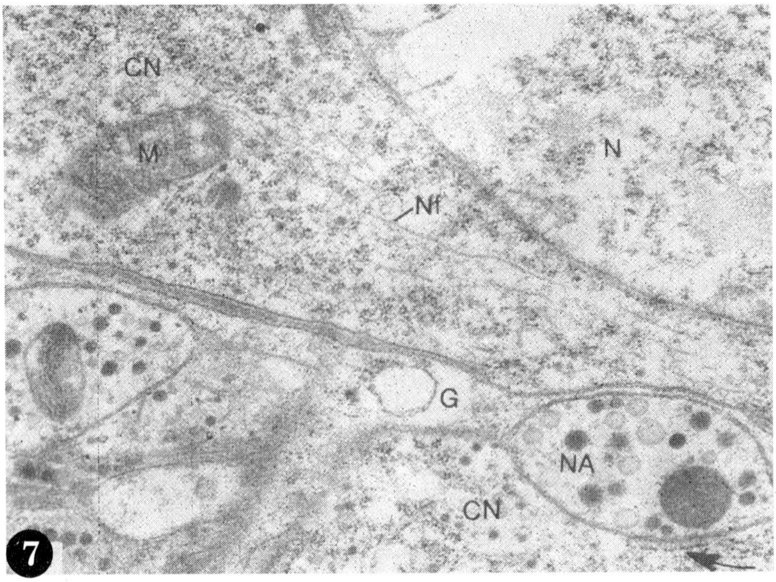

Plate 182. Fig. 7. The nonneurosecretory ganglion cells (CN) in the central region of the corpus cardiacum of *Aedes*. Neurosecretory axons (NA) are found between the parikarya of these cells, making possible axosomatic synapses (arrow). Nf, neurofilaments; N, nucleus; G, glial cell; M, mitochondria. X 24,850. (From Meloa and Lea, 1972).

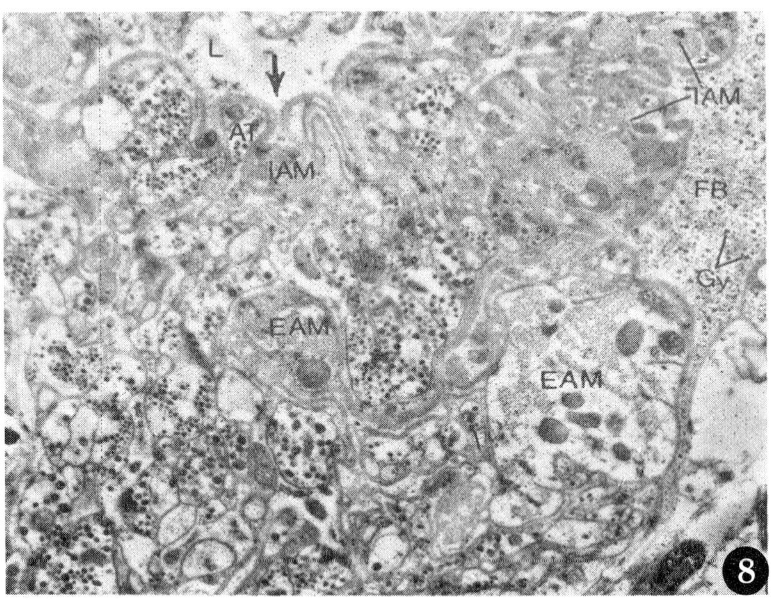

Plate 183. Fig. 8. Transverse section through the aortic neurohaemal site, which extends the length of the corpus cardiacum of *Aedes* showing the interdigitation (arrow) of the neurosecretory axon terminals (AT) and the internal aorta muscle fibres (IAM) around the lumen (L) of the aorta. EAM, external aorta muscle; Gy, glycogen granules; FB, fat body. X 6900. (From Meloa and Lea, 1972).

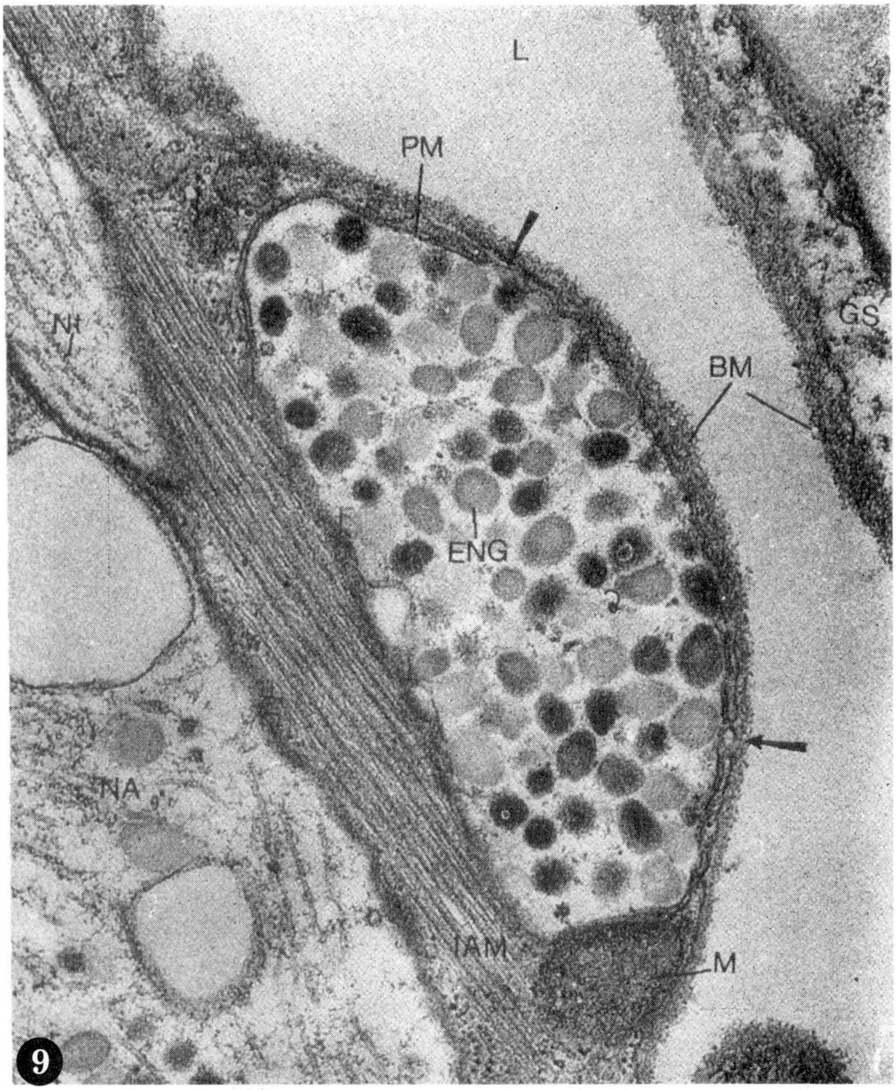

Plate 184. Fig. 9. A neurosecretory axon terminal projecting into the lumen in the aortic neurohaemal site, after passing between fibres of the internal aorta muscle (IAM). Only the plasma membrane (PM) of the axon and a thin basement membrane (BM) separate the elementary neurosecretory granules (ENG) from the haemolymph. Arrows point to discontinuities in the axolemma where ENG appear to be released. Glial sheath (GS) of the recurrent nerve which forms the ventral wall of the aortal lumen is seen at upper right. Nt, neurotubules; M, mitochondria in muscle; NA, neurosecretory axon. X 63,000 (*Aedes*) (From Meola and Lea, 1972).

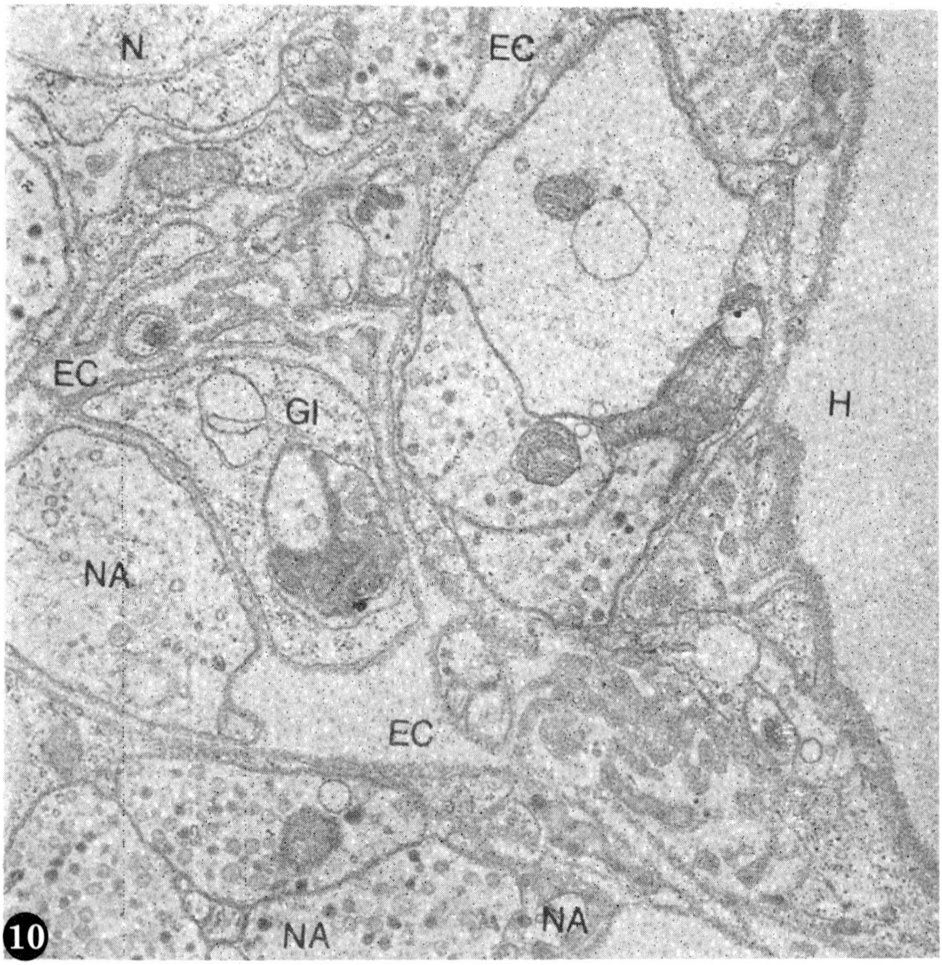

Plate 185. Fig. 10. An enlargement of part of the lacunar neurohaemal site of *Aedes* showing the loosely knit organization of this area due to the ramifications of the extracellular channels (EC). The association of the lacunar glial cell (Gl) with the neurosecretory axons (NA) and their terminals is seen. The nuclei (N) of the pituicyte-like glial cells lack the dark chromatin masses of the ``typical'' sheathing glial cells found elsewhere in this system. H, haemocoel. X 5000 (From Meola and Lea, 1972).

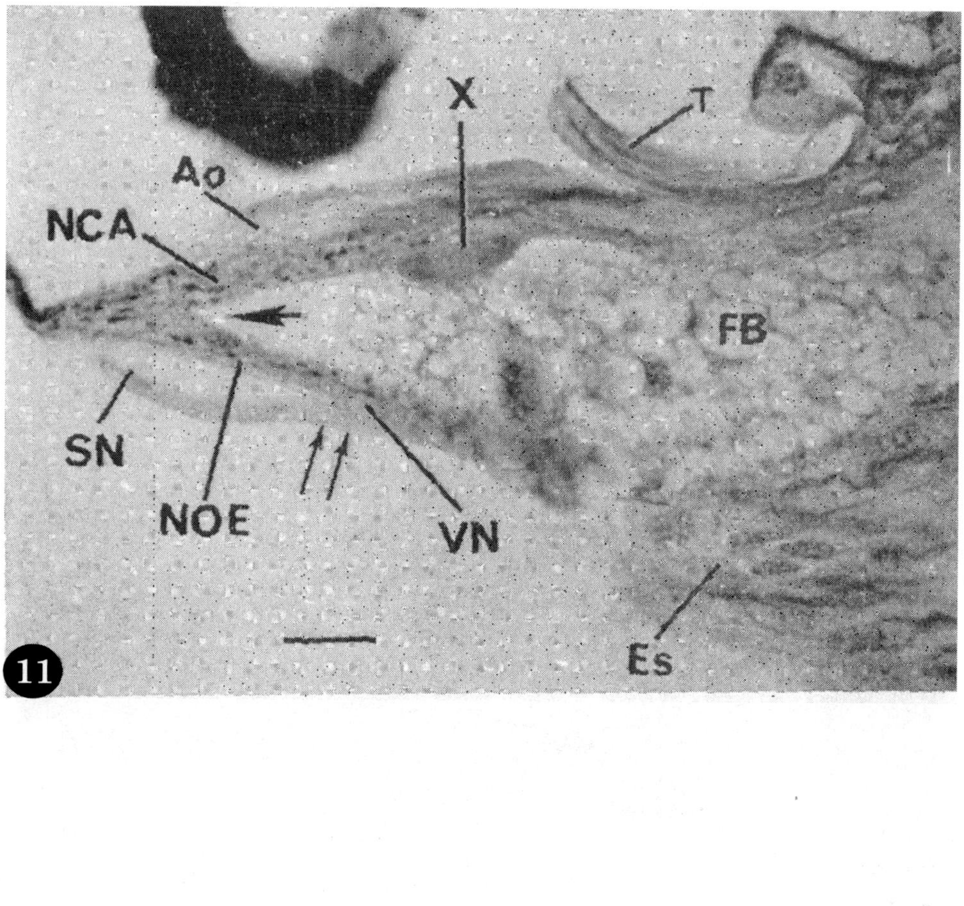

Plate 186. Fig. 11. Sagittal section showing bifurcation (arrow) of one of the pair of neurosecretory tracts that extend posteriorly from the corpus cardiacum forming the nervi corporis allati (NCA) and nervi oesophagei (NOE). The stomatogastric nerve (SN) joins (double arrow) the NOE to form the ventricular nerve (VN). Ao, aorta, T, trachea; X, secretory cells of the corpus allatum complex; Es, oesophagus; FB, fat body. Bouin's fixative, PF stain, Scale line, 20 μ. (*Aedes*). (From Meola and Lea, 1972).

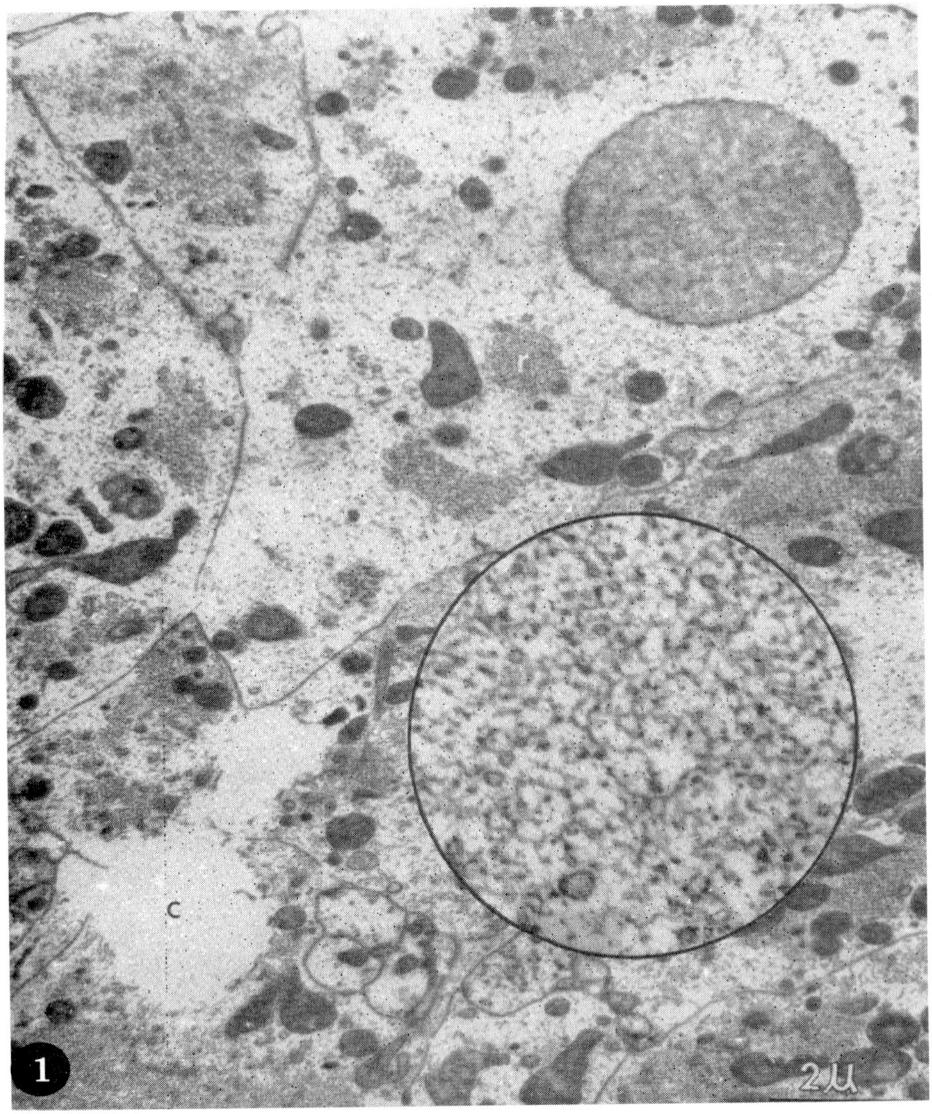

Plate 187. Fig. 1. An electron micrograph of a portion of the corpus allatum of a three day old female *Drosphila.* Note that the cytoplasm contains transparent areas within which are nests of reticulate material (r). The inset shows some of this reticulate material at a three times higher magnification. The reticulum appears to consist of both tubules and solid filaments. However, tubules lying in the plane of the section may be mistaken for filaments, since their diameters are 2-3 times smaller than the thickeness of the sections. Note that the palsma membranes are broken at many places and that the cells appear to be discharging their secretions into centrally located lacunae (c). (From King *et al.*, 1966a)

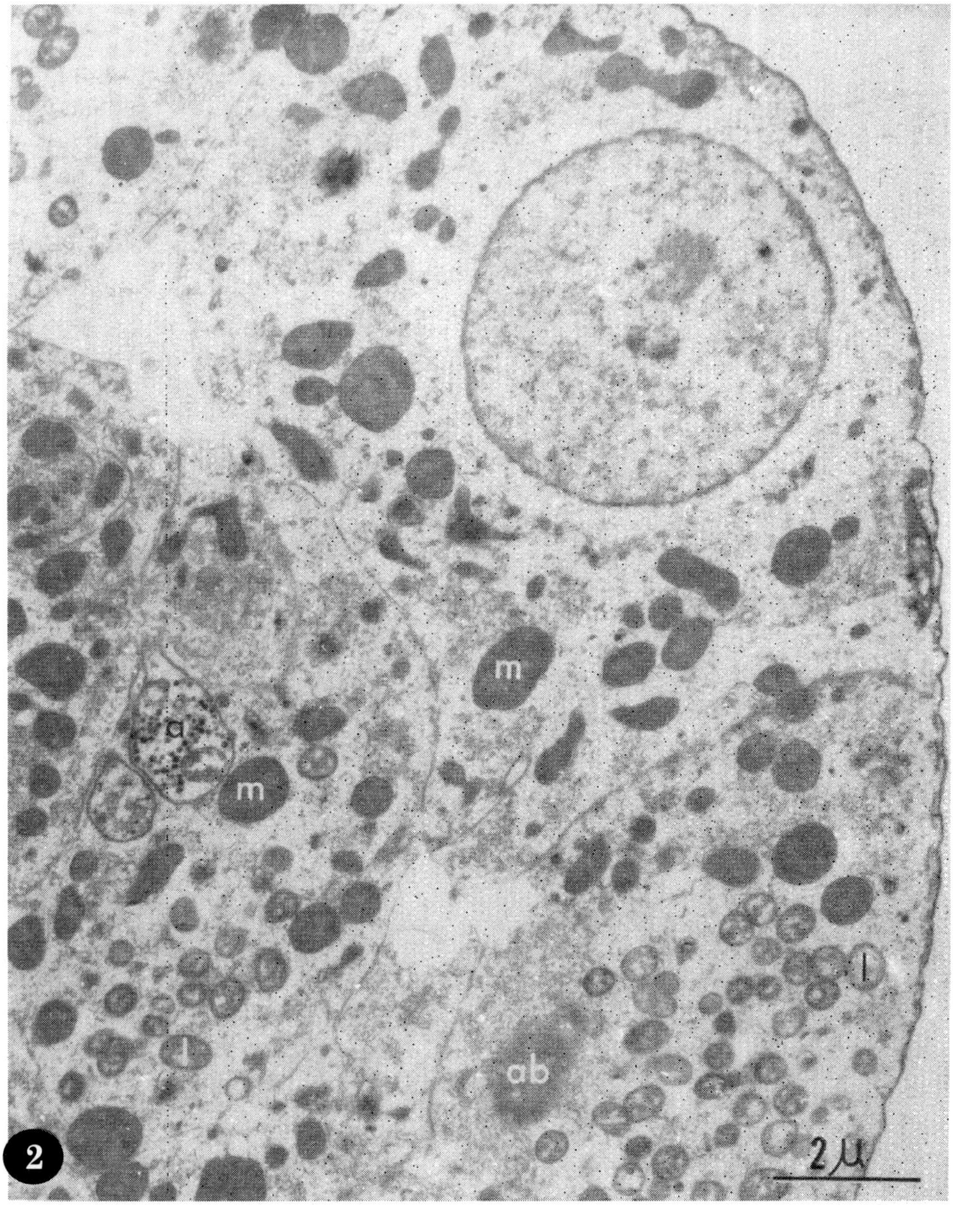

Plate 188. Fig. 2. An electron micrograph of a portion of the corpus allatum of a 24-day adult female *Drosophila melanogaster*. An interface between five adjacent corpus allatum cells is shown. Two sectioned axons (a) containing neurosecretory spheres are included. The cytoplasm of the corpus allatum cells contains dense mitochondria (m), a reticulum of tiny tubules, lysosomes (l), and suspended polyribosomes. The lower right cell also contains an allatum body (ab). (From King *et al.* 1966a).

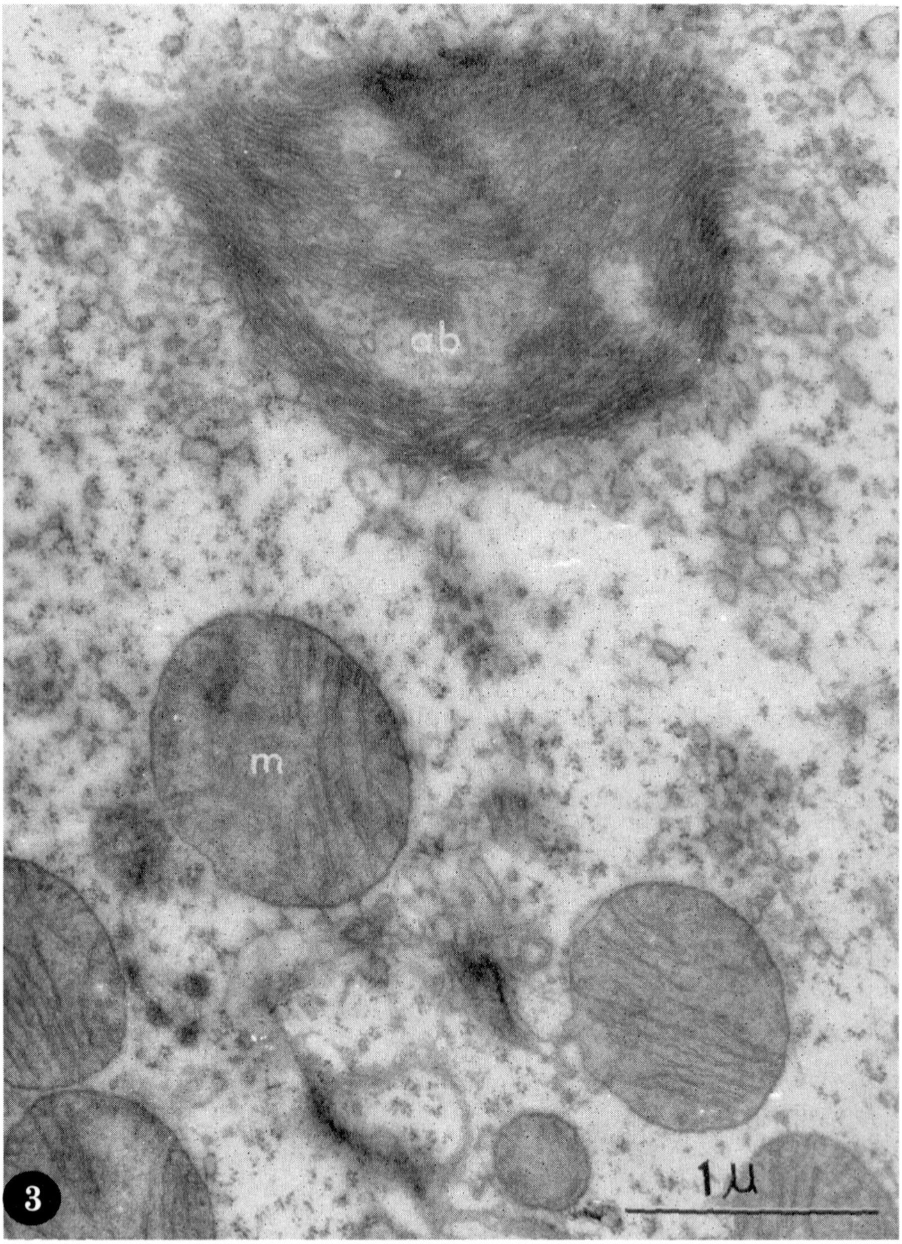

Plate 189. Fig. 3. A higher power electron micrograph of some of the organelles characteristic of corpus allatum cells. The cross-sectioned allatum body (ab) consists of stacks of smooth surfaced cisternae which seem to be budding off vesicles from their tips. The denseness of the mitochondria (m) is due to flocculent material contained between their cristae. (*Drosophila*) (From King *et al.*, 1966a).

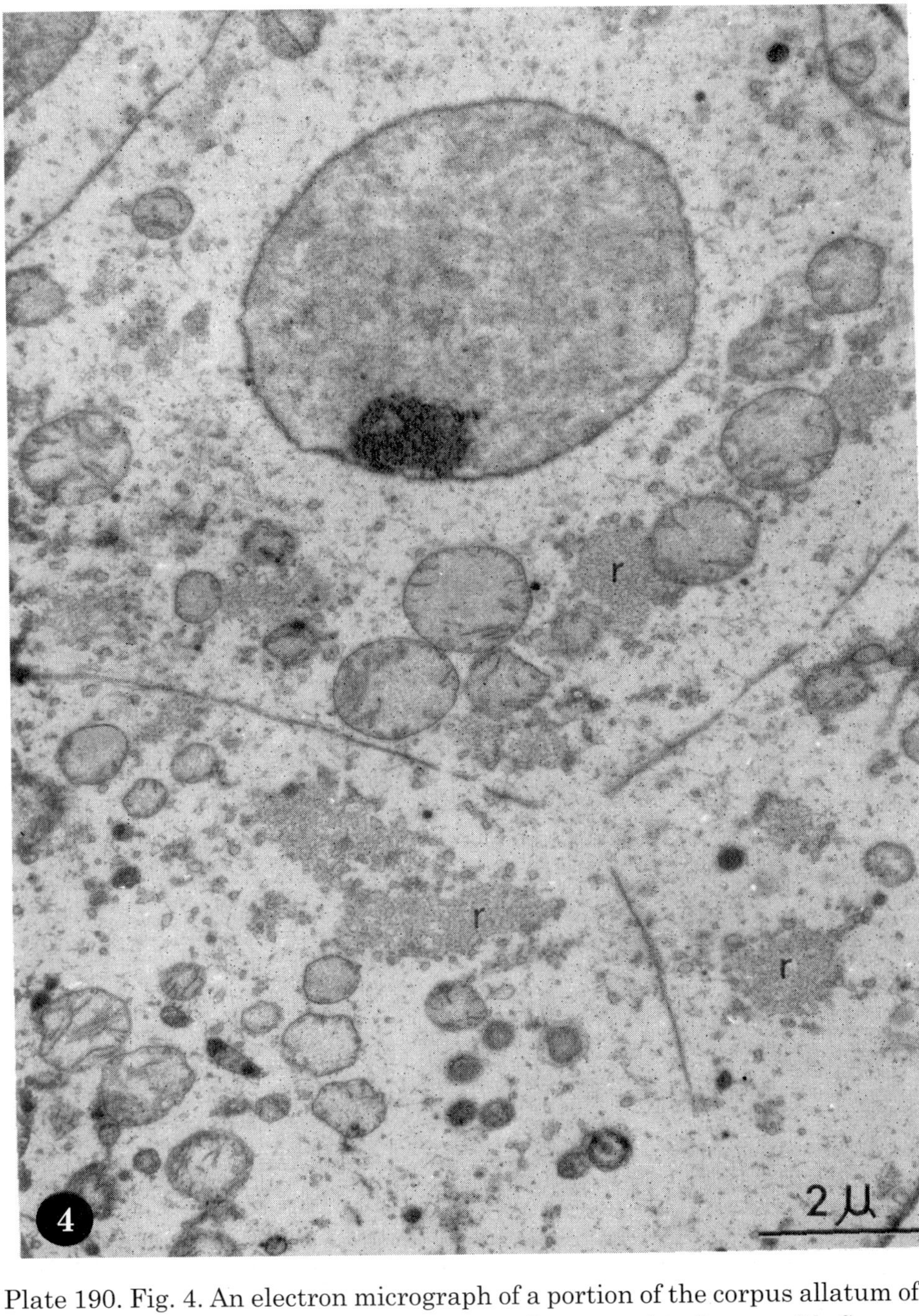

Plate 190. Fig. 4. An electron micrograph of a portion of the corpus allatum of a six-day-old *fes* female *Drosophila*. Contrast this figure with figure 1. The pathological swelling of the mitochondria is evident. r, reticulum. (From King *et al.*, 1966a)

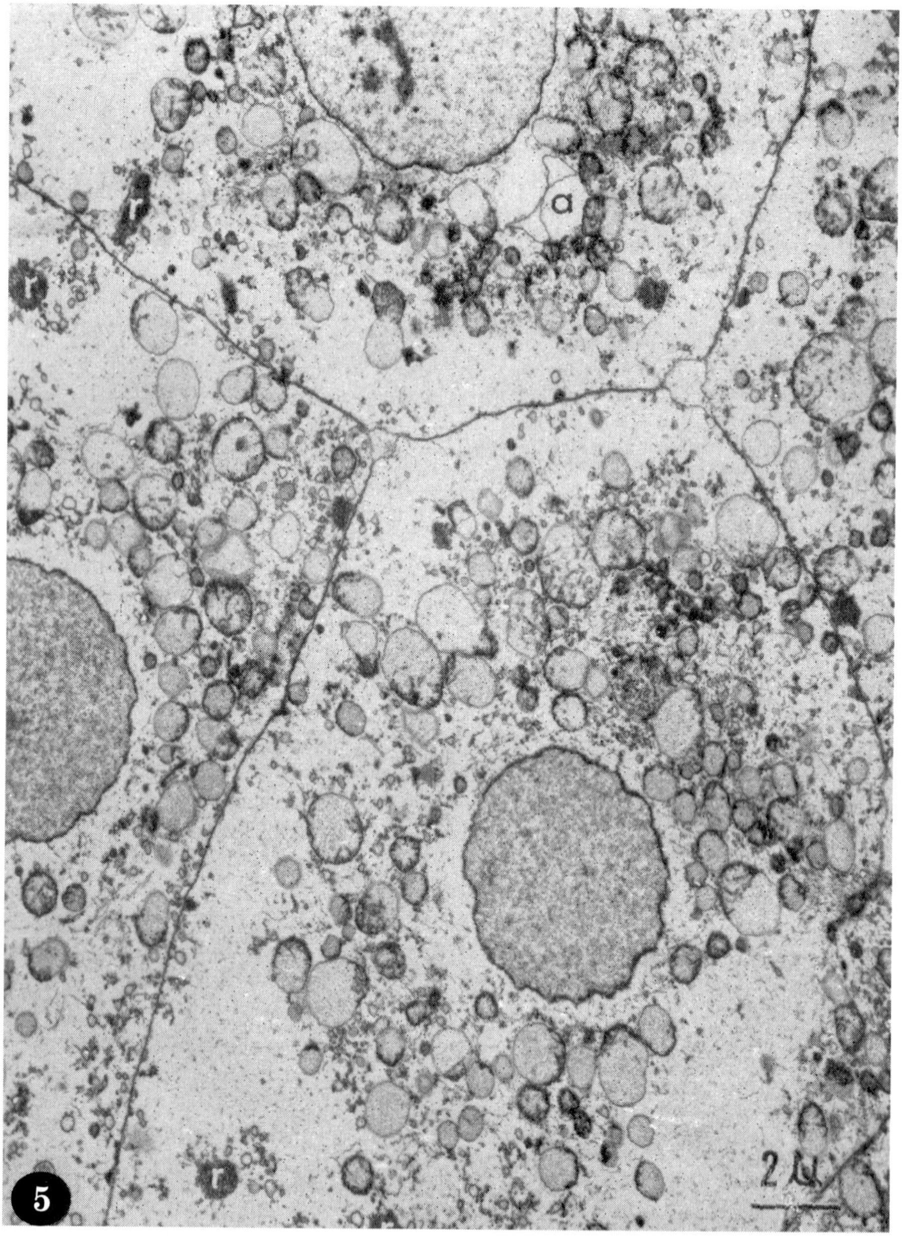

Plate 191. Fig. 5. An electron micrograph of a portion of the corpus allatum of 24-day-old *fes* female *Drosophila*. Contrast this figure with figure 4. Note the swollen mitochondria, the crenate nuclei, and the tangled cytoplasmic filaments. The islands of reticulate material (r) seem compressed as compared to those of the six-day series. a, collapsed axon. (From King *et al.*, 1966a).

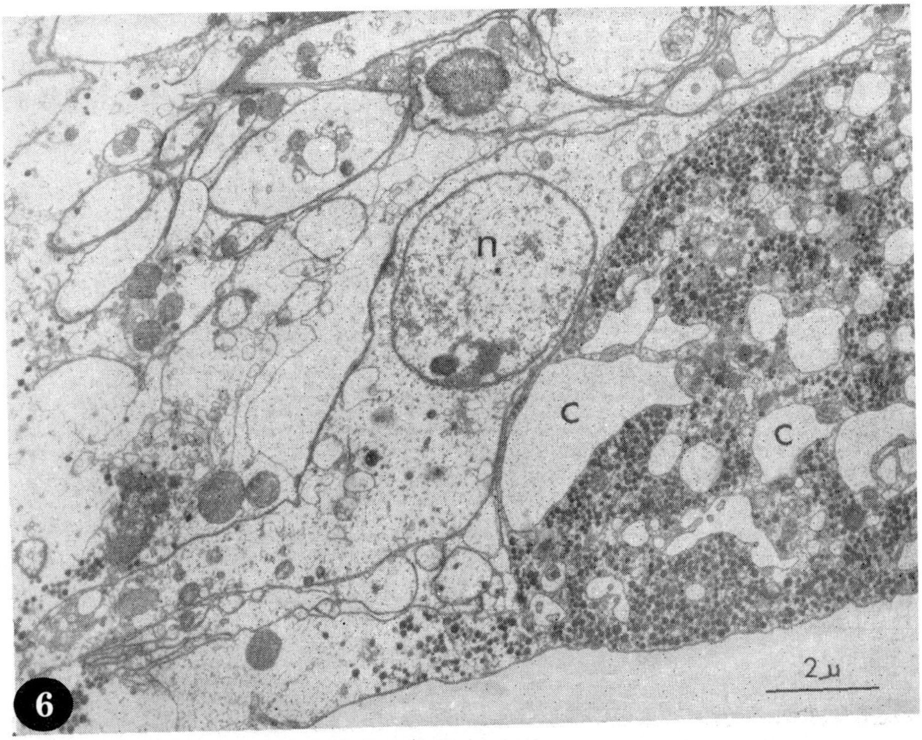

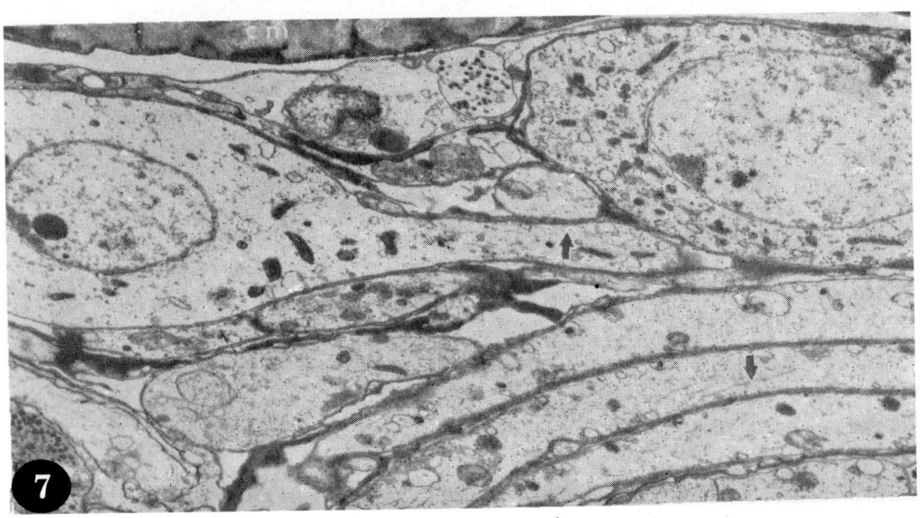

Plate 192. Fig. 6. An electron micrograph of a bipolar neuron in the corpus cardiacum of *Drosophila* c, cisternae of cortical cell; n, nucleus of neuron. Fig. 7. An electron micrograph of unipolar neurons of the hypocerebral ganglion of *Drosophila*. cm, cardial muscle. Arrows point to neurofilaments. (From King *et al.*, 1966a).

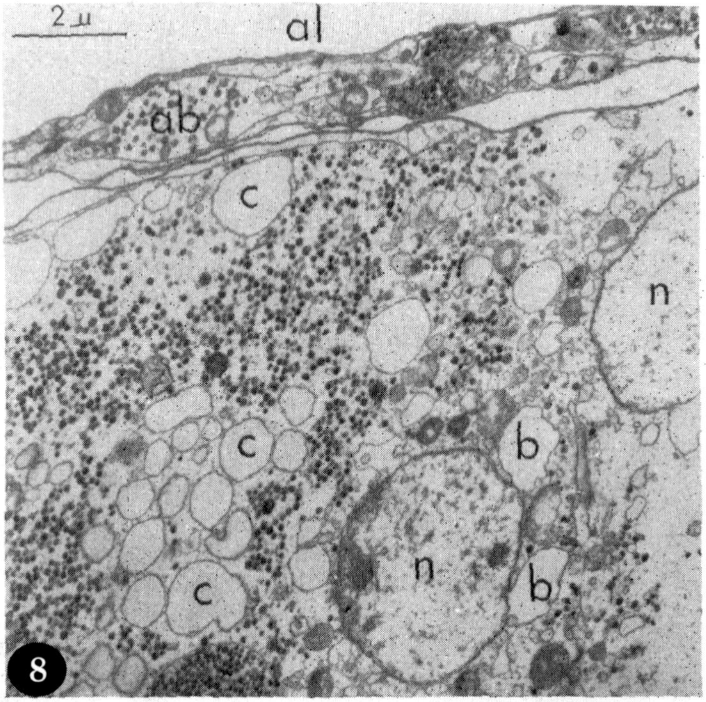

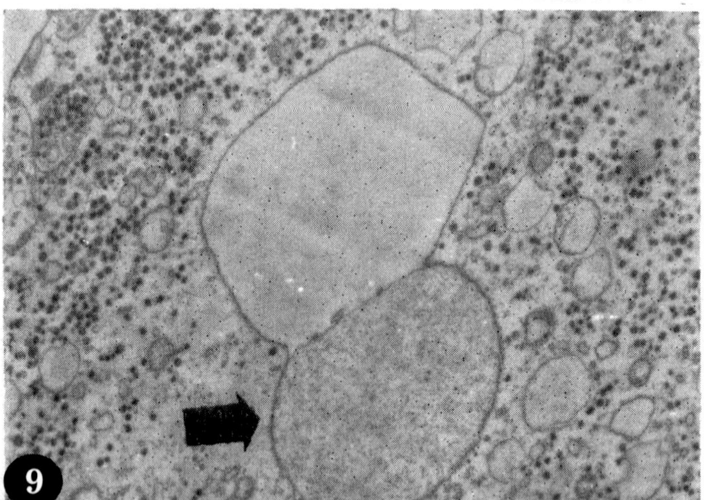

Plate 193. Fig. 8. An electron micrograph of a portion of the aorta and corpus cardiacum ab, axonal bulge in wall of aorta; al, aortal lumen; b, nuclear blisters; c, cisternae; and n, nuclei sharing a common cytoplasm. Fig. 9. An electron micrograph of cortical cells from the corpus cardiacum of a 24-day-old *fes* female *Drosophila*. The arrow points to a blistered nucleus. (From King *et al.*, 1966a).

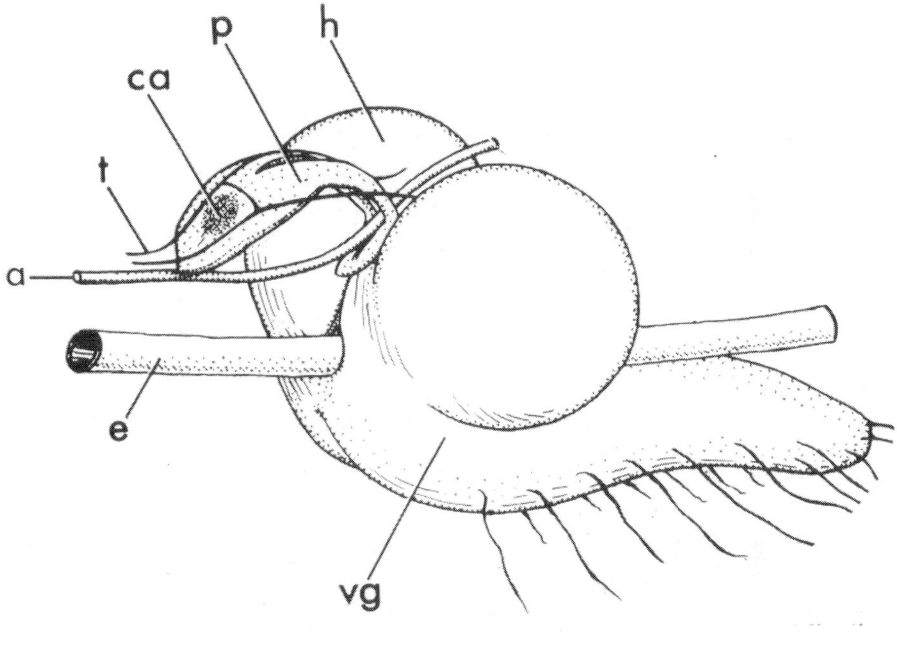

Plate 194. Fig. 1. Diagram of the central nervous system, ring gland, and portions of the circulatory and digestive systems of a third instar larva of *Drosophila melanogaster,* a aorta; ca corpus allatum portion of the ring gland; e oesophagus; h brain hemispher; p left limb of prothoracic gland; t trachea; vg ventral ganglion (From King *et al.,* 1966b)

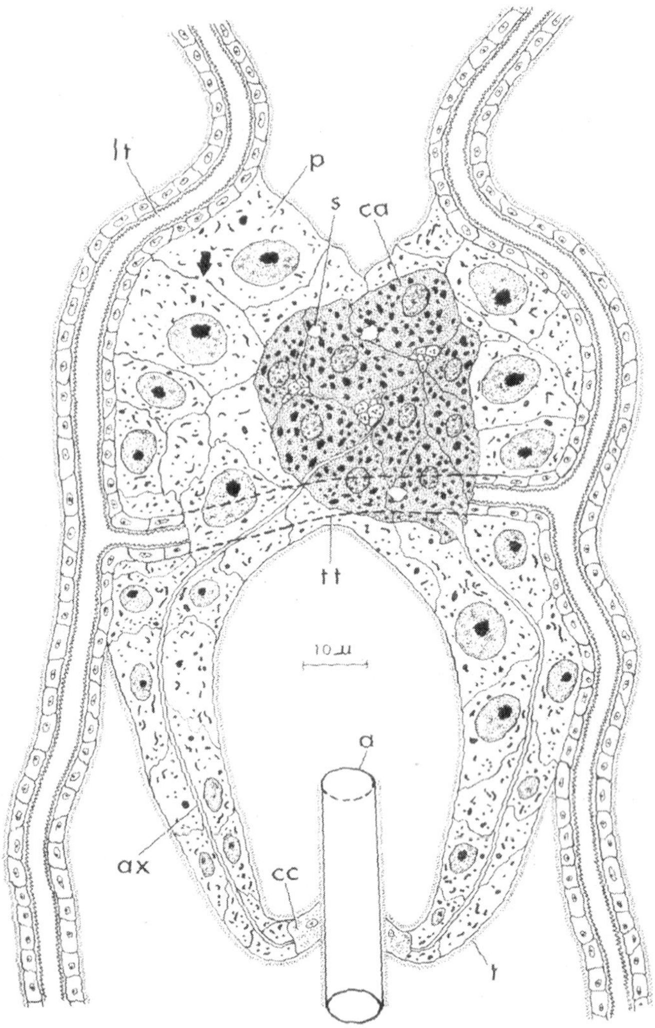

Fig. 2

Plate 195. Fig. 2. Diagram of an ideal transverse section through the ring gland of a third instar larva of *Drosophila*. The corpus allatum is stippled to demonstrate its greater electron density. The solid filaments and short rods drawn in the cytoplasm of the corpus allatum and prothoracic gland cells represent mitochondria. The particles drawn in the cytoplasm of the corpus cardiacum cells and in the axoplasm of the swollen axonal terminations seen in the corpus allatum represent single and clustered neurosecretory spheres. The tracheal tubes are surrounded by an epithelium of tiny polygonal cells. The intima secreted by these cells contains spiral ridges or taenidia which project toward the air-filled lumen (see Fig 10. The path of the transverse trachea through the gland is shown by parallel dashed lines. a aorta; ax axon from corpus cardiacum to corpus allatum (see Fig. 11); ca corpus allatum cell; cc corpus cardiacum cell; lt left lateral trachea; p cell of prothoracic gland; s secretory deposit; t acellular tunica; tt transverse trachea. The arrow points to a pore connecting adjacent prothoracic gland cells (From King *et al.*, 1966b).

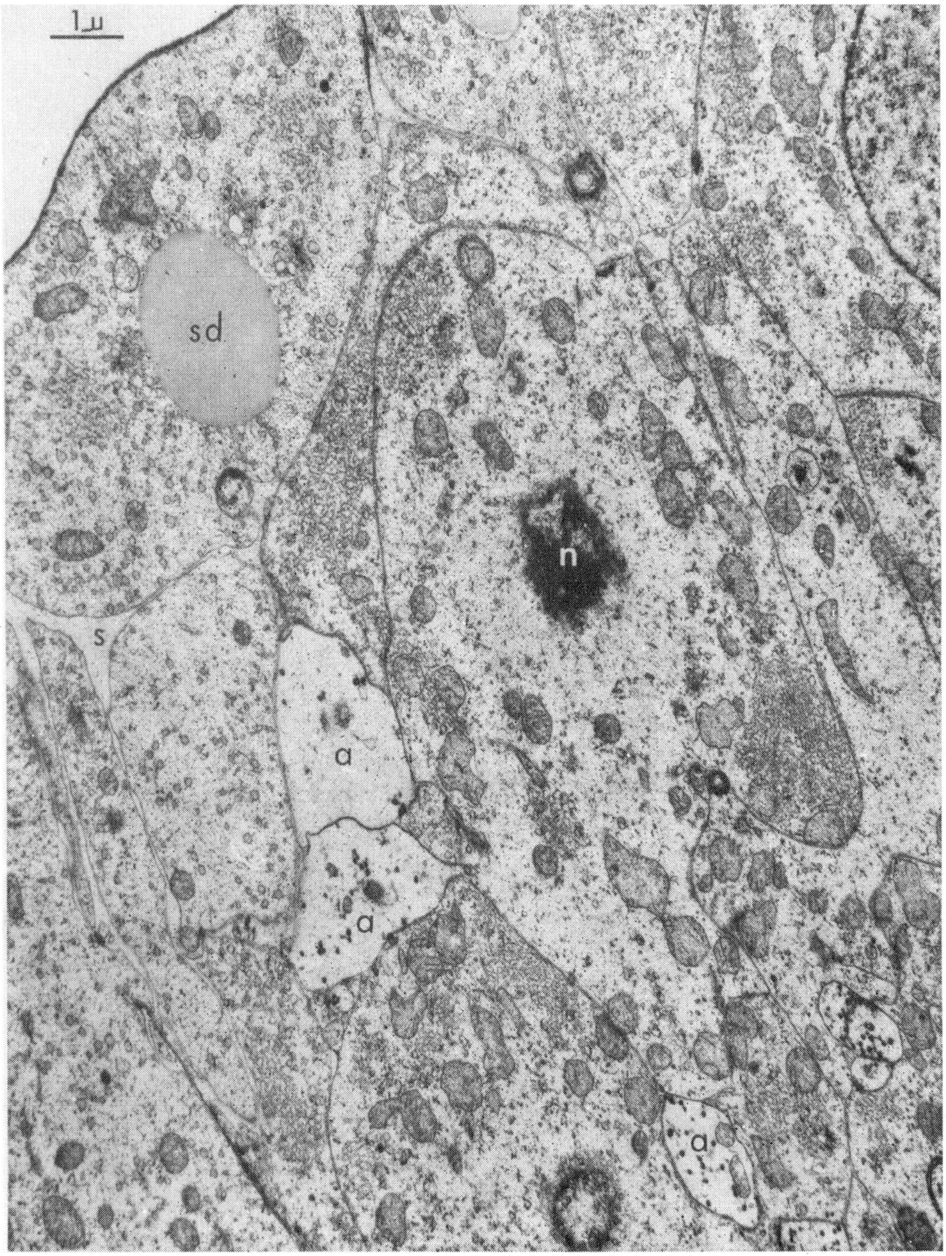

Plate 196. Fig. 3. Electron micrograph of a section through cells in the anterior dorsal portion of the ring gland of a 46 hour, second instar, female larva of *Drosophila,* a sectioned axons containing neurosecretory spheres; n surface of the nucleus of a corpus allatum cell; s intercellular fluid-filled space; sd secretory droplet in prothoracic gland cells (From King *et al.,* 1966b)

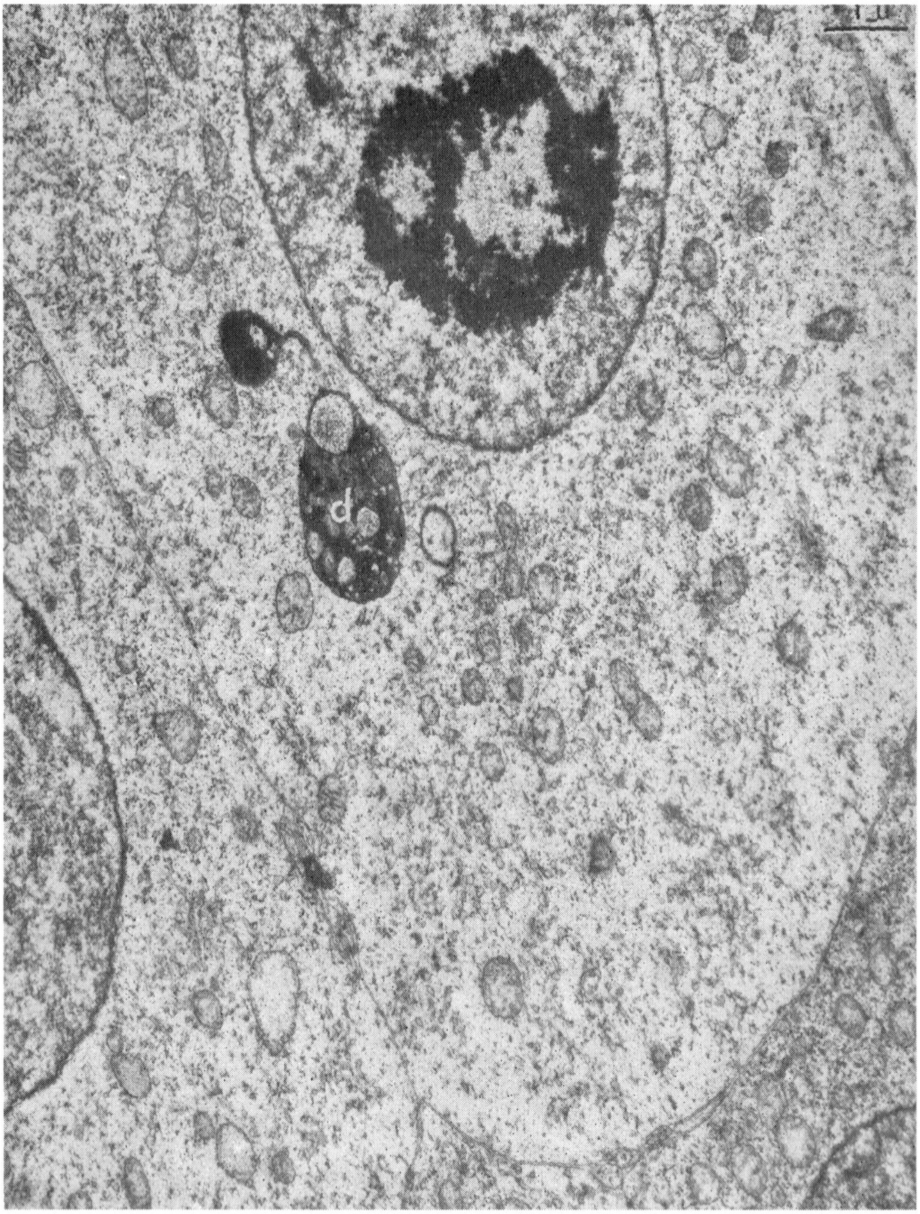

Plate 197. Fig. 4. Electron micrograph of an area in a section of the prothoracic gland of a 46 hour, second instar larva of *Drosophila*. Prothoracic gland cells are characterized by large nuclei with prominent nucleoli. Their cytoplasm contains scattered mitochondria; tiny, smooth-surfaced vesicles; and minute, clustered ribosomes. d digestive vacuole (From King *et al.*, 1966b).

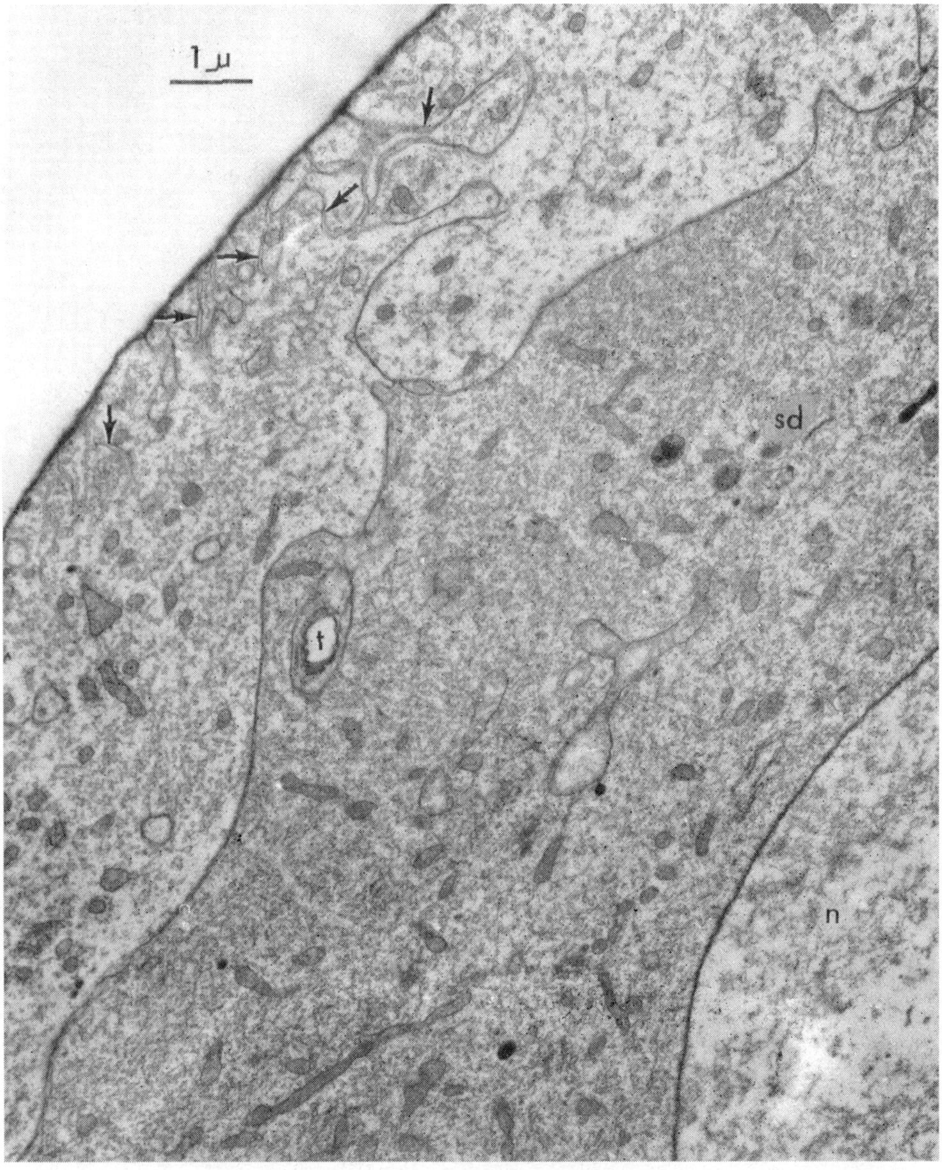

Plate 198. Fig. 5. Section of the prothoracic gland of a 97 hour, third instar larva of *Drosophila*. Contrast this figure with Fig. 4. The concentration of smooth-surfaced vesicles has greatly increased and most mitochondria are longer and thinner than in the 46 hour cell. Arrows point to tubular connections to the surface of the cell which are thought to facilitate the discharge of the secretion. n nucleus; sd secretory droplet; t tracheole (From King *et al.*, 1966b).

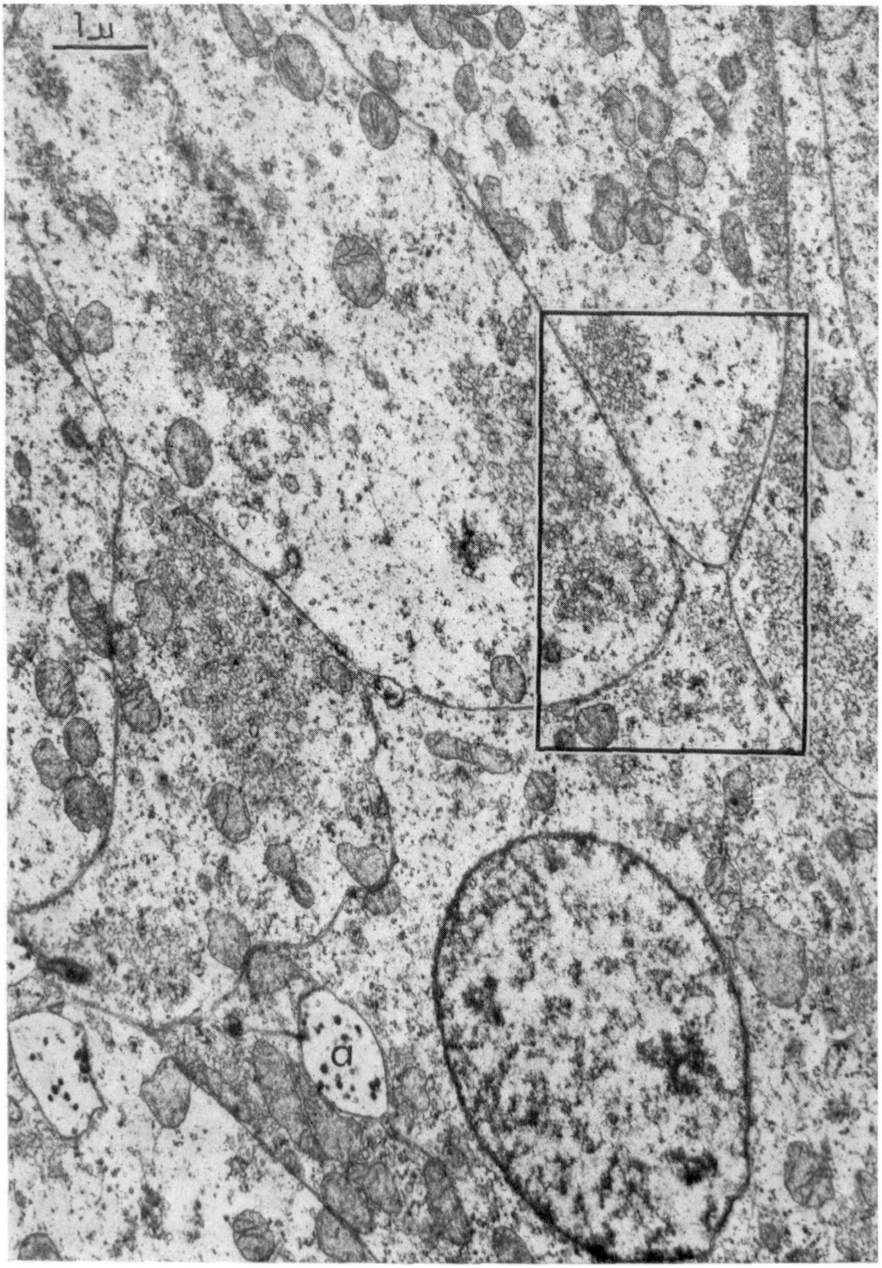

Plate 199. Fig. 6. Section of the corpus allatum of a 46 hour, second instar larva of *Drosophila*. Cells contain nests of smooth-surfaced vesicles, scattered mitochondria, and clusters of ribosomes. The area enclosed in the rectangle is shown at higher magnification in Fig. 8, a sectioned axon containing neurosecretory spheres (From King *et al.,* 1966b)

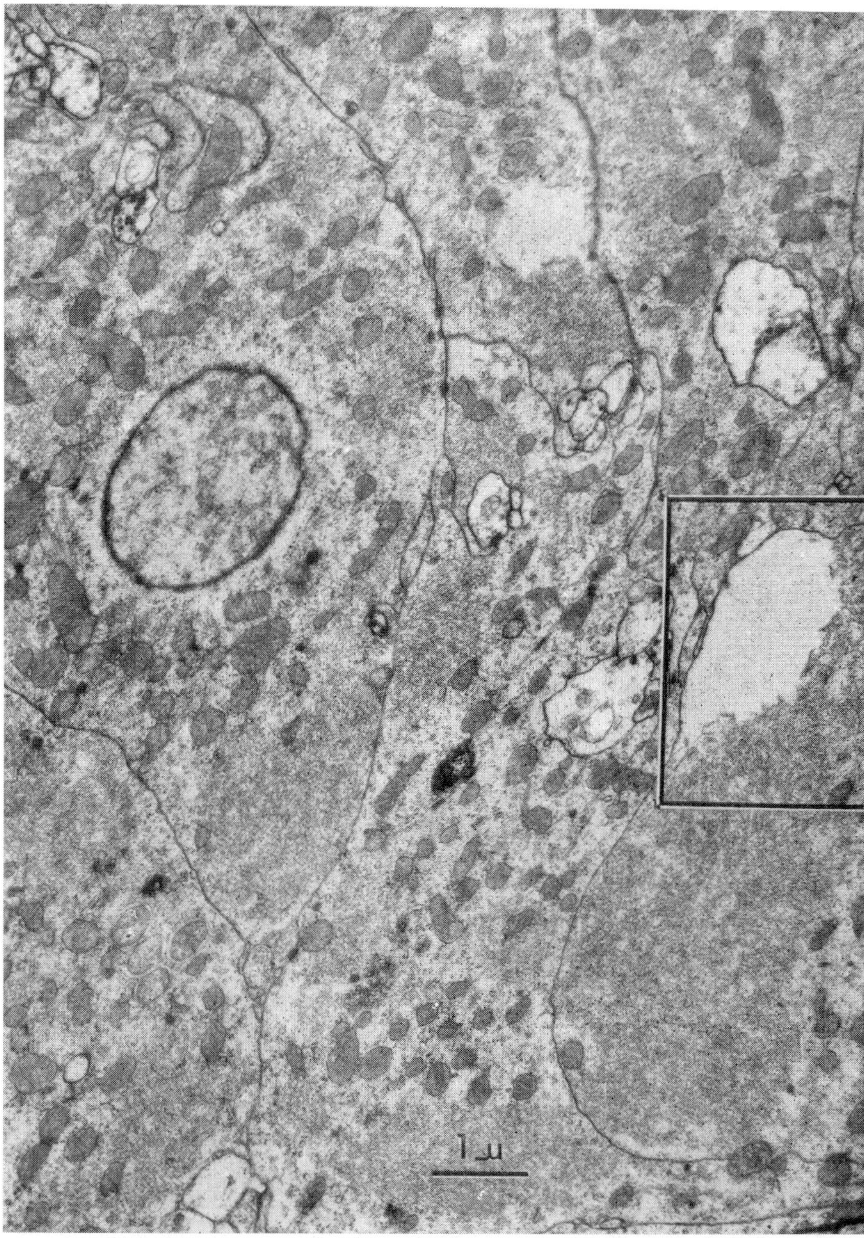

Plate 200. Fig. 7. Corpus allatum of a 97 hour, third instar larva of *Drosophila*. The concentration of smooth-surfaced membranous material has greatly increased. Mitochondria contain far more cristae than in the 46 hour cell. Neurosecretory spheres are seen in sectioned axons. The area enclosed in the rectangle is shown at higher magnification in Fig. 9. (From King *et al.*, 1966b).

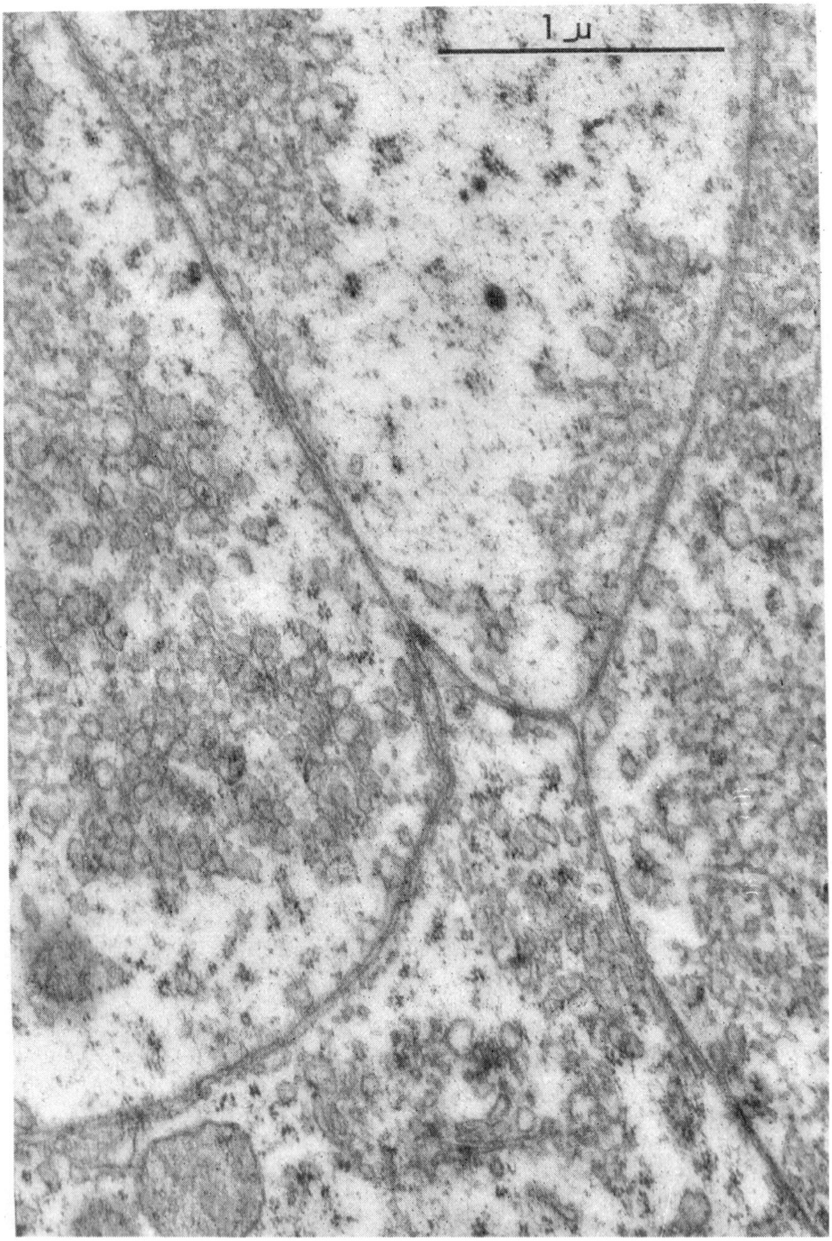

Plate 201. Fig. 8. The area enclosed in the rectangle in Fig. 6 shown at higher magnification. The cytoplasm contains a network of smooth-surface tubules (of diameter ~ 20 mμ) and vesicles (of diameter ~ 100 mμ) and clustered ribosomes. The apposed plasmalemmas of four adjacent cells produce an H-shaped configuration. The space between these membranes is about 10 mμ wide (From King *et al.* 1966b).

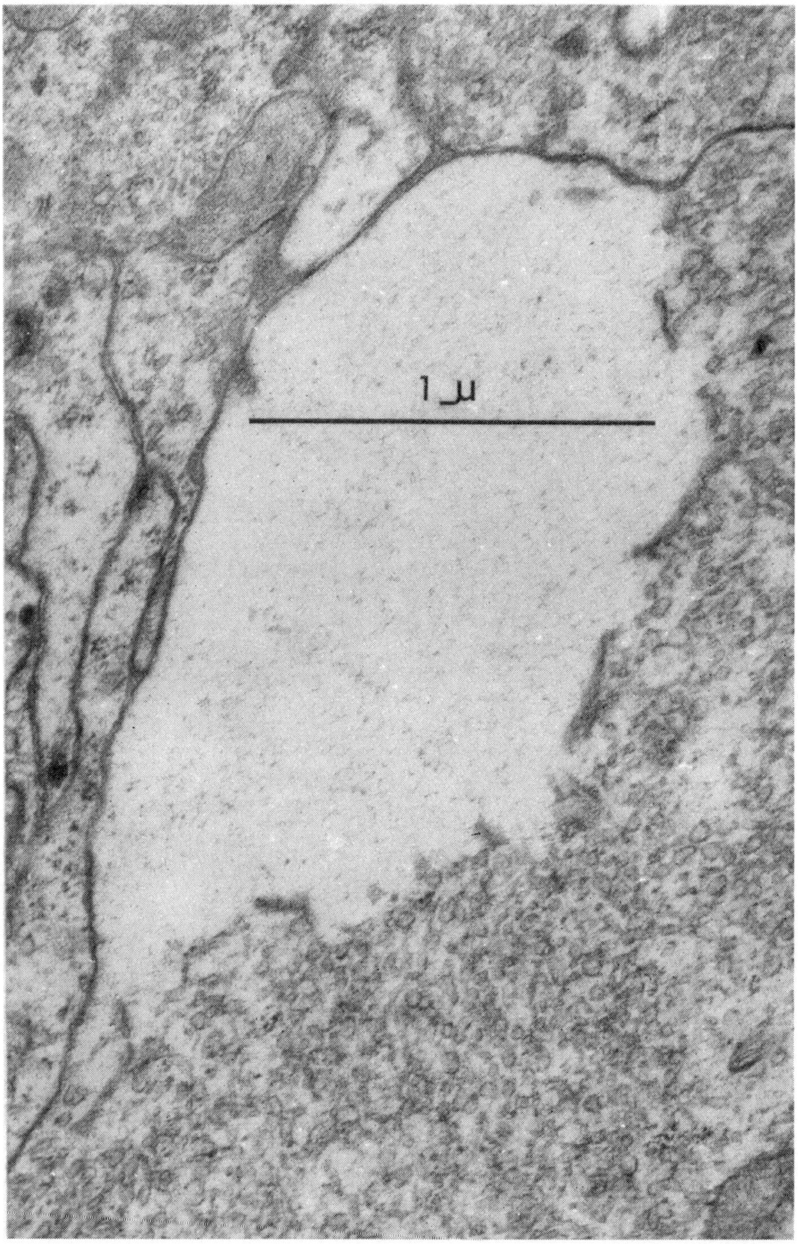

Plate 202. Fig. 9. The area enclosed in the rectangle in Fig. 7 shown at higher magnification. The membranous component consists of closely packed, smooth-surfaced spheres (each about 30 mμ in diameter). A pocket of secretory material seems to have formed against the cell membrane) (From King *et al.*, 1966b).

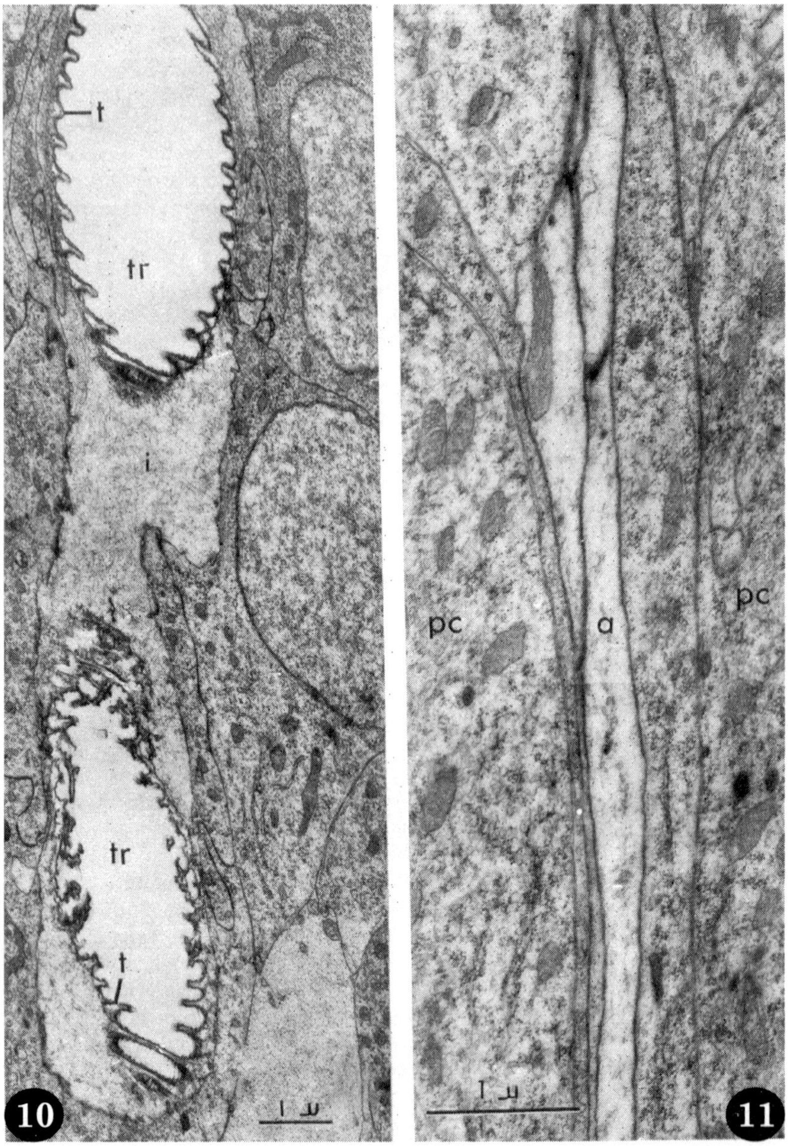

Plate 203. Fig. 10. Section through a lateral trachea of the ring gland of a third instar larva (see Fig 2). The nuclei of two adjacent tracheal cells are evident in the upper and centre right. The cytoplasm of such cells contains mitochondria, scattered sheets of rough-surfaced endoplasmic reticulum, and clustered ribosomes. The intima (i) secreted by these cells has a felt-like apperance. Its external surface is thrown into spiral ridges called taenidia (t). Adjacent taenidia are connected by cross bars which can be seen when sectioned tangentially. The section has passed into the tracheal lumen (tr) at two points. Fig. 11. Longitudinally sectioned paired axons (a) passing between prothoracic gland cells (pc) in a lateral limb of the ring gland of a third instar larva of *Drosophila* (see Fig 2, ax) (From King *et al.*, 1966b).

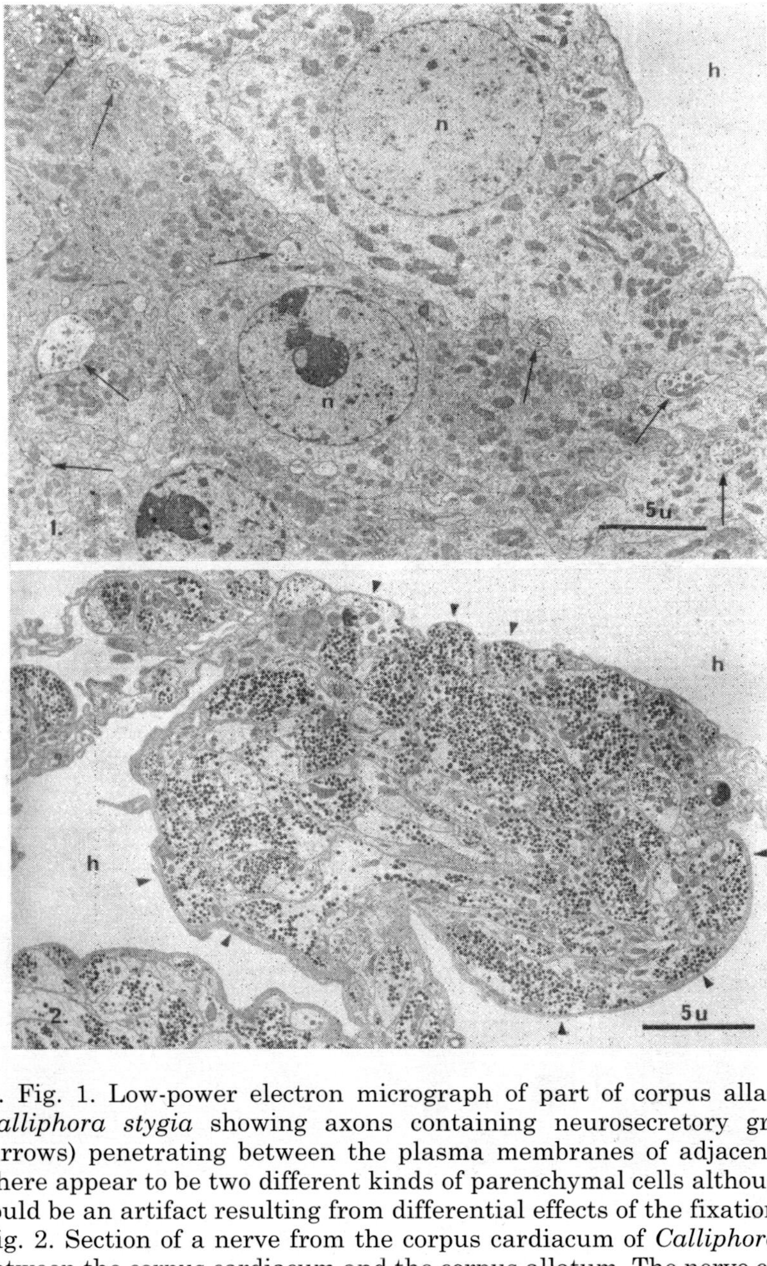

Plate 204. Fig. 1. Low-power electron micrograph of part of corpus allatum of *Calliphora stygia* showing axons containing neurosecretory granules (arrows) penetrating between the plasma membranes of adjacent cells. There appear to be two different kinds of parenchymal cells although this could be an artifact resulting from differential effects of the fixation fluid. Fig. 2. Section of a nerve from the corpus cardiacum of *Calliphora* lying between the corpus cardiacum and the corpus allatum. The nerve consists entirely of glial processes and numerous axons containing neurosecretory granules of different sizes. Some of the axons on the periphery of the nerve contain vesicle clusters from which it is presumed that secretory material is released through the basement membrane into the blood. Arrows indicate the positions at which vesicle clusters were found in this section and two adjacent sections. (From Johnson, 1966)

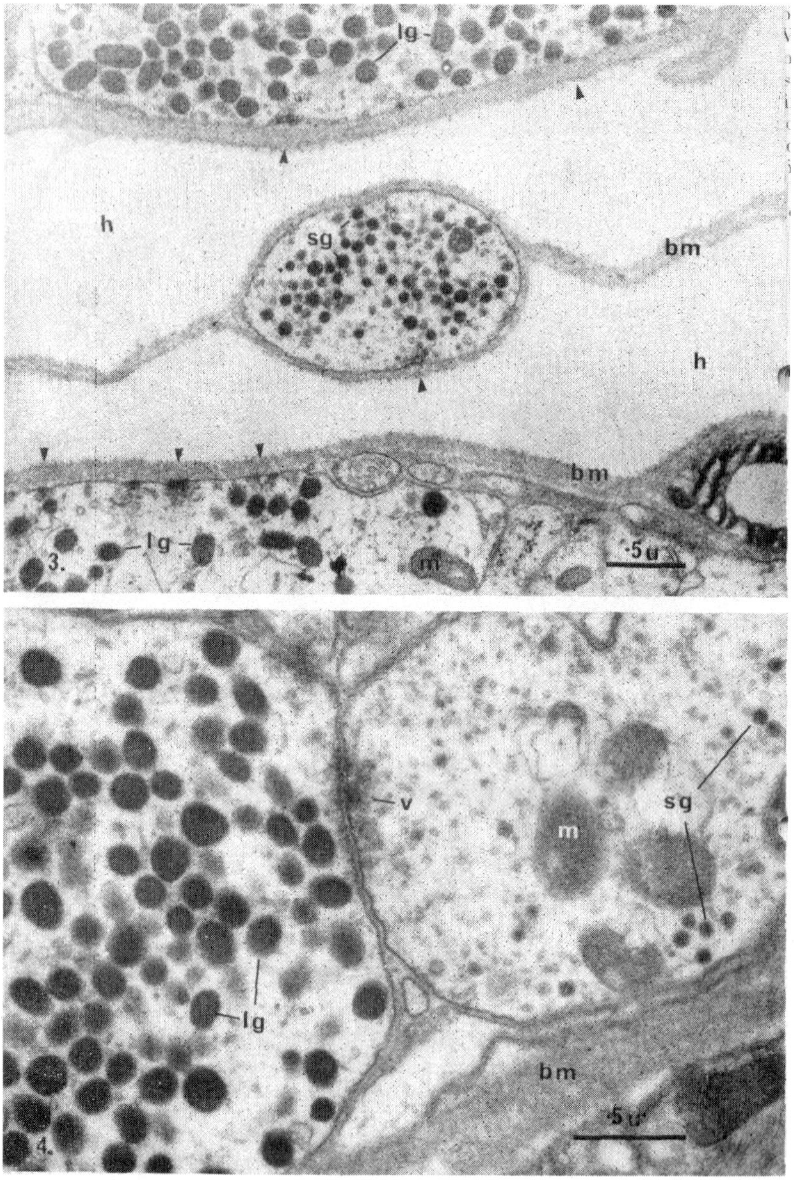

Plate 205. Fig. 3. Section through a folded area on the periphery of the corpus cardiacum of *Calliphora* showing axons containing large granules (lg) and a single axon containing small granules (sg). Vesicle clusters in both types of axon are indicated by the arrows. Fig. 4. Axons containing neurosecretory granuls in the corpus cardiacum of *Calliphora* a synaptic type of contact between an axon containing small granules and one containing large granules is shown. In the micrograph the axon profiles lie adjacent to the heart wall (lower right) but are separated from it by a double layer of basement membrane. (From Johnson, 1966).

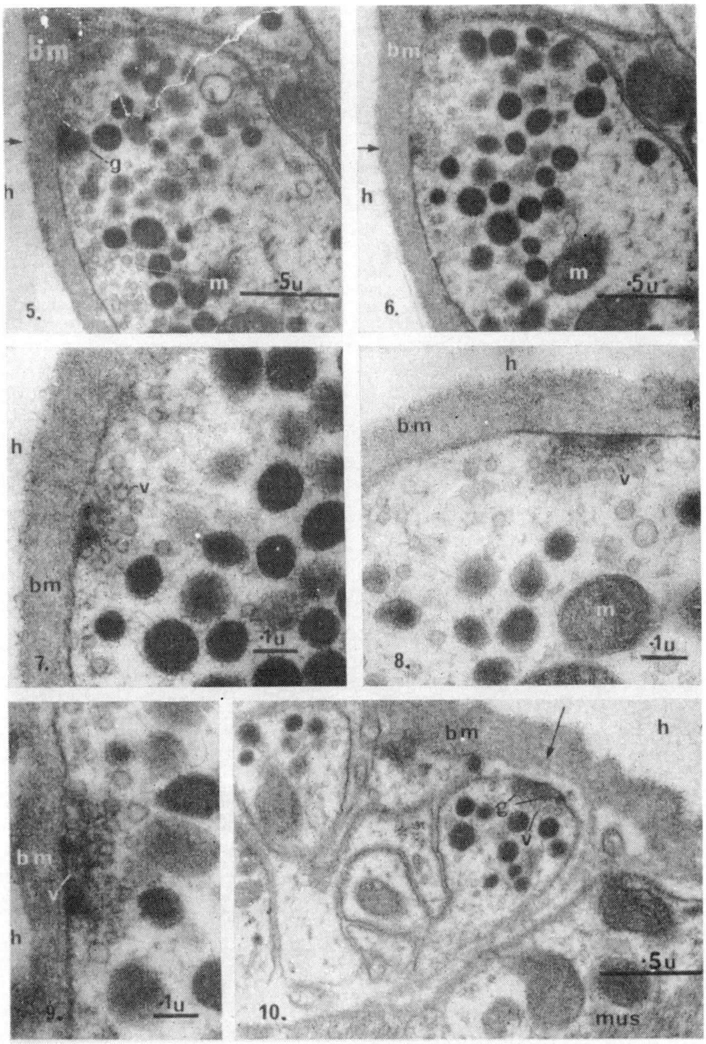

Plate 206. Fig. 5. Axon of *C. stygia* adjacent to the basement membrane (bm) and containing large neurosecretory granules and small vesicles. The vesicles are aggregated along the plasma membrane beneath the basement membrane. A large granule (g) is present just beneath the plasma membrane and has vesicles and dense areas on either side of it (arrow). Fig. 6. Adjacent section to Fig. 5 showing a cluster of small vesicles and two dense areas in the same position in the axon profile as the dense granule seen in Fig. 5 (arrow). Fig. 7. High magnification of vesicle cluster seen in Fig. 6. Fig. 8. Axon containing small neurosecretory granules with a vesicle cluster and dense material adjacent to the basement membrane. Fig. 9. Vesicle cluster immediately beneath the basement membrane in an axon containing large granules. Note the variation in size of the vesicles and the large vesicle (v) surrounded by dense material within the cluster. Fig. 10. Axon containing neurosecretory granules in the wall of the heart. The axon lies beneath the outer layer of basement membrane separating the heart from the haemocoel. A cluster of small vesicles (v) is present around two granules (g) which are appressed to the plasma membrane. The axon is adjacent to a muscle cell (mus). (From Johnson, 1966).

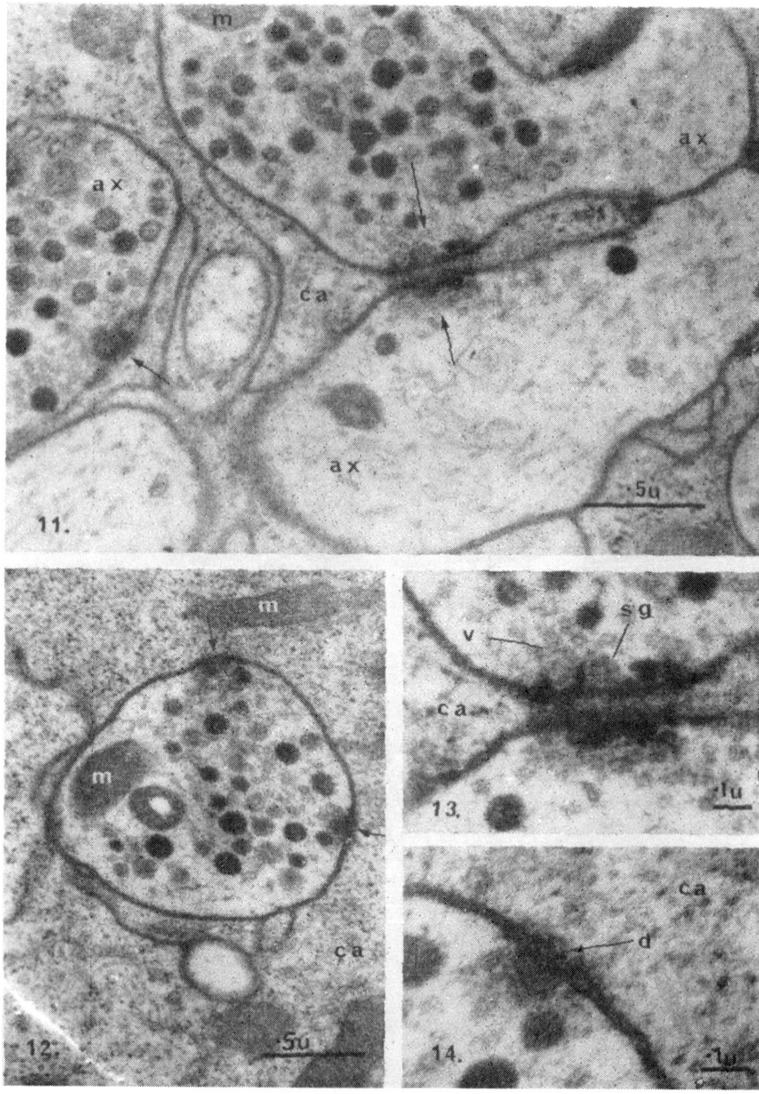

Plate 207. Fig. 11. Several axons (ax) lying within the corpus allatum of *Calliphora*. Three areas of probable release of secretory material are shown (arrows); two of them lie opposite one another and are separated by a narrow process from a corpus allatum cell (ca). Fig. 12. Neurosecretory cell axon lying between corpus allatum cells of *Calliphora* and showing two probable release sites (arrows). There is a dense layer of cytoplasm in the corpus allatum cell opposite each of the vesicle clusters. Note the proximity of neurosecretory granules to the vesicle clusters and the dense areas within them. Fig. 13. Enlargement of probable release sites in Fig. 11. Fig. 14. Enlargement of one of the probable release sites in Fig. 12 to show the dense layer in the corpus allatum cell adjacent to the vesicle cluster. (From Johnson, 1966).

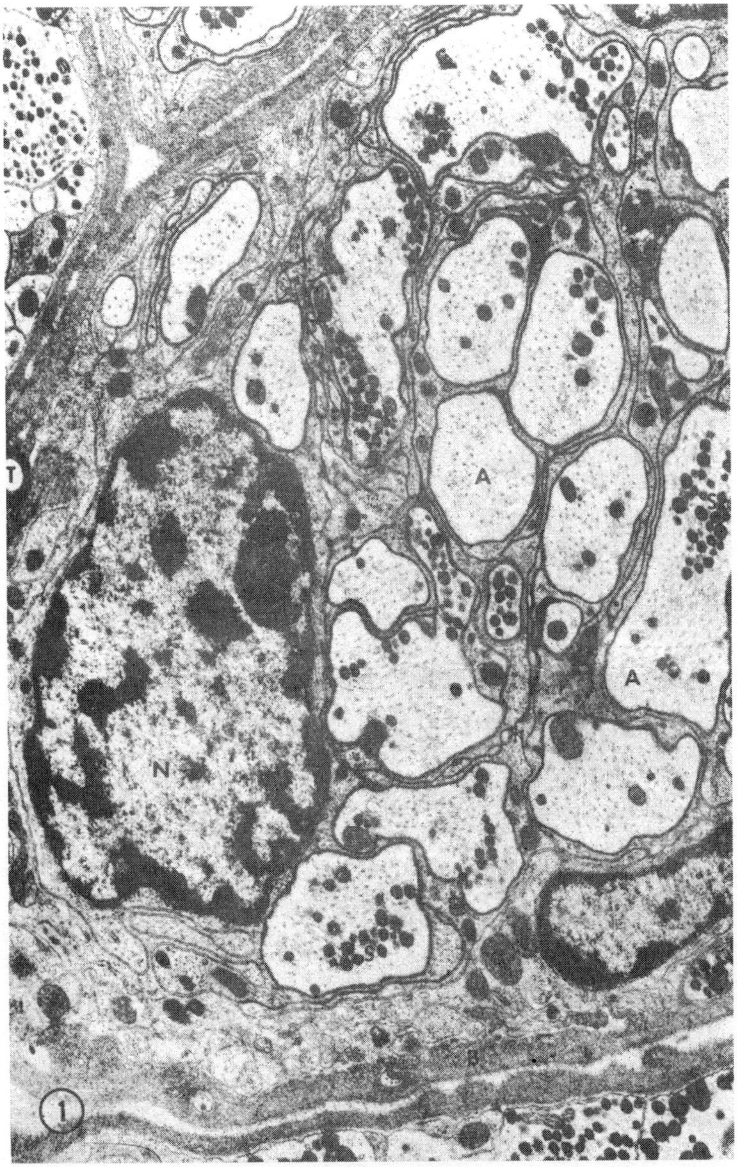

Plate 208. Fig. 1. Electron micrograph of a transverse section of the main nerve entering the corpus cardiacum of *Hypera postica*. The nerve trunk contains numerous axons (A) ensheathed in glial processes limited by irregular and branching mesaxons (*). Note the glial nuclei (N) and areas of glial cytoplasm; also the basement membrane (B) surrounding the trunk. The axons contain microtubules, and, in many instances, groups of electron-opaque neurosecretory droplets. (S). A transverse profile of a tracheole is included at (T). X 19000. (From Tombes and Smith, 1970).

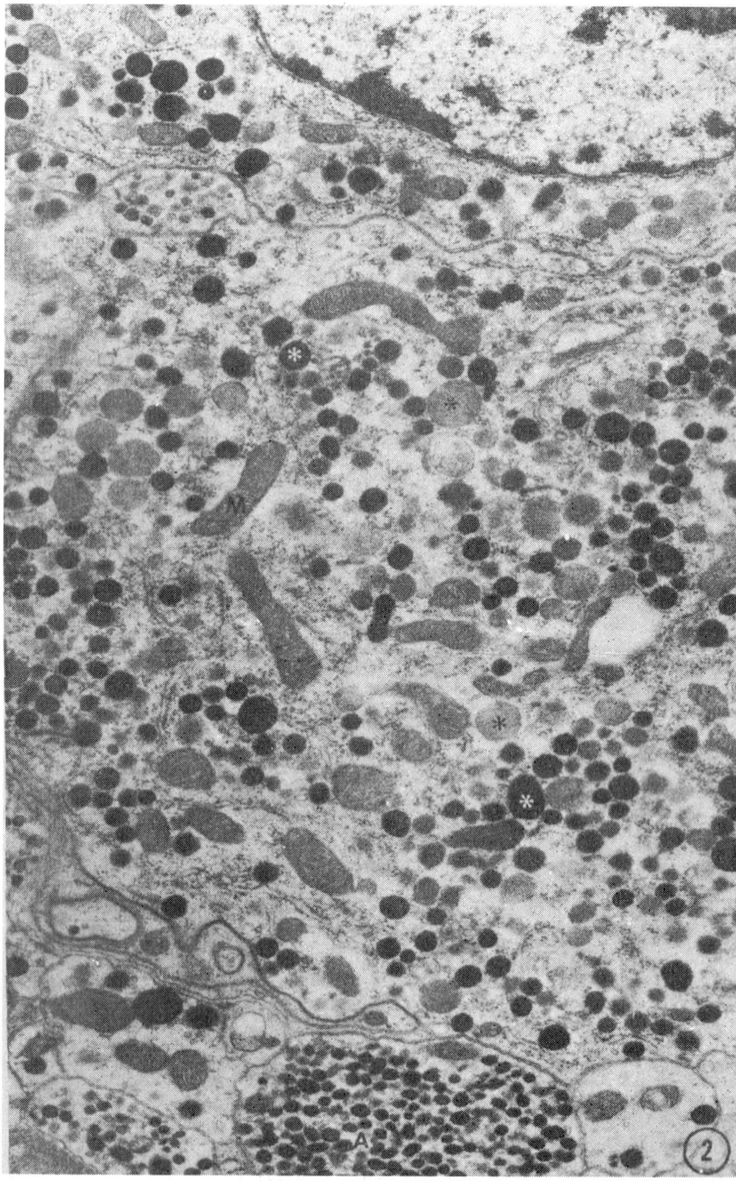

Plate 209. Fig. 2. Illustrating the organization of the intrinsic secretory cell of the corpus cardiacum of *Hypera postica*. The cytoplasm is extensive and contains elongated mitochondria (M), and secretion droplets of varying degrees of electron-opacity (black and white asterisks) which form in association with the cisternae of the dictyosomes or Golgi bodies. Profiles of several small axons, probably of extrinsic origin (A) are closely apposed to the surface of the intrinsic cells. These contain microtubules and neurosecretory droplets. X 27000 (From Tombes and Smith, 1970).

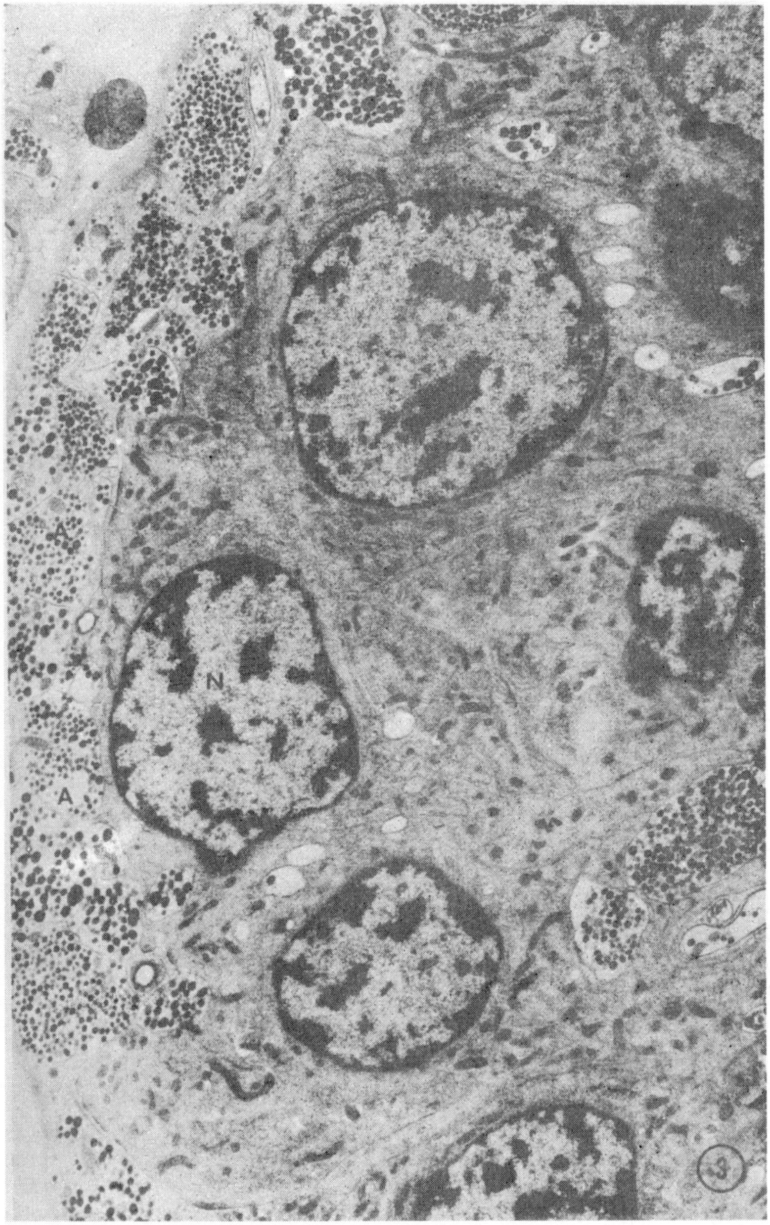

Plate 210. Fig. 3. Survey electron micrograph of a group of corpus allatum cells of *Hypera* : note the nuclei of these cells (N) and the surrounding cytoplasm. The cells of the c. allatum are, in this insect, closely associated with axon processes (A), probably originating in the brain. These axons, which are generally filled with secretory droplets form an almost complete investment over the surface of the c. allatum and also penetrate between the c. allatum cells (c.f. figs. 4.5). X 11000. (From Tombes and Smith, 1970).

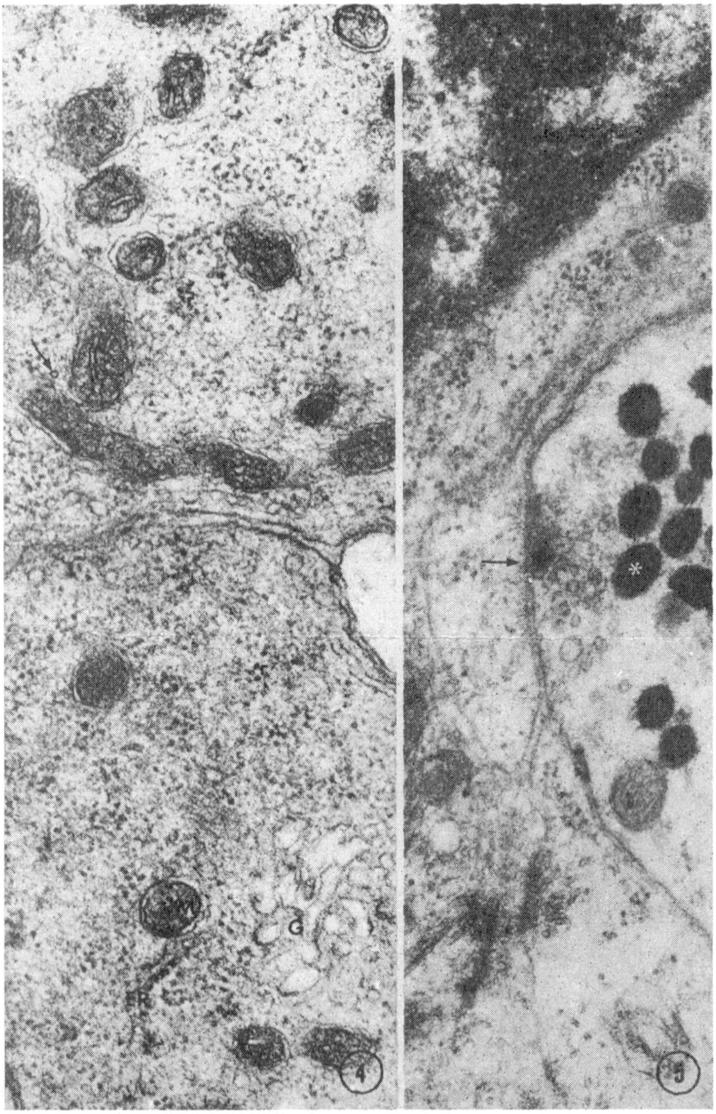

Plate 211. Fig. 4. Micrograph illustrating the general cytoplasmic features of the *Hypera* corpus allatum cell. Small mitochondria (M) are numerous, and these are accompanied by prominent Golgi bodies (G) and infrequent cisternae of the rough surfaced endoplasmic reticulum (ER). Un attached ribosomes (r) are abundant, and microtubules (arrows) are often more widespread than in this field. X 96000. Fig. 5. A field similar to the last, illustrating an axon in close association with the surface of a corpus allatum cell. The axoplasm contains electron opaque neurosecretory droplets (*), and a focal cluster of small synaptoid structures aggregated around a dense `core' (opposite arrow) a configuration possibly representing the site of neurosecretion release from the axon onto the surface of a c. allatum cell. X 70000. (From Tombes and Smith, 1970).

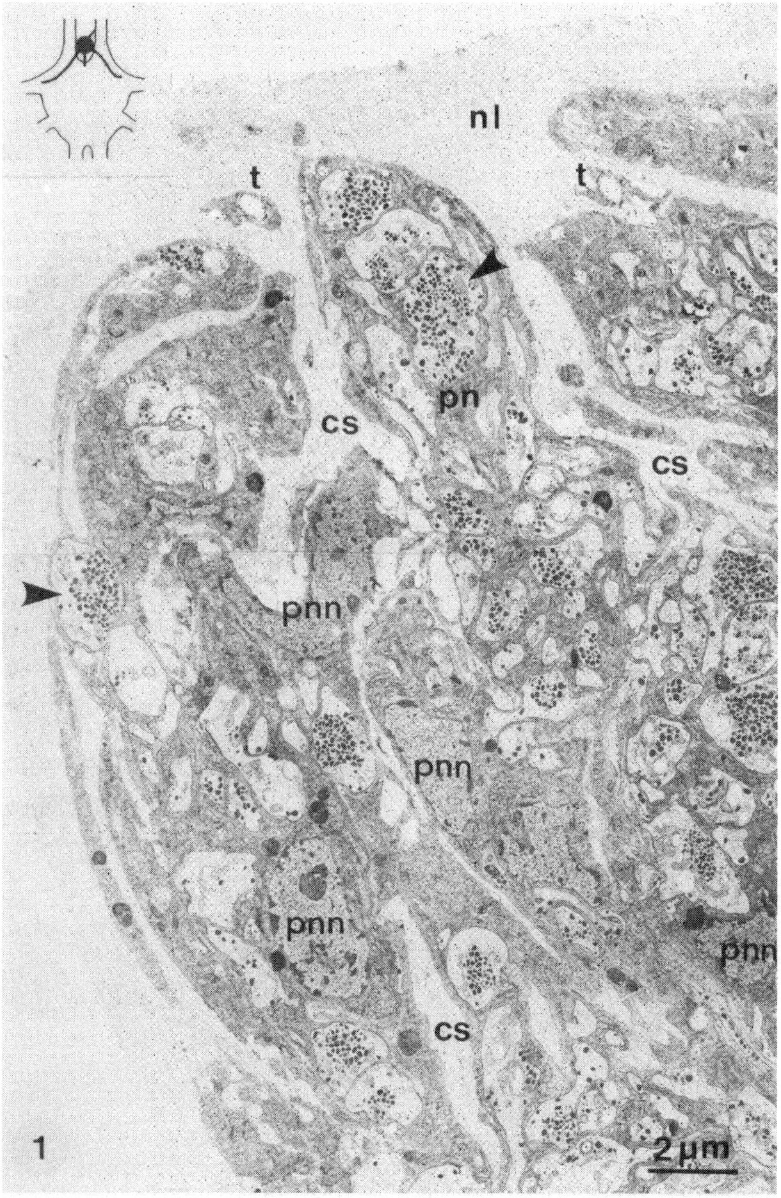

Plate 212. Fig. 1. Peripheral area of the intermediate external median PO in *Chrysocarabus auronitens*. Deep connective stroma (cs) invaginations penetrate from the neural lamella (nl) into this part of the organ. Many electron dense perineurial cells (pn) ensheath all the neurosecretory endings (arrowheads) which are filled with a single type of granule. Pnn, perineurial cell nucleus; t, tracheole. Inset : drawing of an abodminal ganglion and its associated PO. X 7000. (From Provansal-Baudez and Baudry-Partiaoglou, 1983).

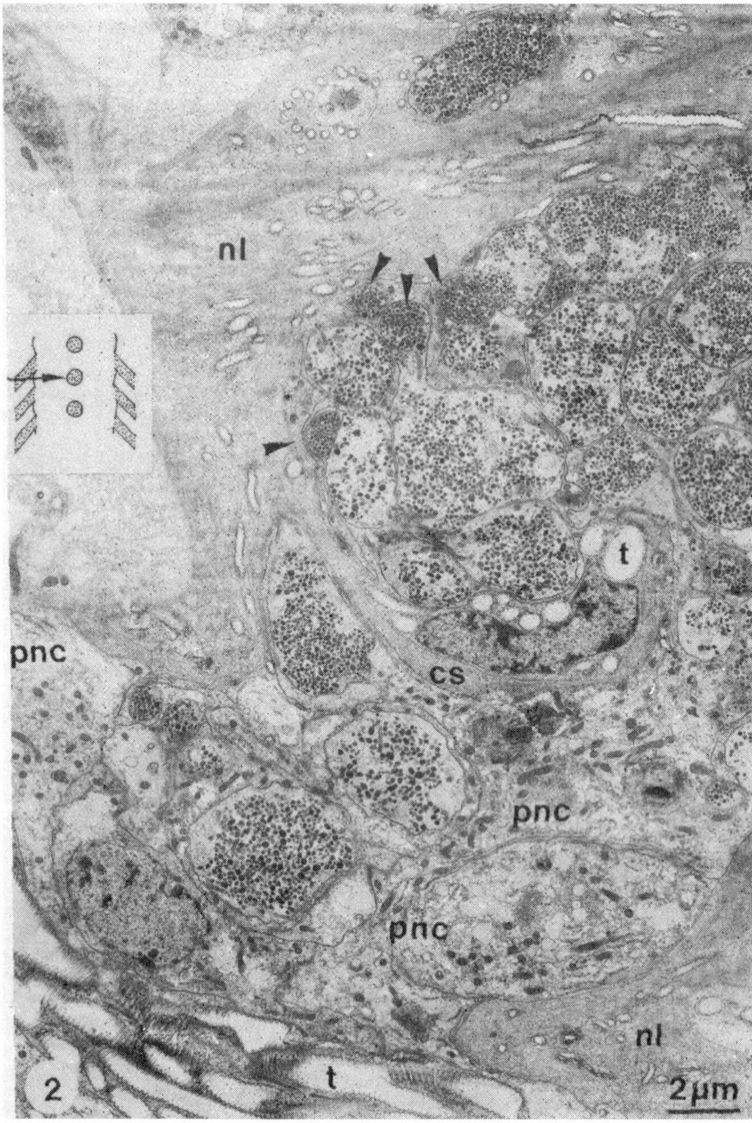

Plate 213. Fig. 2. Transverse semisection through an advanced interal median PO of *Oryctes rhinoceros*. The organ lies in a depression on the dorsal side of the abdominal neuromeres (see inset). Perineurial cells have a large cytoplasmic area (pnc) at the periphery of the organ and send out narrow processes around all the neurosecretory axons and their endings. The latter are recognizable by their heavy load of neurosecretory granules (arrowhead). As in *C. auronitens*, there is only one type of neurosecretory axon. Note the abundance of tracheoles (t) in the neural lamella (nl) which entirely surrounds the PO and sends out a few connective stroma invaginations (cs). X 5600. (From Provansal-Baudez and Baudry-Partiaoglou, 1983).

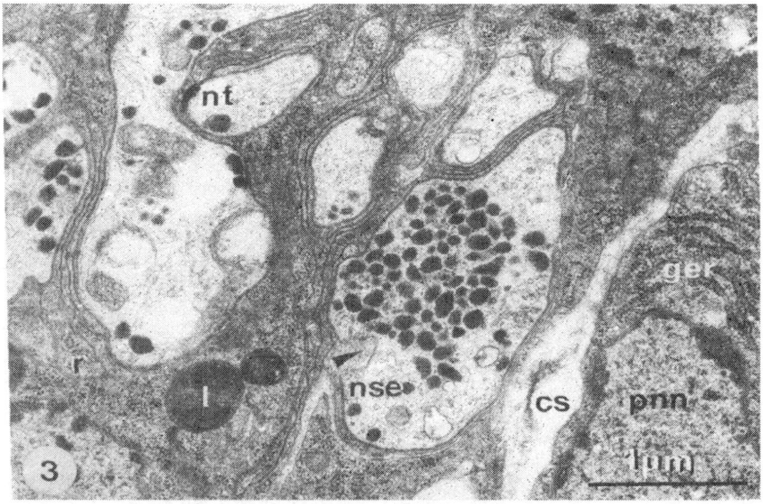

Plate 214. Fig. 3. Median PO of *C. auronitens*. High magnification of a central area illustrated in Fig. 1, showing details of perineurial cytoplasm which contains numerous free ribosomes (r), granular endoplasm reticulum (ger), and lysosomes (1). A neurosecretory ending (nse) ensheathed by perineurial cells had limited access to intercellular spaces (arrowhead). cs. connective stroma; pnn, perineurial cell nucleus; nt, neurotubule. X 25,000. (From Provansal-Baudez and Baudry - Partiaoglou, 1983).

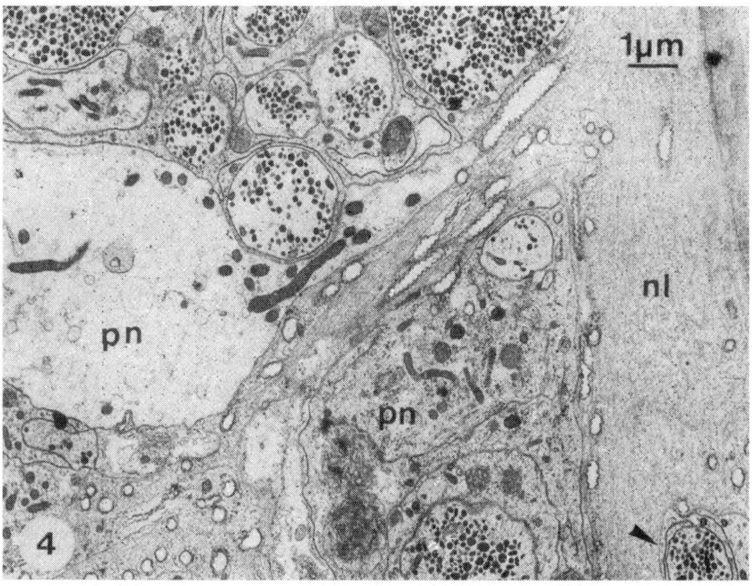

Plate 215. Fig. 4. Median PO of *O. rhinoceros*. Peripheral region of the organ showing different aspects of the same type of perineurial cell (pn). Even the small neurosecretory endings (arrowhead) isolated in the neural lamella (nl) have a glial sheath. X 8000. (From Provansal-Baudez and Baudry-Partiaoglou, 1983).

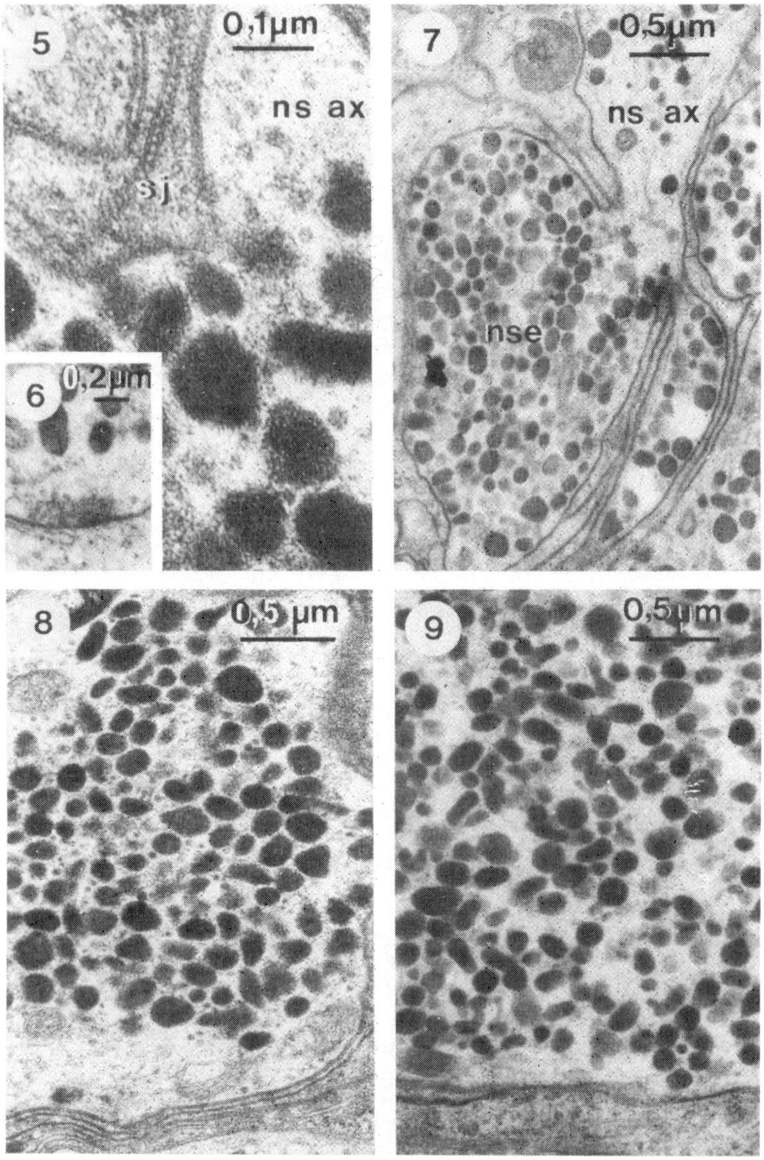

Plate 216. Fig. 5. Septate junction (sj) between glial processes ensheathing a neurosecretory axon (ns ax) in the median PO of *C. auronitens*. X 133,000. Fig. 6. Synaptoid site facing a glial cell in the median PO of *O. rhinoceros*. X 30,000. Fig. 7. Neurosecretory axon (ns ax) sprouting a neurosecretory ending (nse) in the median PO of *O. rhinoceros*. Note the difference in the amount of neurosecretory granules in the axon and its ending. X 25,000. Figs. 8 and 9. Comparison of the types of neurosecretory axon in the median PO of *C. auronitens* (Fig. 8) and *O. rhinoceros* (Fig. 9). In both species the largest neurosecretory granules measure 160 X 250 nm. X 30,000. (From Provansal – Baudez and Baudry-Partiaoglou, 1983).

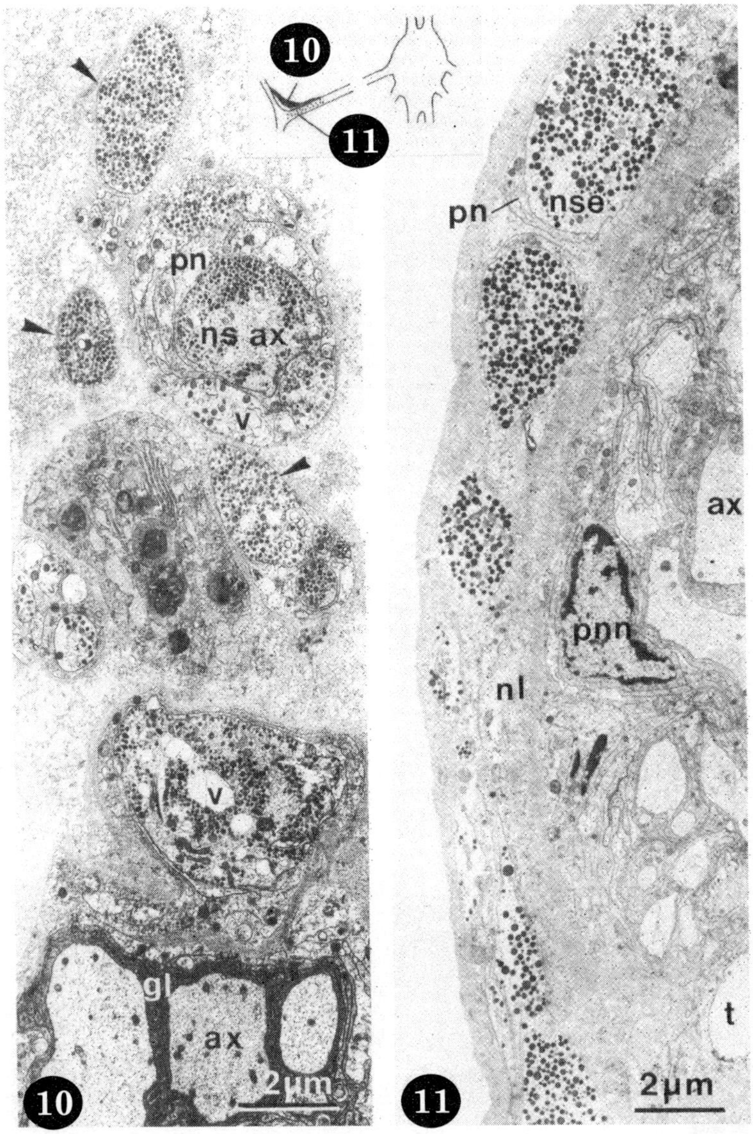

Plate 217. Figs. 10 and 11. Advanced internal transverse PO of *Tenebrio molitor*. The organ, located along segmental nerves, is composed of two regions (inset) : a neurohaemal strip (Fig. 10), and a neurohaemal area at the periphery of the segmental nerve (Fig. 11). Fig. 10. The neurohaemal strip forms a very loose tissue in which neurosecretory axons (ns ax) are ensheathed by perineurial cells (pn), but the scattered axon endings (arrowhead) have no perineurial sheath. Note also the numerous vacuoles (v) in these axons. X 8000. Fig. 11. The peripheral neurohaemal area lies in the neural lamella (nl), but makes smaller swellings than in *O. rhicoceros*. ax, axon; pn, perineurial cell process; pnn, perineurial cell nucleus; nse, neurosecretory ending; t, tracheole. X 7000. (From Provansal-Baudez and Baudry-Partiaoglou, 1983).

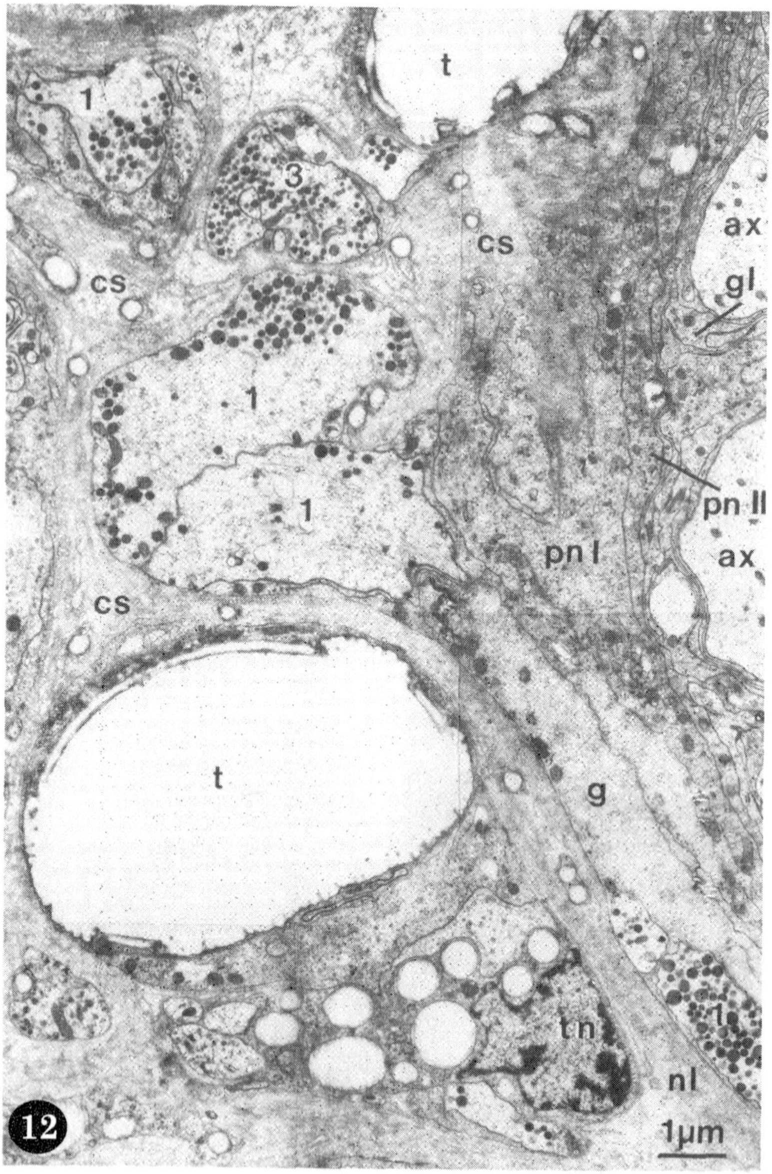

Plate 218. Fig. 12. The advanced transverse PO in *O. rhinoceros*. This nerohaemal swelling is located between two segmental nerves close to their emergence from the neuromere. Two types of neurosecretory axon can be identified (1 and 3), they lie in a loose connective stroma (cs) and are almost entirely sheathless. Type I perineurial cells (pn I) contain large deposits of glycogen rosettes (g). gl, glial cells surrounding axons (ax); nl, neural lamella; pn II, type II perineurial cell; t, trachea; tn, tracheal cell nucleus. X 10,400 (From Provansal – Baudez and Baudry-Partiaoglou, 1983).

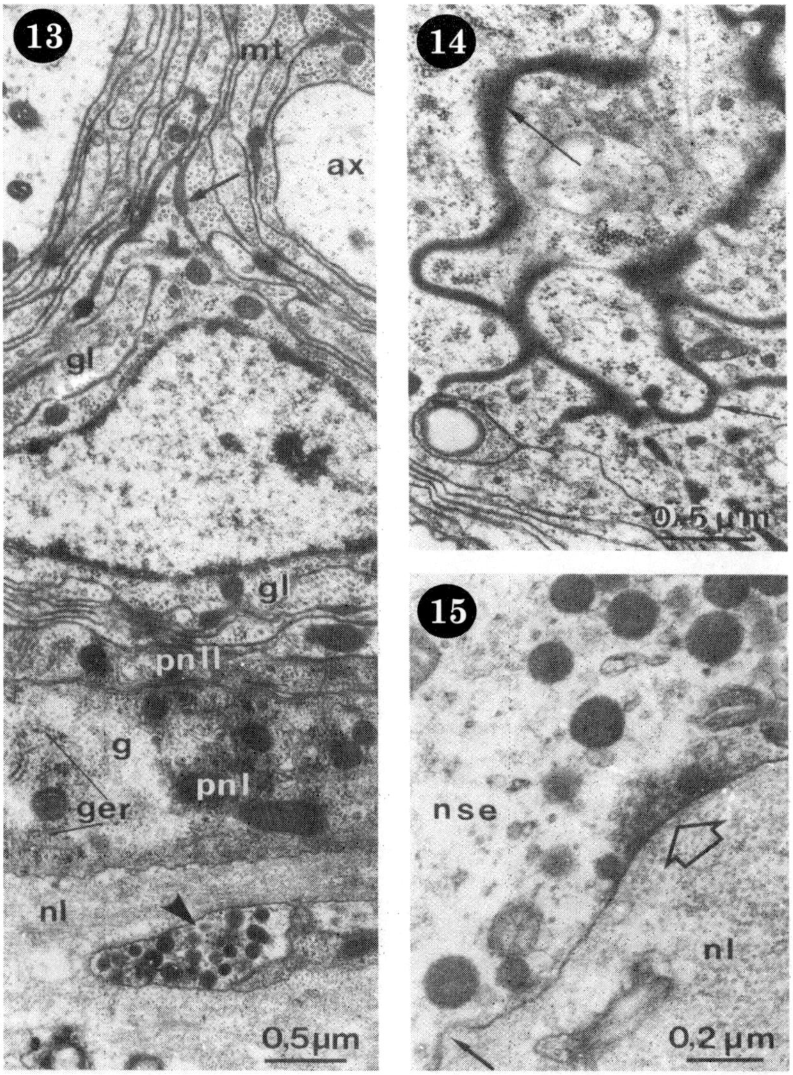

Plate 219. Figs. 13-15. Fig. 13. Peripheral portion of a segmental nerve in *O. rhinoceros* showing the acellular neural lamella (nl) containing a naked neurosecretory ending (arrowhead), two underlying types of perinuerial cells (pn I and pn II), and glial-ensheathed axons (ax). Dense material lies in the intercellular spaces between adjacent glial cells (arrow). g, glycogen; ger, granular endoplasmic reticulum; gl, glial cell; mt, orientated bundles of microtubules. X 25,000. Fig. 14. Part of a glial cell of an abdominal neuromere in *O. rhinoceros* showing many interdigitations forming enlarged intercellular spaces always filled with dense material (arrows). These interdigitations are more pleated and the cytoplasmic area is larger than in the segmental nerve of the same species. X 30,000. Fig. 15. Neurosecretory ending (nse) in the transverse PO of *O. rhinoceros*. Synaptoid site (white arrow) and omega shaped profile (black arrow) in a neurosecretory ending (nse) in direct contact with the neural lamella (nl). X 60,000. (From Provansal-Baudez and Baudry-Partiaoglou, 1983).

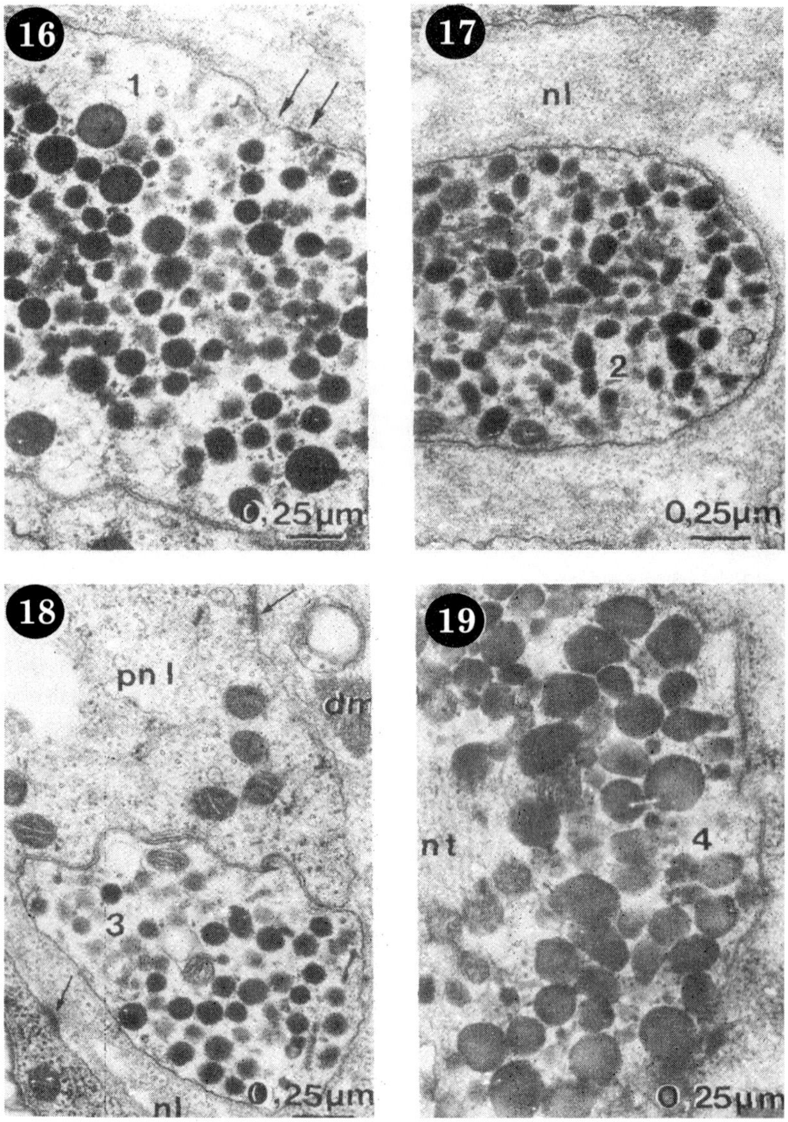

Plate 220. Figs. 16-19. These electron micrographs show different types of neurosecretory axons in the transverse PO of *O. rhinoceros*. X 37,000. Fig. 16. Spherical dense elementary granules of a type 1 axon (1). Arrows indicate probable exocytosis. Fig. 17. Ovoid or irregularly shaped dense elementary granules in a type 2 axon (2) of an isolated neurosecretory ending in the neural lamella (nl). Fig. 18. Spherical dense elementary granules of a type 3 axon (3). The neurosecretory axon is in direct contact with the neural lamella (nl) along most of its surface. The neural lamella (nl) frequently contains patches of dense material (dm). In Type I perinuerial cells (pn I) hemidesmosomes often occur facing the neural lamella (arrows). Fig. 19. Irregular gray elementary granules of a type 4 axon (4) whose neurotubules (nt) are visible. (From Provansal-Baudez and Baudry-Partiaoglou, 1983).

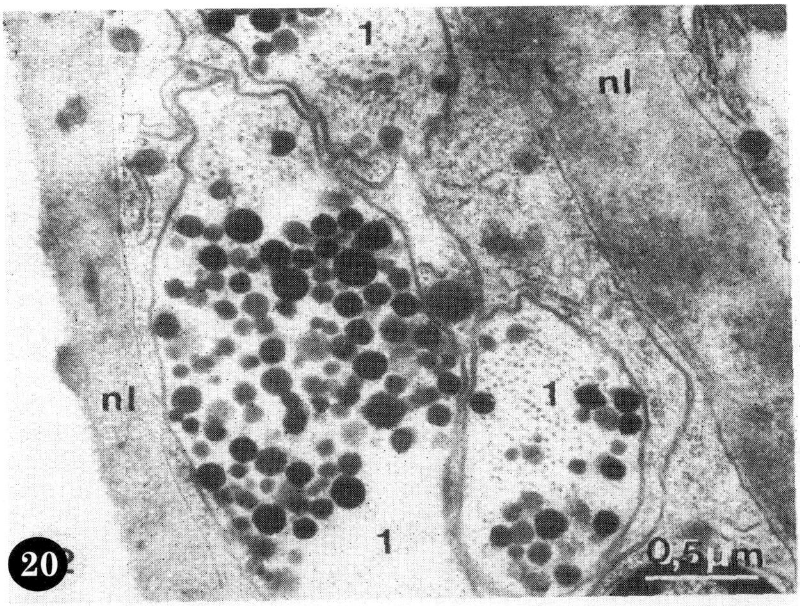

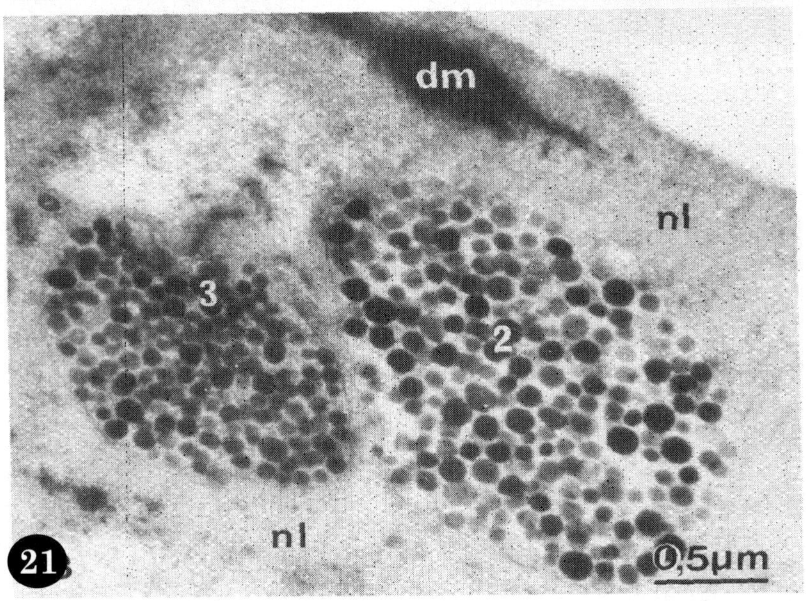

Plate 221. Figs. 20 and 21. These electron micrographs show different types of neurosecretory axons in the transverse PO of *T. molitor*. All the elementary granules are dense and shperical, and only differ in their size. X 30,000. Fig. 20. The neurohaemal area at the periphery of the segmental nerve forms a muff in the neural lamella (nl). (1) neurosecretory type 1 axons. Fig. 21. Neurosecretory endings of type 2 and 3 axons (2) (3) in the neurohaemal strip. As in *O. rhinoceros,* patches of dense material (dm) are visible in the neural lamella (nl). (From Provansal-Baudez and Baudry-Partiaoglou, 1983).

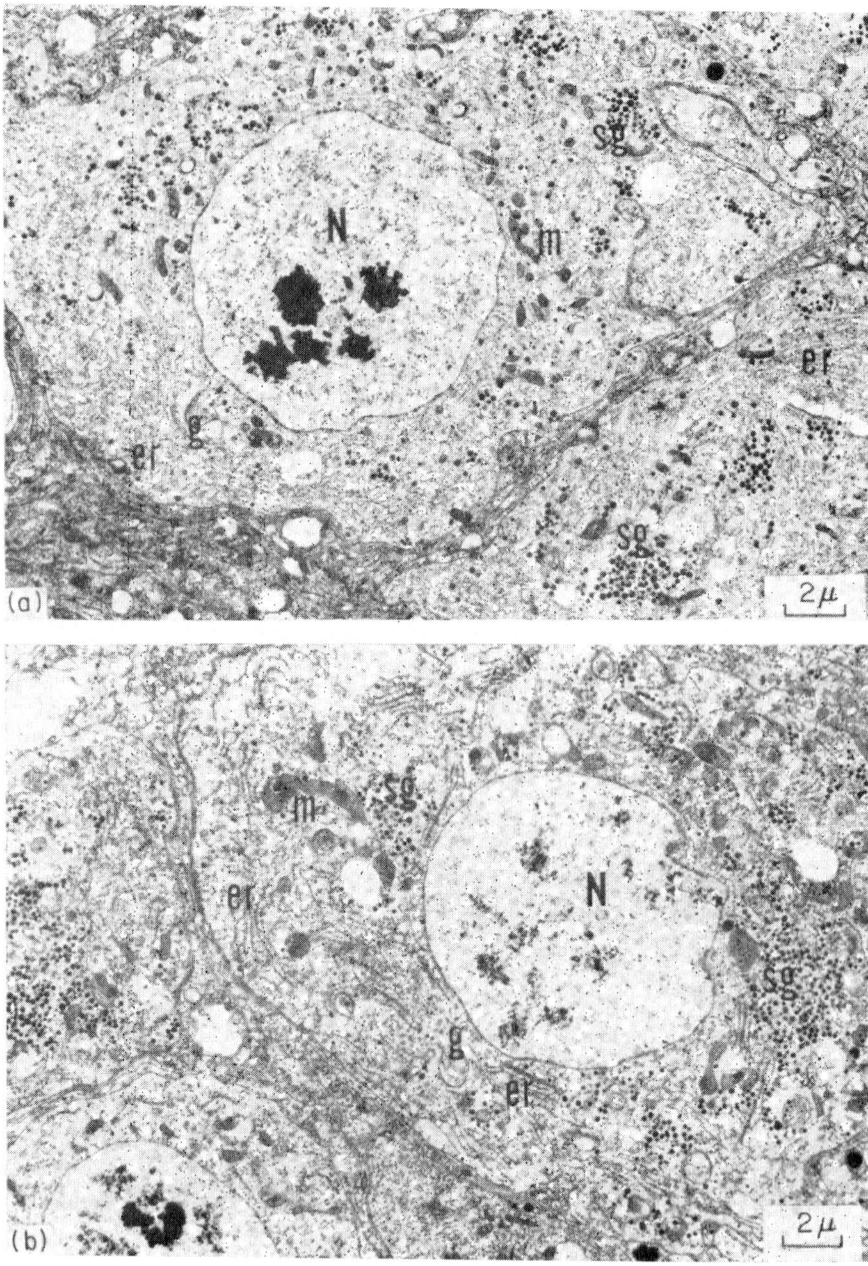

Plate 222. Fig. 1. Electron micrographs of NS-II cells 24-22 hr before the larval-pupal ecdysis of *Pieris rapae crucivora*, showing the differences in the appearance of the er and other cell structures between non-diapause (a) and diapause individuals (b). N, nucleus; er, endoplasmic reticulum; g, Golgi body; m, mitochondria; sg, secretory granule. (From Kono, 1973).

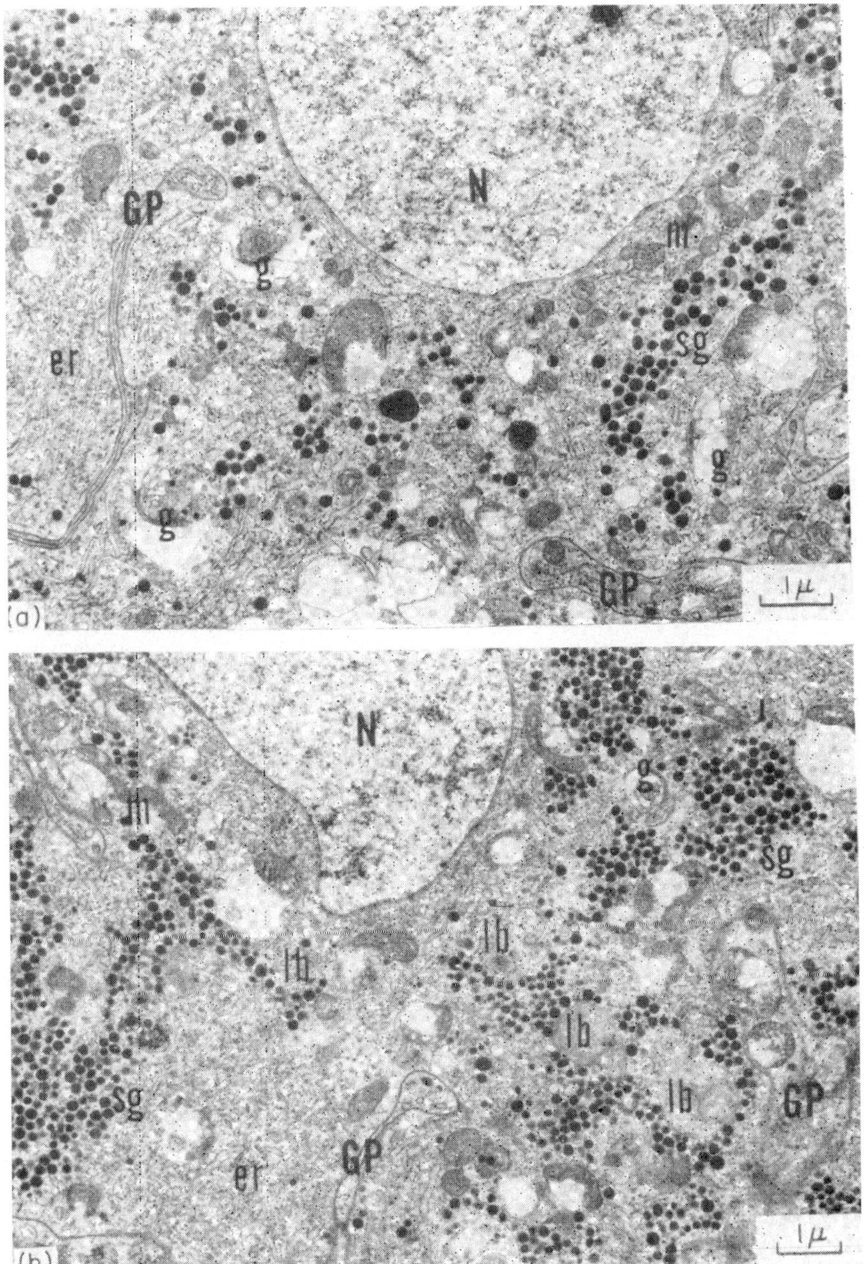

Plate 223. Fig. 2. Comparison of NS II cell appearances between non-diapause (a), and diapause individuals (b) 8-6 hr before larval-pupal ecdysis (*Pieris*). GP, Glial process; N, nucleus; er, endoplasmic reticulum; g, Golgi body; lb, lamellar body; m, mitochondria; sg, secretory granule. (From Kono, 1973).

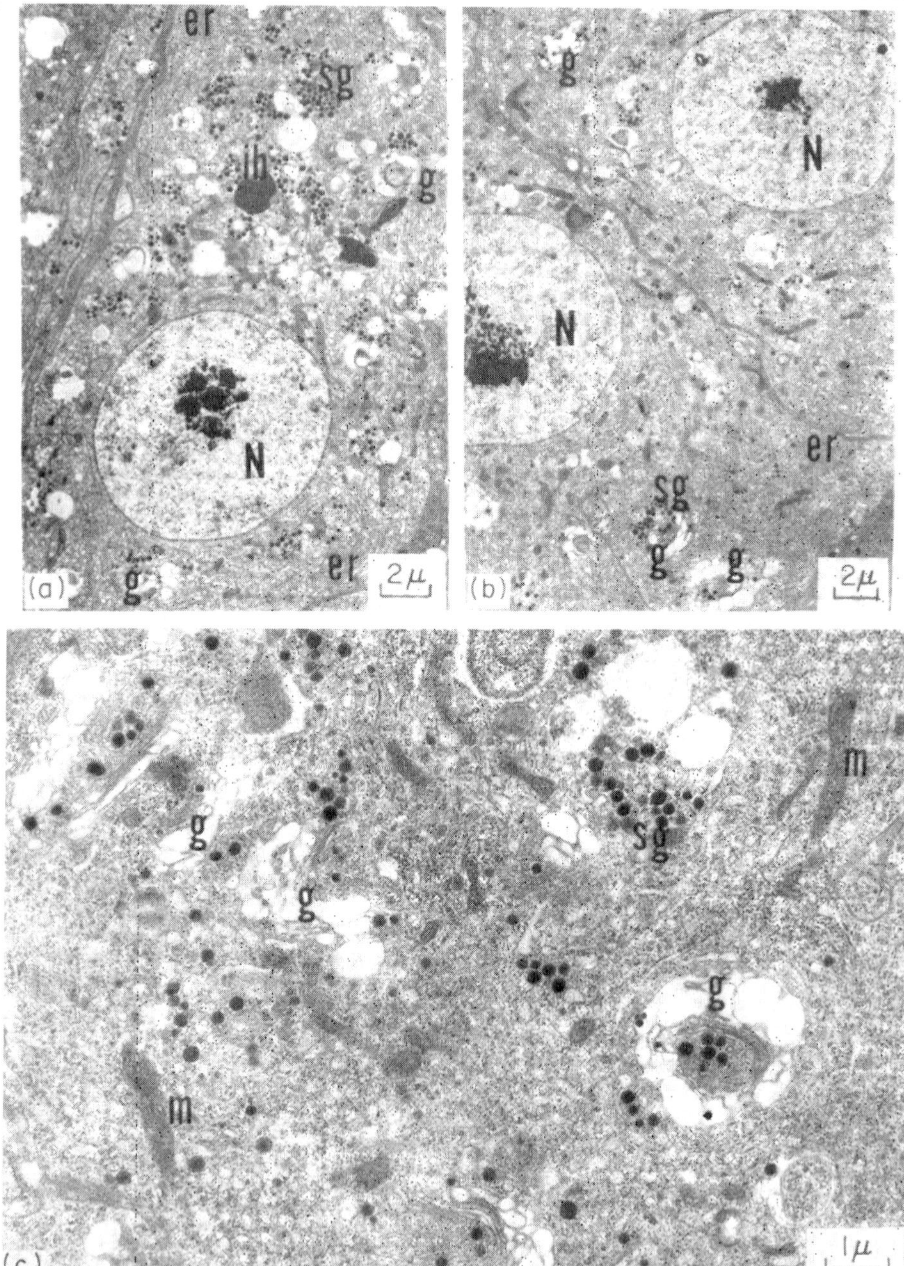

Plate 224. Fig. 3. Electron micrographs of NS-II cells of non-diapause individuals 0-4 hr after larval-pupal ecdysis (a) of *Pieris* and 8-12 hr after larval-pupal ecdysis (b,c). N, nucleus; er, endoplasmic reticulum; g, Golgi body; ib, inclusion body; m, mitochondria; sg, secretory granule. (From Kono, 1973).

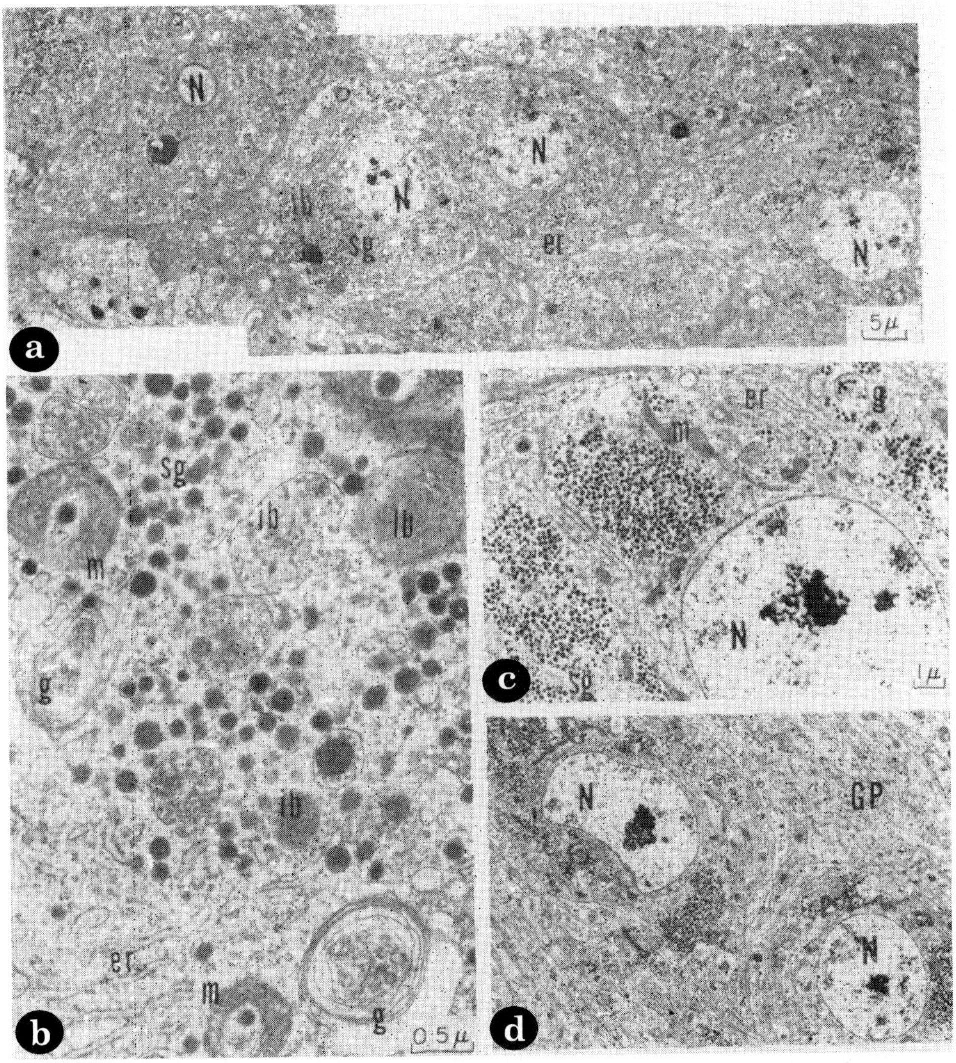

Plate 225. Fig. 4. Electron micrographs of NS-II cells of diapause individuals 0-4 hr after the larval-pupal ecdysis (a, c) of *Pieris* and 8-12 hr after the larval pupal ecdysis (b,d). NS-II cells of these ages include large aggregates of neurosecretory granules. GP, Glial process; N, nucleus; db, dense body; er, endoplasmic reticulum; g, Golgi body; ib, inclusion body; lb, lamellar body; m, mitochondria; sg, secretory granule. (From Kono, 1973).

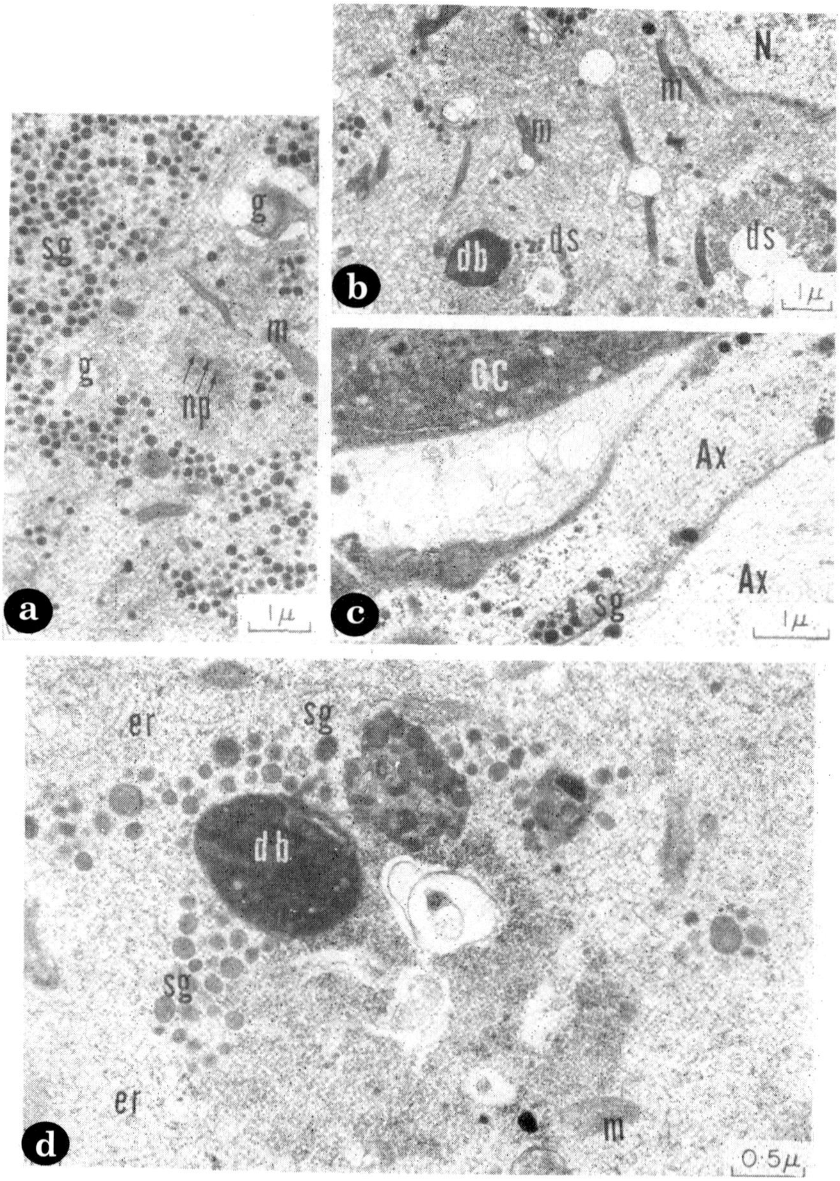

Plate 226. Fig. 5. NS-II cell of diapause individuals (*Pieris*) 2 days after the larval pupal ecdysis (a) and 7 days after the larval-pupal ecdysis (b) and its axon 7 days after the ecdysis (c). Fig. (d) shows the characteristic structure (ds in Fig. b) appeared in the cytoplasm of certain NS II cells 7 days after ecdysis to the pupa. Ax, axon; N, nucleus; db, dense body; ds, structure suggesting the site of destruction of granules; er, endoplasmic reticulum; g, Golgi body; m, mitochondria; np, nuclear pore; sg, secretory granule. (From Kono, 1973).

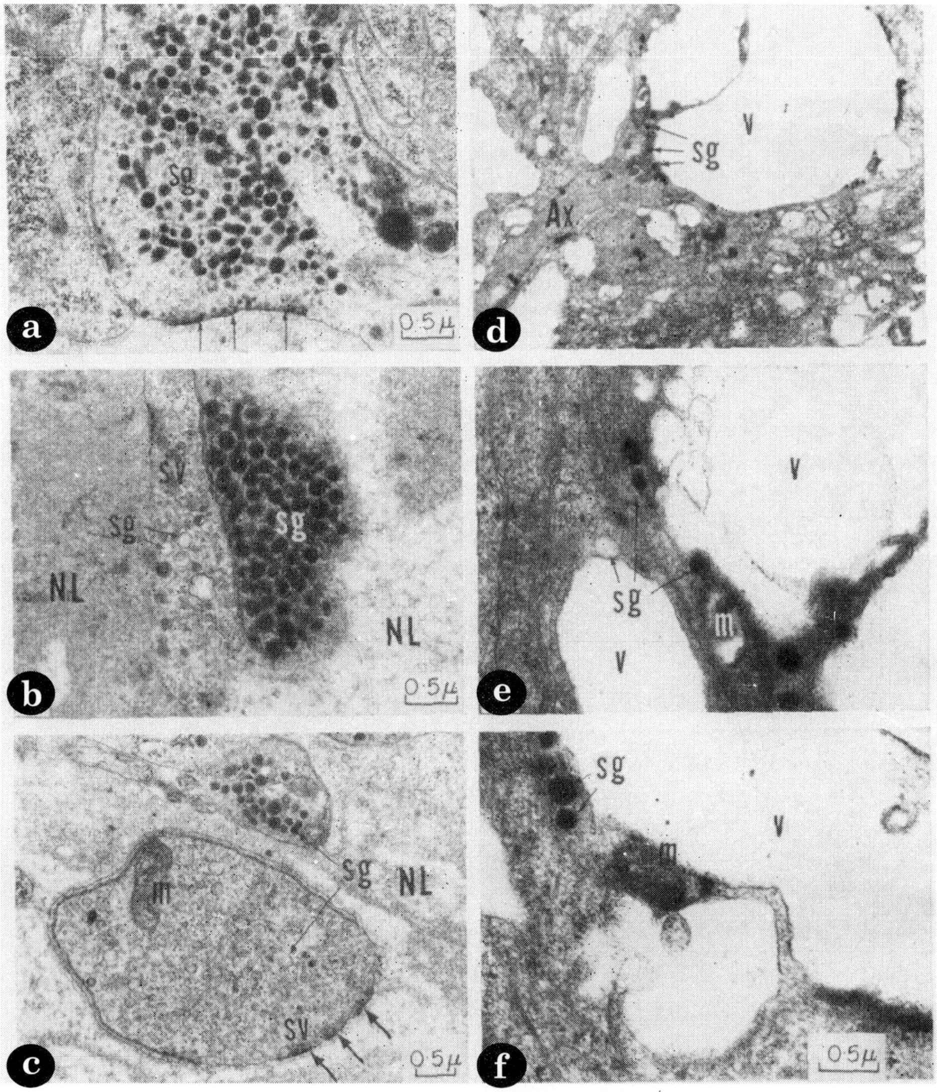

Plate 227. Fig. 6. Endings and peripheral parts of axons of NS-II cell of *Pieris*. (a) (Diapause pupa 8-12 hr after larval-pupal ecdysis) : Aggregation of synaptic vesicles along the axon membrane (arrows), and electron dense secretory granules can be observed. (b) (Non-diapause pupa 8-12 hr after ecdysis). (c) (Diapause pupa 8-12 hr after ecdysis) : Electron transparent granules are observed in accompaniment with the electron dense granules. (d) (Non-diapause pupa 0-4 hr after ecdysis): Secretory granules are arranged around the vacuoles in the peripheral part of axon. (e and f). Enlargement of Fig. (d) showing the `empty' granules and granules budded off in the vacuole. Ax, axon; NL, neural lamella; m, mitochondria; sg, secretory granule; sv, synaptic vesicle; v, vacuole. (From Kono, 1973).

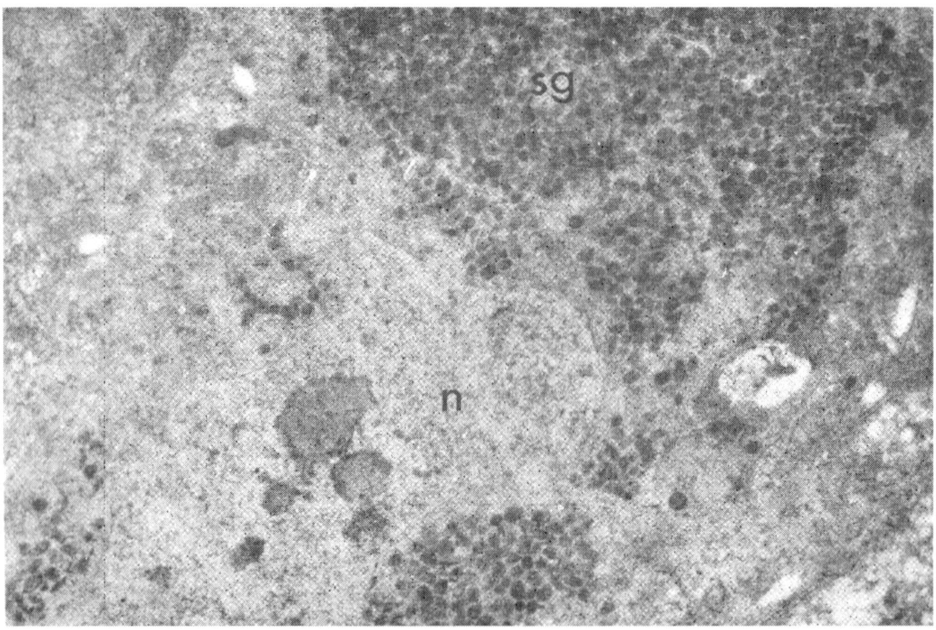

Plate 228. Fig. 1. NS-II cell of diapause pupae 20 days after pupation without chilling (*Pieris*). Cytoplasm is filled with secretory granules. Diameter of secretory granule in NS II cell is 220-240 nm. (From Kono, 1977).

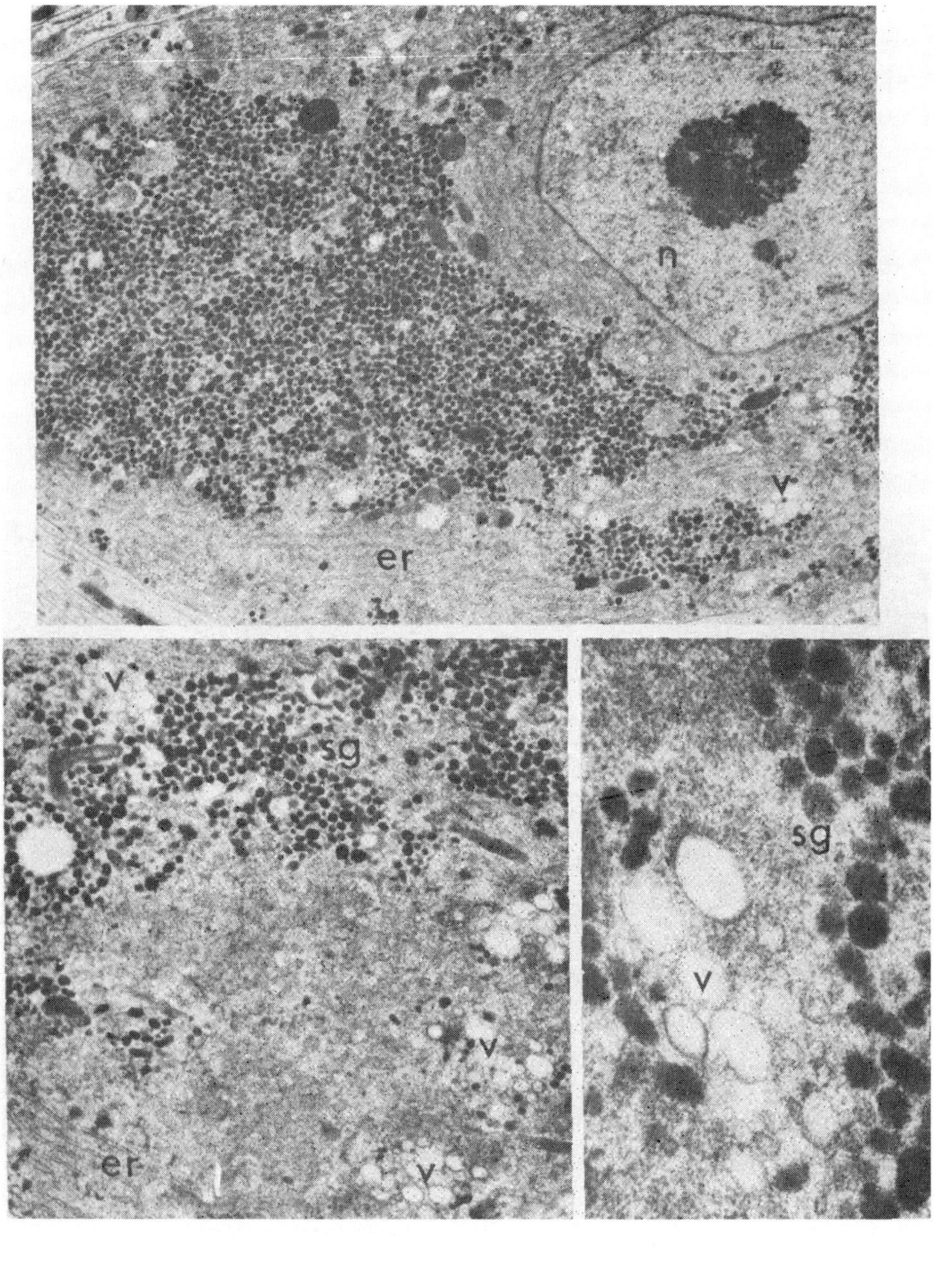

Plate 229. Fig. 2. NS-II cell of diapause pupae of *Pieris* after 20 days of chilling. The cytoplasm is also occupied with large aggregates of secretory granules. Lower photographs show the groups of vesicles in the cytoplasm. (From Kono, 1977).

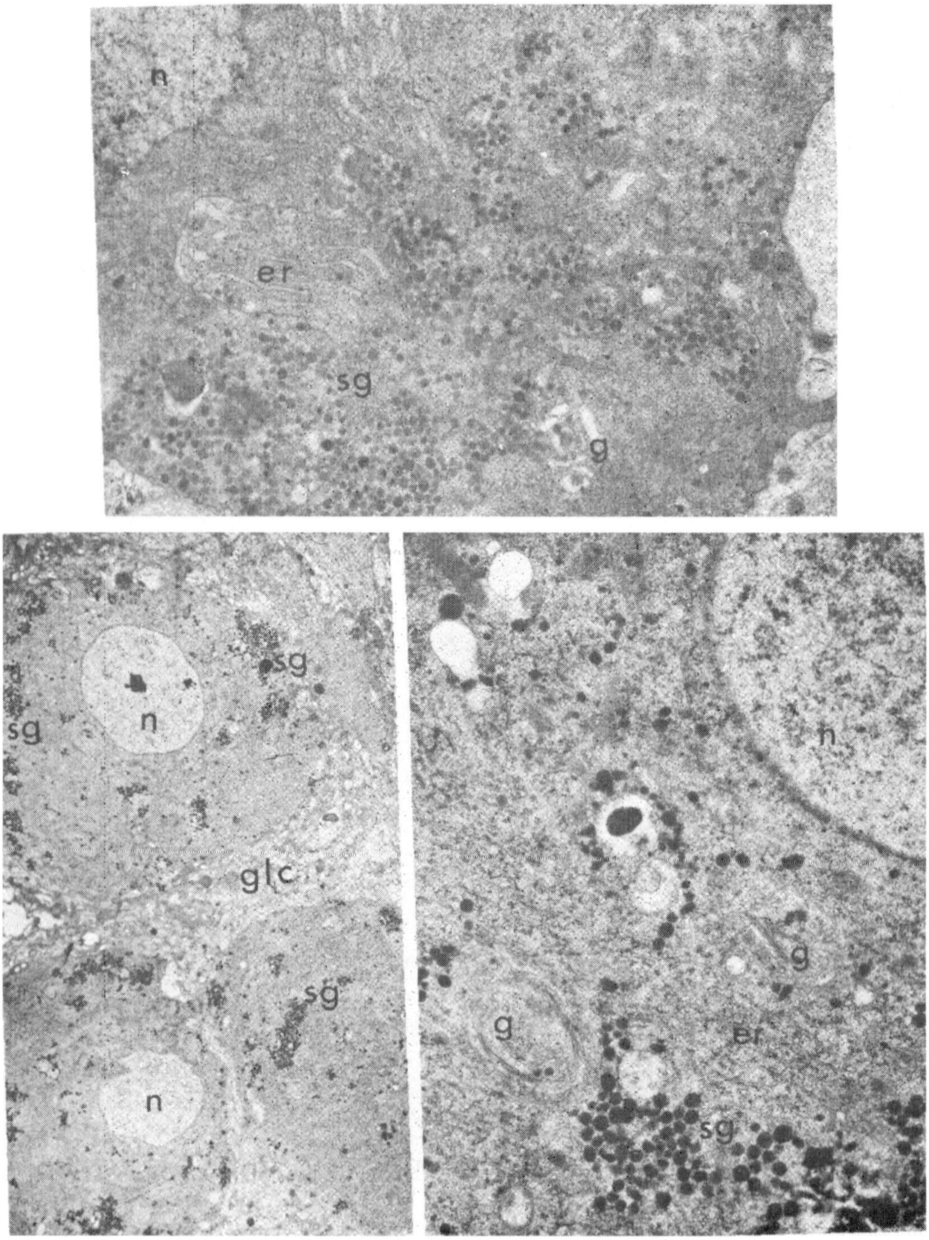

Plate 230. Fig. 3. NS-II cell of *Pieris* after 35 days of chilling. Top, before post-incubation. Secretory granules moved to the marginal zone of cytoplasm. Lower left, after 6-8 hr incubation following the chilling. Secretory granules decreased in number and aggregates of granules situated in the marginal zone. Lower right, after 6-8 hr incubation. Showing restored Golgi bodies. (From Kono, 1977).

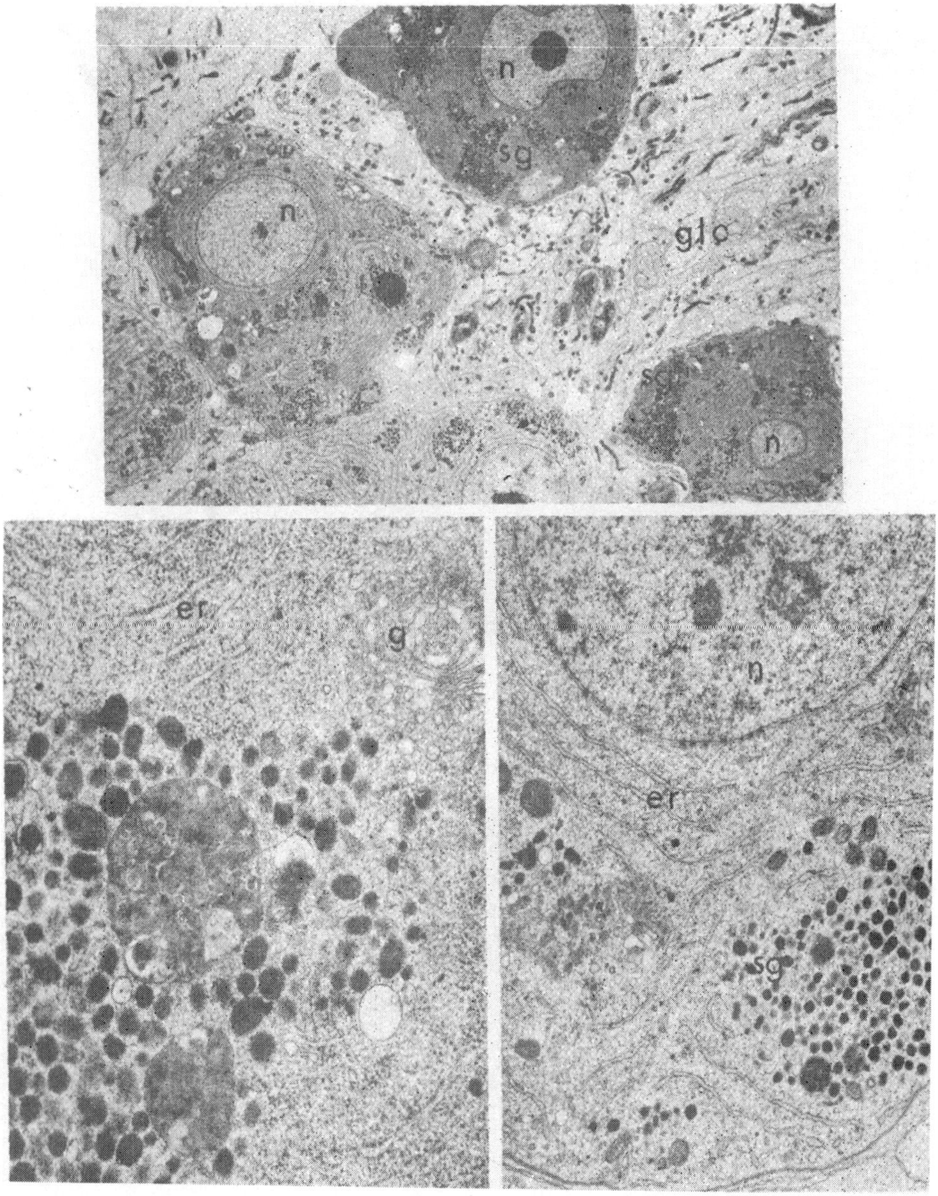

Plate 231. Fig. 4. NS-II cell of *Pieris* after 75 days of chilling without incubation. Low magnification (top) shows aggregates of secretory granules in the marginal zone of cytoplasm, and high magnifications (lower photographs) show Golgi bodies and dulated rough ER. (From Kono, 1977)

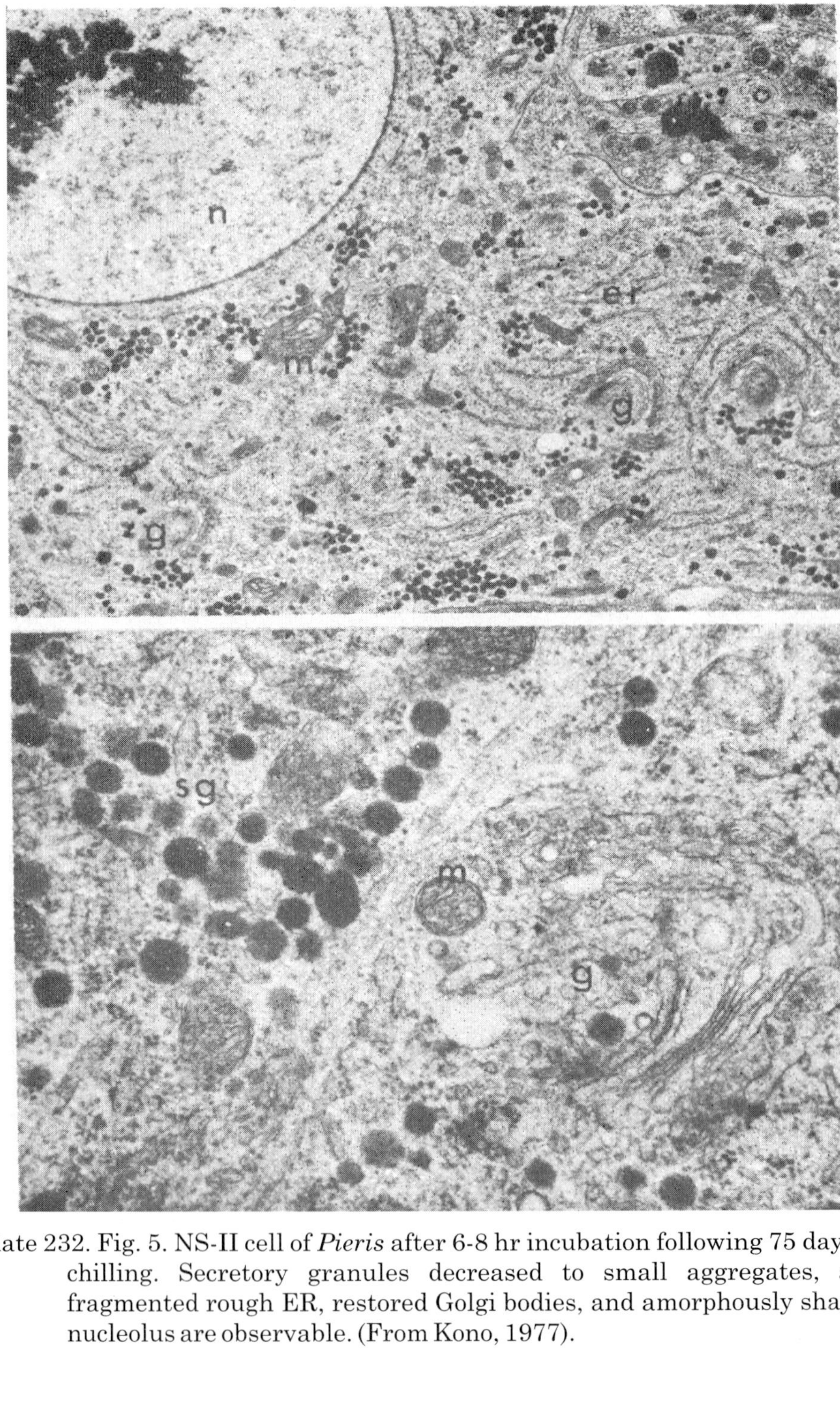

Plate 232. Fig. 5. NS-II cell of *Pieris* after 6-8 hr incubation following 75 days of chilling. Secretory granules decreased to small aggregates, and fragmented rough ER, restored Golgi bodies, and amorphously shaped nucleolus are observable. (From Kono, 1977).

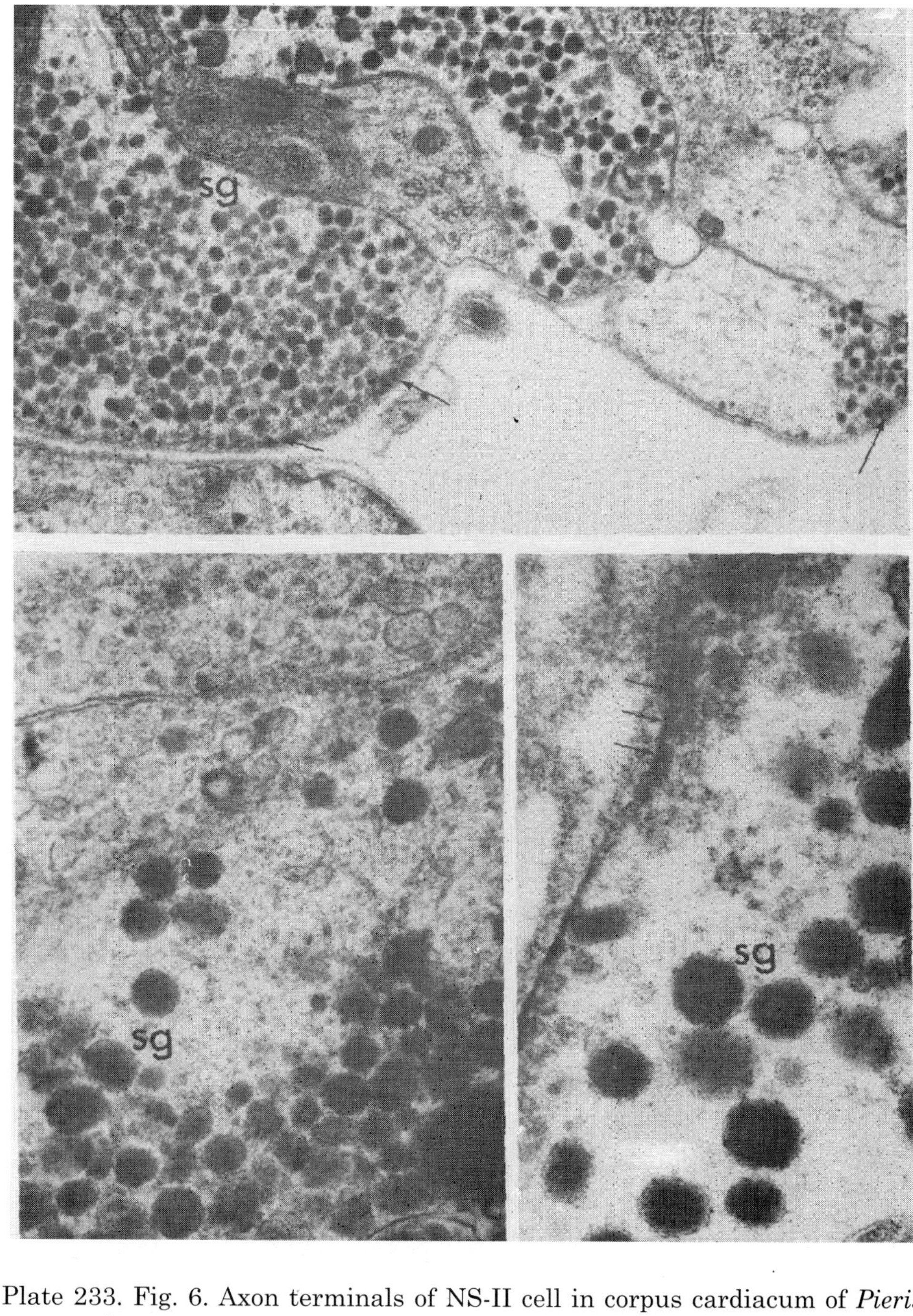

Plate 233. Fig. 6. Axon terminals of NS-II cell in corpus cardiacum of *Pieris* showing release of secretory materials (arrows). Empty granules are observable. (From Kono, 1977)

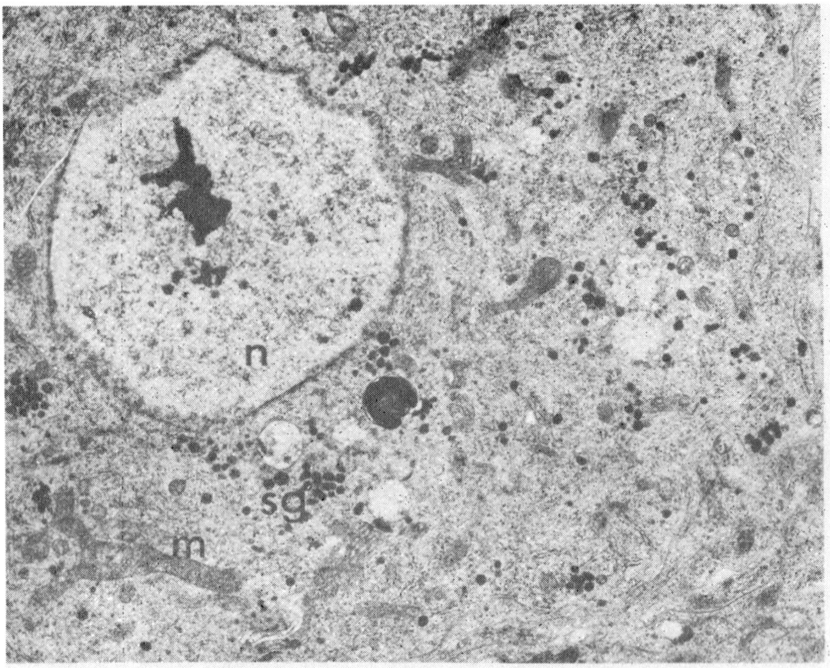

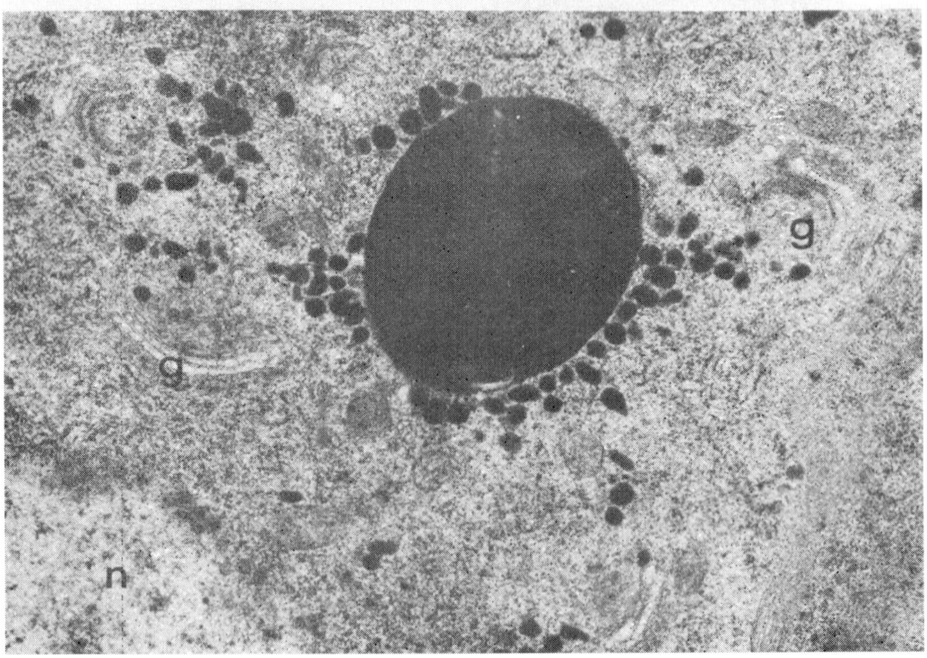

Plate 234. Fig. 7. NS-II cell of *Pieris* after 24 hr incubation following 75 days of chilling. Secretory granules reduced to small number, and branched mitochondria are observable. (From Kono, 1977).

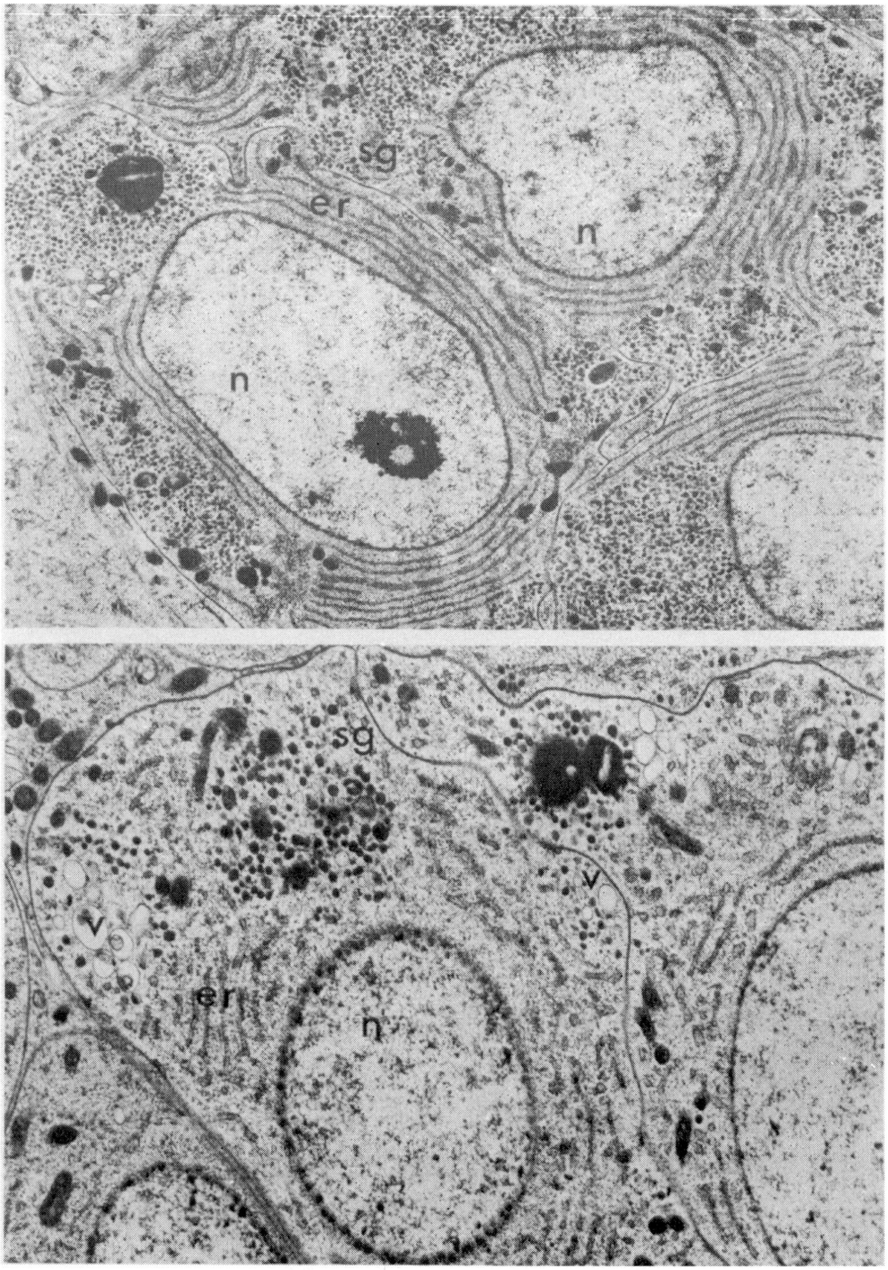

Plate 235. Fig. 8. (Upper), NS IV cell of *Pieris* after 20 days of chilling showing large aggregates of secretory granules and paralleled plate like rough ER. (Lower), NS-IV cell after 75 days of chilling without incubation showing decreased secretory granules and fragmented rough ER. Diameter of secretory granule in NS-IV cell is approxi. 120 nm. (From

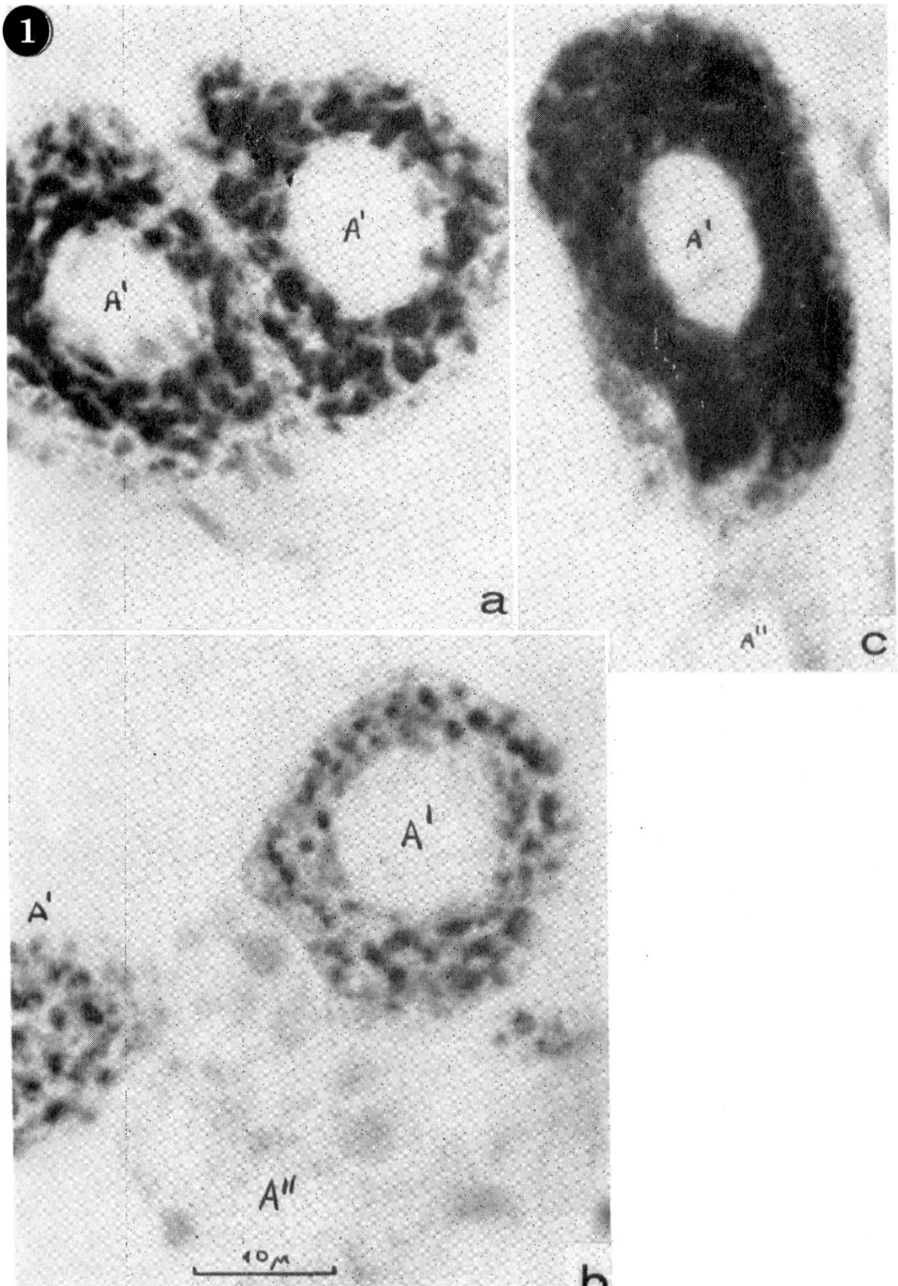

Plate 236. Fig. 1. A'-cells (A') and A''-cells (A'') of *Bombyx mori* caterpillars at the beginning (a), and the end (b) of the fourth instar, and after several days of starvation (c). Paraldehyde-thionin staining. X 1,850. (From Bassurmanova and Panov, 1967).

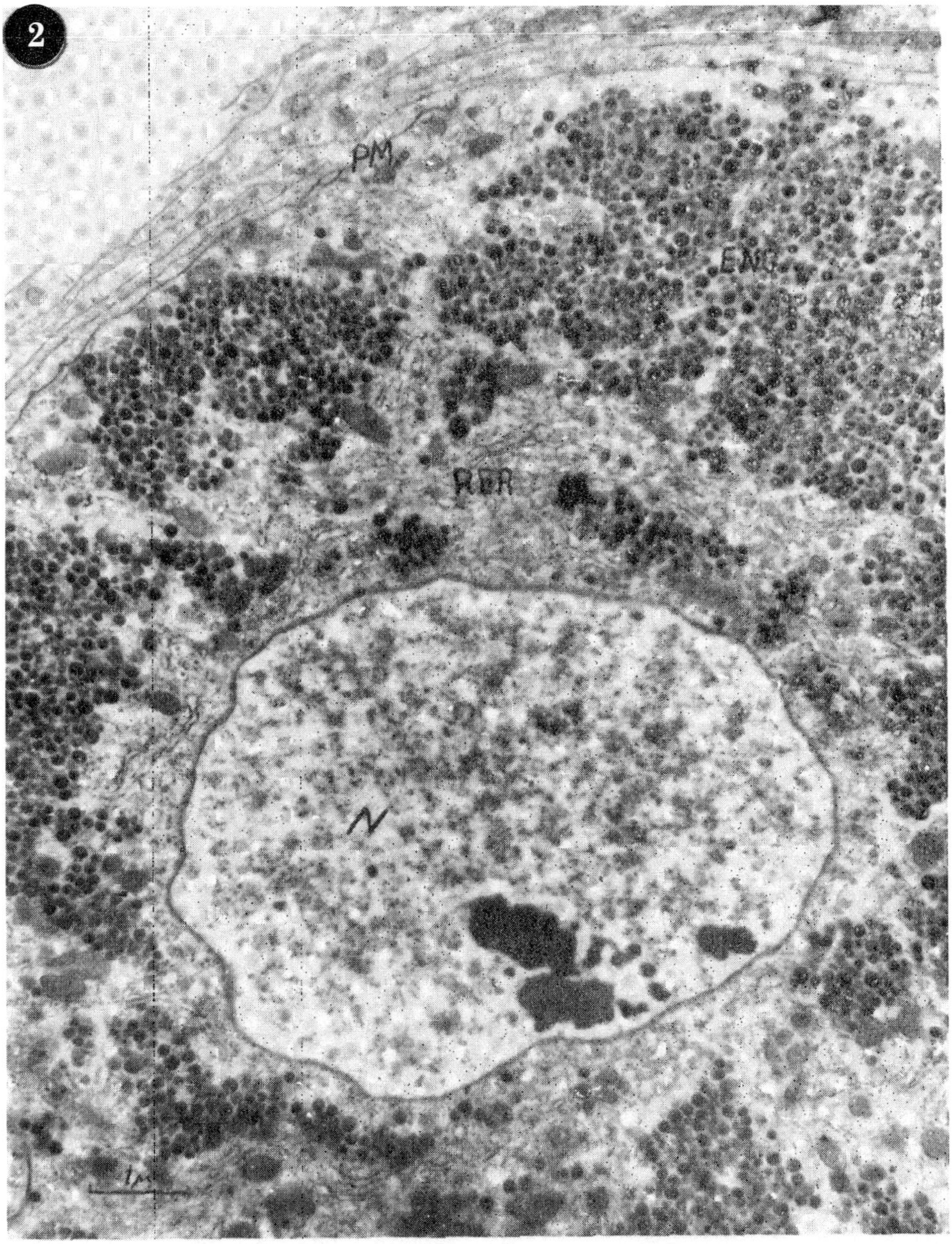

Plate 237. Fig. 2. Area of the A''-neurosecretory cell of *Bombyx* containing only electron-dense elementary neurosecretory granules (ENG). N, nucleus; RER, granular endoplasmic reticulum; PM, plasma membrane. X 11,000 (From Bassurmanova and Panov, 1967).

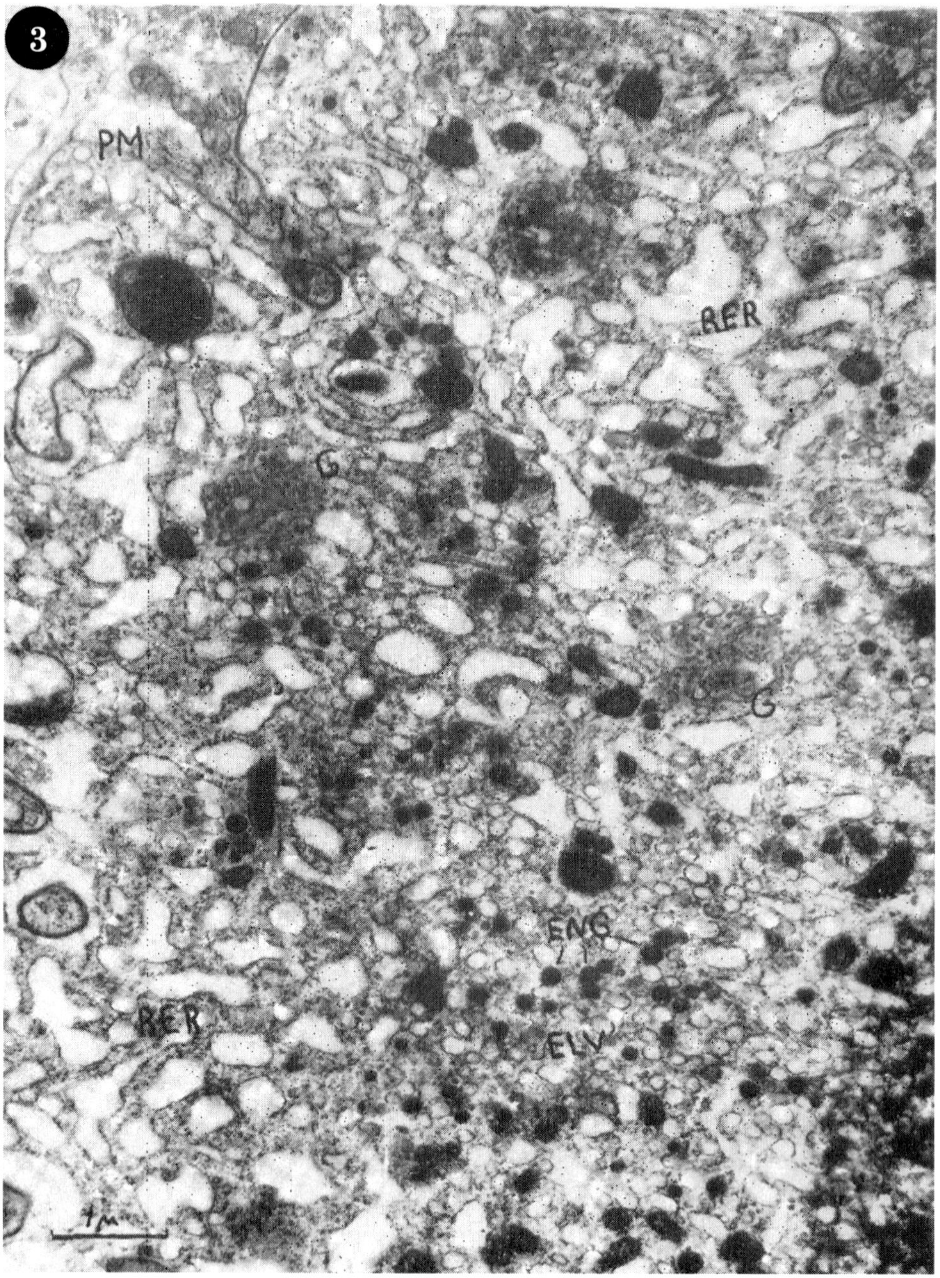

Plate 238. Fig. 3. Peripheral area of A'-cell body in the middle of the fifth instar caterpillar of *Bombyx*. Endoplasmic reticulum (RER) is intensively developed. ELV, electron-lucent vesicles; ENG, electron-dense neurosecretory granules; G, Golgi apparatus; PM, plasma membrane; X 13,000. (From Bassurmanova and Panov, 1967)

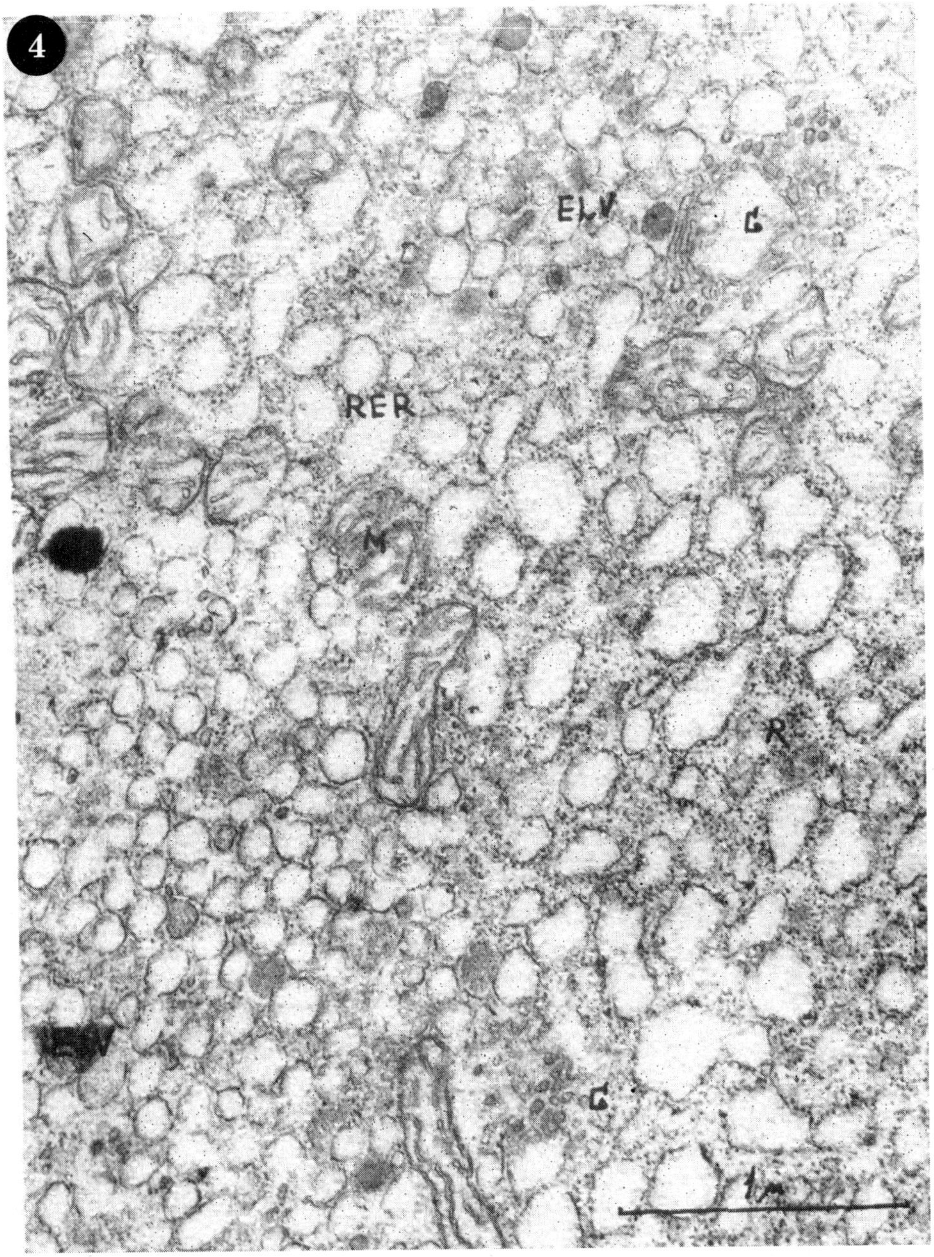

Plate 239. Fig. 4. Peripheral area of caterpillar A'-cell body at the beginning of the fourth instar. A layer of endoplasmic reticulum (RER) divides two clusters of electron-lucent vesicles (ELV). At the edge of one there is a row of mitochondria (M). In cytoplasm numerous free ribosomes (R) are observed. G. Golgi apparatus. X 36,000. (From Bassurmanova and Panov, 1967).

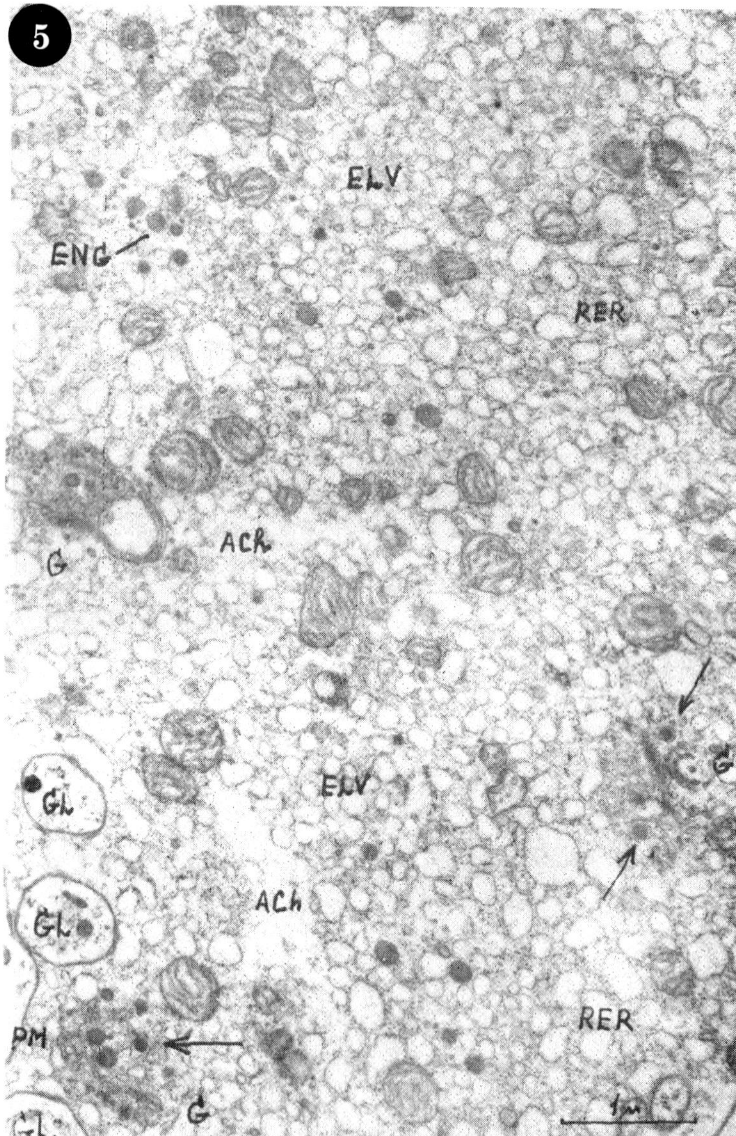

Plate 240. Fig. 5. A'-cell of a starving fifth instar caterpillar of *Bombyx*. Peripheral zone. Cisternae of endoplasmic reticulum (RER) are greatly reduced. A large number of single and grouped electron-lucent vesicles (ELV). Electron-dense neurosecretory granules (ENG) occur singly. In Golgi zone (G) there is a large number of electron-dense granules (an arrow indicates to the intermediate stage of neurosecretory granule formation). PM, plasma membrane; GL glial intrusions; Ach, axoplasmic intrusions in cell body (``axoplasmic channels''). X 22,000. (From Bassurmanova and Panov, 1967).

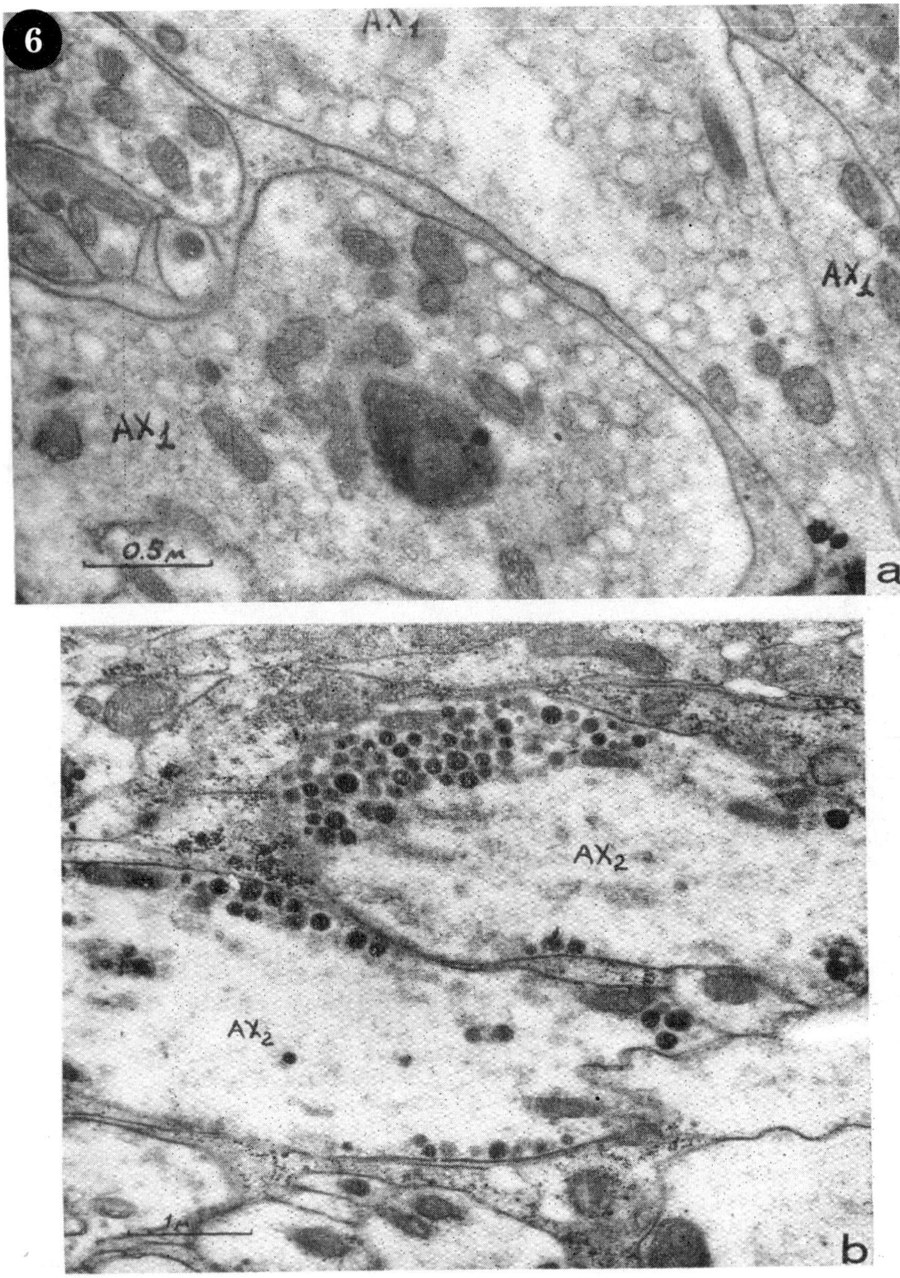

Plate 241. Fig. 6. Micrographs of neurosecretory axons from neuropile of the caterpillar brain of *Bombyx*. a, axons of A'-cells containing numerous electron-lucent vesicles and single electron-dense neurosecretory granules (AX_1); b, axons belonging probably to A"-cells (AX_2). a, X 36,000; b, X 17,000 (From Bassurmanova and Panov, 1967).

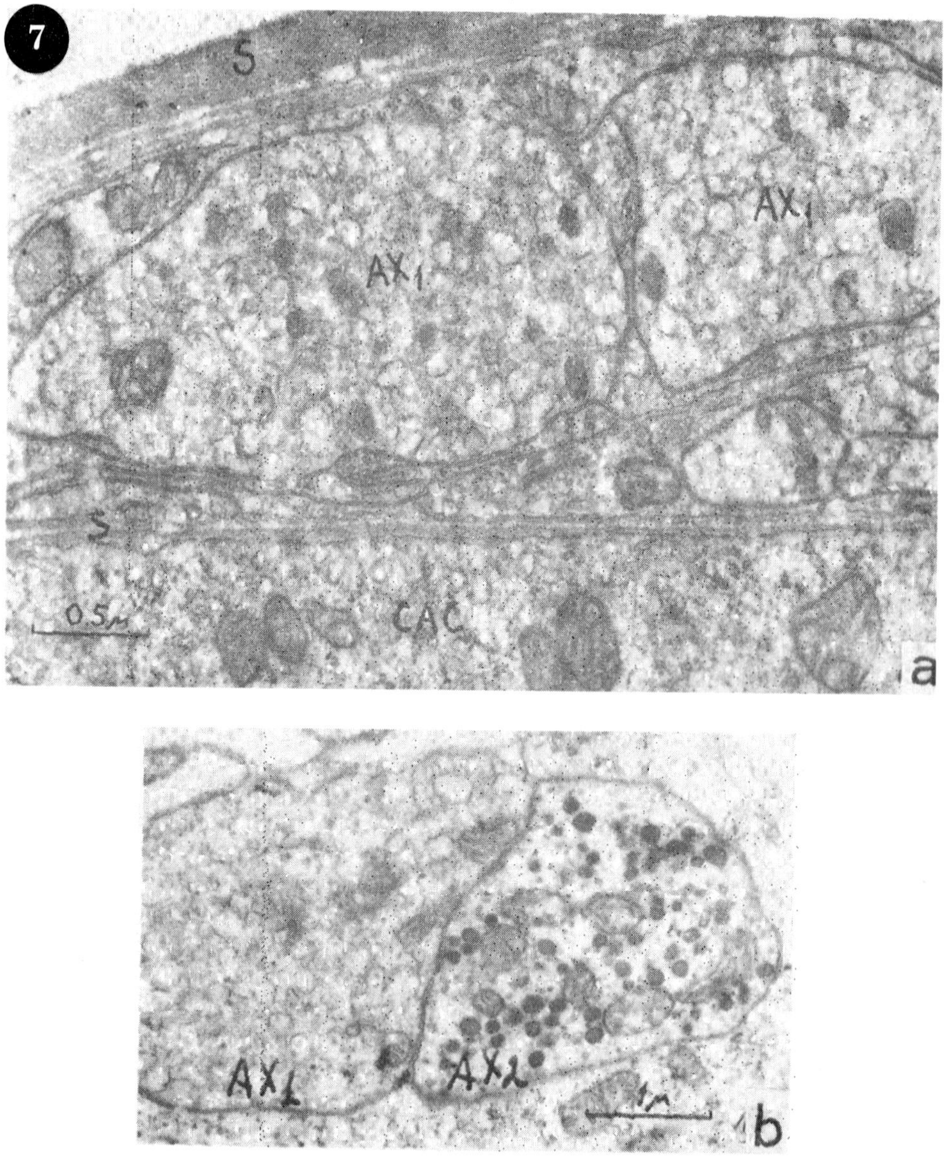

Plate 242. Fig. 7. Secretory axons in corpus allatum (a) and corpus cardiacum (b) of *Bombyx* AX_1, axons of A'-cells; AX_2, axon with electron-dense neurosecretory granules; CAC, corpus allatum cell; S, corpus allatum sheath and its intrusion between the secretory axons and corpus allatum cell. a. X 29,000; b. X 16,000. (From Bassurmanova and Panov, 1967)

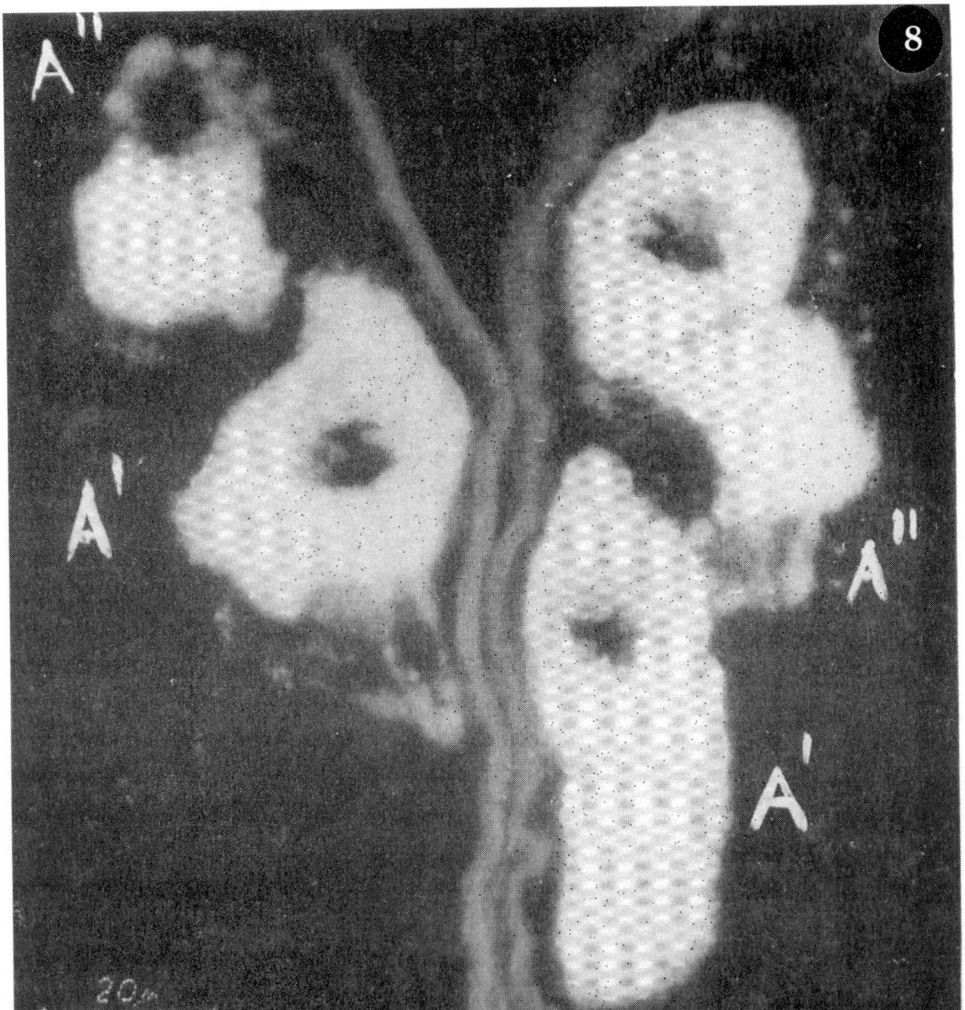

Plate 243. Fig. 8. Neurosecretory cells of the first medial group in the caterpillar brain of *Bombyx mori*. Bouin, fluorescence microscopy. Pseudoisocyanin staining after potassium permanganate oxidation. A',. A'-cells; A'', A''-cells; a double connective-tissue sheath of the brain, (neural lamella) goes along the micrograph's middle line. X 950. (From Bassurmanova and Panov, 1967).

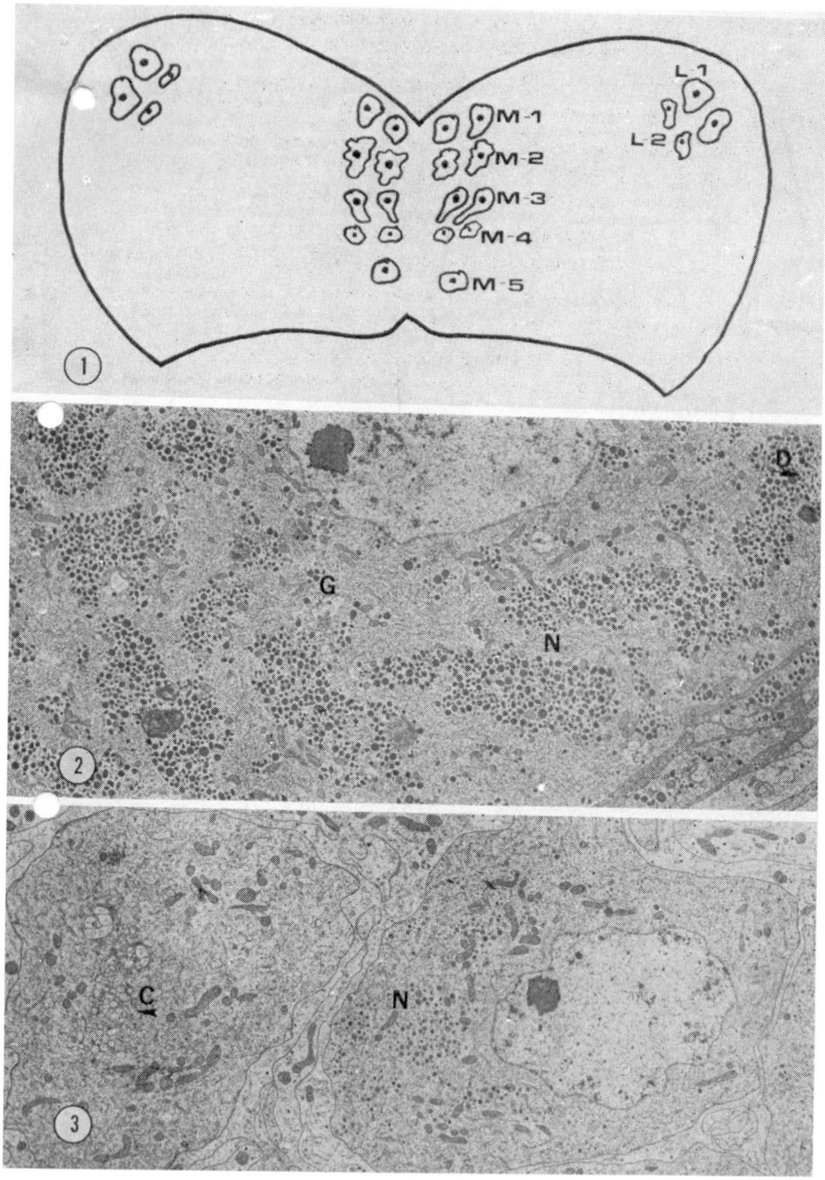

Plate 244. Fig. 1. Schematic illustration of a transverse section through the brain showing the location of the different types of NSC in the brain of diapausing pupae of *Manduca sexta*. M = Medial; L= Lateral. Fig. 2. The morphology of the M-5 cell which was similar to the M-1 or M-2 cell types, showing accumulation of neurosecretory granules (N), Golgi apparatus (G), and deposits of glycogen (D). X 7500. Fig. 3. Low magnification showing the M-3 (N) and M-4 (C) cell types. The M-4 cell shows the electron lucent granules (arrow). X 7000. (From Borg and Bell, 1977).

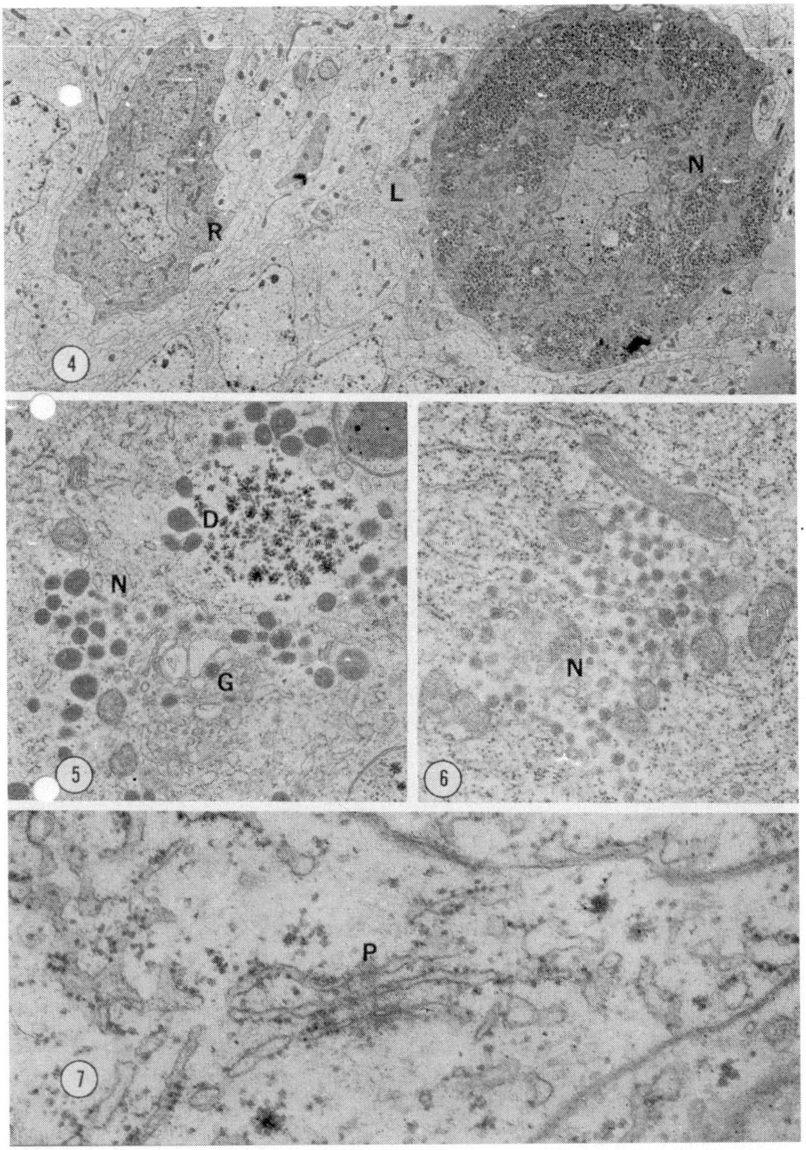

Plate 245. Fig. 4. Low magnification of the lateral NSC of *Manduca* with the extra cellular lipid (L), the L-1 cell type with accumulation of neurosecretory granules (N), and the smaller L-2 cell type (R). X 4200. Fig. 5. High magnification of the L-1 cell type showing glycogen deposit (D), Golgi apparatus (G), and neurosecretory granules (N). X 24,000. Fig. 6. High magnification of L-2 cell type showing the abundant rough endoplasmic reticulum and neurosecretory granules (N). X 35,000. Fig. 7. A dispersing (or forming) annulate lamellae showing the rough endoplasmic reticulum and the pore complex (P). X 75,000. (From Borg and Bell, 1977).

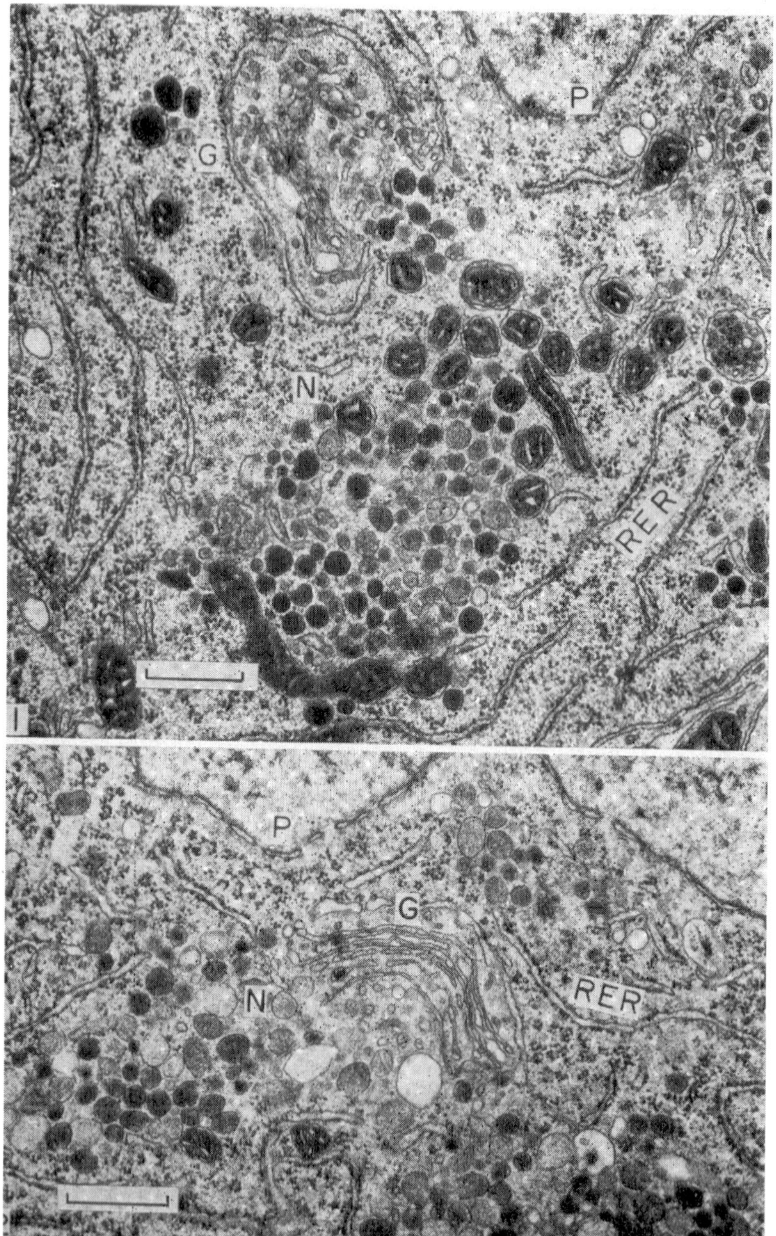

Plate 246. Fig. 1. Median NSC *in vivo* of *Manduca sexta* showing the accumulation of neurosecretory granules (N), rough endoplasmic reticulum (RER), Golgi apparatus (G), and nuclear pores (P). Fig. 2. Median NSC after 4 days *in vitro* showing neurosecretory granules (N), rough endoplasmic reticulum (RER), Golgi apparatus (G), and nuclear pores (P). (From Borg and Marks, 1973). Scale of Figs. 1-5 = 0.5 μm.

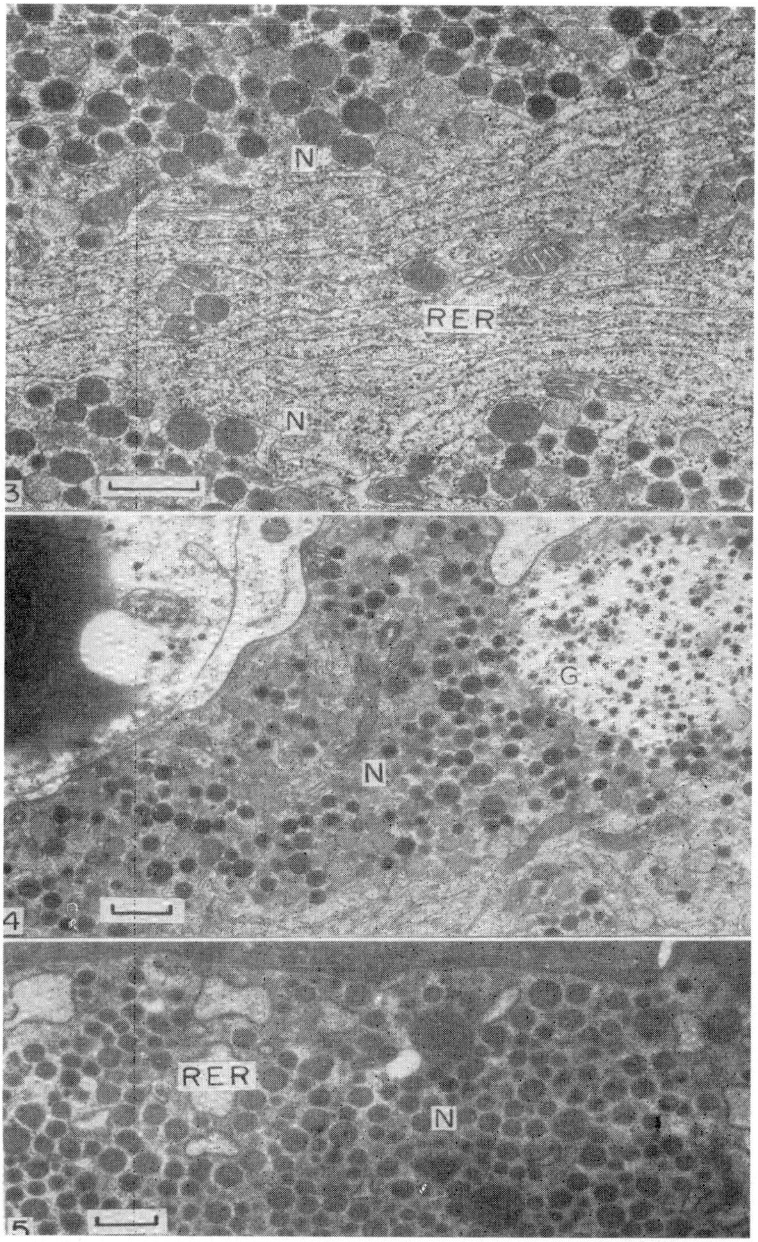

Plate 247. Fig. 3. Accumulation of neurosecretory granules (N) in the median NSC after 4 days *in vitro* of *Manduca* located between parallel arrays of rough endoplasmic reticulum (RER). Fig. 4. Neurosecretory granules (N) in the median NSC after 4 days *in vitro* showing glycogen (G) and lipid droplets. Fig. 5. Neurosecretory granules (N) after 10 days *in vitro* showing the swollen and vacuolated rough endoplasmic reticulum (RER). (From Borg and Marks, 1973).

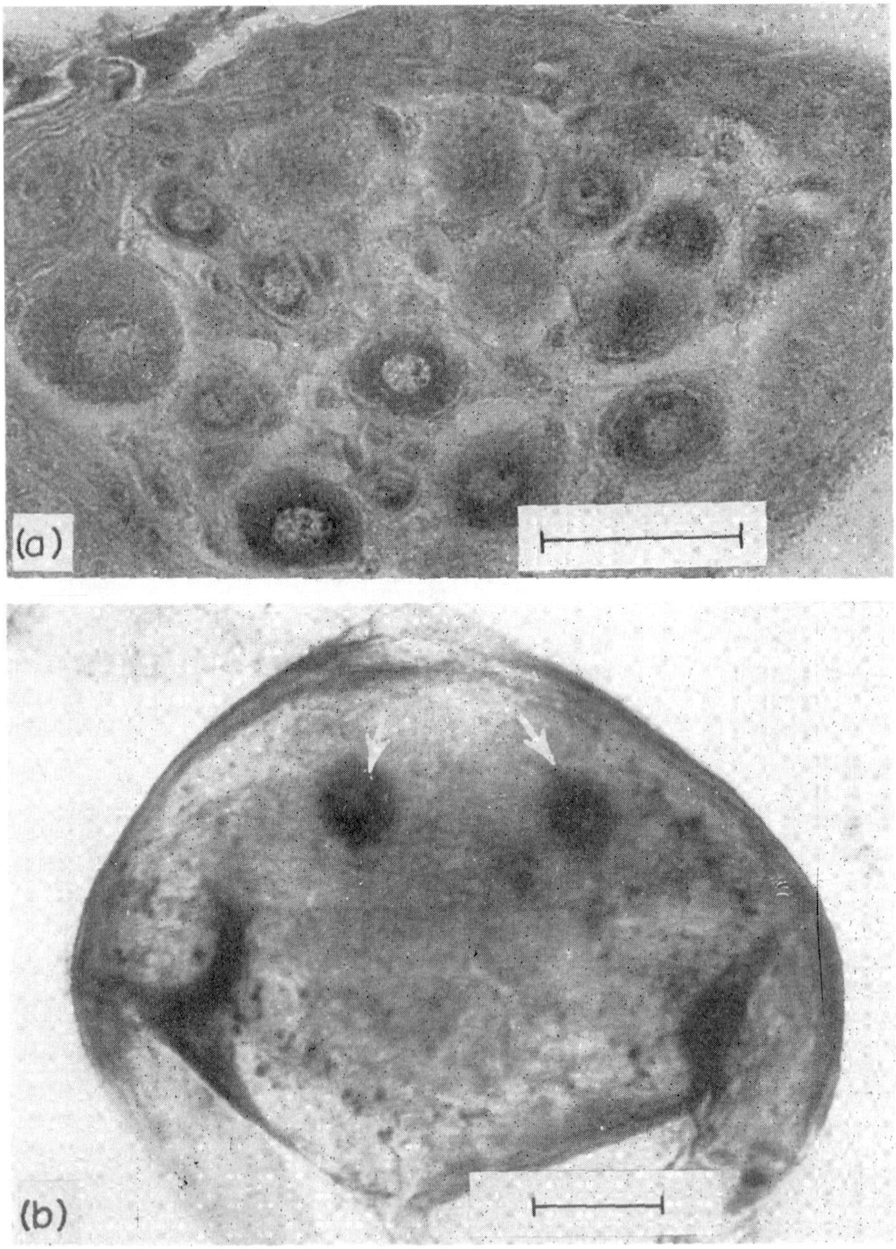

Plate 248. Fig. 1. (a) Transverse section of a frontal ganglion of *M. sexta* showing numerous large neurons (25-50 μm dia.) stained with alcian blue-phloxine; (b) appearance of the NS cells (indicated by arrows) in the frontal ganglion of a diapause pupa. Fixed 21 days after the larval-pupal ecdysis and stained *in situ* with PF (scale = 50 μm). (From Bell *et al.*, 1974).

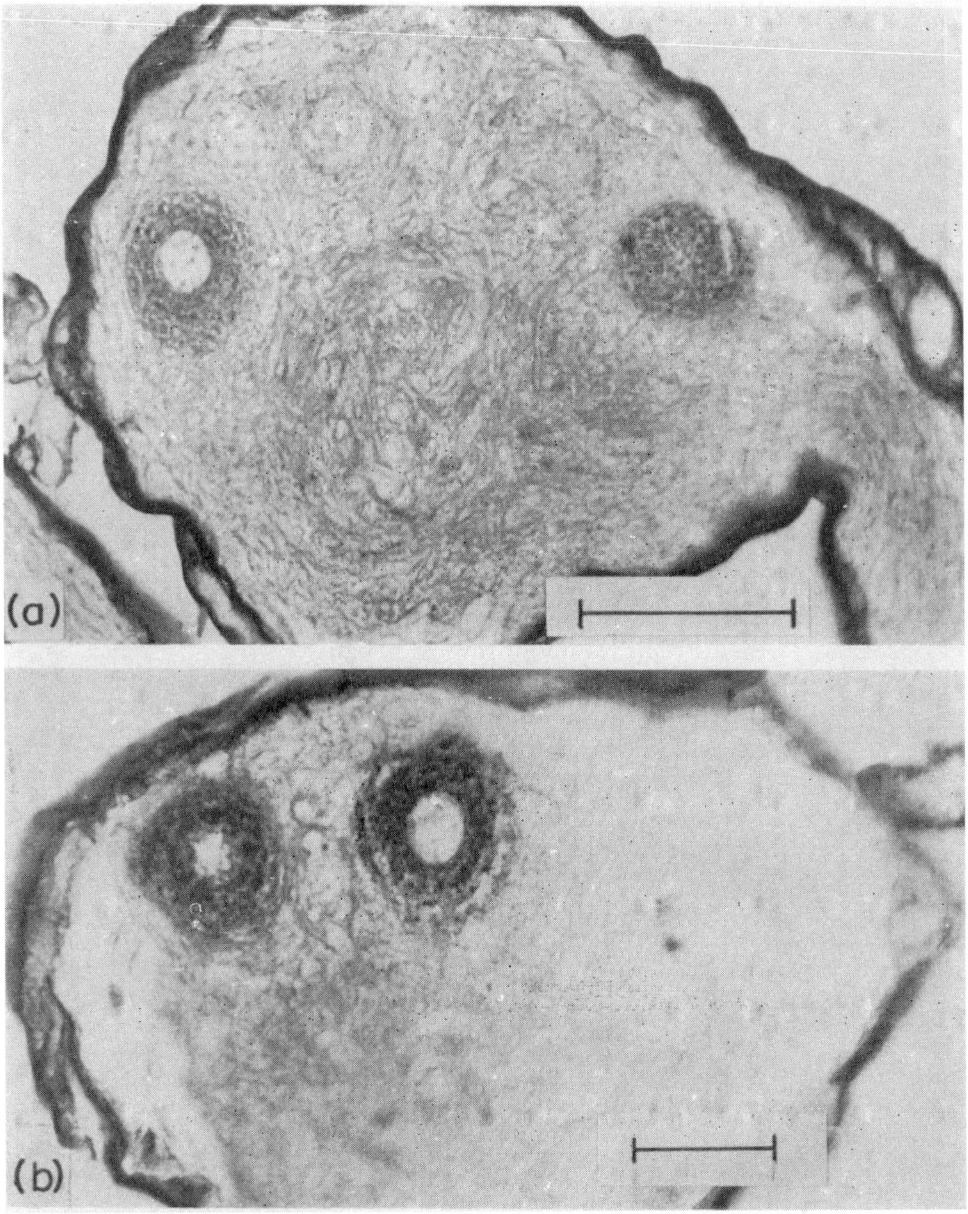

Plate 249. Fig. 2. A comparison of the neurosecretory cells in the frontal ganglion of a feeding (a) vs. a starved (b) tobacco hornworm larva. Paraldehyde-fuchsin + Halmi's counterstain. Note the greater accumulation and clumping of the cyto-plasmic inclusions in the starved specimen (scale = 50 μm). (From Bell *et al.*, 1974).

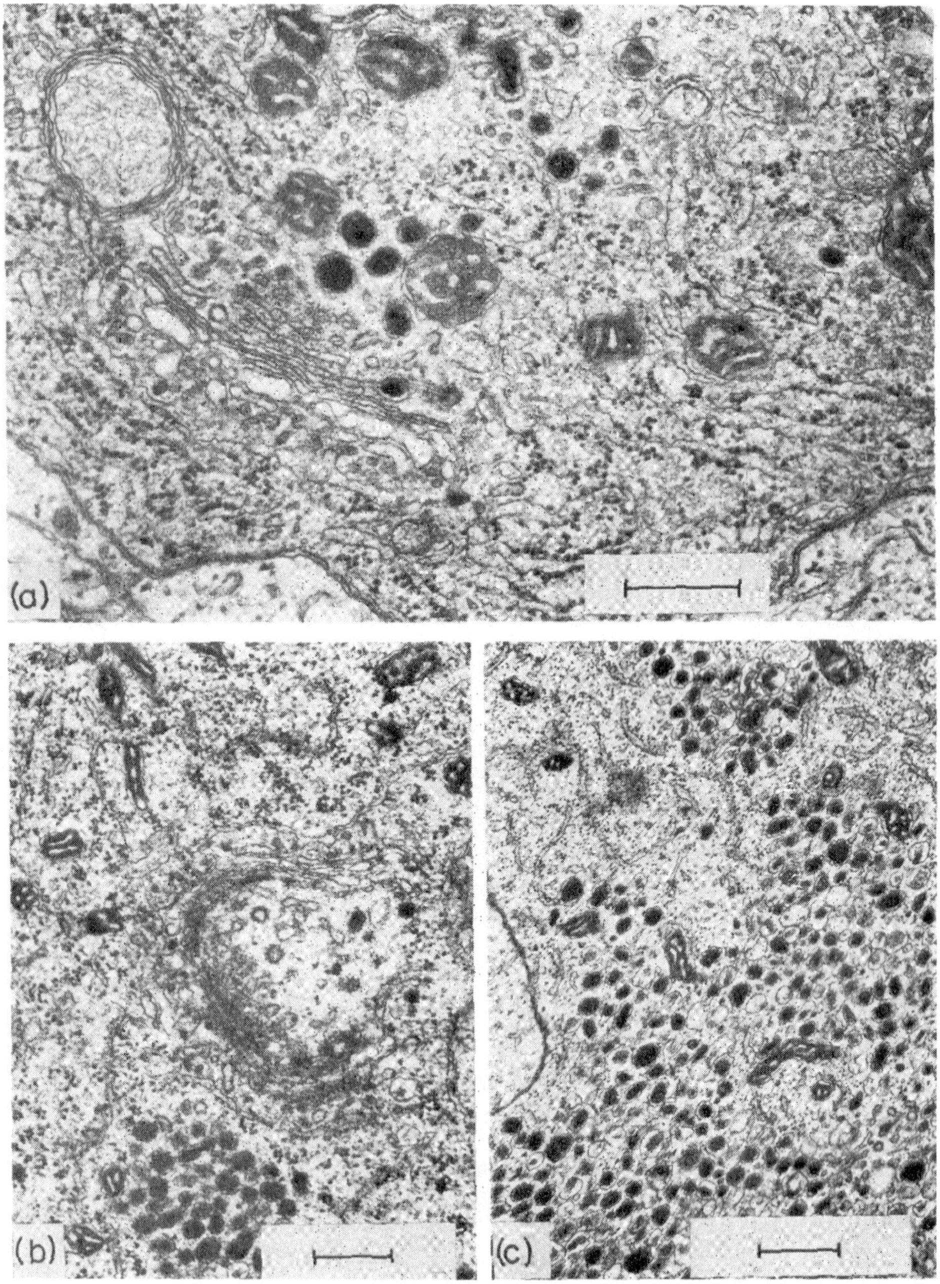

Plate 250. Fig. 3. Electron micrographs showing typical accumulations of neurosecretory granules in the NS cells of frontal ganglia of *Manduca* during (a) larval feeding, (b) pharate adult development, and (c) pupal diapause. Note the large accumulation of granules in the inactive (diapause) stage (scale = 0.5 μm). (From Bell *et al.*, 1974).

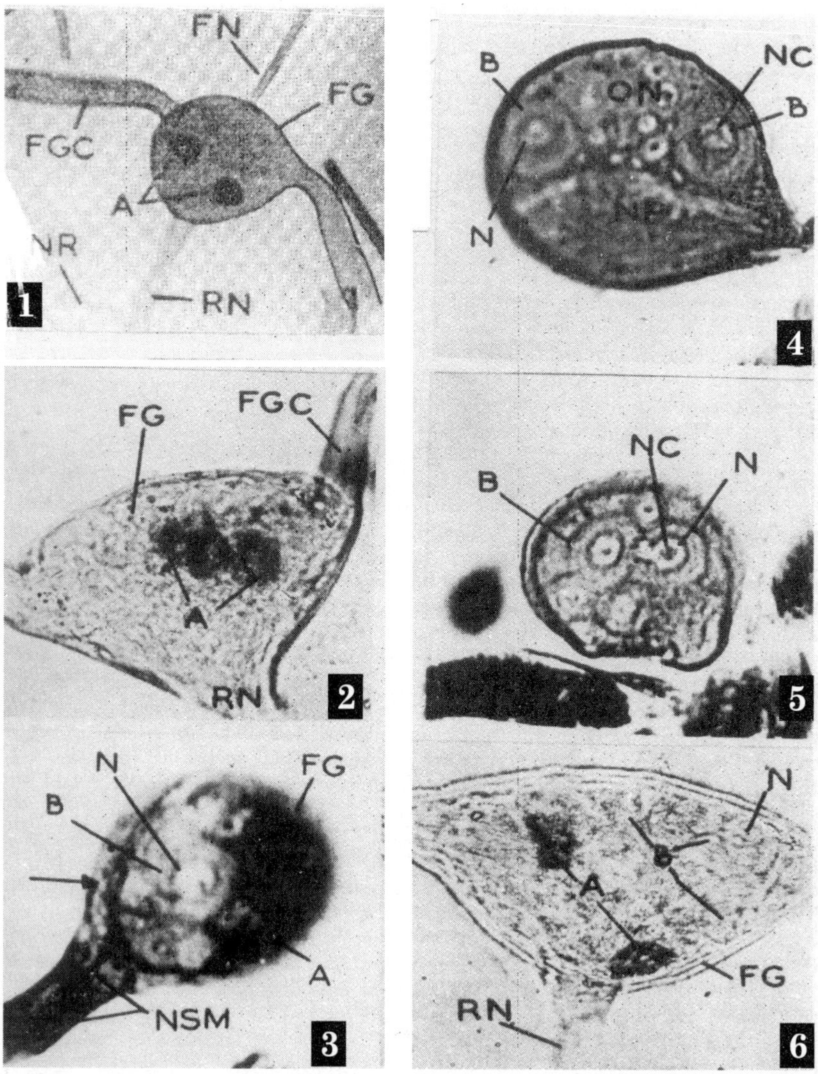

Plate 251. Fig. 1. Whole mount of frontal ganglion of *Achoea janata* showing 2 A-cells of a mature 5th instar larva and associated nerves of the ganglion. PARF X 180. Fig. 2. Whole mount of the frontal ganglion of the mature larva of *Philosamia ricini* showing 3 closely located A-cells packed with NSM. PARF X 540. Fig. 3. Frontal section of the FG of *Achoea janata* showing both A- and B- cells. PARF X 800. Fig. 4. Frontal section of FG of a mature larva of *Philosamia ricini* showing 2-B-cells packed with green-granular material. Ordinary nerve cells are also seen. PF X 800. Fig. 5. T.S. of the frontal ganglion of *Achoea janata* showing B-cells. The nucleolus is very distinctly seen in these cells. PF X 800. Fig. 6. Whole mount of the frontal ganglion of *Philosamia ricini* showing 2 A-cells, one close to the root of the frontal ganglion connective of a side and other to the recurrent nerve. Lightly stained B-cells are also seen. PARF X 500. (From Awasthi, 1980).

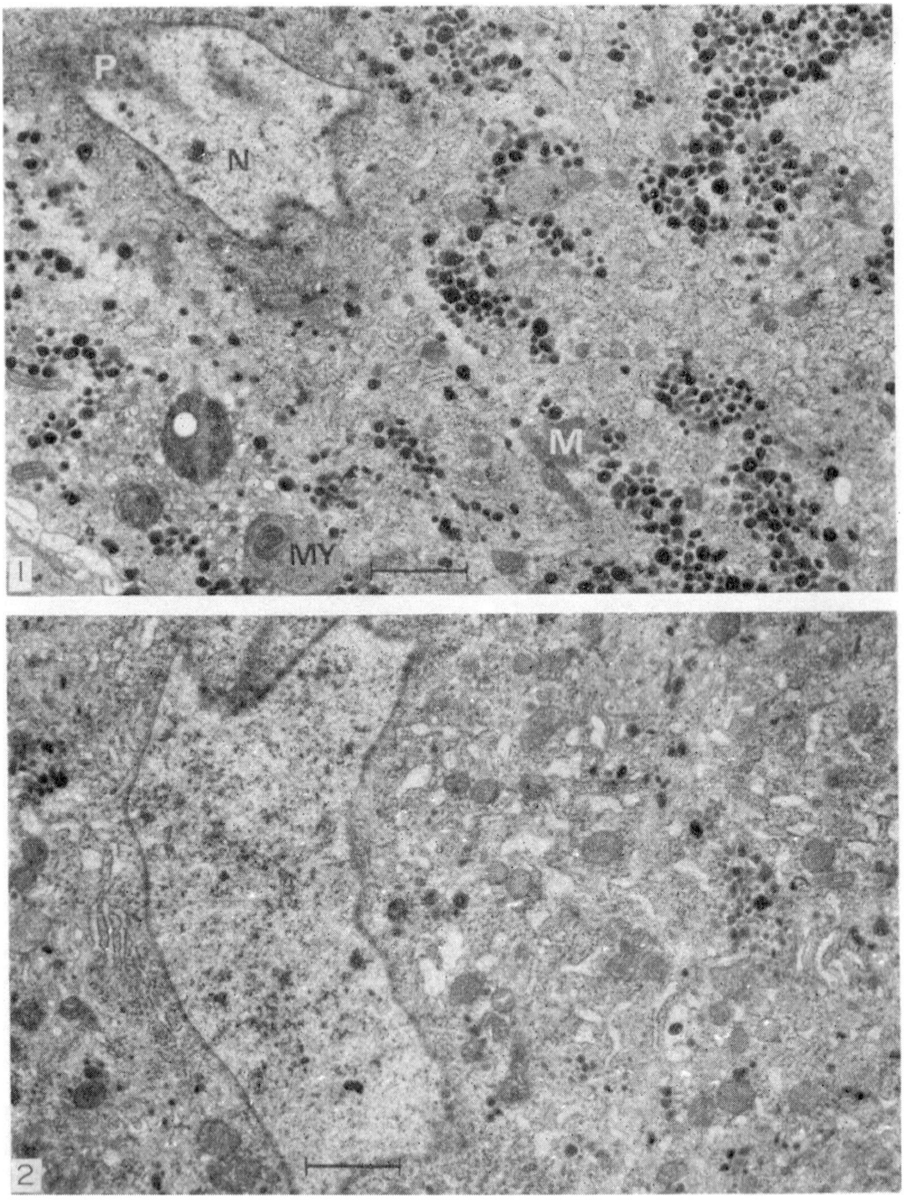

Plate 252. Fig. 1. Neurosecretory cell located in the diapause suboesophageal ganglion of *Bombyx*. My, myelin-like structure; M, mitochondria; N, nucleus; P, nuclear pore. Note the presence of many cytoplasmic granules. The suboesophageal ganglion of the larva soon after the completion of spinning throughout. Scale = 2μ. Fig. 2. Neurosecretory cell located in the non-diapause suboesophageal ganglion. Note a slight swelling endoplasmic reticulum and sparsely distributed granules. Scale = 2μ. (From Park and Yoshitake, 1971).

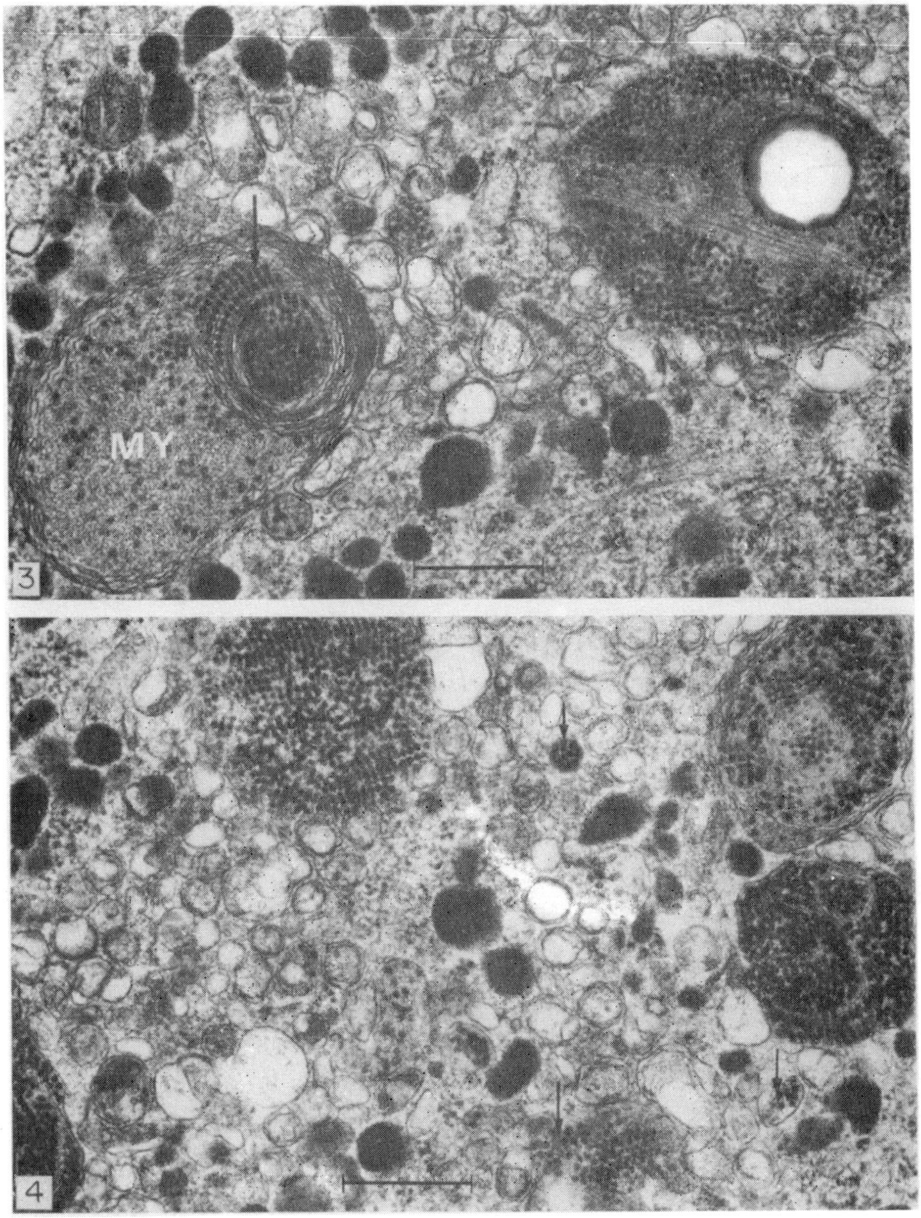

Plate 253. Fig. 3. Neurosecretory cell located in the diapause suboesophageal ganglion of *Bombyx*. Myelin-like structure (MY) is surrounded by many vacuolar structures. Highly dense particles come into line along the myelin sheath (see arrow). Scale = 0.5 μ. Fig. 4. Neurosecretory cell located in the diapause suboesophageal ganglion of *Bombyx*. Note that the particles are discharged from the myelin-like structure and reach the vacuole (see arrow). Scale = 0.5 μ. (From Park and Yoshitake, 1971).

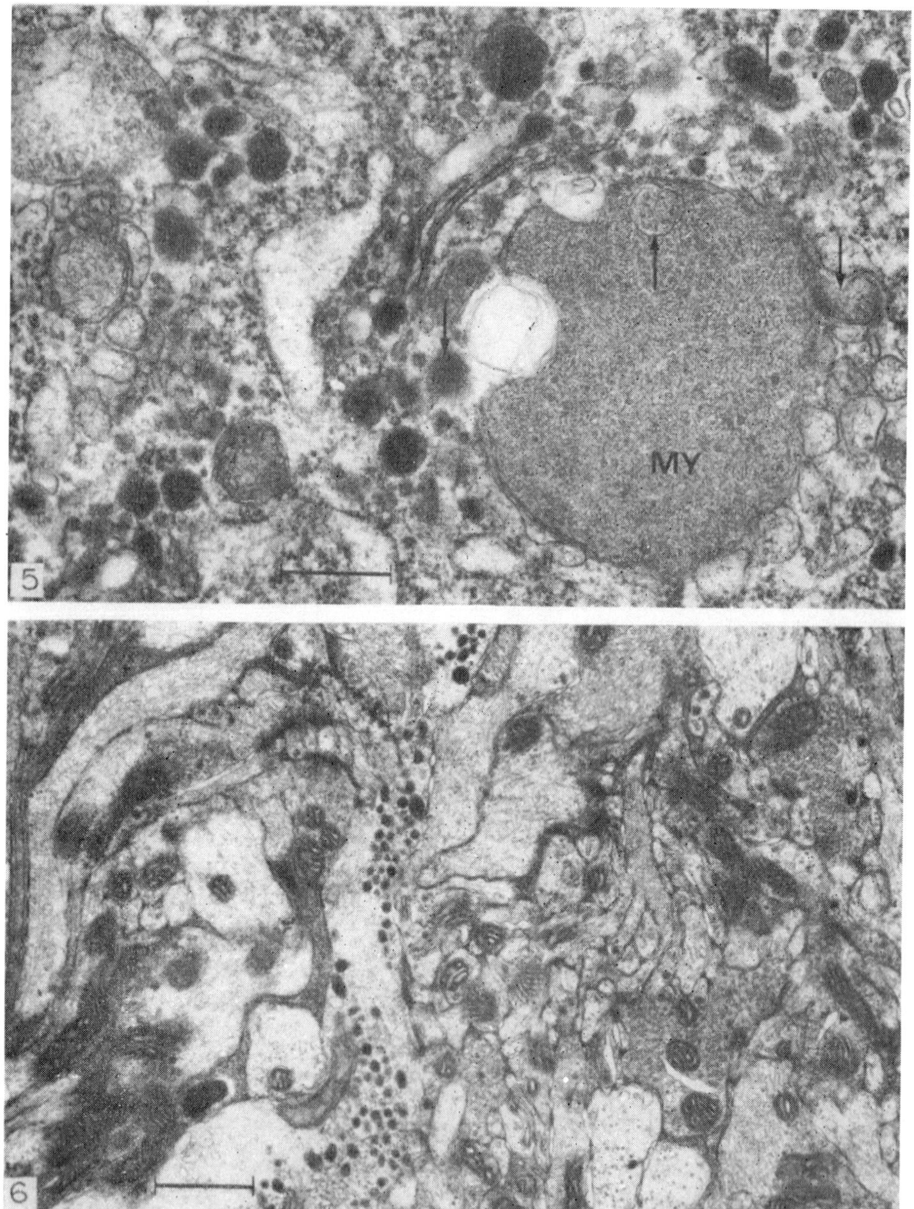

Plate 254. Fig. 5. Neurosecretory cell located in the non-diapause suboesophageal ganglion of *Bombyx*. Note that there is no small dense particle formation in the myelin-like structure (MY) in which a part is either dispersed into the vacuoles or surrounded directly by a membrane (see arrow). Scale = 0.5 μ. Fig. 6. The cytoplasmic granules were released into axon in the non-diapause suboesophageal ganglion of *Bombyx*. Scale-2μ. (From Park and Yoshitake, 1971).

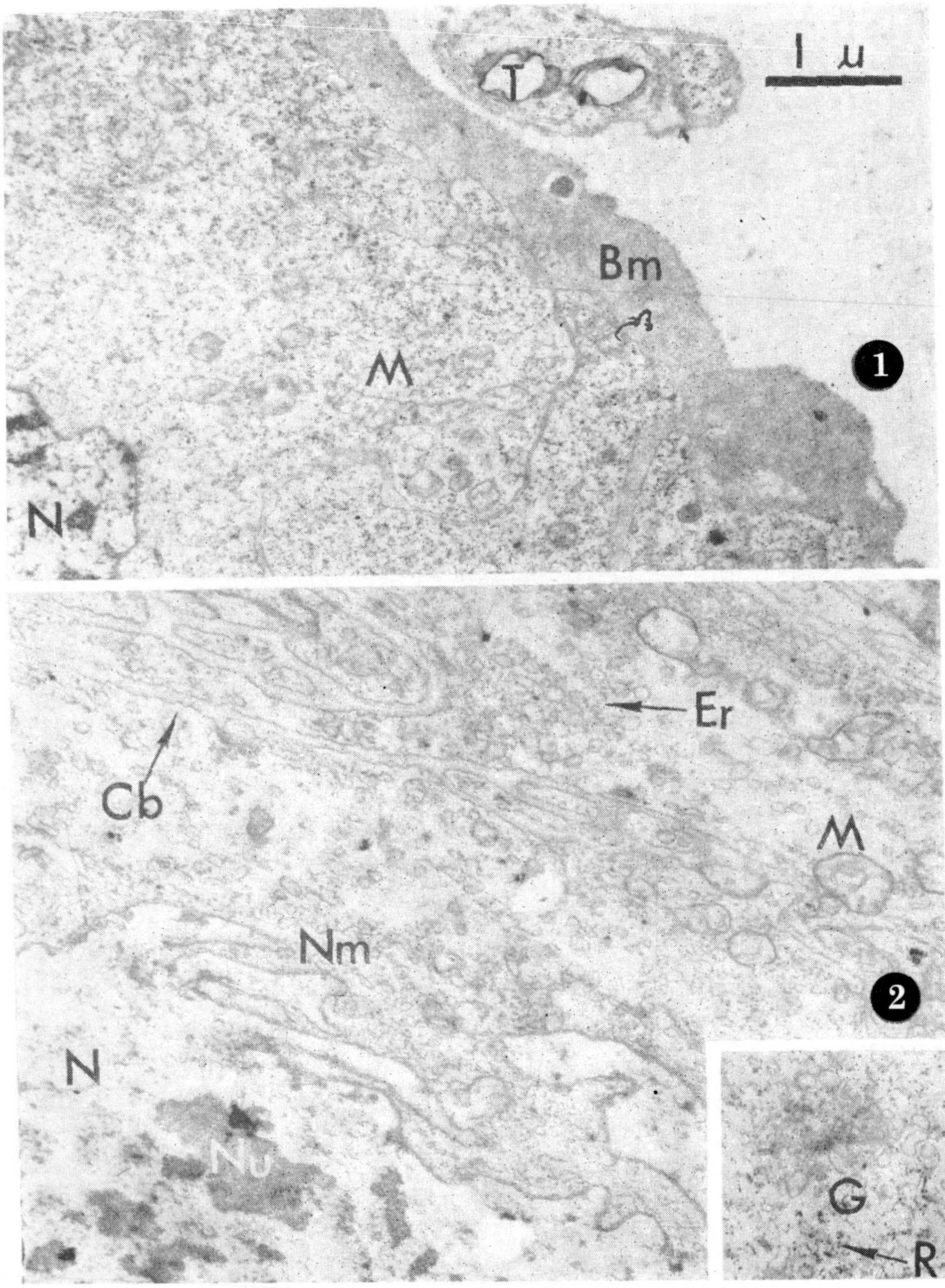

Plate 255. Fig. 1. Portion of a peripheral cell of the corpus allatum of a male diapausing pupa of *Hyalophora*. Note lack of structure in basement membrane. Fig. 2. Portion of cell from the interior of the corpus allatum of a diapausing, male pupa. Inset shows Golgi apparatus. (From Waku and Gilbert, 1964).

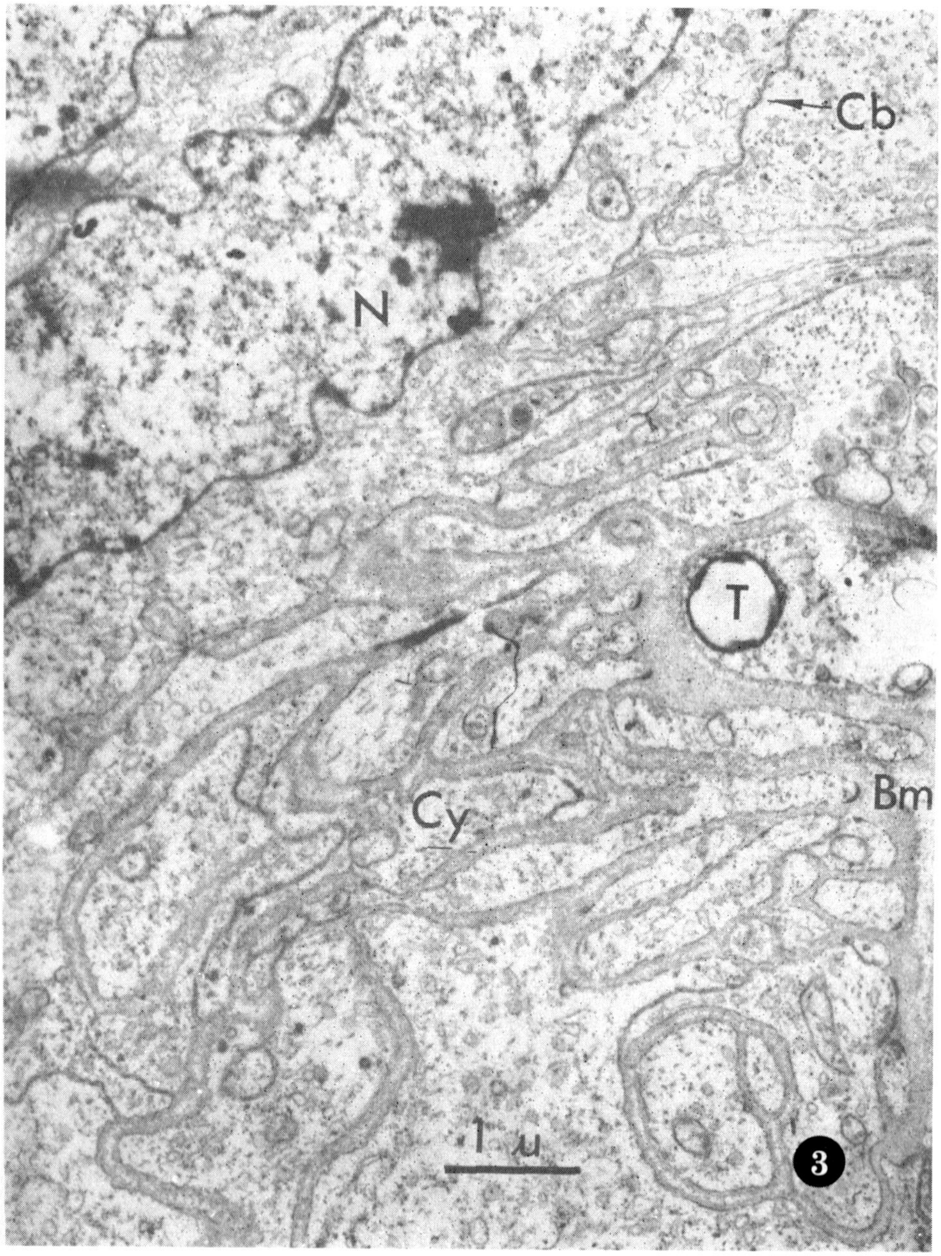

Plate 256. Fig. 3. Portion of a peripheral cell of the corpus allatum of a diapausing, male pupa of *Hyalophora*. Note involution of basement membrane resulting in compartmentalization of the cytoplasm. (From Waku and Gilbert, 1964).

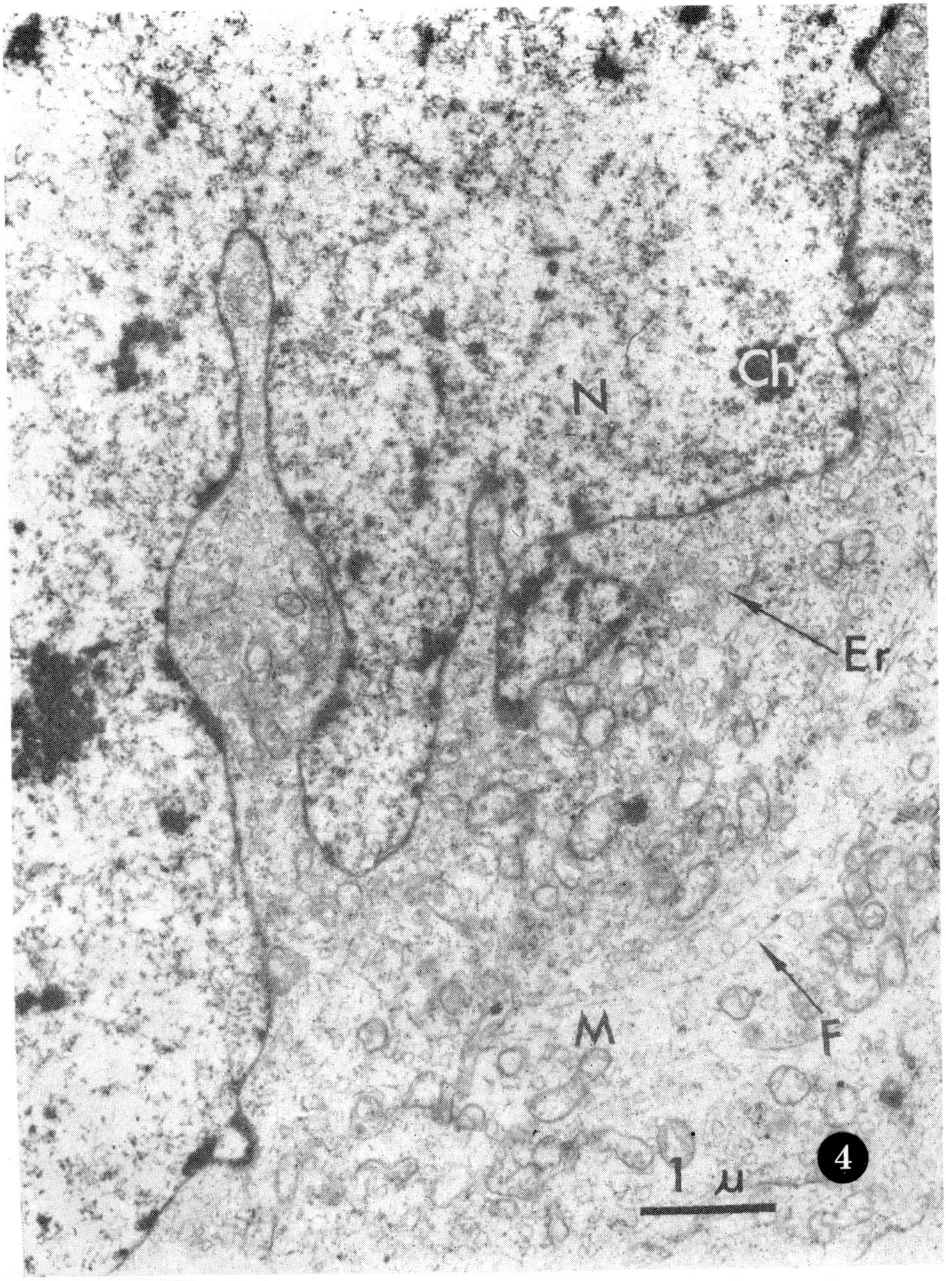

Plate 257. Fig. 4. Cytoplasm bordering nucleus of a corpus allatum cell from diapausing, male pupa of *Hyalophora*. Note heavy concentration of mitochondria and endoplasmic reticulum around nuclear envelope and invasion of nucleoplasm by cytoplasmic processes. The significance of the filamentous structures is unknown. (From Waku and Gilbert, 1964).

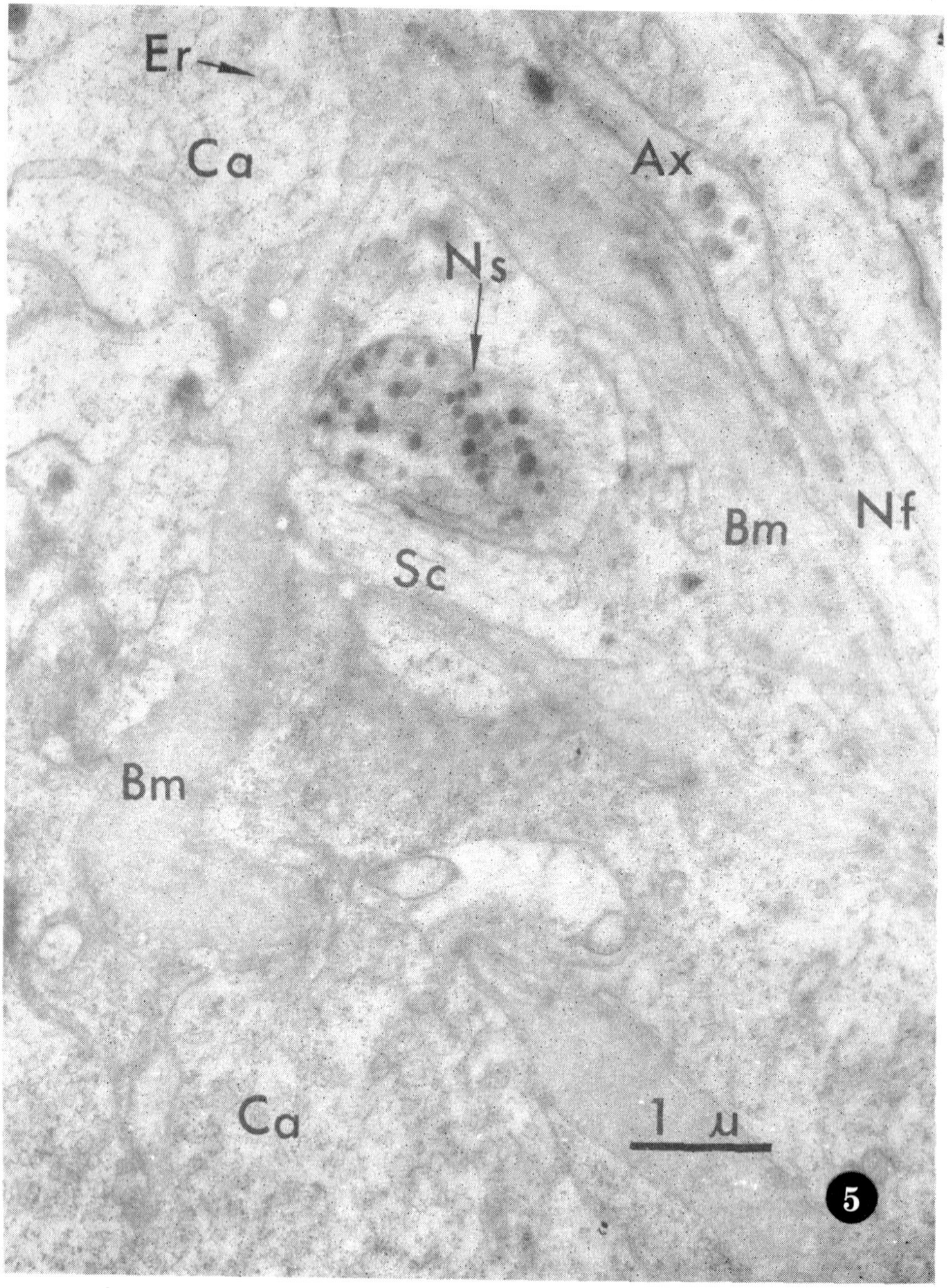

Plate 258. Fig. 5. Cross section of "neurosecretory" neuron adjacent to a peripheral cell of the corpus allatum of a diapausing male pupa of *Hyalophora*. Note continuity of basement membrane surrounding corpus allatum and nervous tissue. (From Waku and Gilbert 1964).

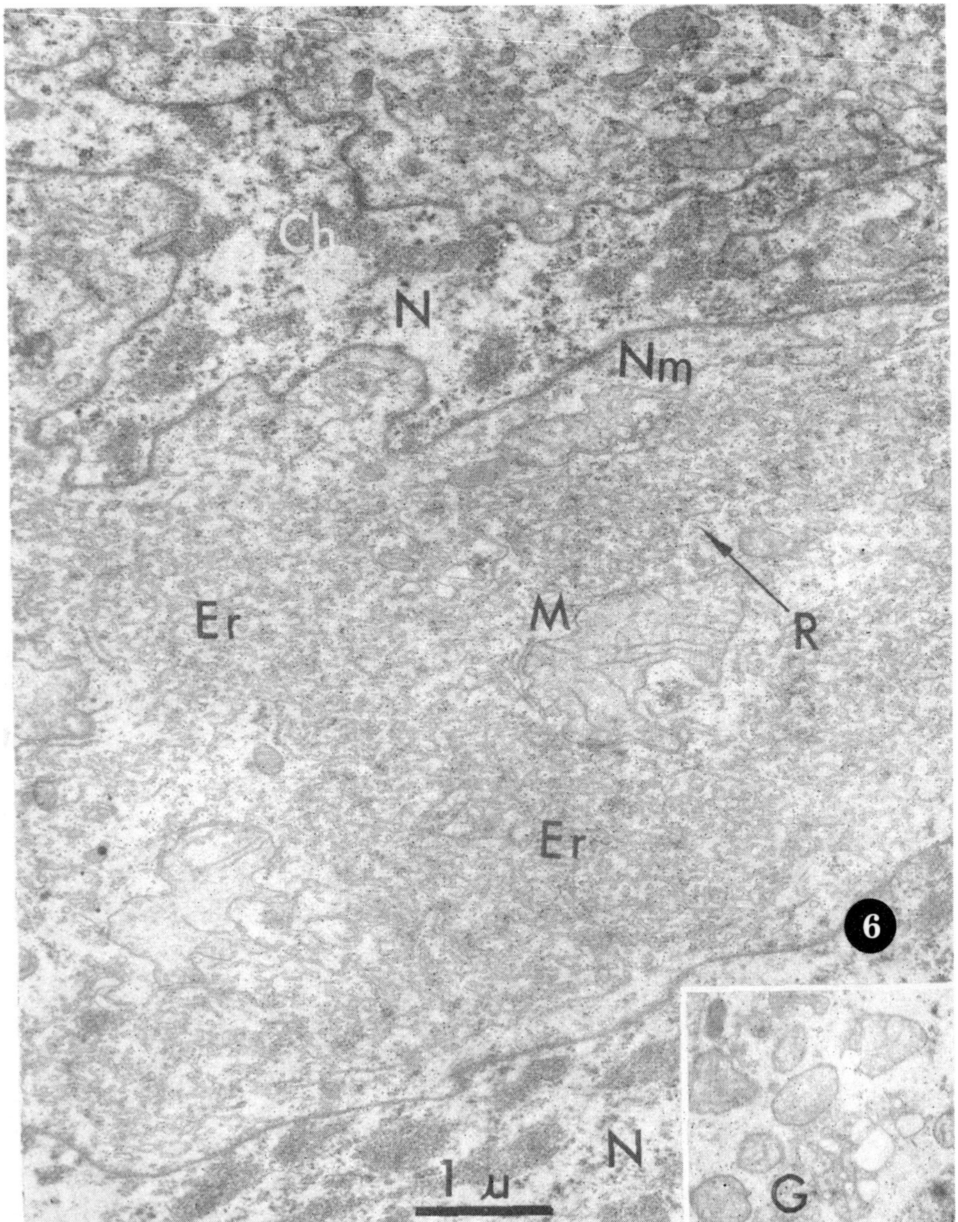

Plate 259. Fig. 6. Nucleus and cytoplasm of a cell from the corpus allatum of a young, male adult of *Hyalophora*. Note convoluted nucleus with much chromatin material, extremely dense endoplasmic reticulum and large mitochondria. Inset shows Golgi apparatus. (From Waku and Gilbert, 1964).

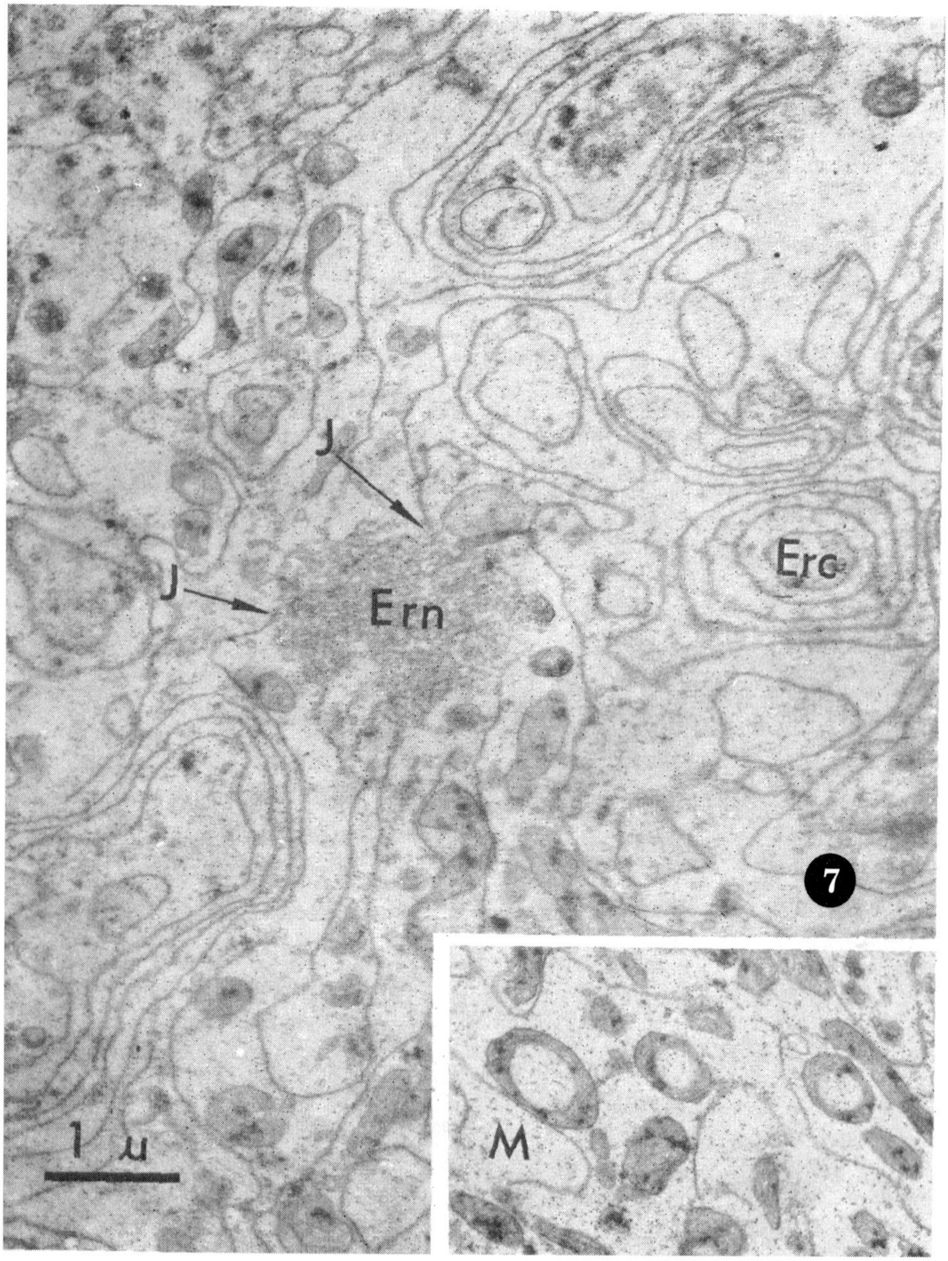

Plate 260. Fig. 7. Cytoplasm of a cell from the corpus allatum of a one day old, male adult of *Hyalophora*. Note scarcity of ribosomes when compared to previous figure and apparently ``indented" mitochondria with numerous intramitochondrial granules (inset). ERC circular ER, ERN net like ER, J junction between circular ER and net like ER. (From Waku and Gilbert, 1964).

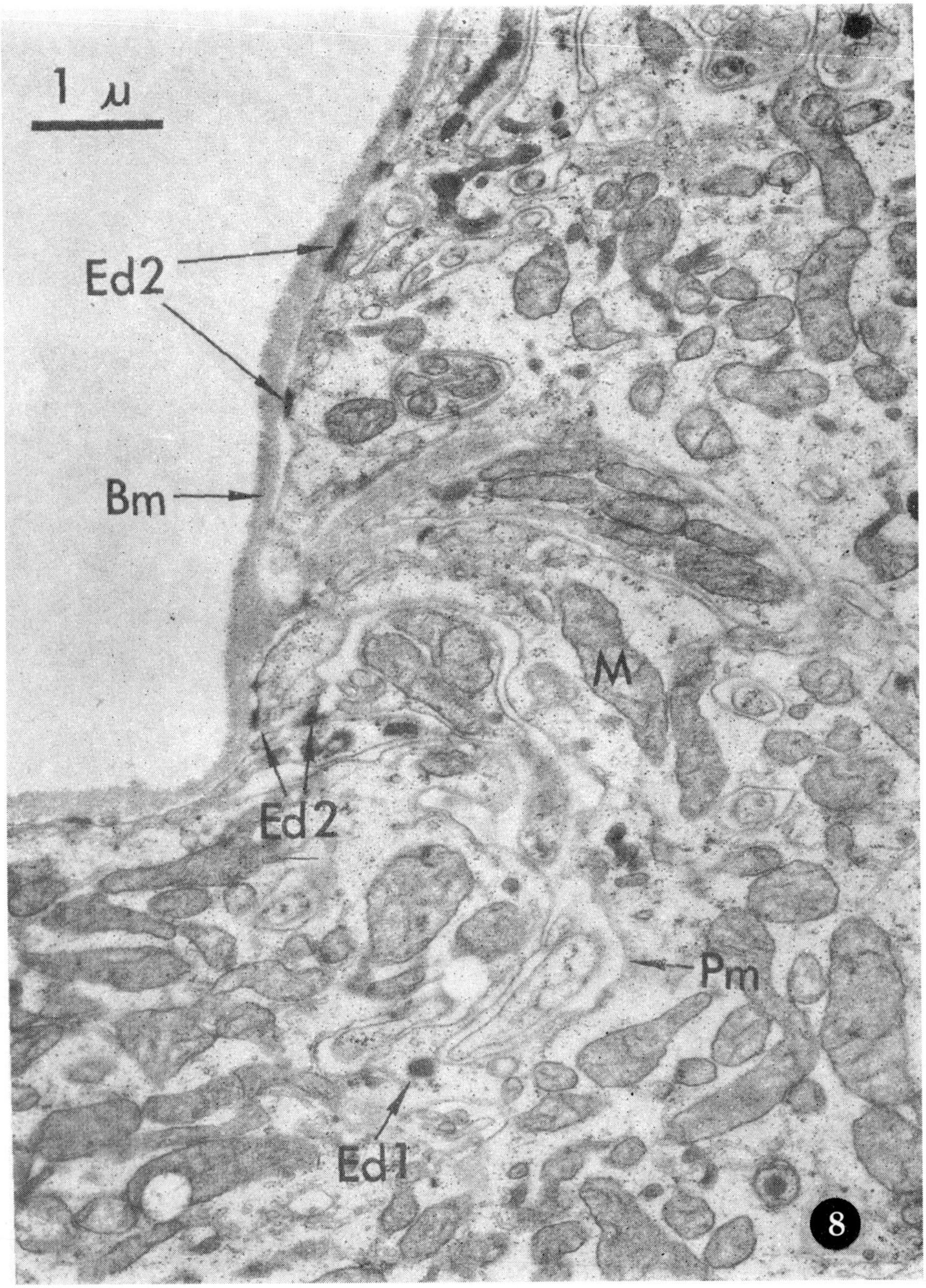

Plate 261. Fig. 8. Portion of a peripheral cell from the corpus allatum of a young, male adult of *Hyalophora*. Note high concentration of mitochondria and electron-dense bodies (Ed1, Ed2). (From Waku and Gilbert, 1964).

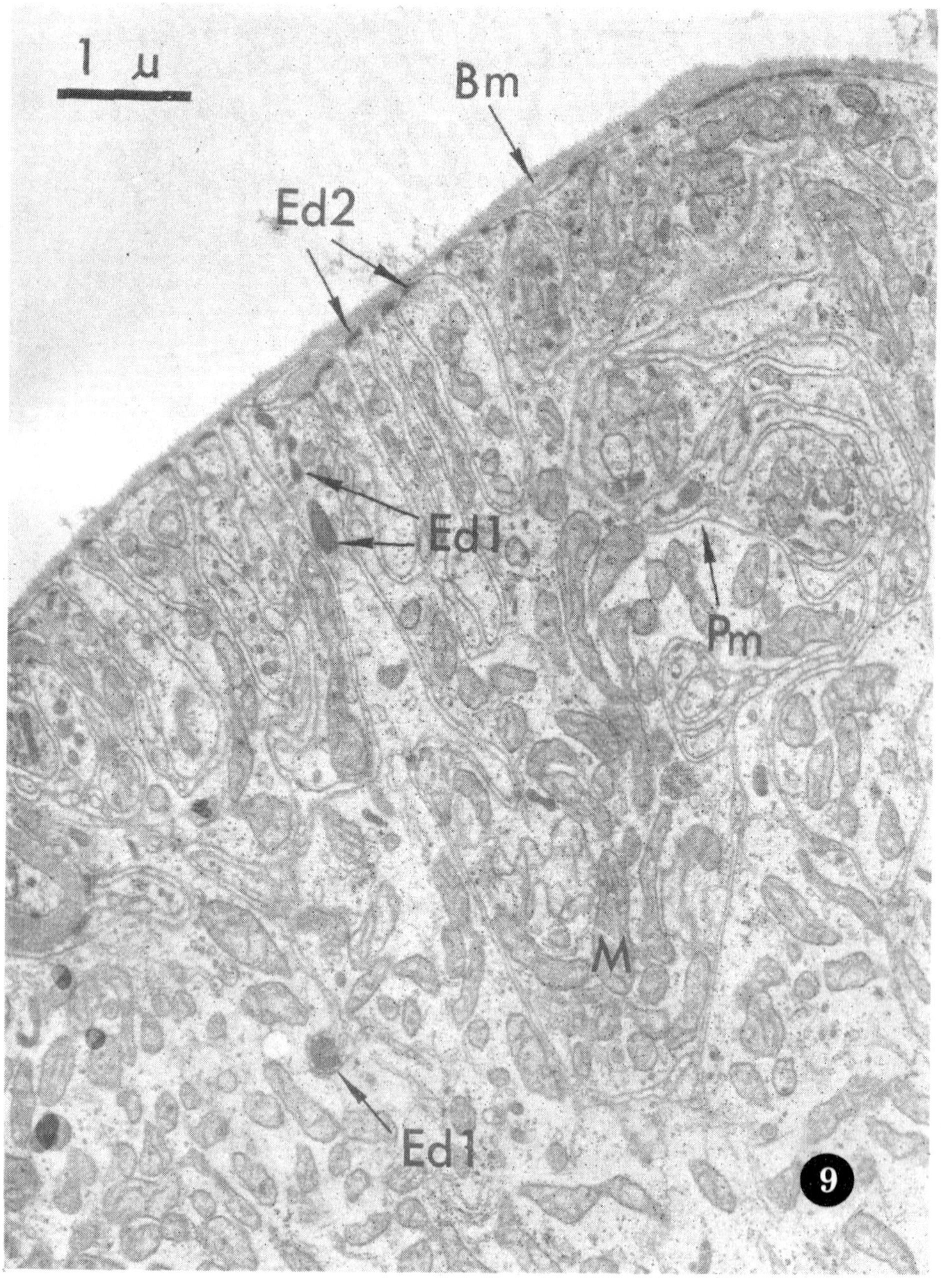

Plate 262. Fig. 9. Portion of a peripheral cell from the corpus allatum of a young, male adult of *Hyalophora*. Note extremely convoluted plasma membrane and electron-dense bodies 2 (Ed2) just below the basement membrane. Double arrows to Ed 1 suggest continuum between Ed1 and Ed 2. (From Waku and Gilbert, 1964).

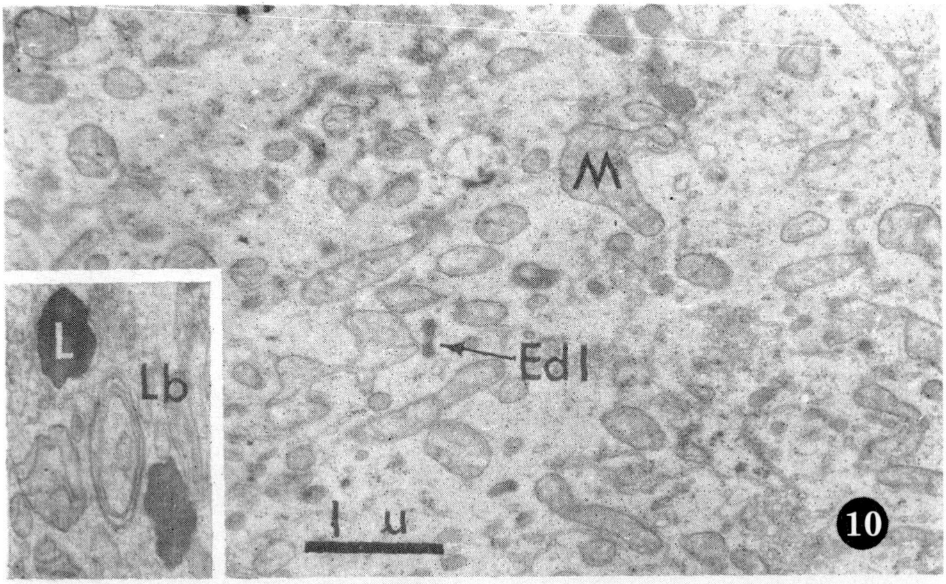

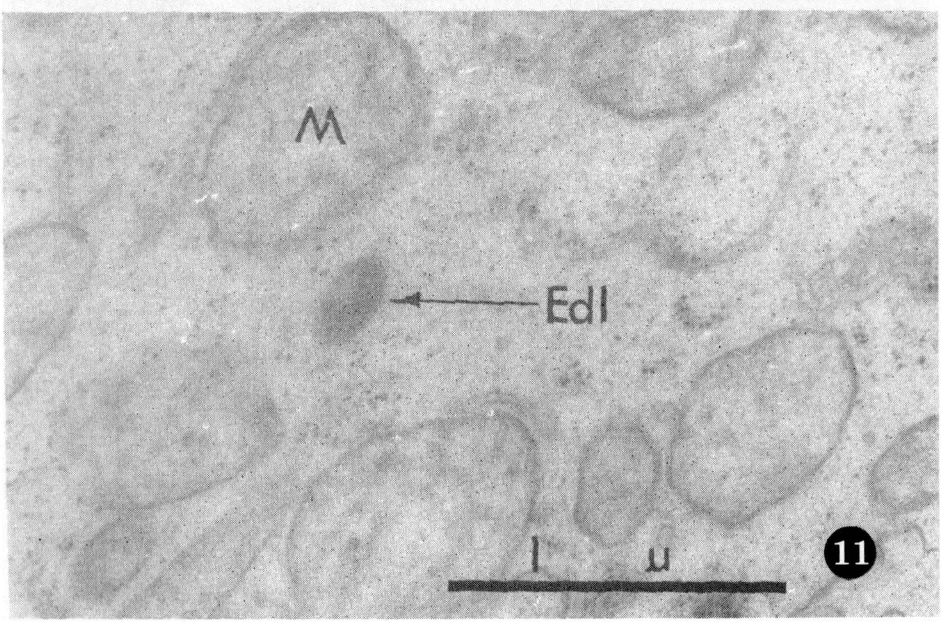

Plate 263. Fig. 10. Portion of a cell from the interior of the corpus allatum of a young, male adult of *Hyalophora*. Note proximity of Ed 1 to the mitochondria. The laminated body (Lb) in the inset is seen in many cells but its nature is yet unknown. Fig. 11. Higher magnification of portion of a cell from the interior of the corpus allatum of a young, male adult of *Hyalophora*. Note electron-dense body 1. (Ed1) (From Waku and Gilbert, 1964).

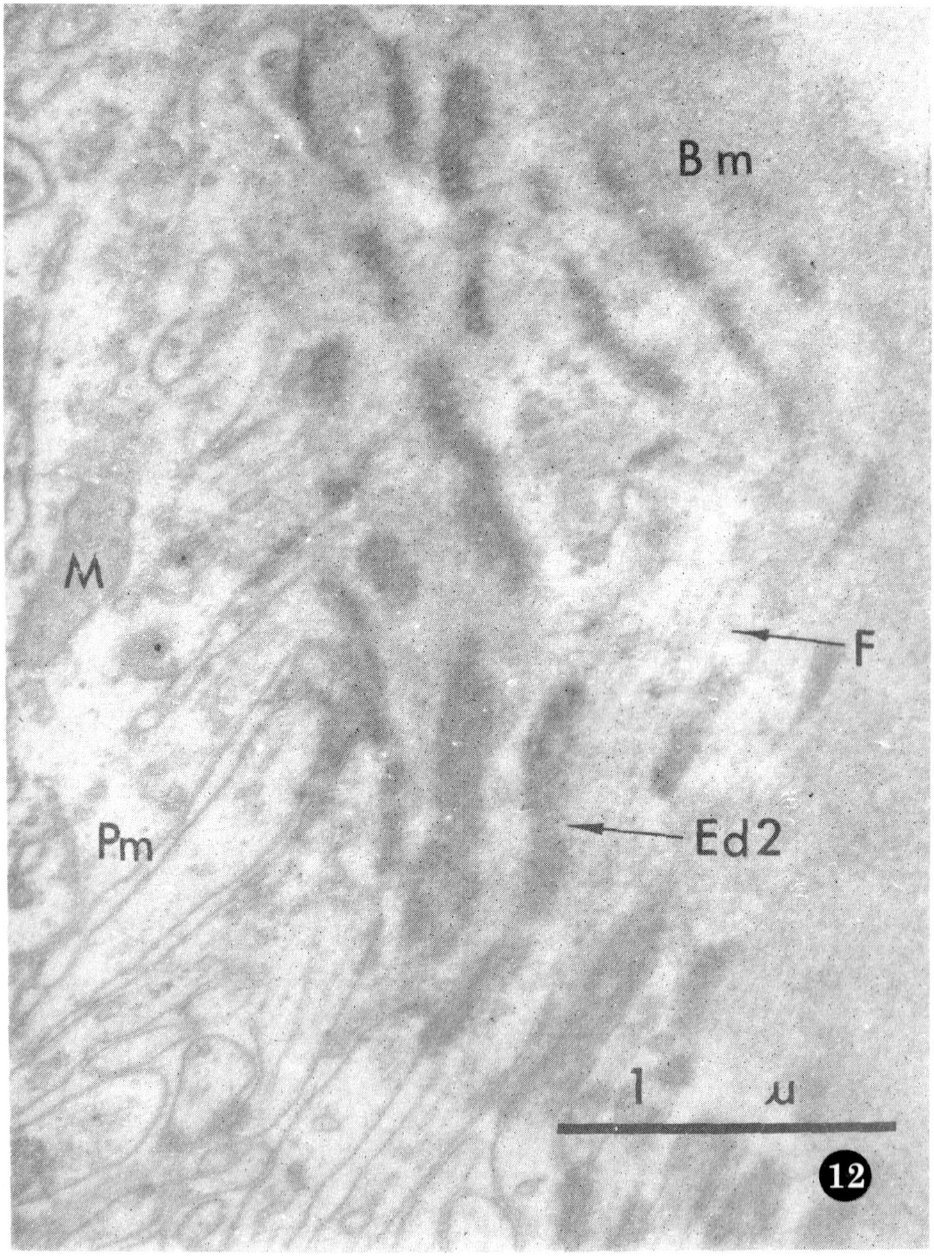

Plate 264. Fig. 12. Higher magnification of a portion of a cell from the periphery of the corpus allatum of a young, male adult of *Hyalophora*. Note the electron-dense bodies 2 (Ed2) protruding from the cell surface into the basement membrane. Filamentous structures are seen within the basement membrane. The extreme convolution of the plasma membrane can also be seen. (From Waku and Gilbert, 1964).

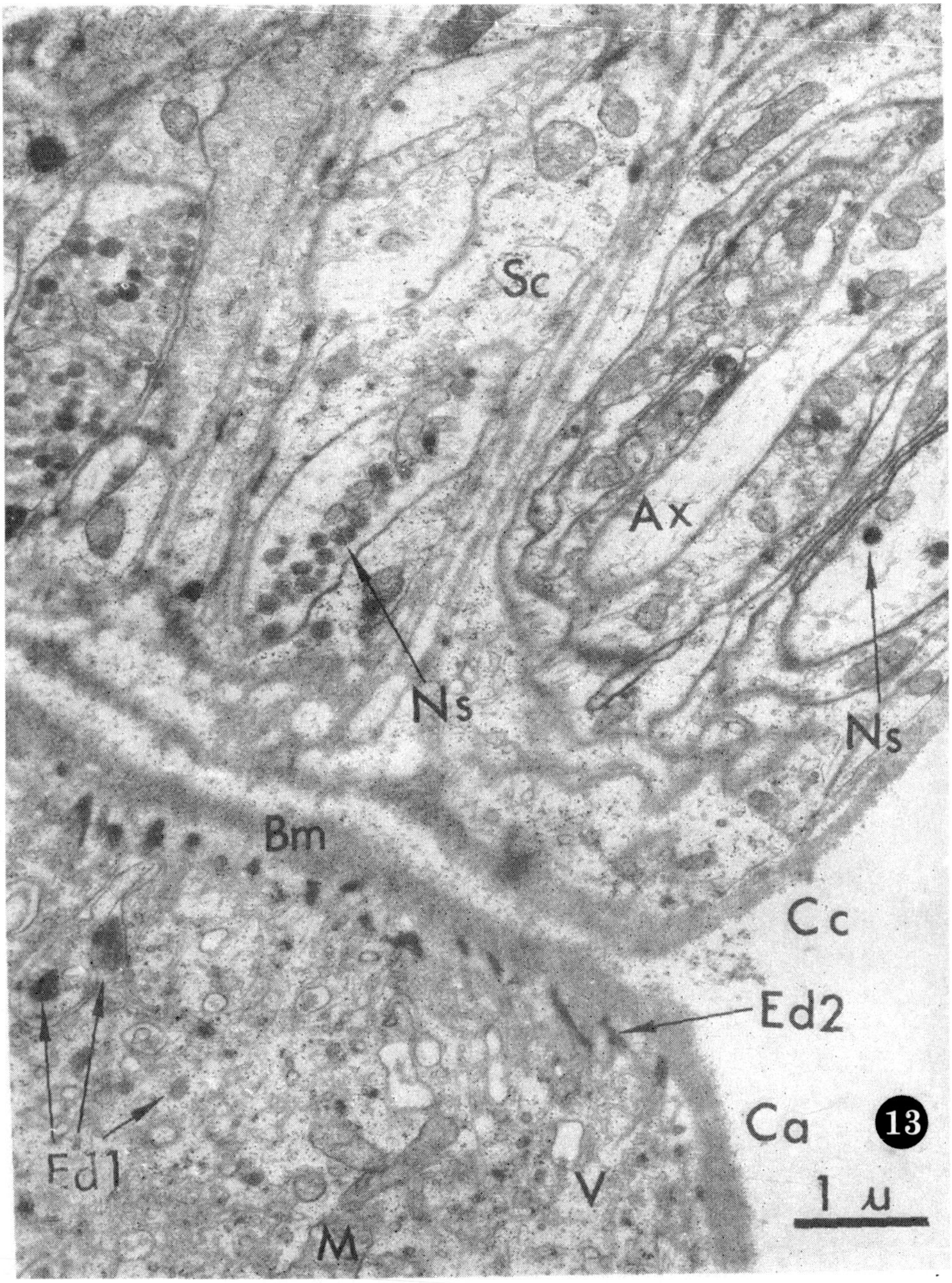

Plate 265. Fig. 13. Peripheral region of the corpus allatum (Ca) and corpus cardiacum (Cc) of a young, male adult of *Hyalophora*. Note ``neurosecretory'' axons in corpus cardiacum. Electron-dense bodies 2 (Ed2) are prominent in the corpus allatum but absent from the corpus cardiacum. (From Waku and Gilbert, 1964).

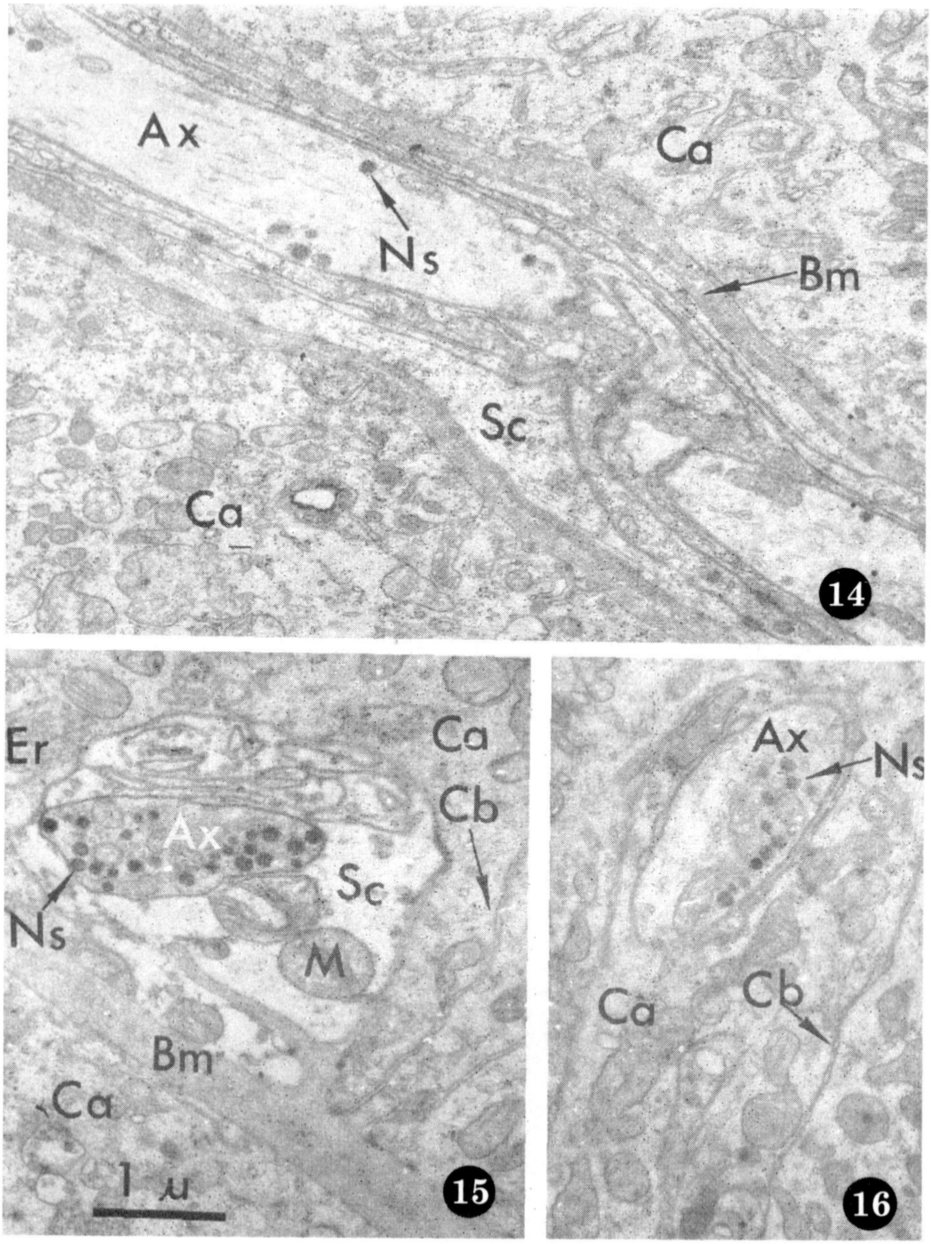

Plate 266. Fig. 14. Nerve fibre innervating a cell from the corpus allatum (CA) of a young, male adult of *Hyalophora*. Note Schwann cell (Sc) and basement membrane covering the neuron. Fig. 15. Similar to figure 14, but in this case the basement membrane covering the axon is absent. Fig. 16. Similar to figures 14, 15 but in this case both the basement membrane and Schwann sheath are absent leaving a ``bare'' axon. (From Waku and Gilbert, 1964).

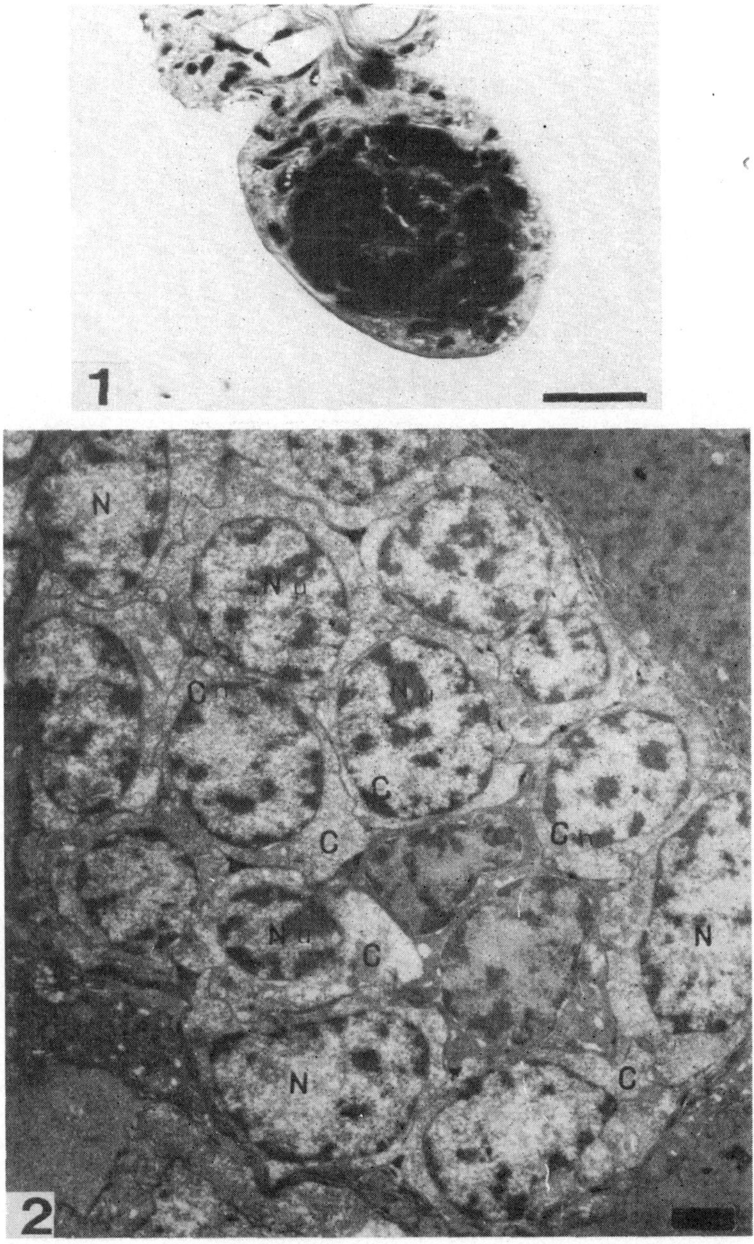

Plate 267. Fig. 1. General appearance of pre-diapausal corpus allatum of *Monema flavescens*. The gland is small and cells are packed in it compactly. Scale 10 µm. Fig. 2. Group of cells of corpus allatum in pre-diapause. Note relative paucity of cytoplasm. The nucleo-cytoplasmic ratio is high. N, nucleus; Ch, chromatin; C, cytoplasm. Scale 2 µm. (From Takeda, 1977).

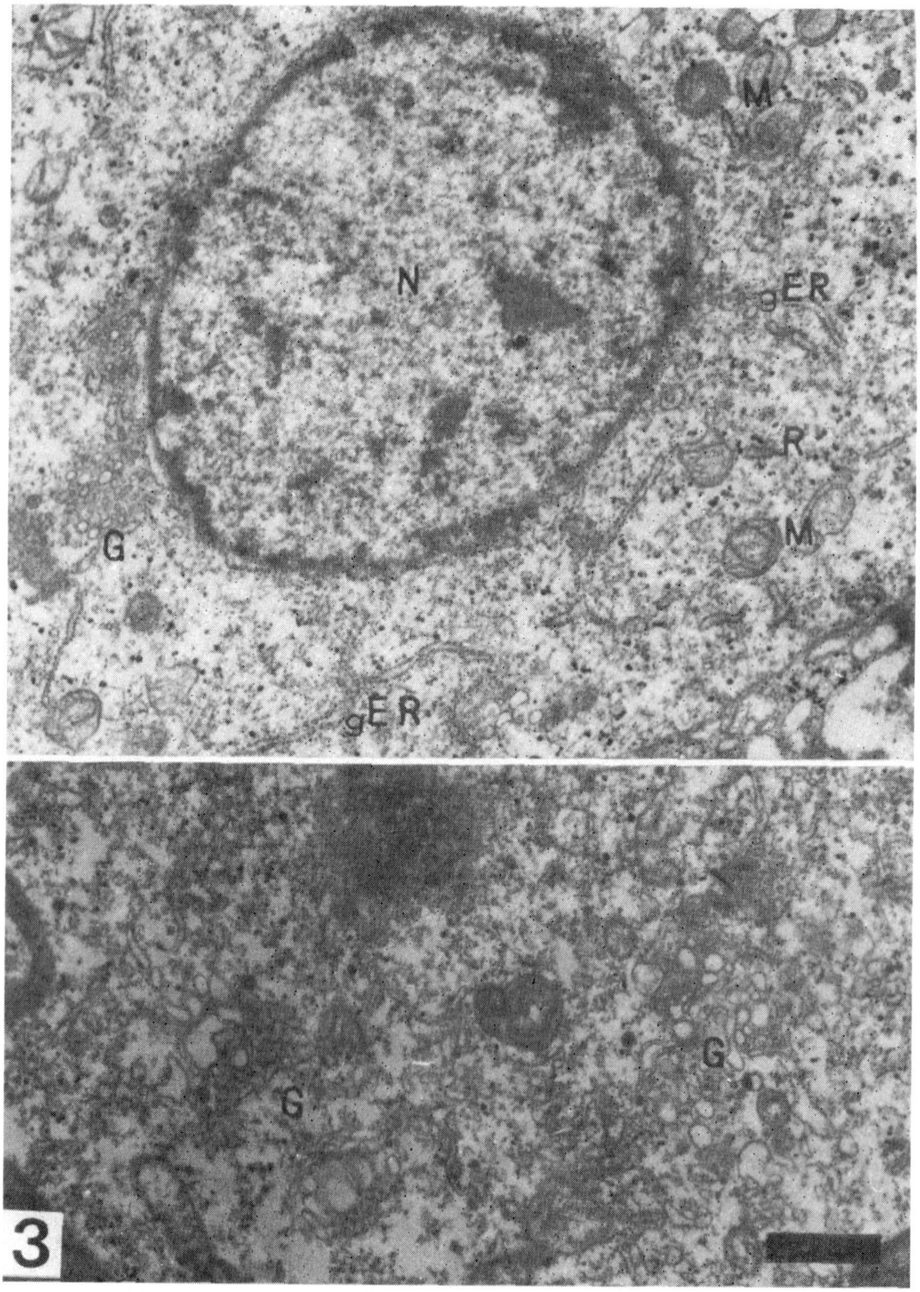

Plate 268. Fig. 3. Higher magnification of gradually activating cells of corpus allatum of *Monema* at the onset of diapause. Note the enlargement and enrichment of cytoplasm and the well developed Golgi apparatus (G). N, nucleus; M, mitochondrion; gER, granular endoplasmic reticulum; R, ribosome. Scale 1 μm. (From Takeda, 1977).

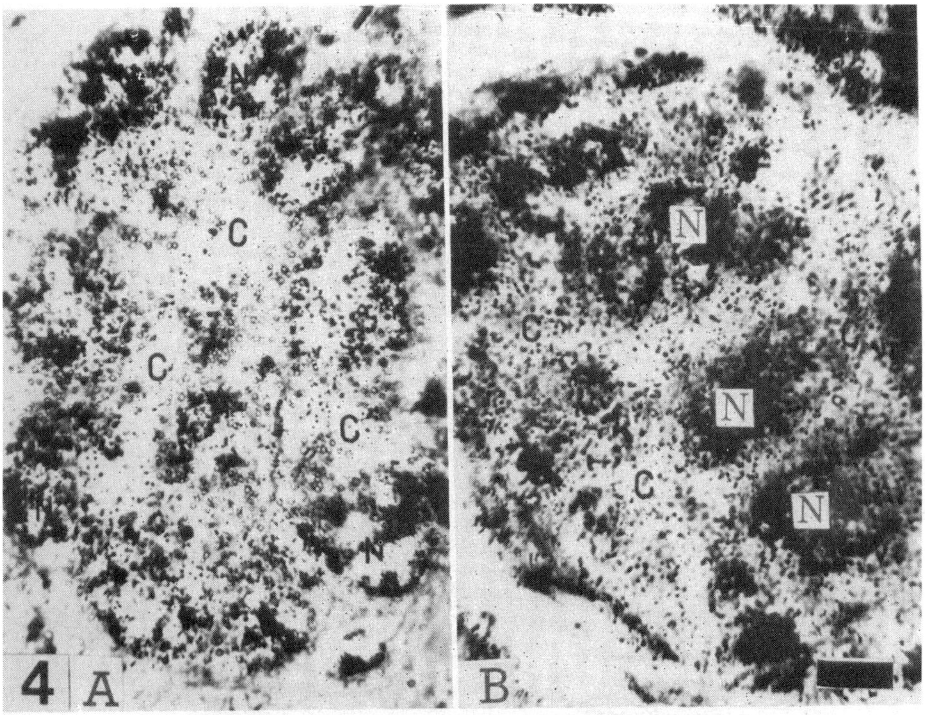

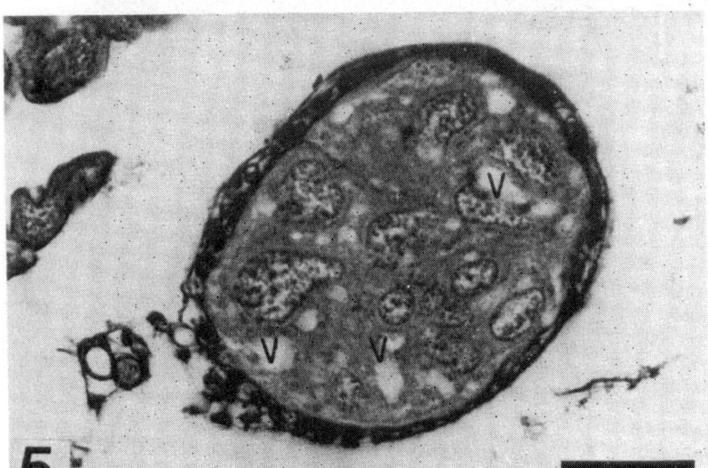

Plate 269. Fig. 4. Autoradiographs of corpus allatum of *Monema* in diapausing prepupa injected with tritium-uridine. Incorporation is found first in the region of nucleolus of each nucleus (A) and then in both nucleus and cytoplasm (B). N, nucleus; C, cytoplasm. Scale 5 μm. Fig. 5. General appearance of corpus allatum of *Monema* in diapause. Cells contain some unstained vacuoles of various sizes and the gland has enlarged. V, vacuole. Scale 10 μm. (From Takeda, 1977).

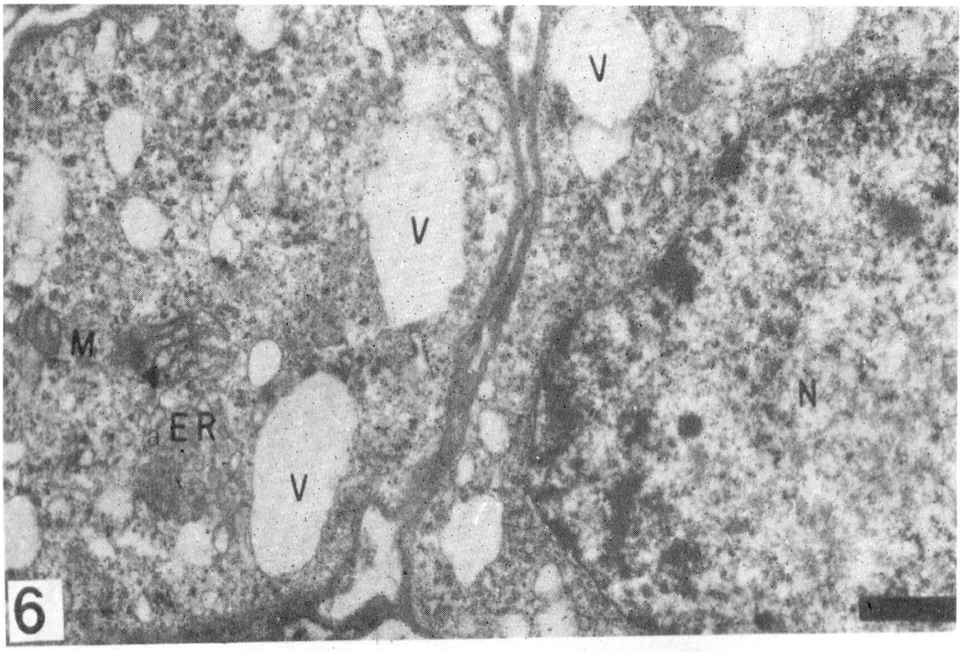

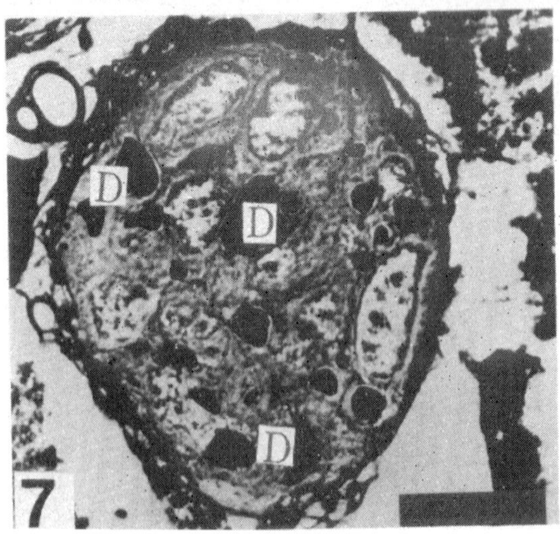

Plate 270. Fig. 6. Higher magnification of corpus allatum in diapause *Monema*. Cytoplasm is mainly filled with a large number of vacuoles. N, nucleus; V. vacuole, M. mitochondria, aER, agranular endoplasmic recticulum. Scale 1 μm. Fig. 7. Frozen section of corpus allatum in diapuase *Monema*. Cells contain some droplets of lipoidal nature stained with sudan black B. D. droplet. Scale 10 μm. (From Takeda, 1977).

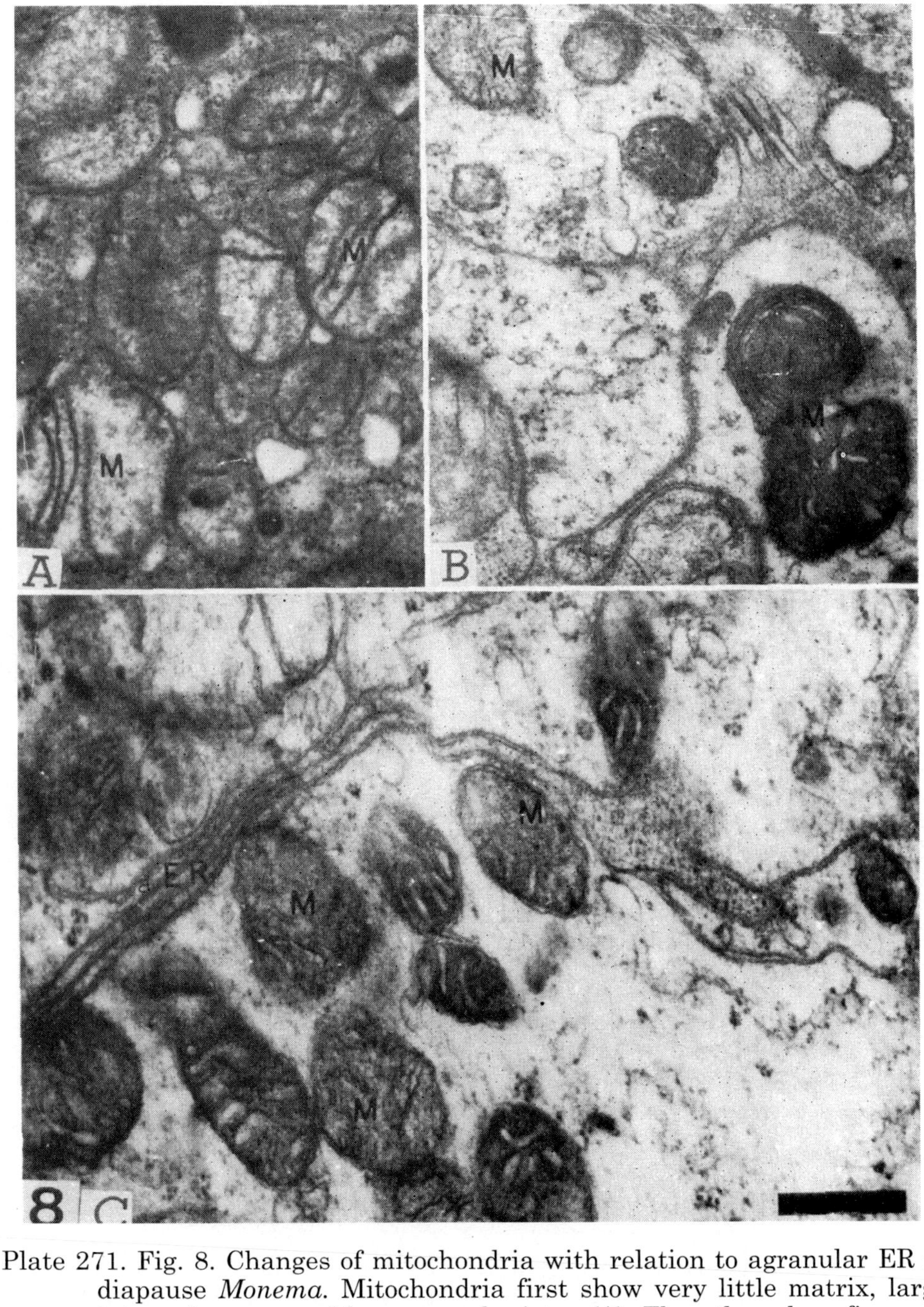

Plate 271. Fig. 8. Changes of mitochondria with relation to agranular ER in diapause *Monema*. Mitochondria first show very little matrix, large internal spaces and long spread cristae (A), Then they show fine well developed and complicated cristae (B) and finally they are closely contacted to some agranular ER (aER) (C), M. mitochondrion. Scale 0.5 μm (From Takeda, 1977).

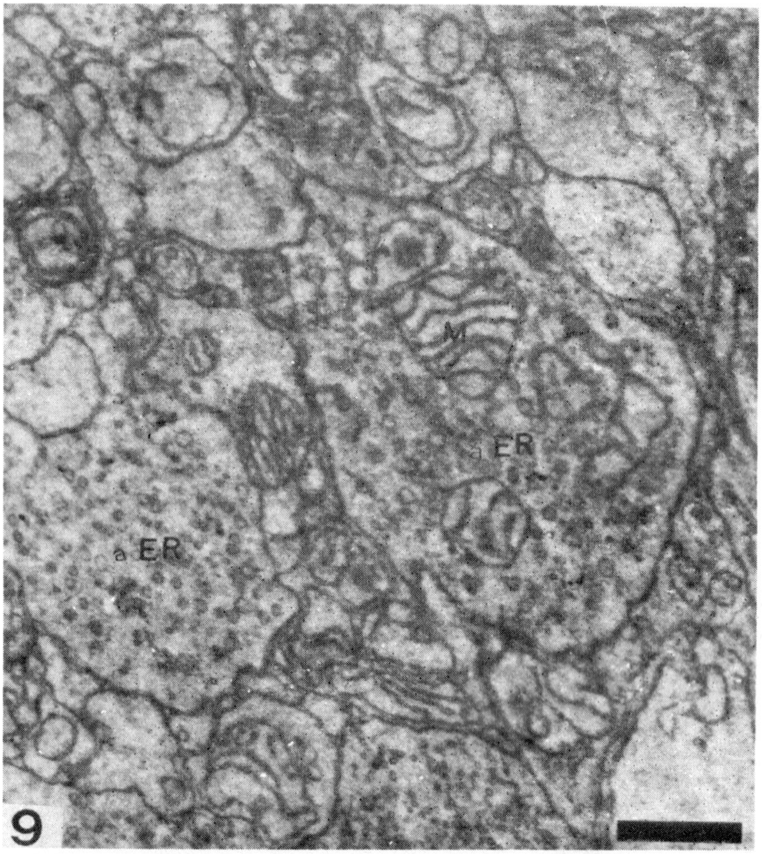

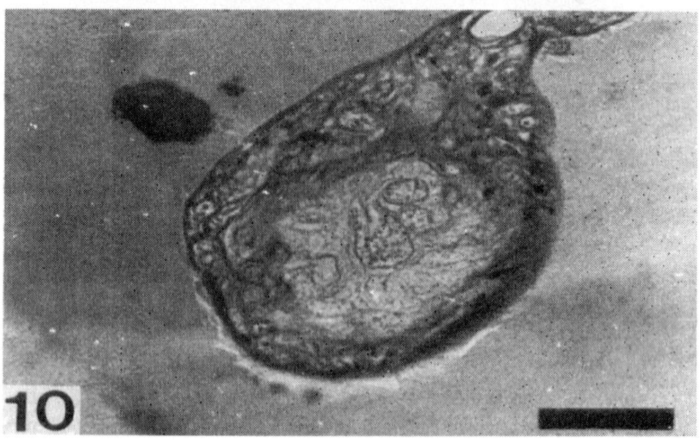

Plate 272. Fig. 9. Well developed agranular ER (aER) found in corpus allatum of *Monema* of higher diapause intensity. M, mitochondrion. Scale 0.5 μm. Fig. 10. General appearance of post-diapausal corpus allatum of *Monema*. Its appearance gradually becomes similar to that seen before diapause. Scale 10 μm. (From Takeda, 1977).

APPENDIX

The morphocological organisation of the neuroendocrine system of some common insects.

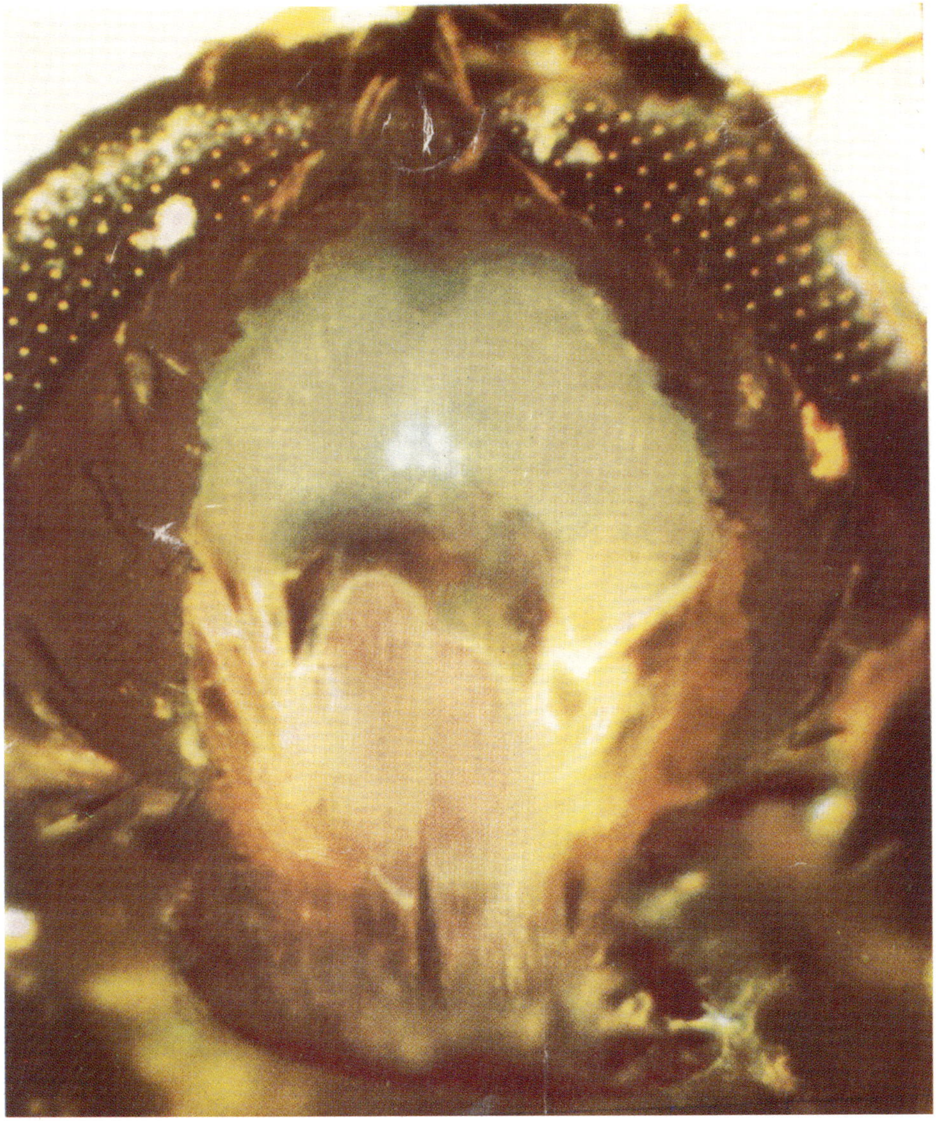

Plate 1 Fig.1 Occiput opened exposing the protocerebrum of a mosquito with the paired groups of bluish medial neurosecretory cells. (From Lea, 1967)

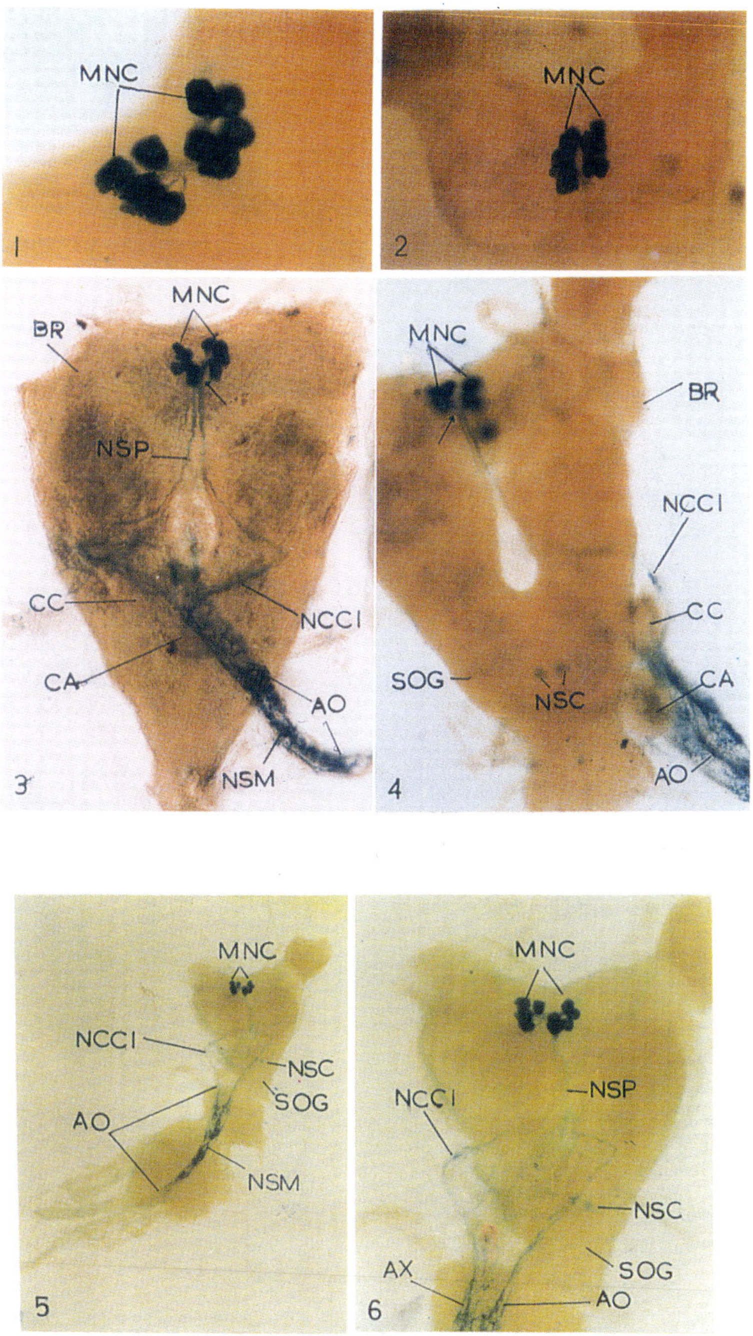

Plate 2. Figs. 1-6. The neuroendocrine system of *Nezara viridula* showing 5 + 5 medial neurosecretory cells and other components (for abbreviations see: plate 4) (Awasthi).

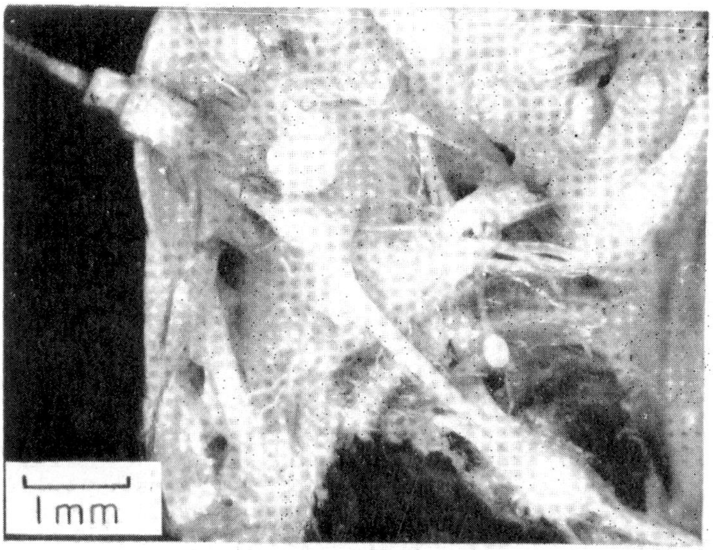

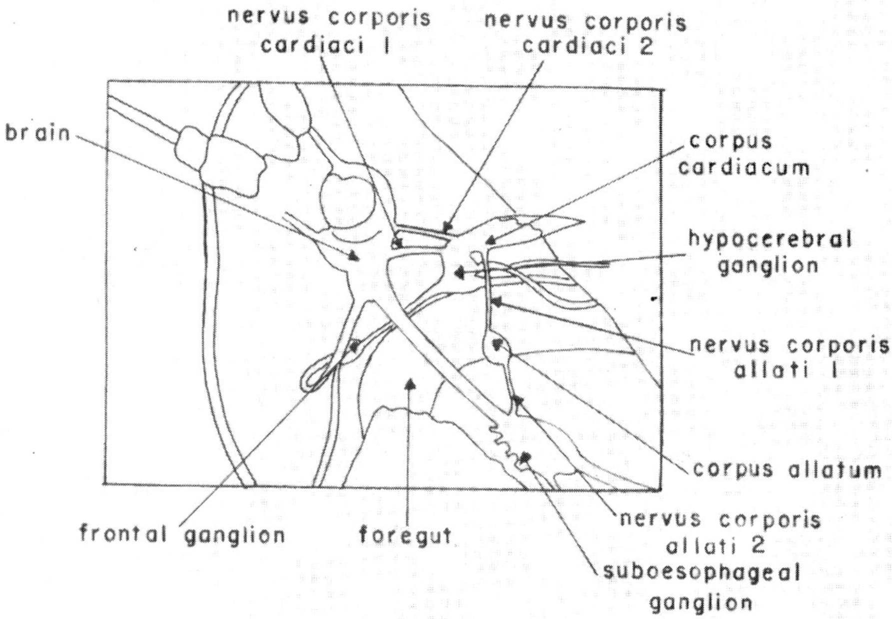

Plate 3. Fig. 1. Dissection of the head capsule of *Schistocerca* sp. showing the anatomical relationships between the central nervous system, the stomatogastric nervous system, and the endocrine glands. (From Strong, 1965).

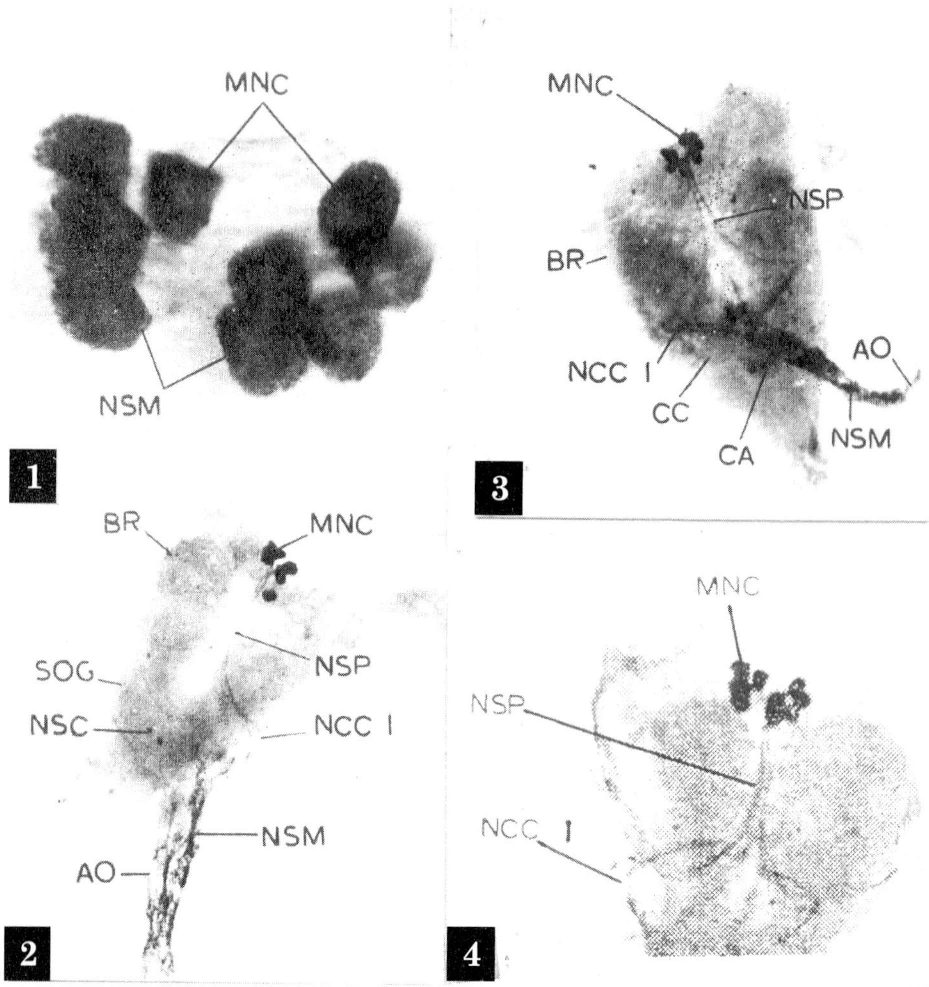

Plate 4. Fig. 1. Dissected pars intercerebralis of the protocerebrum of *N. viridula* showing 5+ 5 A type of medial neurosecretory cells (MNC) which are heavily loaded with neurosecretory material (NSM). (PAVB, *in situ*) X 455. Fig. 2. Whole mount of the brain-retrocebral endocrine glands showing medial neurosecretory cells (MNC). The nervi corporis cardiaci-I (NCC-I) innervates the aorta wall (AO) from lateral aspects. The two A-type of neurosecretory cells (NSC) in the suboesophageal ganglion (SOG) are clearly marked out. (PAVB, *in situ*) X 110. Fig. 3. Whole mount of the neurosecretory system and retrocerebral endocrine glands of *N. viridula* showing medial neurosecretory cells (MNC) and neurosecretory pathways (NSP) within the brain (BR). The two nervi corporis cardiaci-I (NCC-I) situated out side the brain are seen entering the aorta wall (AO). The NSM and fibres of NCC-I are confined upto the middle part in the aorta. (PAVB *in situ*) X 95. Fig. 4. Whole mount of the anterior part of the neurosecretory system showing 5 + 5 medial neurosecretory cells and other components viz MNC, NSP and NCC-1. (From Awasthi, 1972a).

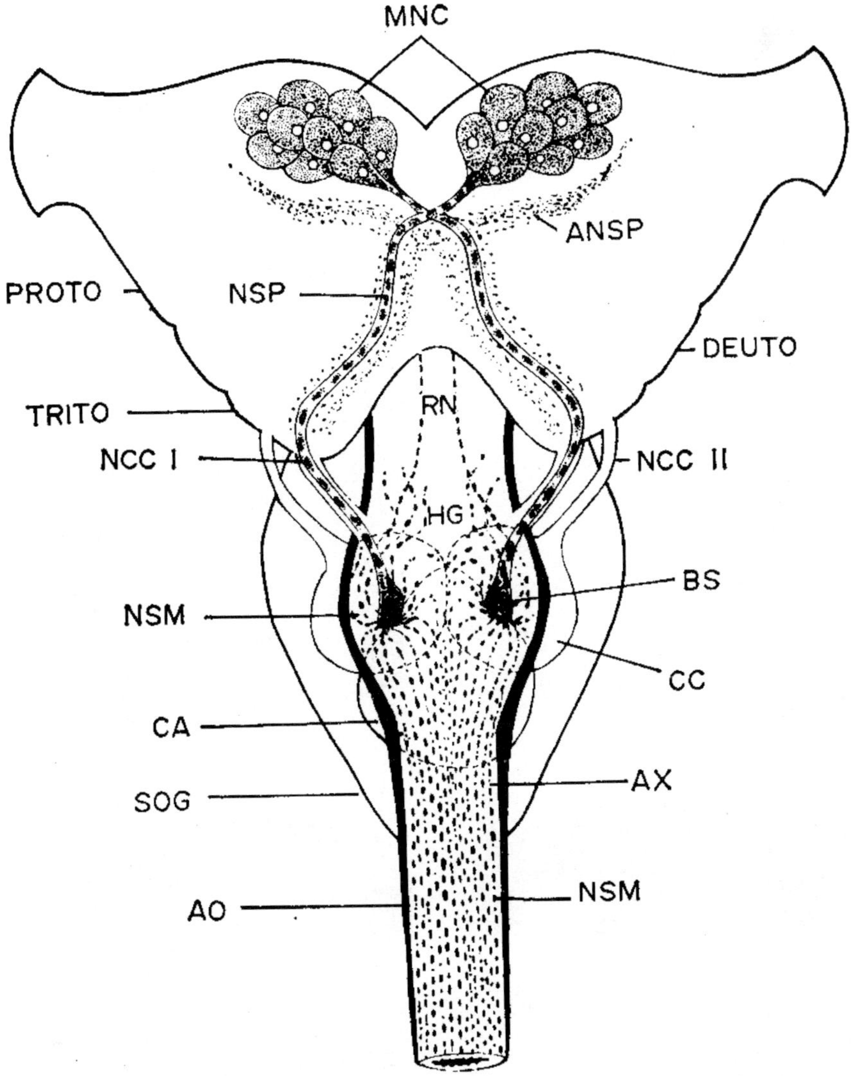

Plate 5. Fig. 1. Schematic diagram of the neurosecretory system of *Metochus uniguttatus* Thunb. Aorta (AO), axons (AX), associated neurosecretory pathway (ANSP), corpora cardiaca (CC), corpus allatum (CA), deutocerebrum (DEUTO), hypocerebral ganglion (HG), medial neurosecretory cells (MNC), neurosecretory pathways (NSP), nervi corporis cardiaci I and nervi corporis cardiaci II (NCC-I and NCC-II), neurosecretory material (NSM), Protocerebrum (PROTO), recurrent nerve (RN), suboesophageal ganglion (SOG), tritocerebrum (TRI). Further, note the origin of secondary axons from the bulbous swelling (BS). (From Awasthi, 1973c).

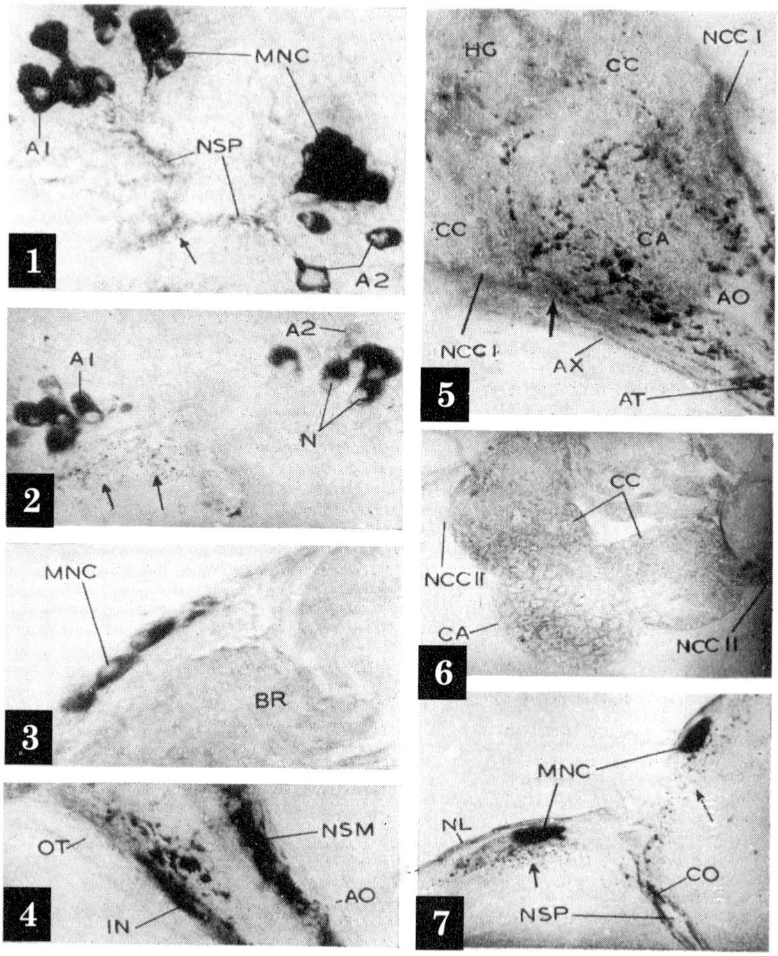

Plate 6. Fig.1. Dissected pars intercerebralis of protocerebrum, of *Metochus,* showing 2 groups of medial neuro-secretory cells (MNC), each of them composed of 9 cells, 5 A-1 and 4 A-2 cells. Neurosecretory pathway (NSP) of each side crosses over at arrow. X 390. Fig.2. Dissected pars intercerebralis of the protocerebrum. A-1 cells full and A-2 cells with very small amount of NSM, thus showing a metachronous rhythm. Oozed NSM from cell bodies stored below (arrow). Note axonal position of nuclei in the cells. X 300. Fig. 3. T.S. of brain of nymph. Medial neurosecretory cells (MNC) contain lesser amount of NSM. The pathway is devoid of NSM X 300. Fig. 4. Whole mount of aorta (AO). Note presence of neurosecretory material (NSM) in inner layer (IN) only, outer (OUT) being devoid of NSM. X 400. Fig. 5. Whole mount of the retrocerebral endocrine glands with intact aorta (AO). Aorta lies dorsally and endocrine glands ventral to it. Note bulbous swelling of NCC-1 (arrow) that gives rise to numerous secondary axons (AX). Bulbous swelling (↑), secondary axons and terminals are packed with NSM. Two corpora cardiaca (CC) and a medial corpus allatum (CA) are also visible. Some secondary axons are also seen extending up to hypocerebral ganglion (HG). X 450. Fig. 6. T.S. of retrocerebral endocrine glands. Corpora cardiaca (CC) and corpus allatum (CA) are without NSM. Further distinguish nervi corporis cardiaci II (NCC-II) terminating in the CC and are devoid of NSM. X 390. Fig. 7. T.S. of brain showing 2 medial neurosecretory cells (MNC) and the pathways (NSP) that cross over each other (CO). Note oozed NSM from cell bodies deposited below MNC (arrow). X 390. (From Awasthi, 1973c).

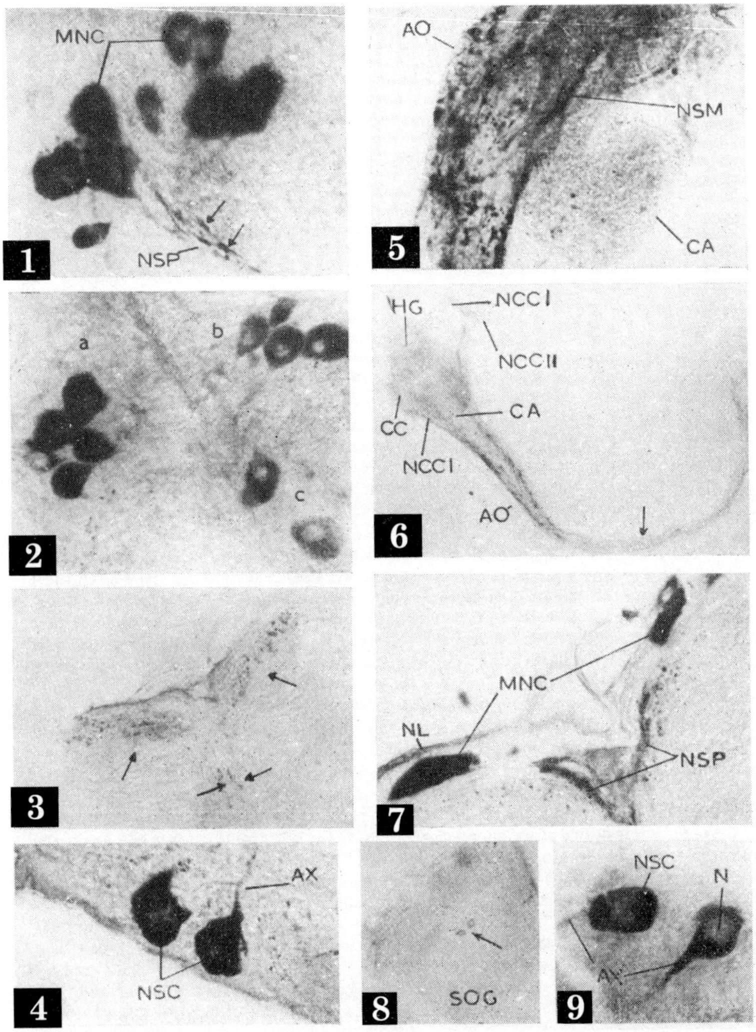

Plate 7. Fig. 1. Dissected pars intercerebralis of protocerebrum of *Metochus* showing one group of medial neurosecretory cells, consisting of 9 cells. Transport of NSM via axonal pathway (NSP) is clearly visible (arrowed). X 430. Fig. 2. Dissected pars intercerebralis of the protocerebrum showing 3 medial groups of neurosecretory cells, viz. a, b, c, consisting of 5, 4, and 2 cells. Nuclei are oval and located towards axonal side. X 400. Fig. 3. T.S. of brain showing associated neurosecretory pathways formed by oozed NSM from cell bodies and the axonal pathways. X 360. Fig. 4. T.S. of suboesophageal ganglion, showing 2 A-type neurosecretory cells (NSC) with plenty of NSM in cells and axons. X 410. Fig. 5. Whole mount of aorta (AO) showing abundant NSM in the axons. Corpus allatum is devoid of NSM. X 410. Fig. 6. Whole mount of retrocerebral endocrine glands with dorsal aorta (AO) showing 2 corpora cardiaca (CC) and a corpus allatum (CA) ventral to aorta. NCC-I and NCC-II and hypocerebral ganglion (HG) are also visible. Further, note posterior part of aorta (arrowed) is devoid of NSM. X 80. Fig. 7. T.S. of brain, showing 2 medial neurosecretory cells (MNC) and their pathways packed with plentiful NSM. X 300. Fig. 8. Whole mount of suboesophageal ganglion of a newly-moulted bug. Two neurosecretory cells are visible with small amount of NSM. X 80. Fig. 9. Dissected suboesophageal ganglion showing 2 prominent neurosecretory cells (NSC). Nuclei are concentric in position (in contrast to MNC, where these are axonal in position). Cells and axons are packed with NSM. X 410. (From Awasthi, 1973c).

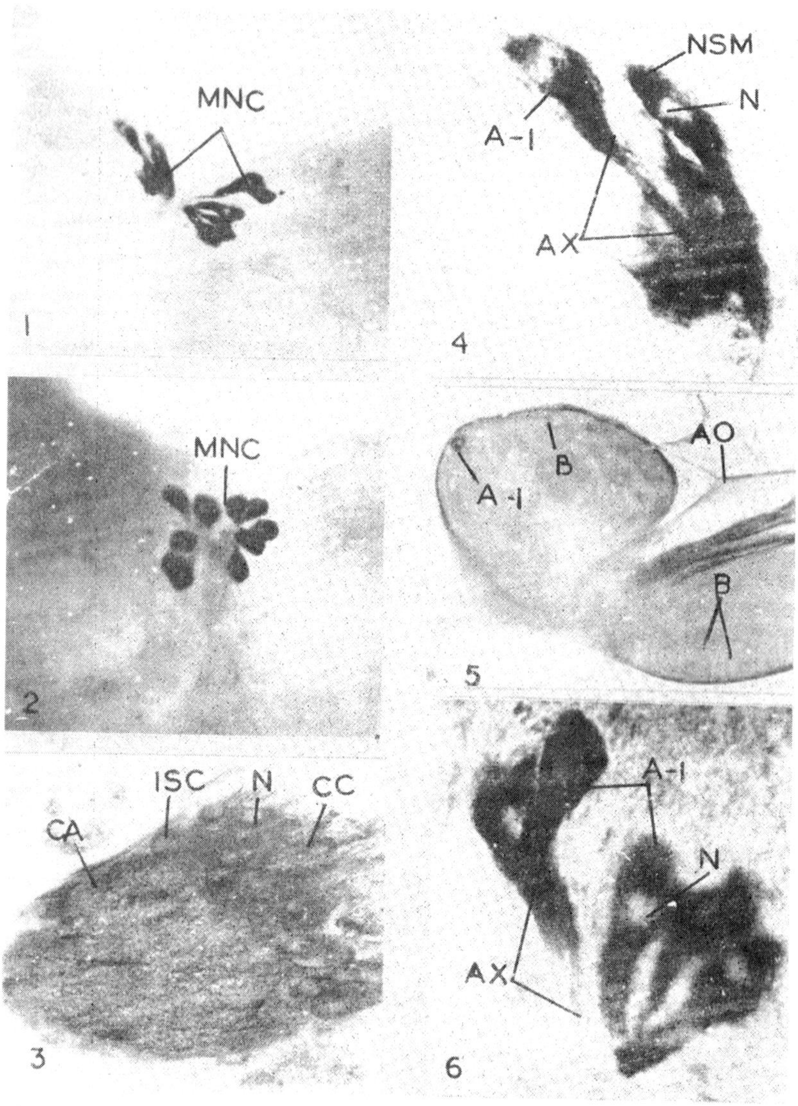

Plate 8. Fig. 1. Dissected pars intercerebralis medialis of the protocerebrum of the milkweed bug. *Lygaeus pandurus* Scop. showing two clusters of medial neurosecretory cells loaded with cysteine and/or cystine rich NSM (PAVB *in situ*) X 160. Fig. 2. Dissected pars intercerebralis of the protocerebrum showing the two groups of cells that lie very close to one and other each cluster consisting of 5 + 5 cells. Note the oval shape of the cells. PAVB (*in situ*) X 240. Fig. 3. Frontal section passing through the CC and CA. Note the two types of chromophil cells in the CC and one type of small cells in the CA. An allatic nerve is absent. PARF X 450. Fig. 4. A magnified cluster of cells of Fig. 1 showing the details of the cells viz. Nuclei (N) are small in proportion to perikaryon. Axons (AX) due to copious amount of NSM are clearly marked out. X 750. Fig. 5. Sagittal section passing through brain and suboesophageal ganglion of a 4 day old bug. Note A and B- cells of the brain. The two B-cells in the suboesophageal ganglion are clearly seen. Note a small amount of NSM in the A-and B-cells. PF X 160. Fig. 6. A magnified part of Fig. 1 showing one medial group of 5 A-1 type of cells. X 750 (From Awasthi, 1980a).

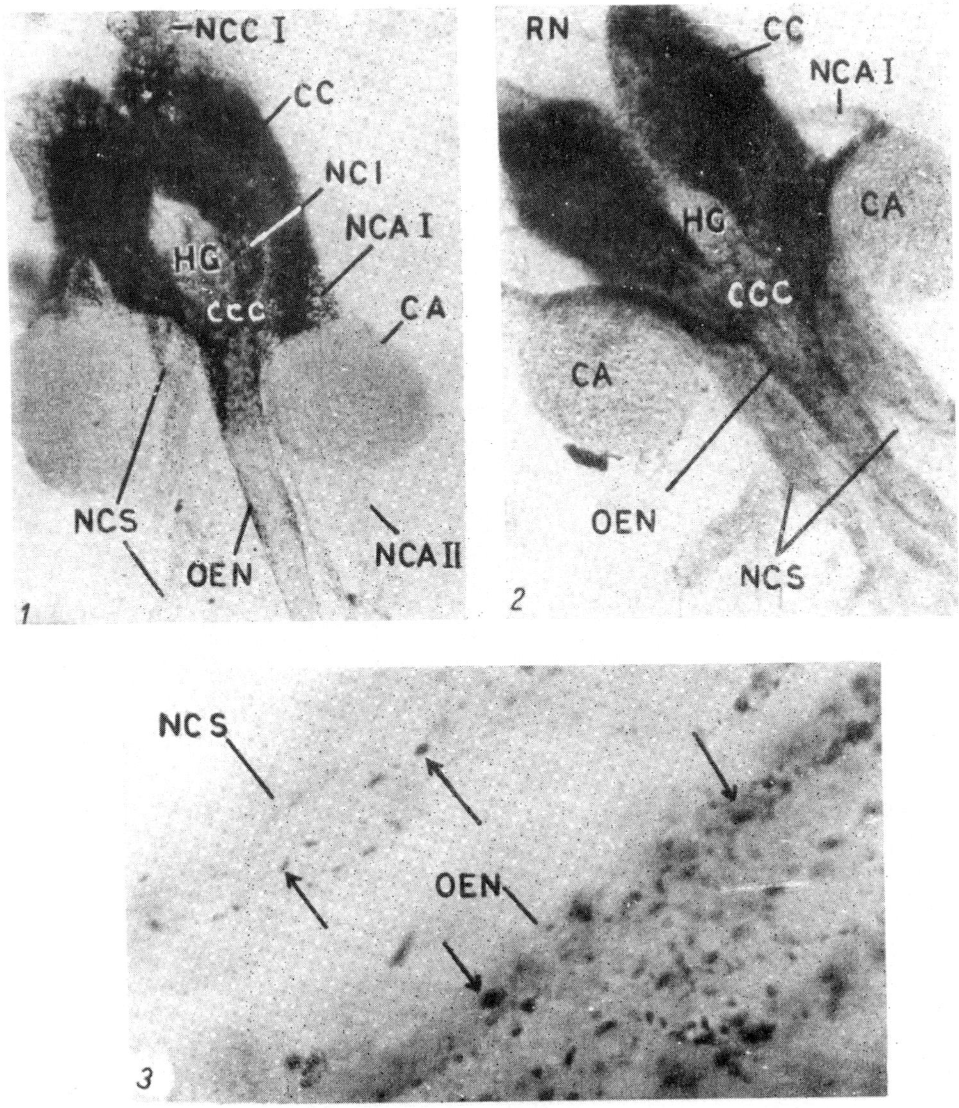

Plate 9. Fig. 1. Whole mount of retrocerebral endocrine glands of *Epilampra* with associated nerves. Note a heavy accumulation of cysteine or cystine rich NSM in the nervi corporis cardiaci I (NCC-I), corpora cardiaca (CC), nervi corporis allati I (NCA-I), nervi cardiostomatogastrici (NCS), nervi corporis cardiaci interni (NCI), commissura corporis cardiaci (CCC) and oesophageal nerve (OEN). The corpora allata CA and nervi corporis allati II (NCA-II) are devoid of ((NSM); (PAVB)). X 145. Fig. 2. Whole mount of retrocerebral endocrine glands of *Epilampra* with associated nerves. The NCA-I, CC, NCI, NCS, HG, CCC, and OEN are loaded with cysteine or cystine rich NSM. Note the mode of attachment of CA with the CC. The CA are devoid of NSM. The posterior part of recurrent nerve (RN) contains a small amount of NSM (PAVB). X 145. Fig. 3. Part of the whole mount of OEN and NCS. Note an accumulation of cysteine or cystine rich NSM (arrows) in both the nerves (PAVB). X 590. (From Awasthi, 1975b).

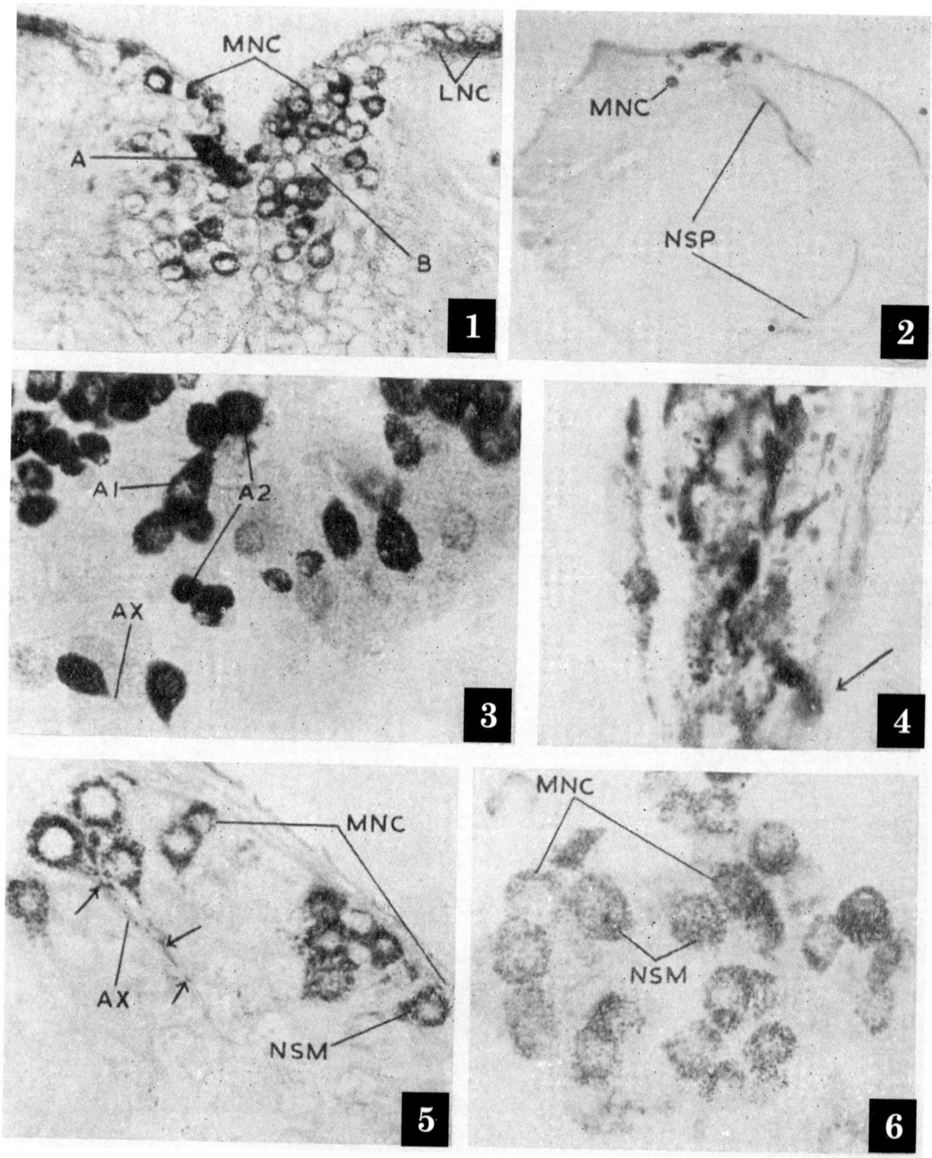

Plate 10. Fig. 1. T.S. of the brain of *Gryllodes sigillatus* showing medial (MNC) and lateral (LNC) neurosecretory cells. Note the deeply stained A-cells and the lightly stained B-cells, (PF). X 400. Fig. 2. L.S. of the brain, showing medial neurosecretory cells (MNC) and the neurosecretory pathway (NSP). (PAVB). X 85. Fig. 3. Dissected pars intercerebralis of the protocerebrum showing A-1 and A-2 cells. (PAVB- *in situ*). X 400. Fig. 4. A portion of the corpus cardiacum cut longitudinally showing abundant neurosecretory material in the axons and its diffusion through the wall of the corpus cardiacum (arrow). (PAVB). X 1500. Fig. 5. H.S. of the brain. Note the two medial groups of neurosecretory cells (MNC) and the movement of neurosecretory material along the axon (AX. Arrows). (PAVB), X 400. Fig. 6. Dissected pars intercerebralis of the protocerebrum. Note the granular nature of neurosecretory material (NSM) in the medial neurosecretory cells (MNC). (PF *in situ*) 400. (From Awasthi, 1969a).

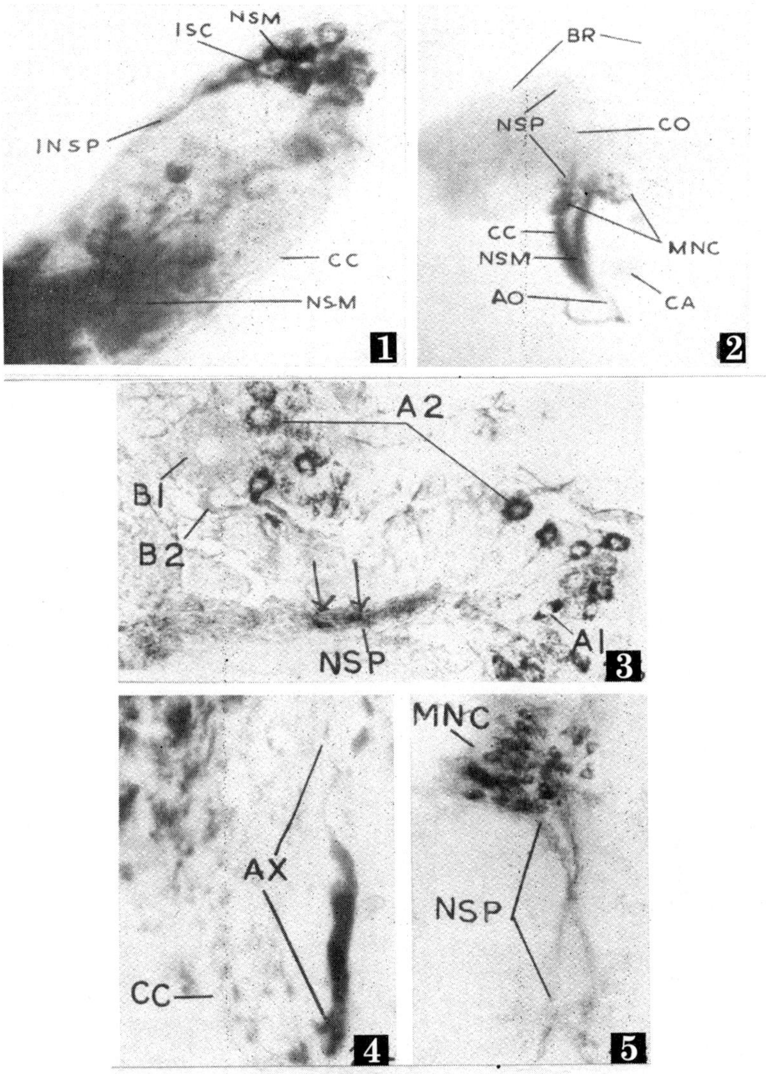

Plate 11. Fig. 1. Whole mount of one lobe of the corpus cardiacum of *Gryllodes* showing intrinsic secretory cells (ISC), alongwith neurosecretory material (NSM) and intrinsic neurosecretory pathway (INSP). (PAVB *in situ*). X 500. Fig. 2. Whole mount of the brain and the postcerebral endocrine glands. Note the two medial groups of neurosecretory cells (MNC), neurosecretory pathways (NSP), crossing over (CO), corpora cardiaca (CC) with neurosecretory material (NSM). The corpus allatum (CA) of one side is also seen. Note the absence of NSM in the aorta (AO). (PAVB- *in situ*) X 65. Fig. 3. T.S. of the brain showing A-1, A-2 and B-1, B-2 cells. Note abundant NSM in the neurosecretory pathway (NSP). (PF) X 300. Fig. 4. L.S. of the corpus cardiacum (CC). Note the bulbous axon terminal (AX) owing to heavily loaded with NSM. (PAVB) X 1500. Fig. 5. Dissected pars intercerebralis of the protocerebrum of a immature individual. Note loosely arranged medial neurosecretory cells (MNC), neurosecretory pathways are without NSM (PAVB *in situ*) X 85. (From Awasthi, 1969a).

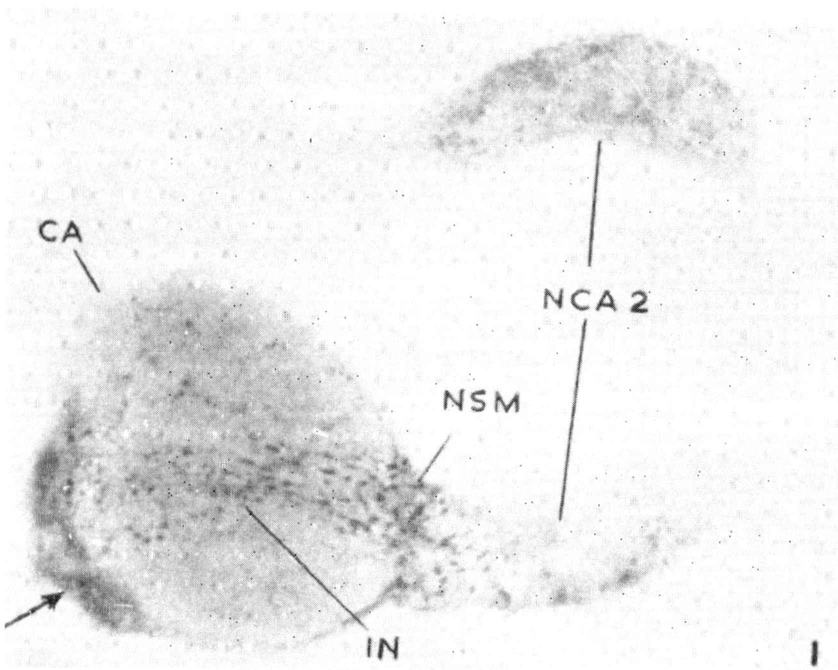

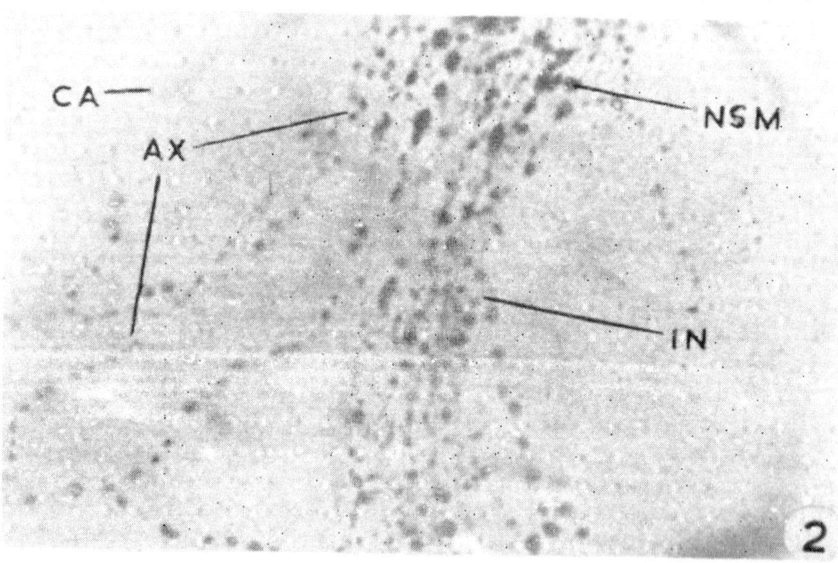

Plate 12. Fig.1. Whole-mount preparation of corpus allatum of *Gryllodes* showing superficial innervation (IN). Note accumulated NSM at the point of entry of NCA 1 (arrow) and the place of emergence of axons over corpus allatum. X 500. Fig. 2. A magnified view of Fig. 1 showing the pattern of innervation and NSM in the axons. X 950. (From Awasthi, 1968).

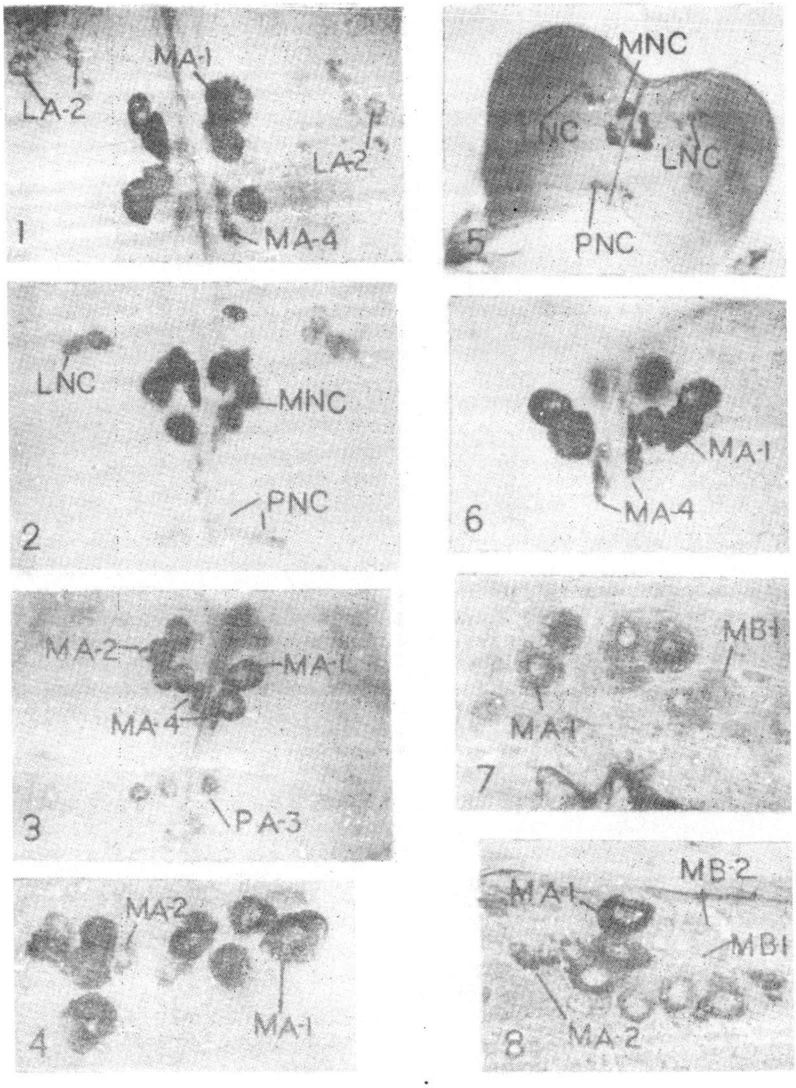

Plate 13. Fig. 1. Whole mount of the brain of one-day old fifth instar larva of *Utetheisa pulchella* showing medial (MA-1 and MA-4) and lateral (LA-2) cells. X 300. Fig. 2. Whole mount of the brain of a prepupa of *Utetheisa* showing medial (MNC), lateral (LNC) and posterior (PNC) neuroserectory cells. X 300. Fig. 3. Whole mount of the brain of a three-day old fifth instar larva showing medial (MA-1, MA-2 and MA-4) and posterior (PA-3) cells. X 280. Fig. 4. Whole mount of the brain of an adult female (at the time of egg laying) showing moderately filled medial (MA-1, MA-2) cells. X 350. Fig. 5. Whole mount of the brain of a fifth instar larva (before prepua) showing medial (MNC), lateral (LNC) and posterior (PNC) cells. X 330. Fig. 6. Whole mount of the brain of a fifth instar larva of *Utetheisa* showing the large granulated medial (MA-4) cells, the MA-1 cells are also seen. X 330. Fig. 7. F.S. of brain of a newly moulted adult male showing the medial (MA-1 and MB-1) cells. X 330. Fig. 8. F.S. of brain of a newly moulted adult female *Utetheisa* showing deeply stained medial MA-1 and MA-2 cells. The medial MB-1 and MB-2 cells are also seen. X 300. (From Awasthi and Singh, 1985).

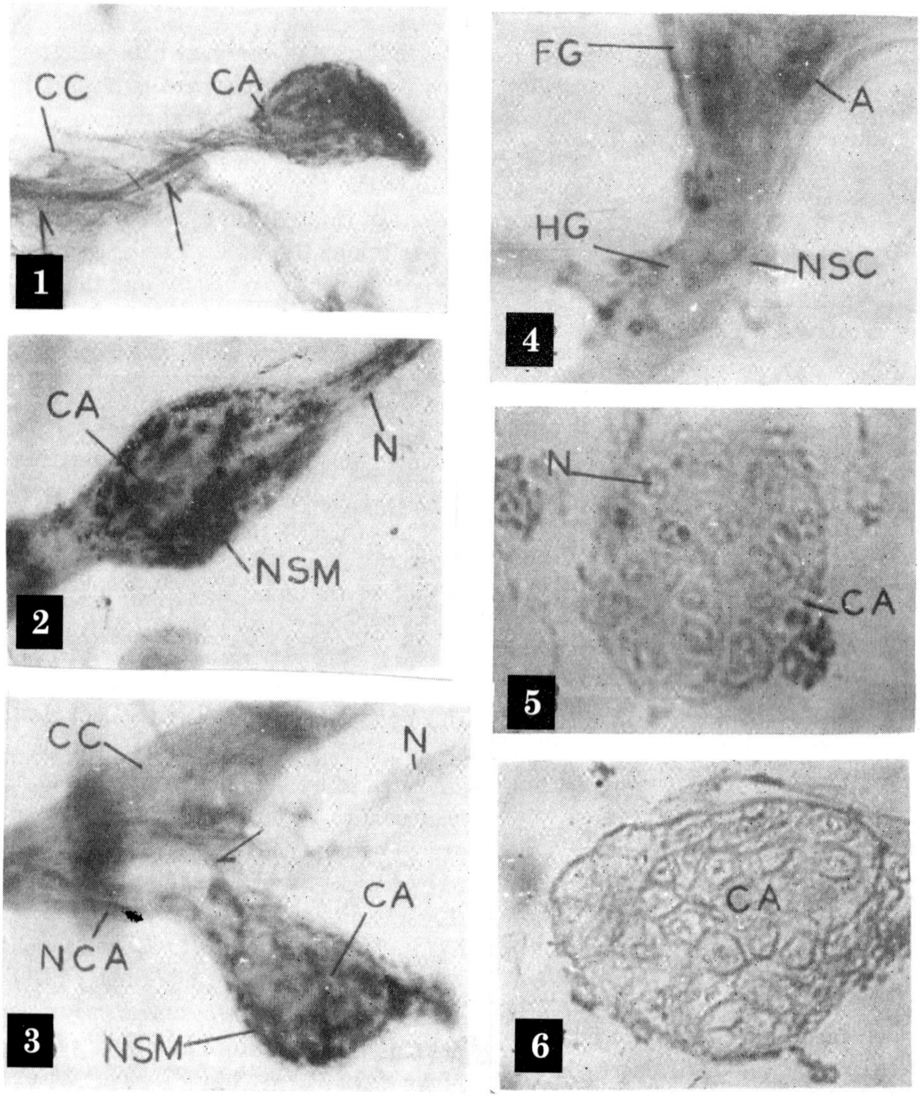

Plate 14. Fig. 1. W.M. of retrocerebral complex of *Amsacta collaris* showing CC devoid of NSM and CA is heavily loaded with NSM. Arrows indicate the path of the axons of NCC-1 which innervate the CA. X 250. Fig. 2. W.M. of corpus allatum (CA) of *Amsacta* the gland is heavily loaded with NSM. The post allatic nerve (N) also contains NSM. X 325. Fig. 3. W.M. of retrocerebral complex. The CC are devoid of NSM and the axons innervating the CA superficially are provided with a large amount of NSM. The axons enters the CA via CC as well as some of them come out side the CC and then enter to CA (arrow). X 380. Fig. 4. W.M. of frontal (FG) and hypocerebral ganglion (HG) showing neurosecretory cells of A-type. X 425. Fig. 5. T.S. of corpus allatum of diapausing pupa. Note absence of NSM from the entire gland. X 400. Fig. 6. T.S. of corpus allatum of a one month old diapausing pupa. The gland is totally devoid of NSM. X 350. (Singh and Awasthi, 1981b).

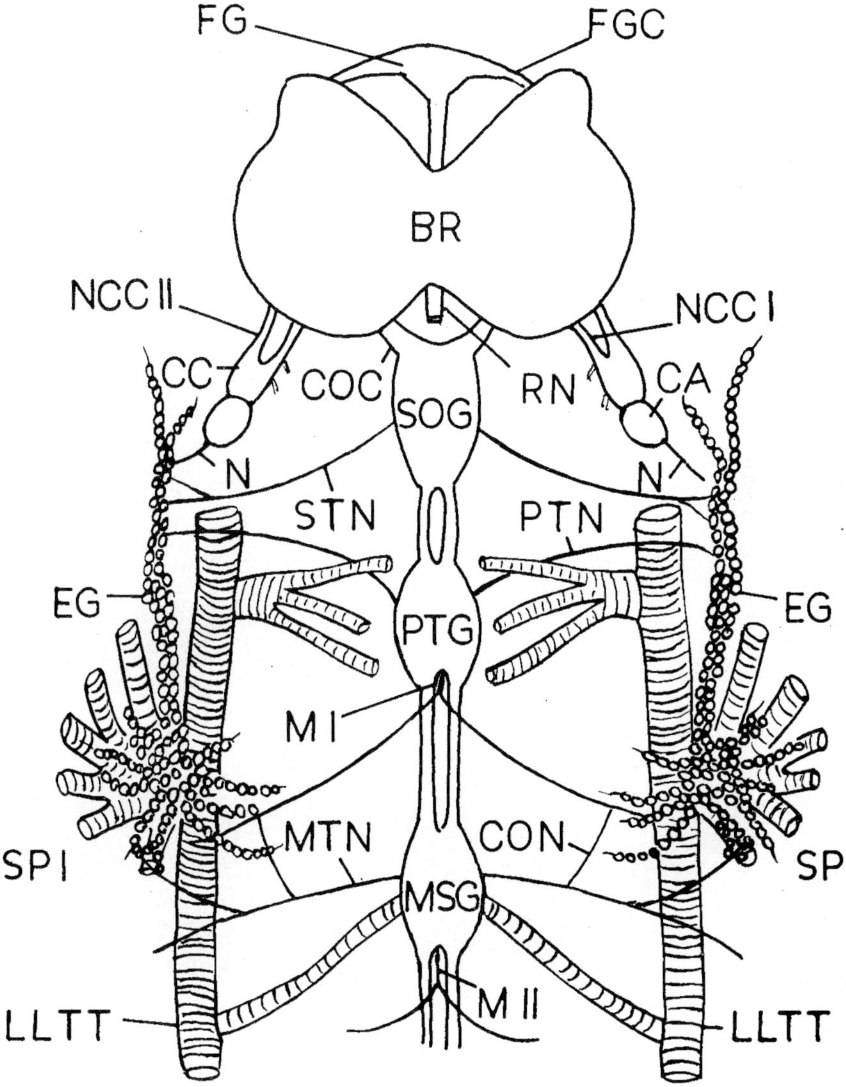

Plate 15. Fig. 1. Diagrammetic representation (not drawn to scale) of the various components of the neuroendocrine-system of a mature larva of *Amsacta collaris,* BR Brain, CA corpora allata; CC corpora cardiaca; COC circum oesophageal connective; CON connective; EG ecdysial gland; FG frontal ganglion; FGC frontal ganglion connective; LLTT lateral longitudinal tracheal trunk M-1 medial-1; M-II medial-II; MSG mesothoracic ganglion; MTN mesothoracic transverse nerve; N nerve; NCC-1, nervi. corporis cardiaci I; NCC-II nervi corporis cardiaci II; PTG prothoracic ganglion; PTN prothoracic transverse nerve; RN recurrent nerve; SOG suboesophageal ganglion; SP-1 spiracle I; STN suboesophageal transverse nerve. (From Awasthi and Singh, 1983).

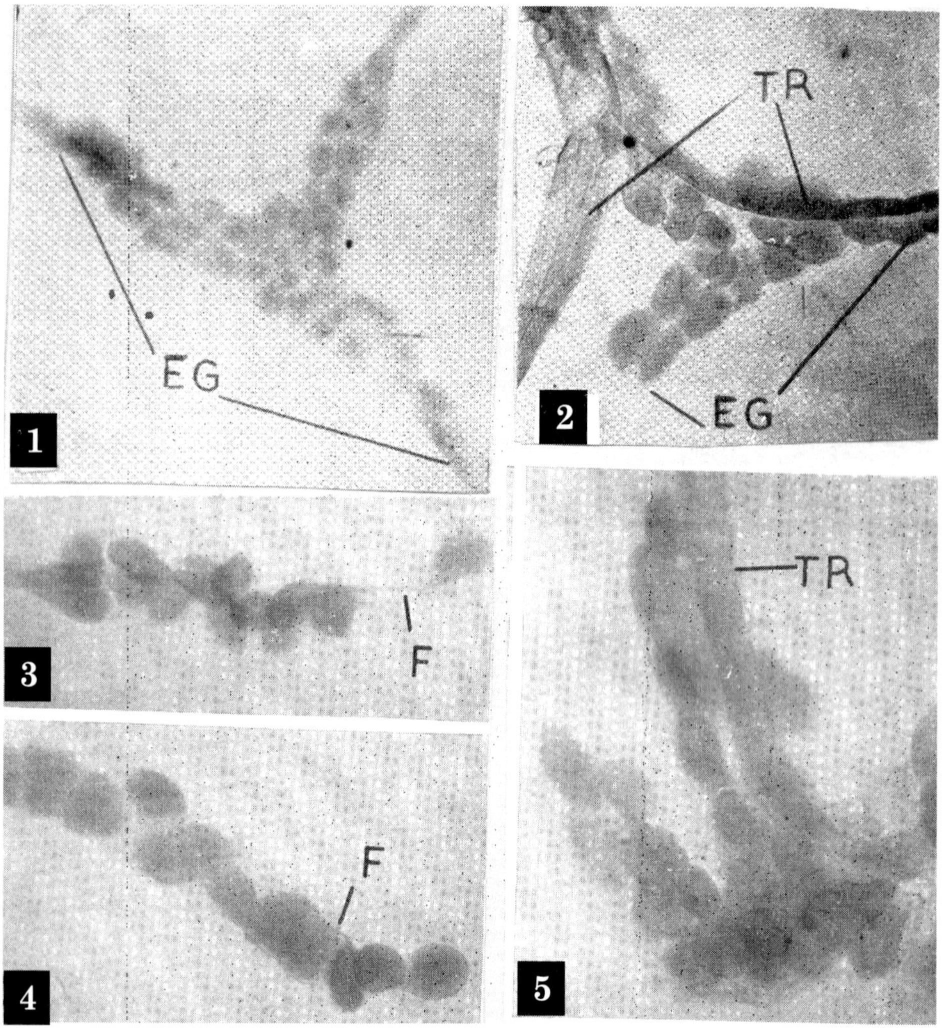

Plate 16. Figs. 1-2. Ecdysial glands of 5th instar larvae of *Achoea janata* Fig. 1 X 265, Fig. 2 X 220 (From Singh and Awasthi, 1981a). Figs. 3 - 5. Ecdysial glands of *Amsacta collaris*. Fig. 3. EG of the 6 th instar larva (before moulting). Note alternate arrangement of cells. Filamentous structure (F) is clearly seen. X 275. Fig. 4. EG of the 6 th instar larva. Note one row of oval cells. X 280. Fig. 5. EG of 6 th instar larva (before moulting) showing the main cluster and the tracheal supply (TR). X 250. (From Awasthi and Singh, 1983).

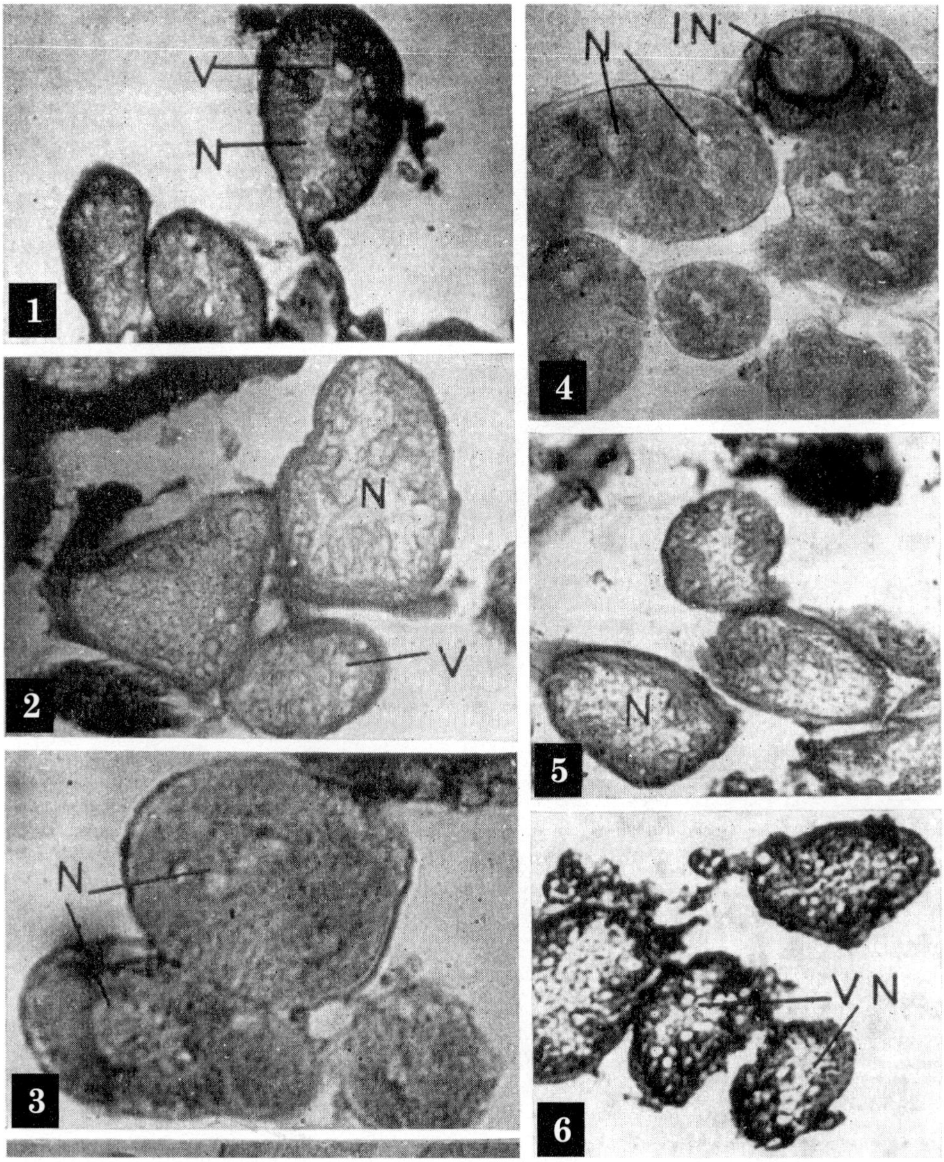

Plate 17. Fig. 1. EG of 2 d old 6th instar larva of *Amsacta* showing convoluted cytoplasm with a few vacuoles (V). X 245. Fig. 2. EG of 2 d old 6th instar larva showing numerous vacuoles (V) at the periphery of the nucleus (N) as well as in the cytoplasm. X 250. Fig. 3. EG of an old male moth showing beginning of distortions. X 360. Fig. 4. EG of male moth showing nuclear (N) inclusions (lN). X 250. Fig. 5. EG of 6 d old, 6th instar larva showing a few vacuolus in the cytoplasm. X 245. Fig. 6. EG of 3 d old, 6th instar larva showing numerous vacuoles (V) in the nucleus (N) as well as in the cytoplasm. X 333. (Awasthi and Singh, 1983).

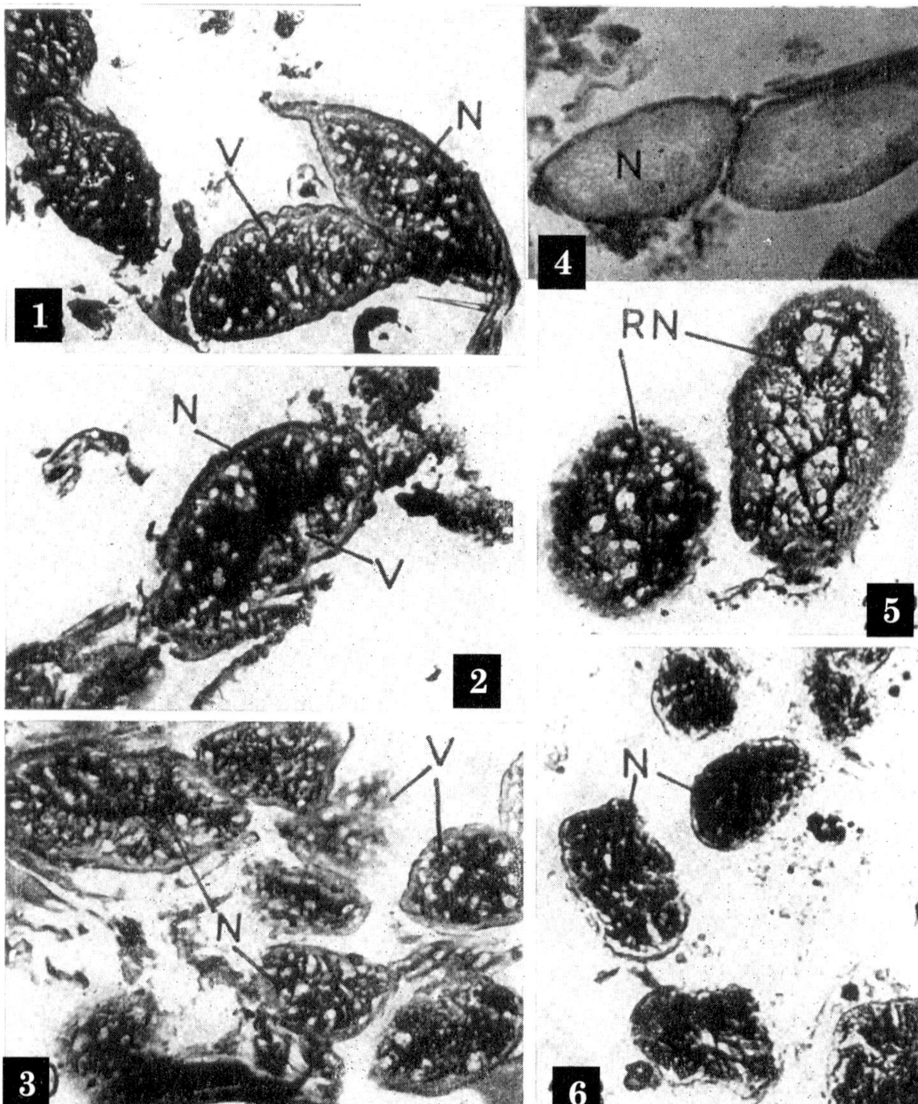

Plate 18. Fig. 1. EG of a 4 d old. 6th instar larva of *Amsacta* showing highly reticulate nucleus with numerous vacuoles in the cytoplasm. X 310. Fig. 2. EG of 5 d old (early), 6th instar larva, vacuoles are not present in the nucleus (N) but they are in the cytoplasm. X 320. Fig. 3. EG of a 5 d old. 6th instar larva. Note presence of a reticulate nucleus (N) and numerous vacuoles (V) in the cytoplasm. X 310. Fig. 4. EG just after boring (before pupation) vacuoles are absent in the gland X 250. Fig. 5. EG of 3 d old 6th instar larva showing reticulate nucleus (RN) and numerous vacuoles in the cytoplasm. The section has passed through the lateral side of the cells. X 350. Fig. 6. EG of a 3 month old diapause pupa. Note large and compact nucleus (N) with thin cytoplasm devoid of vacuoles. X 350. (From Awasthi and Singh, 1983).

INDEX